WAVES IN FLUIDS

WAVES IN FLUIDS

JAMES LIGHTHILL

Lucasian Professor of Applied Mathematics
University of Cambridge

CAMBRIDGE UNIVERSITY PRESS

Cambridge
London   New York   Melbourne

Published by the Syndics of the Cambridge University Press
The Pitt Building, Trumpington Street, Cambridge CB2 1RP
Bentley House, 200 Euston Road, London NW1 2DB
32 East 57th Street, New York, NY 10022, USA
296 Beaconsfield Parade, Middle Park, Melbourne 3206, Australia

First published 1978
First paperback edition 1979

Printed in Great Britain
at the University Press, Cambridge

Library of Congress Cataloguing in Publication Data
Lighthill, M J Sir.
Waves in fluids.

Bibliography: p.
Includes indexes.
1. Waves. 2. Fluid dynamics. 3. Wave-motion,
Theory of. I. Title
QC157.L53   532'.0593   77–8174
ISBN 0 521 21689 3 hard covers
ISBN 0 521 29233 6 paperback

CONTENTS

Chapter 3. Water waves

Chapter 4. Internal waves

Epilogue

Bibliography
(indexed as pages A to Q)

PREFACE

The aims of this book are set out in the prologue. The main subject matter is developed in chapters 1–4. Several further topics are sketched briefly in the epilogue.

Although no references are included in the text, an annotated bibliography is designed to take the reader through the book's subject matter, indicating where he or she can read more about each topic mentioned. This is followed by a notation list showing the meanings of the principal symbols used.

Pages 470 to 486, which constitute the bibliography, have subsidiary page designations A to Q, which are used for bibliographical references throughout the Author Index and Subject Index.

Within each chapter, mathematical equations are numbered consecutively: (1), (2), (3), etc. The numbering then *begins again from* (1) in the next chapter, or in the epilogue. When, in any chapter, we refer to a numbered equation, we mean the equation of that number *in the same chapter*.

By contrast, figures are numbered continuously (from 1 to 117) throughout the book. Exercises for the reader are given at the end of each chapter.

Cambridge   J A M E S   L I G H T H I L L
1978

PROLOGUE

This book is designed as a comprehensive introduction to the science of wave motions in fluids (that is, in liquids and gases); an area of knowledge which forms an essential part of the dynamics of fluids, as well as a significant part of general wave science; and, also, has important applications to the sciences of the environment and of engineering. The subject's extent and variety are enormous: the different types of waves in fluids, the different fundamental ideas that have been developed to interpret their properties, and the different applications of those properties are so extensive that a comprehensive introduction in one volume demands very careful selection.

The design adopted for the book, in four chapters and an epilogue, has two principal aims. First, as the chapter titles suggest, it allows an analysis in depth of four important and representative types of waves in fluids (sound waves; one-dimensional waves in fluids; water waves; internal waves) to precede brief descriptions of some other important types in the epilogue. At the same time, the subject matter of the four chapters is chosen so that, as far as possible, all the most generally useful fundamental ideas of the science of waves in fluids can be developed at length, one after another. The main exceptions are certain very difficult, advanced ideas which could not, even in a comprehensive introduction, be treated so fully; they are merely sketched, with references to more extensive treatments, in the epilogue.

Thus, each chapter is designed *both* to analyse the main type of wave system named in its title *and* to develop an important body of fundamental ideas of general application to waves in fluids. The ideas developed in each chapter are especially important for the wave system of that chapter, but are applicable to wave science generally, and to other systems of waves in fluids in particular. Therefore, later chapters include some applications of the ideas they have developed to the wave systems of earlier chapters; as when methods developed in chapter 2 are used to analyse the generation and propagation of supersonic booms, or when methods developed in chapter 4 are used to analyse the wavemaking resistance of ships.

Practical applications are continually indicated: to noise-abatement research in chapters 1 and 2, to areas as diverse as hydraulics and circulation physiology in chapter 2, to oceanography and ocean exploitation in chapters 3 and 4, and to numerous parts of atmospheric science in chapter 4. Some even more diverse applications are indicated in the epilogue.

Within wave science as a whole, the nature of waves in fluids is characterised especially by their ability to interact with complex fluid flow fields. Such interactions are therefore described at length in this book (see, in particular, sections 1.10, 2.14, 3.9, 3.10, 4.6, 4.7 and 4.12).

Some ideas of fundamental importance treated early in chapter 1 are the property of linearity (direct linear superposability of different wave motions); the concept of energy transport by waves; and the differing character of propagation in one, two and three dimensions. Next, two quite distinct sets of ideas (complementary of their use) are developed: (i) for sources small in comparison with the length of the waves generated ('compact sources'), and (ii) for fluid systems on a scale large compared with the wavelength; both are applied to noise-source problems. In later chapters, both sets of ideas are taken still further; see especially section 4.9 for compact sources, and section 4.5 for the general ray-tracing technique applicable to systems with properties varying gradually on a scale of wavelengths.

In the meantime, sections 1.11 and 1.12 describe an intermediate régime. Finally, chapter 1, like every chapter in the book, includes an account of processes associated with wave *attenuation*: processes involving dissipation in the body of the fluid (section 1.13); or dissipation near either a solid boundary (section 2.7), or a free boundary (section 3.5); or the generation of steady streaming motions by wave attenuation (section 4.7).

Chapter 2 resumes in detail the theme of one-dimensional propagation, and shows how a common treatment is possible for a wide range of seemingly quite different systems; including propagation of sound in ducts, of the blood pulse in arteries, and of 'long' water waves in open channels. Fundamental ideas treated in the first half of chapter 2 include (i) the different effects on a wave of *discontinuous* or *gradual* changes in the properties of the containing tube or channel, (ii) the application of that knowledge to propagation in branching systems, and (iii) the study of a variety of types of resonance which can occur.

The second half of chapter 2 gives an extended treatment of nonlinear effects; and, in particular, of those effects which generate a local steepening of waveforms. Shock waves and other essentially discontinuous waves involving a balance between steepening and dissipative effects are described at length, and methods are outlined for tracing the development of complex signals including shock waves. Finally, one-dimensional nonlinear theory is applied to the propagation of signals along those abstract 'ray tubes' whose properties were first encountered in chapter 1.

The subject of water waves, broached in chapter 2 only as far as 'long' waves (waves of length far exceeding the water depth) and their associated discontinuities (the 'hydraulic jumps') are concerned, is pursued much further in chapter 3. Beyond the basic dynamics of surface waves with gravity or surface tension as the restoring force, chapter 3 is concerned to introduce also the special properties of 'dispersive' waves. The fundamental distinction between phase velocity and group velocity is developed for general dispersive systems that are isotropic (with propagation properties independent of direction, although varying with wavelength). The subject is approached from three complementary standpoints (sections 3.6–3.8) and then applied to the analysis of surface waves generated by storms, or by obstacles in a stream, or by the motion of a ship through water.

Similarly, chapter 4 is primarily concerned to explain wave dispersion in systems that are *not* isotropic, including those internal gravity waves in stratified fluids which give the chapter its title. These are systems for which the group velocity and the phase velocity may be quite different in direction as well as in magnitude, with many important consequences. Several other fundamental ideas are also treated at length: 'trapped waves', caustics, wave-flow interaction, travelling wave sources in general and, finally, waveguides.

On the other hand, two groups of fundamental ideas of a higher order of difficulty are postponed to the epilogue, where indeed they are only sketched (with references). These include theories of the interaction between dispersive effects and nonlinear effects, and theories of the development of statistical assemblages of waves through nonlinear interactions.

Readers approaching this book are likely to possess basic knowledge of *dynamics*, including the elementary dynamics of vibrations, and the elementary dynamics of fluids. However, for these and for all other matters on which prior knowledge may be desirable, a selection of suitable texts is suggested in the bibliography.

The science of waves in fluids is here approached quantitatively, and with the aim of outlining, wherever possible, techniques of quantitative analysis. Subject to this aim, however, the extent of mathematical development has been kept to a minimum. No mathematics has been included for its own sake; furthermore, all the mathematical analyses which have been included have to the maximum possible extent been given clear physical interpretations. Nevertheless, the subject of waves seems to demand the use of complex variables; accordingly, the elementary theory of functions of a complex variable, and (for similar reasons) the elementary properties of Fourier integrals, are among the mathematical knowledge which our readers are either assumed to possess or else (perhaps) assisted to acquire through direction to suitable texts.

The most fundamental waves in fluids are sound waves (chapter 1), because they can exist in a fluid without any external force field needing to be present. Readers familiar with the elementary dynamics of vibrations know that a wave or any other vibrating system involves a balance between a restoring force and the inertia of the system. Most of the waves treated in this book involve external restoring forces; especially gravity (chapters 2, 3 and 4) but also surface tension (section 3.4) or tube elasticity (section 2.2). Other external forces, important for wave systems treated in the epilogue, include magnetic force fields and the Coriolis force felt by rotating fluids.

Sound waves propagate, however, independently of external forces. The restoring force balancing the fluid's inertia is provided entirely by the fluid's own compressibility. Because the compressibility properties of the fluid are the same in all directions, sound propagation is isotropic.

By contrast, most wave motions due to an *external* restoring force are anisotropic, and this is why the general theory of anisotropic propagation given in chapter 4 is so important. Waves on a horizontal water surface are an exception, because they are subject only to two-dimensional propagation in horizontal directions; and, evidently, a vertical external force such as gravity can make no distinction between different horizontal directions. On the other hand, when the source of those waves is a moving ship, they are made effectively anisotropic by the Doppler effect (section 4.12).

Those extensive parts of the dynamics of fluids that are strongly influenced by the properties of waves in fluids include many large and important fields of current research. These are found, for example, in modern aeronautical engineering, and other branches of engineering where flow noise is important, and in those parts of naval architecture and offshore-structure technology that interact with the wave properties of the sea surface.

Such research fields include, furthermore, the analysis of tides and surges in oceans and seas and estuaries; and the study of numerous ocean-current patterns of a wavelike nature. They include the analysis of several atmospheric propagation phenomena of great importance, from small-scale 'clear air turbulence' to large-scale wavelike wind patterns; and many properties of the air–sea interaction. Other active areas of geophysical study include wave propagation in the ionosphere, and in the liquid core of the earth, while astrophysical observations constantly reveal wavelike gaseous motions suitable for analysis by similar methods. This book, designed as a comprehensive introduction to waves in fluids, is intended to prepare readers to be able to enter any of these active research fields, by giving them that wide background of fundamental ideas in terms of which the specialised literature of such a field can be readily understood.

1. SOUND WAVES

1.1   The wave equation

As remarked in the prologue, it is a balance between the compressibility and the inertia of a fluid that governs the propagation of sound waves through it. The linear theory of this propagation is described in chapter 1.

Use of a linear theory, for waves of any kind, implies that we consider disturbances so weak that in equations of motion we can view them as small quantities whose products are neglected. Such products of small quantities occur, for example, in the well-known expression for the acceleration of a fluid element:

$$\partial\mathbf{u}/\partial t + \mathbf{u}\cdot\nabla\mathbf{u}, \tag{1}$$

where $\mathbf{u}$ is the vector velocity field. In this expression (significant whenever inertia is important, as it is for practically all waves in fluids) the linear term $\partial\mathbf{u}/\partial t$ represents the local rate of change of $\mathbf{u}$ at a fixed point, while the nonlinear term $\mathbf{u}\cdot\nabla\mathbf{u}$ describes how the element's velocity changes owing to its changing position in space. This 'convective rate of change' of $\mathbf{u}$ involves products of its spatial gradients with components of $\mathbf{u}$ itself, and so is neglected in a linear theory.

In this chapter, then, disturbances are supposed weak enough for such nonlinear contributions to inertial effects, together with nonlinear terms in the restoring forces (here, those associated with compressibility), to be neglected. Investigations of just *how* weak disturbances need to be for the theory to be reasonably good, and of what detailed effect on stronger disturbances the nonlinear terms may have, are postponed to chapter 2.

In this section, taking into account compressibility and inertia but no other properties of the fluid, we obtain the linearised equations of the theory of sound in their simplest form, a very useful one. We postpone consideration of how sound waves are influenced by effects neglected here (especially viscosity, heat conduction, external forces including gravity, and inhomogeneities such as stratification) to section 1.2 and later parts of the book.

The inertial nature of a fluid of density ρ is expressed when we apply to a small fluid element Newton's second law of motion. This demands that the product of the mass per unit volume ρ and of the acceleration (1) is the force on the element per unit volume, which in the absence of external forces is due solely to those internal stresses through which neighbouring fluid acts on it. When viscous stresses are neglected, this force per unit volume is simply minus the gradient ∇p of the fluid pressure p; thus

$$\rho(\partial \mathbf{u}/\partial t + \mathbf{u}\cdot\nabla\mathbf{u}) = -\nabla p. \tag{2}$$

Compressibility implies that the density of a fluid element may change, in accordance with the well-known equation of continuity:

$$\partial\rho/\partial t + \mathbf{u}\cdot\nabla\rho + \rho\nabla\cdot\mathbf{u} = 0. \tag{3}$$

The first two terms in (3) make up the total rate of change of ρ for the element. Thus, the *divergence* $\nabla\cdot\mathbf{u}$ of the velocity field is identified by (3) as the rate of increase of volume of an element moving in that velocity field, divided by the volume; in other words (since the element's mass is conserved) *minus* the rate of increase of density divided by the density. At the same time, an alternative interpretation of equation (3), based on grouping the second and third terms together as $\nabla\cdot(\rho\mathbf{u})$, is also possible and is used later (section 1.10).

We linearise these equations by regarding as small quantities all departures from a state in which the fluid has uniform density ρ_0 and is at rest. In the absence of external forces this implies that the pressure also takes a uniform value, say p_0.

Equations (2) and (3), with products of small quantities neglected, become the linearised equations of momentum

$$\rho_0\,\partial\mathbf{u}/\partial t = -\nabla p \tag{4}$$

and of continuity

$$\partial\rho/\partial t = -\rho_0\nabla\cdot\mathbf{u}. \tag{5}$$

These forms result from the neglect in (2) of $\mathbf{u}\cdot\nabla\mathbf{u}$ as already discussed, and the similar neglect in (3) of $\mathbf{u}\cdot\nabla\rho$, both involving products of small velocities with small gradients. At the same time the factor ρ in one term of each equation is replaced by ρ_0, the error being the product of a small quantity $(\rho - \rho_0)$ with another small quantity $(\partial\mathbf{u}/\partial t$ or $\nabla\cdot\mathbf{u})$. From this there result local rates of change of velocity $\mathbf{u}$ and density ρ directly proportional to pressure gradient and to velocity divergence, respectively.

One quantity that on the linear theory of sound behaves extremely simply is the vorticity; that is,

$$\mathbf{\Omega} = \nabla\times\mathbf{u}, \tag{6}$$

the curl of the velocity field (for general properties of vorticity, see texts on fluid dynamics). In fact, equation (4) implies that

$$\partial\mathbf{\Omega}/\partial t = 0, \tag{7}$$

since the curl of ∇p vanishes. Thus, the vorticity field is independent of time: vorticity 'stays put' on the approximations involved in the linear theory of sound, however much other quantities may be propagated.

This conclusion may astonish a reader familiar with Helmholtz's theorem that 'vortex lines move with the fluid'; all such changes due to convection are neglected in a linear theory. This is reasonable, however, in a theory of sound, which predicts that changes in other quantities (such as pressure) propagate at hundreds of metres per second, compared with which convection by relatively small flow velocities appears negligible.

That *rotational* part of the velocity field which is 'induced' by the vorticity field $\mathbf{\Omega}$ must, according to equation (7), be independent of time. The remaining part of the velocity field is irrotational and so can be written as the gradient $\nabla\phi$ of a 'velocity potential' ϕ. Only this part exhibits the fluctuations associated with sound propagation.

From this point on, we write

$$\mathbf{u} = \nabla\phi, \tag{8}$$

so that $\mathbf{u}$ is taken as the irrotational part of the velocity field (what is left after subtracting the velocities induced by the steady vorticity field). On linear theory, this irrotational propagating velocity field shows *no interaction* with any steady rotational flow field. Actual propagation of sound fields across vortex lines that actually move with the fluid is studied later (section 4.6); when flow velocities are very small compared with the sound speed the interaction is found to involve at most very gradual changes.

Equations (4) and (8) imply that

$$p - p_0 = -\rho_0\,\partial\phi/\partial t, \tag{9}$$

because the *gradients* of both sides of (9) are everywhere equal, and because both sides vanish in undisturbed parts of the fluid provided that the velocity potential, as is usual, is taken as the solution of (8) vanishing in such parts. Equation (9) differs from the well-known 'Bernoulli equation' for unsteady irrotational flows by omission of the term $-\frac{1}{2}\rho_0(\nabla\phi)^2$ (which on a linear theory is negligible) on the right-hand side.

Equations (5) and (8) express the rate of change of density as

$$\partial\rho/\partial t = -\rho_0\nabla^2\phi \tag{10}$$

in terms of the Laplacian $\nabla^2\phi$, but no further progress beyond (9) and (10) can be made until the compressibility properties of the fluid have been used to infer an explicit relationship between changes of pressure and density. The character of such a relationship is discussed below (section 1.2) but here we simply assume a functional dependence $p = p(\rho)$. Linearising this means expanding in a Taylor series about $\rho = \rho_0$ as

$$p = p(\rho_0) + (\rho - \rho_0)p'(\rho_0) + \dots \tag{11}$$

and neglecting all terms beyond those shown as involving squares and higher powers of $\rho - \rho_0$. Then

$$\partial p/\partial t = p'(\rho_0)\,\partial\rho/\partial t, \tag{12}$$

whence we deduce, substituting for p on the left from (9) and for $\partial\rho/\partial t$ on the right from (10), that

$$\partial^2\phi/\partial t^2 = c^2\nabla^2\phi, \tag{13}$$

where the constant c (with the dimensions of velocity) is defined by the equation

$$c^2 = p'(\rho_0). \tag{14}$$

Most readers will recognise equation (13) as 'the wave equation': an equation characteristic of any phenomena, with energy conserved, involving propagation through a homogeneous medium at a single wave speed c, independent of waveform or direction of propagation. It is satisfied, for example, by components of electromagnetic fields in free space with c as the velocity of light $3 \times 10^8\,\mathrm{m\,s^{-1}}$. We find, however (section 1.2), that the sound speed c given by (14) is smaller by several orders of magnitude.

A simple solution of (13), representing a 'plane wave' travelling in the positive x-direction, is

$$\phi = f(x - ct). \tag{15}$$

Here $f(x)$ is the waveform at time $t = 0$, while the waveform at a later time t has identical shape but is shifted a distance ct in the positive x-direction. The wave is 'longitudinal' in the sense that the velocity field $\mathbf{u} = (u, v, w)$, satisfying

$$u = f'(x - ct), \quad v = w = 0, \tag{16}$$

is parallel to the direction of propagation.

For such a travelling plane wave, equation (9) gives

$$p - p_0 = \rho_0 c u, \tag{17}$$

a simple proportionality between excess pressure and the component of fluid velocity in the direction of propagation. This proportionality arises because, at a point in a travelling wave where the pressure is increasing, the gradient of pressure in the direction of propagation takes the negative value $-c^{-1}\partial p/\partial t$ and so *accelerates* the fluid, whose corresponding component of acceleration $\partial u/\partial t$ (see (4)) takes the positive value $(\rho_0 c)^{-1}\partial p/\partial t$. Note that (17) gives much bigger pressure variations for given u than would be obtained in steady flows through $\frac{1}{2}\rho_0 u^2$ terms.

Equation (15) is not the only solution of the wave equation which depends on just the two variables x and t; another such solution is

$$\phi = g(x+ct), \tag{18}$$

and the general solution is the sum of (15) and (18) with the functions f and g arbitrary. Equation (18) represents a plane wave travelling in the negative x-direction; the velocity field $\mathbf{u} = (u, v, w)$ satisfies

$$u = g'(x+ct), \quad v = w = 0 \tag{19}$$

and the excess pressure is $\quad p-p_0 = -\rho_0 cu, \tag{20}$

but of course the ratio of excess pressure to the velocity component in the direction of propagation $(-u)$ is still $+\rho_0 c$.

The general plane wave, travelling in the direction of the vector (ξ, η, ζ), is

$$\phi = h(\xi x + \eta y + \zeta z - ct), \tag{21}$$

which satisfies (13) provided that $\xi^2 + \eta^2 + \zeta^2 = 1$. Equations (8) and (9) show that

$$\mathbf{u} = (\xi, \eta, \zeta)(\rho_0 c)^{-1}(p - p_0), \tag{22}$$

again signifying longitudinal waves with velocity component $(\rho_0 c)^{-1}(p - p_0)$ in the direction of propagation.

The independence of the wave speed c on the direction (ξ, η, ζ) as well as on the shape h of the waveform, is a simplifying feature of 'the wave equation' that disappears in many other problems of waves in fluids (see below, beginning with chapters 3 and 4).

1.2 The speed of sound

Although the formula (14) for the speed of sound was known already to Newton, he failed to obtain good agreement between its indications and the results of observations of sound speed. Boyle's experiments on gases had shown that moderate pressure increases so decrease gas volume that pressure is closely proportional to density at a fixed temperature, suggesting that (14) can be written $c^2 = p_0/\rho_0$, which in atmospheric air at $20\,°C$ gives $c = 290\,\mathrm{m\,s^{-1}}$, significantly lower than the observed value of $340\,\mathrm{m\,s^{-1}}$.

Not till more than a century later did Laplace explain this shortfall as due to the inappropriateness of using data obtained at a fixed value of the temperature. Wherever in a sound wave a fluid element is being compressed, neighbouring fluid is doing work upon it and this 'work of compression' adds internal energy to the element and so raises its temperature. Experiments such as those of Boyle on gases are performed in a container of large heat capacity with which the gas is allowed to exchange heat after compression, the volume change being measured only after reaching a steady state associated with return to the initial temperature.

By contrast, for the local compressions inside a sound wave there is no such restraint upon the temperature rise, whose value and whose effect on the speed of sound we now calculate for those gases which do to close approximation satisfy Boyle's law. These are the so-called 'perfect gases', including atmospheric air and any gas whose density is very small compared with that of the same substance in a condensed phase. Then the pressure takes to good approximation the value

$$p = RT\rho, \tag{23}$$

where T is the absolute temperature in kelvins (that is, 273 plus the temperature in °C) and where the formula

$$R = \frac{8314 \, \text{m}^2 \text{s}^{-2} \text{K}^{-1}}{\text{mean molecular weight}} \tag{24}$$

gives RT in m^2s^{-2} in terms of the mean molecular weight of the gas or gas mixture.

Physically, a perfect gas is one in which at each instant only a very small proportion of the molecules are sufficiently close to others to be interacting with them. Its pressure is close to the value $RT\rho$, associated with momentum transfer by the random translational motions of molecules, because contributions proportional to higher powers of ρ from intermolecular forces are negligible. Its internal energy E per unit mass is to close approximation a function $E(T)$ of the temperature alone, proportional to the average energy (translational, rotational and vibrational) of an isolated molecule, because contributions from the potential energy associated with intermolecular forces are again negligible.

Whenever the volume of an element of gas is unchanging, any rise in temperature dT demands, per unit mass, a heat input equal to the required increase $E'(T)dT$ in the internal energy. For this reason $E'(T)$ is written c_v, the *specific heat* (heat input per unit mass per unit increase in temperature) *at constant volume*.

There is an additional rise in internal energy (due to input of *work* rather than heat) when a fluid element is compressed, so that its volume, say V, changes by an amount dV which is negative. The total work done on it by the pressure p of adjacent elements is then $p(-dV)$, since every adjacent small element does a quantity of work (that is, the *force* with which it acts *times the displacement* in the direction of that force) per unit reduction in volume equal to the force divided by the area of application; that is, to the pressure p. But the volume V per unit mass is ρ^{-1}, and so the energy change per unit mass due to work of compression is

$$dE = p(-d\rho^{-1}) = p\rho^{-2}d\rho. \tag{25}$$

When this is the only source of internal energy change, as in sound waves when both *conduction* of heat and its generation by *dissipation* of mechanical energy can be neglected, the corresponding temperature rise dT is given by

$$p\rho^{-2}d\rho = dE = c_v\,dT = (c_v/R)(\rho^{-1}dp - p\rho^{-2}d\rho), \tag{26}$$

where (23) has been used to relate temperature changes to changes in pressure and density.

The square of the velocity of sound, c^2, given by (14) as $dp/d\rho$, is now seen, for the pressure–density relations (26) characteristic of sound waves in perfect gases, to take the value

$$c^2 = \gamma p/\rho = \gamma RT, \tag{27}$$

where for perfect gases $\qquad \gamma = (R+c_v)/c_v. \tag{28}$

Expression (27) is Newton's value p/ρ multiplied by γ, which for atmospheric air takes the value 1.40. We obtain the corresponding value of R from (24), using a mean molecular weight of 29.0, and infer at a temperature $T = 293$ K (corresponding to 20 °C) $c = 340$ m s^{-1} to two significant figures, as observed.

For perfect gases in general, the quantity (28) can always be written

$$\gamma = c_p/c_v, \tag{29}$$

where c_p, the specific heat *at constant pressure*, is necessarily equal to $c_v + R$. That is because gas, on being heated at constant pressure (for example, under a given column of mercury), *expands*, doing work on adjacent fluid equal to $pd\rho^{-1}$ per unit mass (namely, minus the work (25) done by adjacent fluid on the element), and this is RdT for changes at constant pressure by (23). Both this work RdT and the rise $c_v\,dT$ in internal energy must be supplied by the total heat input $c_p\,dT$.

Typical values of γ range from a maximum of $\frac{5}{3}$ for monatomic gases, whose internal energy is simply the translational energy of the molecules (contributing an amount $\frac{3}{2}R$ to c_v), through $\frac{7}{5}$ for diatomic gases (possessing an additional contribution of R to c_v from the energy of molecular rotation), to values as low as 1.2 or even 1.1 for polyatomic gases at high temperature (with a large further contribution to c_v from molecular vibrations). In all these cases the formula (27), stating that the square of the sound speed, c^2, exceeds its Newtonian value by the factor γ, is well borne out by observations.

For liquids, or for gases that are too dense to be regarded as 'perfect gases', the simple equation of state (23) and functional relationship between E and T have to be replaced by more complicated relationships

$$p = p(\rho, T), \quad E = E(\rho, T), \tag{30}$$

but c^2 is still given as the value of $dp/d\rho$ in changes satisfying (25). Rather surprisingly, it still turns out that

$$c^2 = \gamma c_N^2, \tag{31}$$

where the 'Newtonian' value c_N^2 is $\partial p/\partial \rho$ keeping T constant, and where equation (29) still specifies γ as the ratio of the two specific heats.

This is because any small temperature change dT at constant pressure produces a density change

$$d\rho = -\frac{\partial p/\partial T}{\partial p/\partial \rho}\, dT, \tag{32}$$

and so the heat input $dE - p\rho^{-2}d\rho$ at constant pressure bears to its constant-volume value $(\partial E/\partial T)\,dT$ the ratio

$$\gamma = \left[\frac{\partial E}{\partial T} + \left(\frac{\partial E}{\partial \rho} - p\rho^{-2} \right) \left(-\frac{\partial p/\partial T}{\partial p/\partial \rho} \right) \right] \bigg/ \frac{\partial E}{\partial T}. \tag{33}$$

On the other hand, a density change satisfying (25) implies a temperature change

$$dT = \frac{p\rho^{-2} - \partial E/\partial \rho}{\partial E/\partial T}\, d\rho, \tag{34}$$

and so the alternative definition (31) of γ gives

$$\gamma = \left[\frac{\partial p}{\partial \rho} + \frac{\partial p}{\partial T} \left(\frac{p\rho^{-2} - \partial E/\partial \rho}{\partial E/\partial T} \right) \right] \bigg/ \frac{\partial p}{\partial \rho}, \tag{35}$$

which evidently is identical with (33).

A more fundamental insight into the nature of the sound speed, however, is given by general thermodynamic theory through the concept of 'entropy'.

For a full account of entropy and its properties, see texts on thermodynamics and statistical physics; here, only those properties most needed for the study of waves in fluids are briefly summarised.

Entropy is a quantity that remains constant in any 'reversible' process, like the changes postulated above as occurring in sound waves. The internal energy E per unit mass changes exactly by the amount (25) in a reversible process; there is no extra change due to dissipation of kinetic energy into heat or due to transfer of heat from outside; also, while the process takes place, the fluid continues to satisfy the same relationships (30) as would characterise equilibrium conditions. Thus, a reversible process is one which so avoids *abrupt gradients* in both space and time that (i) viscous dissipation of energy into heat and conduction of heat produce negligible effects and (ii) the distribution of heat energy as between different modes of molecular motion remains always close to an equilibrium distribution. It is reversible because an equal and opposite change through the same set of equilibrium states restores the initial condition.

It follows that, when gradients in a sound wave are not too abrupt, the entropy per unit mass S remains constant. Accordingly, if a relationship

$$p = p(\rho, S) \tag{36}$$

for a particular fluid in equilibrium conditions can be derived, then

$$c^2 = \partial p / \partial \rho, \tag{37}$$

where this partial derivative now signifies a derivative keeping S constant.

The entropy, however, is not just any quantity constant in a reversible process (a process which in particular satisfies (25)); it and the absolute temperature T are so defined that any departure from (25) in a small change between equilibrium states satisfies

$$dE - p\rho^{-2}d\rho = TdS; \tag{38}$$

in other words, the net additional heat absorbed is TdS. When all quantities are written as functions of ρ and T as in (30), equation (38) implies that

$$\frac{\partial S}{\partial \rho} = \frac{1}{T}\left(\frac{\partial E}{\partial \rho} - p\rho^{-2}\right), \quad \frac{\partial S}{\partial T} = \frac{1}{T}\frac{\partial E}{\partial T}. \tag{39}$$

Hence, expressing that $\partial^2 S/\partial\rho\partial T = \partial^2 S/\partial T\partial\rho$, we obtain Maxwell's relationship

$$\partial E/\partial \rho = (p - T\partial p/\partial T)\rho^{-2}, \tag{40}$$

which can be used in (33) to give another representation of γ in terms of measurable quantities:

$$\gamma = 1 + (\alpha^2 T c_N^2 / c_v), \tag{41}$$

where it is necessary to explain that α is the fluid's coefficient of expansion, $(-\mathrm{d}\rho/\rho\mathrm{d}T)$ for changes at constant pressure; this implies that

$$\mathrm{d}p = c_{\mathrm{N}}^2(\mathrm{d}\rho + \alpha\rho\mathrm{d}T) \qquad (42)$$

for general changes and thus gives the values of $\partial p/\partial\rho$ and $\partial p/\partial T$ inserted in (33) and (40) to derive (41).

The quantity γ always exceeds 1 but does so for most liquids by a considerably smaller margin than for most gases. The margin is exceptionally small for cold water with its unusually low value of the product αT (a value which actually vanishes at $T = 277$ K). Values of αT more typical of liquids in general are exhibited by hot water; the product rises to 0.27 when $T = 371$ K (that is, 98 °C), giving $\gamma = 1.10$. The sound speed c takes values about $1400\,\mathrm{m\,s^{-1}}$ in water, and values of similar magnitude in most liquids.

The second law of thermodynamics states that the *total entropy* in any thermally isolated system can never decrease. Either the processes occurring are reversible and the entropy remains constant, or they deviate from reversibility and the entropy increases; as, for example, when a heat input $T\mathrm{d}S$ per unit mass results from viscous dissipation of kinetic energy, or when a certain quantity of heat is transferred from a part of the system with higher T to a part with lower T (so that the latter gains more entropy than the former loses). From the point of view of statistical physics, entropy is a measure of the randomness of the molecular organisation of the substance and this measure of randomness remains constant in reversible changes, while, however, its total value for an isolated system can only increase in irreversible changes as the system moves into regions of 'state space' with greater and greater probability....

In sound waves, any irreversible processes neglected in section 1.1, including viscosity and heat conduction, must in this way produce increases in the total entropy, corresponding to a heating of the fluid through which the sound wave passes and a corresponding gradual dissipation of the mechanical energy of the sound wave (a quantity whose meaning is made precise in section 1.3). A quantitative investigation of this dissipation process is given in section 1.13.

At this stage we may also discuss briefly any possible effect on sound propagation of another feature neglected in section 1.1; namely, such an external force field as gravity. Its presence means that the pressure p_0 and density ρ_0 in the undisturbed fluid are not uniform, but rather satisfy the hydrostatic relationship $\qquad \nabla p_0 = \rho_0\,\mathbf{g}, \qquad\qquad (43)$

where $\mathbf{g}$ is the vector acceleration due to gravity.

In the first instance we assume, however, that the entropy per unit mass has a uniform value in the undisturbed fluid, and therefore remains uniform (if dissipation be neglected) during the whole propagation of a sound wave. Then the linearised momentum equation (4), which must now have an extra term $\rho\mathbf{g}$ added onto the right-hand side, must after subtraction of equation (43) take the form

$$\rho_0 \, \partial\mathbf{u}/\partial t + \nabla(p - p_0) = (\rho - \rho_0) \, \mathbf{g} = (p - p_0) \, \mathbf{g}c^{-2}, \tag{44}$$

using the relation (37) between changes of p and ρ at constant entropy.

The ratio of a typical magnitude of the gradient term $\nabla(p - p_0)$ in (44) to a typical magnitude of $(p - p_0)$ itself is $2\pi/\lambda$, where λ is a typical wavelength. It follows that any sound waves *with λ much smaller than c^2/g* are negligibly affected by gravity. For air, the ratio c^2/g is about 12 km, so the effect of gravity is negligible for all ordinary sound waves. In the atmosphere it can at most become significant for propagation of extremely slow pressure fluctuations with periods of several seconds (and therefore wavelengths in kilometres). For water, the ratio c^2/g is as big as 200 km, which practically rules out any conceivable influence of gravity on sound waves even in such large bodies of water as the ocean.

These arguments are extended to atmospheres with entropy *stratified* (increasing upwards) in chapter 4, where the *same* condition is found to suffice, even though the equations of motion now permit the existence of other, much more slowly propagating waves (the so-called 'internal gravity waves'), to remove any possible influence of gravity upon sound propagation in general and upon the speed of sound in particular.

1.3   Acoustic energy and intensity

A characteristic property of waves is that they permit transport of energy without the need for any net transport of material. 'Acoustic energy' signifies the part of a fluid's total energy associated with the presence of a sound wave, while 'acoustic intensity' means rate of transport of acoustic energy; two ideas made precise in this section within the context of the linear theory of sound.

In any linear theory of vibrations or waves the disturbances, as noted in section 1.1, are regarded as small quantities whose squares we neglect *in equations of motion* (which implies also neglect of the product of two of the quantities, since its numerical value cannot exceed the square of one or the other of them). A different rule applies, however, in *expressions for energy* and its rate of change (or of transport), where terms in the first powers of small quantities should be absent and accordingly we *retain squares* of small quantities and products of two of them, and *neglect only their cubes* (implying neglect of the product of three or more such quantities). General theories (see texts on the dynamics of vibrations) indicate why these two practices are exactly consistent with one another.

Treatments of kinetic energy give a simple illustration of the above rules. In a fluid flow with velocity vector

$$\mathbf{u} = (u, v, w) \tag{45}$$

the *kinetic energy density* (kinetic energy per unit volume) is

$$\tfrac{1}{2}\rho(u^2 + v^2 + w^2), \tag{46}$$

where ρ is mass density. For any wavelike disturbances to a fluid in whose undisturbed state $\rho = \rho_0$ and $\mathbf{u} = 0$, the kinetic energy density on linear theory is accordingly taken as

$$\tfrac{1}{2}\rho_0(u^2 + v^2 + w^2), \tag{47}$$

whose departure from (46), involving products of three small quantities ($\rho - \rho_0$ and two velocity components), is to be neglected. General vibration theories suggest that this inertial contribution (47) to the energy density for the wave should be matched by a contribution associated with restoring forces (for example, with the compressibility of the fluid in the acoustic case) which may be called the 'potential' energy density; they suggest also that appropriately *averaged* values of kinetic and potential energy should be equal.

In acoustic theory, a little care is needed to obtain correct values, both for this 'potential energy' contribution and for the rate of energy transport. We consider the latter first, and initially in only the very simple case, governed by equations (15), (16) and (17), of a plane wave travelling in the positive x-direction.

In this case we expect to find a positive transport across any plane (x = constant) of energy in the positive x-direction. This requires that fluid to the left of the plane shall act on fluid to the right with a positive rate of working. Such rate of working is the product of (i) the force acting across the plane in the positive x-direction, which per unit area is equal to the pressure p, with (ii) the velocity component in that direction, u. Equation (17) confirms the expected tendency of this rate of working per unit area to be positive, because it shows that the 'excess pressure' $p - p_0$ (excess over the undisturbed value p_0) has always the same sign as u, so that its product with u can never be negative.

Although the true rate of working per unit area is pu, we define the acoustic intensity as

$$I = (p - p_0)u. \tag{48}$$

This means that we ignore any work done by the undisturbed 'atmospheric' pressure p_0, which perhaps could hardly be expected to play a role in energy transport, and welcome (48) as being precisely quadratic in the disturbances (which, as we have seen, an energy transport ought to be). Adopting for the time being the philosophy suggested by these crude arguments, we proceed at present with (48), but return later in the section to a full justification for the omission of the embarrassing linear term $p_0 u$.

Applying the same philosophy, we determine the 'potential energy' of the fluid as the work done on it by action of only *the excess pressure $p - p_0$* during compression to density ρ from the undisturbed density ρ_0. For each small increase $d\rho$ this work done, per unit volume, is

$$\rho(p - p_0)(-d\rho^{-1}) = (p - p_0)\rho^{-1}d\rho, \tag{49}$$

which differs from (25) in the substitution of the excess pressure $p - p_0$ for p itself and in being multiplied by ρ to change the value per unit mass into one per unit volume. Note that (49) is quadratic in small quantities, as an energy contribution should be. This means that the factor ρ^{-1} in it should be replaced by ρ_0^{-1} with an error of order the cube of small quantities, while by (14) the factor $p - p_0$ can be replaced by $(\rho - \rho_0)c^2$. Hence the total potential energy for fluid compressed to density ρ is

$$\int_{\rho_0}^{\rho} (\rho - \rho_0)c^2\rho_0^{-1}\,d\rho = \tfrac{1}{2}(\rho - \rho_0)^2 c^2\rho_0^{-1} = \tfrac{1}{2}(p - p_0)^2 c^{-2}\rho_0^{-1}; \tag{50}$$

two alternative forms which, again, are identical if cubes of disturbances be neglected.

The suggestion from general theory that 'appropriately averaged' values of kinetic and potential energy should be equal is dramatically borne out in this special case of a travelling plane wave: their densities (47) and (50) are in fact *everywhere* equal, since in this case $v = w = 0$ and equation (17) applies. The total acoustic energy density (kinetic and potential) is then

$$W = \rho_0 u^2, \tag{51}$$

while equation (48) for the intensity may be thrown, using (17), into the form

$$I = \rho_0 c u^2. \tag{52}$$

These values are consistent with an expectation that the travelling sound wave may transport energy with velocity c, since the rate of transport of energy per unit area, I, is c times the energy per unit volume, W.

Encouraged by these conclusions for plane waves, we may now extend to general three-dimensional motions the ideas of acoustic energy and intensity, holding to the philosophy that only work done by 'excess' pressures is to be considered. The derivation of expression (50) for potential energy density remains unaltered, but the total acoustic energy (sum of (47) and (50)) can no longer be simplified to the form (51) since equation (17) does not hold in general. We may instead use equations (8) and (9) to write this acoustic energy or 'wave energy' W in terms of the velocity potential ϕ as

$$W = \tfrac{1}{2}\rho_0[(\nabla\phi)^2 + c^{-2}(\partial\phi/\partial t)^2].\tag{53}$$

In this three-dimensional case we define the acoustic intensity as a *vector* $\mathbf{I}$, whose component $\mathbf{I}\cdot\mathbf{n}$ in the direction of any unit vector $\mathbf{n}$ is the rate at which energy is being transported in the direction of $\mathbf{n}$ across a small plane element at right angles to $\mathbf{n}$, per unit area of that plane element. This energy transport is the effect of a force $p - p_0$ per unit area (if we consider *excess* pressure only) multiplied by the component $\mathbf{u}\cdot\mathbf{n}$ of velocity in the direction of $\mathbf{n}$ to give a rate of working $(p - p_0)\mathbf{u}\cdot\mathbf{n}$ per unit area, and thus as a three-dimensional generalisation of (48) the formula

$$\mathbf{I} = (p - p_0)\mathbf{u},\tag{54}$$

which in terms of the velocity potential ϕ becomes

$$\mathbf{I} = -\rho_0(\partial\phi/\partial t)\nabla\phi.\tag{55}$$

An important check on the *mutual* consistency of equations (53) and (55) is that they satisfy the equation of *conservation of acoustic energy*:

$$\partial W/\partial t = -\nabla\cdot\mathbf{I},\tag{56}$$

which equates the rate of change of acoustic energy in a small region to the total rate of transport into that region (which per unit volume of the region is minus the divergence of the energy transport vector $\mathbf{I}$). Note that any term such as $\mathbf{u}\cdot\nabla W$ that might be included on the left-hand side of (56) to represent convection of acoustic energy by the fluid velocity $\mathbf{u}$ (which may act additionally to any energy changes due to working by fluid pressures) may be neglected here because W is of the order of the squares of small quantities and so $\mathbf{u}\cdot\nabla W$ is of the order of their cubes. Actually, equation (56) as it stands is exactly satisfied by a linear-theory velocity potential ϕ, for (53) gives the left-hand side as

$$\rho_0[(\nabla\phi)\cdot\nabla(\partial\phi/\partial t) + c^{-2}(\partial\phi/\partial t)(\partial^2\phi/\partial t^2)],\tag{57}$$

while (55) gives the right-hand side as

$$\rho_0[(\nabla\phi)\cdot\nabla(\partial\phi/\partial t)+(\partial\phi/\partial t)\nabla^2\phi], \tag{58}$$

and these are equal by the wave equation (13). It is certainly reasonable, then, to regard W as the density of some useful quadratic measure of acoustic amplitude, and $\mathbf{I}$ as the vector flux of the same quantity.

We ought, on the other hand, to enquire why this is so, since by ignoring any work done by the undisturbed atmospheric pressure p_0 we have certainly failed to include the whole energy density in W. Equation (49) shows that the part excluded is

$$W_{\text{ex}} = \int_{\rho_0}^{\rho} p_0\rho^{-1}\mathrm{d}\rho = p_0\ln(\rho/\rho_0), \tag{59}$$

while equation (54) shows that the corresponding part excluded from the rate of energy transport $\mathbf{I}$ is

$$\mathbf{I}_{\text{ex}} = p_0\,\mathbf{u}. \tag{60}$$

Why can terms like these, for example, such a part of $\mathbf{I}$ proportional to $\mathbf{u}$, be excluded?

In answering this question we must avoid the argument that such a term (60) in the energy transport would be unimportant because its average value in some sense would be zero. To be sure, sound is often generated by vibrations, which on linear theory produce fluctuations in the fluid velocity $\mathbf{u}$ about a zero mean. If, however, $\mathbf{I}_{\text{ex}}$ were to be incorporated with the remainder of $\mathbf{I}$, which is of the order of squares of small quantities, then we should have to include in $\mathbf{I}_{\text{ex}}$ contributions to $\mathbf{u}$ of the same order, that could be calculated only by a nonlinear theory and thus might have a nonzero mean (those would be 'rectified' motions often described as 'acoustic streaming' and studied in more detail in section 4.7 below).

The fact that $\mathbf{I}_{\text{ex}}$, and similarly W_{ex}, could properly be retained *only* if values of $\mathbf{u}$ and ρ in them had been calculated to *second* order in this way, is a powerful incentive (along with the others noted earlier) to exclude them, but implies that any arguments to justify their exclusion must use exact and not merely linearised relationships between the variables concerned. Fortunately it is exactly true that these excluded quantities themselves satisfy a conservation equation

$$\partial W_{\text{ex}}/\partial t+\mathbf{u}\cdot\nabla W_{\text{ex}} = -\nabla\cdot\mathbf{I}_{\text{ex}}, \tag{61}$$

stating that total rate of change for a fluid element of the 'potential' energy associated with working by the undisturbed pressure p_0 equals the rate of energy transport into the element due to that component of the rate of working by neighbouring elements which springs from action by them with pressure p_0. This statement is almost self-evident, and in any case equation (61) is seen by (59) and (60) to be an *exact* multiple of the equation of continuity (3).

This makes clear the status of our equation of conservation of acoustic energy (56): it represents a full equation of conservation of energy with a multiple of the equation of continuity subtracted from it. This multiple (61) can be thought of as a separate conservation equation for the part of the total energy excluded from the 'acoustic energy': one which includes the convective term $\mathbf{u} \cdot \nabla W_{ex}$ because W_{ex} includes first powers of small quantities whereas the corresponding term in (56) was omitted because W does not. The definitions for W and $\mathbf{I}$ with W_{ex} and $\mathbf{I}_{ex}$ excluded, that is, equations (53) and (55), are especially valuable because solutions of the linear wave equation (13) can always be used in them if cubes of disturbances be neglected *and* must satisfy exactly the linear-theory conservation equation (56).

Acoustic intensity may be measured in $W\,m^{-2}$ (watts per square metre), but a logarithmic scale of intensity level is appropriate when the subjective effect of sound is considered, because for a given frequency in Hz (cycles per second) the ear senses equal differences in loudness for equal differences in the *logarithm* of the intensity rather than in the intensity itself. The 'decibel' scale defines the intensity level in decibels (dB) as

$$120 + 10 \log_{10}(I/W\,m^{-2}), \tag{62}$$

the bracket here signifying the magnitude I of the intensity vector, measured in $W\,m^{-2}$. A typical minimum intensity level for audibility of sounds at typical 'higher' frequencies (500 to 8000 Hz) is 0 dB, which by (62) means $I = 10^{-12}\,W\,m^{-2}$.

A similar 'threshold of hearing' at lower frequencies, or at extremely high frequency, is greater: for example 20 dB (meaning $I = 10^{-10}\,W\,m^{-2}$) either at 200 Hz or at about 15 000 Hz, and 40 dB (meaning $I = 10^{-8}\,W\,m^{-2}$) either at 100 Hz or at about 18 000 Hz. Hearing disappears below about 20 Hz and above about 20 000 Hz. Sound of most frequencies causes actual pain above about 120 dB (meaning $I = 1\,W\,m^{-2}$).

Estimation of the power output in watts of different types of sources is discussed in later sections, but we may note typical values in a few cases here; those values become intensities in $W\,m^{-2}$, at a distance r in metres from the source, *when divided by* $4\pi r^2$ (neglecting the source's directional nature if any). The power output of the human voice is around $10^{-5}\,W$ in ordinary conversation, rising to 0.03 W in loud singing. On the other hand, acoustic power from engines may be enormously greater, rising to values as great as 10^5 W for the huge rocket motors used for launching large space-craft.

1.4 The simple source

The wave equation (13), which the velocity potential ϕ satisfies in the linear theory of sound, has a vast variety of solutions in addition to those like (15), (18) or (21) that represent plane waves. Further studies founded on essentially one-dimensional solutions like those are postponed to chapter 2, while in the rest of chapter 1 we develop the properties of an important range of essentially three-dimensional solutions. In so doing, however, we occasionally make comparisons with the properties of plane-wave solutions, since it proves specially instructive to contrast the characteristics of sound-generation mechanisms in different numbers of dimensions.

An understanding of highly complex sound-generation mechanisms or 'sources' of sound can often be built up from the study of a few fundamental solutions of the wave equation. Of these, the simplest is one without any directional nature: such a source, radiating sound equally in all directions, is understandably called a 'simple source' (or alternatively, for reasons that become clearer in section 1.5, a 'monopole' source). Its properties are worked out in detail in this section.

To describe such a simple source, lacking directional dependence, we seek a potential ϕ spherically symmetrical about some central point. In fact

$$\phi = \phi(t, r), \tag{63}$$

depending only on the time t and on the distance r from that central point, must take identical values at points equidistant from it in *all* directions at a fixed time. The Laplacian of (63) is easily calculated as

$$\nabla^2\phi = \partial^2\phi/\partial r^2 + 2r^{-1}\partial\phi/\partial r = r^{-1}\partial^2(r\phi)/\partial r^2, \tag{64}$$

and so the wave equation (13) can be written

$$\partial^2(r\phi)/\partial t^2 = c^2\partial^2(r\phi)/\partial r^2. \tag{65}$$

It is advantageous to be studying the wave equation in a universe with *three* spatial dimensions because only in this number of dimensions can the wave equation for a spherically symmetrical potential ϕ be thus reduced to a one-dimensional wave equation (65) for a simple multiple of it (namely

$r\phi$). That equation is seen, by reference to equations (15) and (18) above, to have the general solution

$$r\phi = f(r-ct)+g(r+ct),\tag{66}$$

but since our sole concern is with waves travelling in the direction r increasing (that is, outwards from the source) we consider only the case $g = 0$ in what follows.

It is interesting to compare (66) with the classical spherically symmetric flow of an 'incompressible fluid'; that is, of an idealised fluid whose density is unaltered by pressure changes, so that $c^2 = dp/d\rho$ is infinite and (13) becomes Laplace's equation $\nabla^2\phi = 0$. That classical flow, also described as a simple source (see texts on fluid dynamics), has

$$\phi = -m(t)/4\pi r.\tag{67}$$

Here, the coefficient $m(t)$ signifies the rate of volume outflow from the centre at time t; indeed, for all r the volume flow through the area $4\pi r^2$ of the fluid a distance r from the centre is

$$4\pi r^2 \partial\phi/\partial r = m(t).\tag{68}$$

This flow at distance r responds instantaneously to variation in the outflow $m(t)$ from the central point because the time for information to travel a distance r is zero for an ideal fluid with infinite sound speed.

We can throw the acoustic solution (66) (with $g = 0$) into a form closely analogous to (67) if the function $f(x)$ is written as $-m(-x/c)/4\pi$ to give

$$\phi = -m(t-r/c)/4\pi r.\tag{69}$$

This differs from (67) *only* by incorporating a time-lag r/c for information about changes in volume outflow $m(t)$ to travel a distance r at the sound speed c; an extraordinarily *modest* difference between solutions of Laplace's equation and of the wave equation, which we may note again as exclusive to space of *three* dimensions!

The volume flow through the area $4\pi r^2$ of the fluid a distance r from the centre is now

$$4\pi r^2 \partial\phi/\partial r = m(t-r/c)+(r/c)\dot{m}(t-r/c),\tag{70}$$

where, in contrast to (68), there are two terms in $\partial\phi/\partial r$ because this time-lag r/c varies with r. That increase in complexity disappears as $r \to 0$, however, when the limiting volume flow emerging from the central point is seen from (70) to be $m(t)$ as before.

Note that r/c need not be exceedingly small for (70) to be nearly $m(t)$, since it constitutes the *first two terms* of the Taylor expansion of $m(t)$ about the point $t-r/c$; we may say in fact that the volume flow at moderate distance r 'makes an attempt' to match the instantaneous outflow value $m(t)$ by a linear extrapolation (using the time rate of change $\dot{m}$) through a time-interval r/c. If typical values of $\ddot{m}$ are in the ratio ω^2 to typical values of m (so that ω is a typical radian frequency of the sound), then the relative error in putting (70) equal to $m(t)$ is of order $(\omega r/c)^2$. This suggests, what later sections confirm, that a source of finite radius r producing variable volume outflow $m(t)$ may act very like a simple point source if the ratio $(\omega r/c)$ be small.

The excess pressure $p-p_0$, given by (9) on the linear theory of sound, is

$$p-p_0 = \dot{q}(t-r/c)/4\pi r \qquad (71)$$

for a simple source, where $q(t) = \rho_0 m(t)$ (72)

can be described on linear theory as the rate of *mass* outflow from the source. In acoustics this rate of mass outflow $q(t)$ is often emphasised in preference to the volume outflow because mass is a more physically significant variable: even a compressible fluid satisfies mass conservation exactly where 'sources' of mass outflow are absent. In acoustics, again, excess pressure (and *not* velocity potential) is the quantity most readily measurable. For both these reasons, equation (71) relating $p-p_0$ to $q(t)$ is regarded as the fundamental equation for a simple source.

Note that it is the *rate of change* of mass outflow, $\dot{q}(t)$, whose variation is mimicked by the excess pressure (71) a time r/c later. This is why the time derivative $\dot{q}(t)$ is commonly called the *strength* of the simple source.

We may compare this property of a simple source in three dimensions (rate of change of mass outflow mimicked by pressure fluctuations) with a contrasting situation in one dimension. Plane waves propagating in the x-direction along a tube of uniform cross-sectional area A satisfy equation (17) and so, if they are generated at $x = 0$ by a fluctuating mass outflow

$$q(t) = A\rho_0(u)_{x=0} = Ac^{-1}(p-p_0)_{x=0}, \qquad (73)$$

the excess pressure for $x > 0$ takes the form

$$p-p_0 = cA^{-1}q(t-x/c). \qquad (74)$$

Expression (74) differs from the three-dimensional form (71) not only in the rather natural lack of the spherical attenuation factor $(4\pi r)^{-1}$ but also in depending directly upon the mass outflow $q(t)$ rather than upon its time derivative.

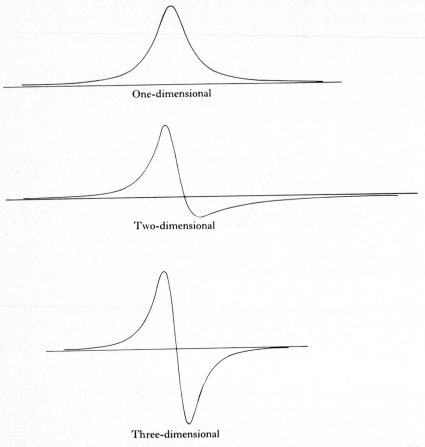

One-dimensional

Two-dimensional

Three-dimensional

Figure 1. Sound pulses generated by a positive pulse of mass outflow (varying with time t like $(t^2 + \tau^2)^{-1}$, where τ is constant) in one-dimensional, two-dimensional and three-dimensional propagation. The pulses (illustrated with arbitrary vertical scales) are proportional to the mass outflow, to its ($\frac{1}{2}$)th derivative and to its first derivative respectively.

Figure 1 shows how a positive pulse of mass outflow generates a proportional positive pulse of pressure excess in such one-dimensional propagation, but generates a pulse of the quite different shape obtained by differentiation in three-dimensional propagation. These contrasts suggest the question: what happens in the intermediate case of two-dimensional propagation?

We can answer this by considering sound generation by an infinite line of sources producing a uniform mass outflow $h^{-1}q(t)$ per unit length. To obtain the excess pressure at a point whose perpendicular distance from that line is s, note that the combined length of those two parts of the line on which distance from that point takes values between r and $r+dr$ is given by simple trigonometry as $2r(r^2-s^2)^{-\frac{1}{2}}\,dr$ if $r > s$, but that there are no such parts if $r < s$, which means that we need to replace $\dot{q}(t-r/c)$ in (71) by $2r(r^2-s^2)^{-\frac{1}{2}}h^{-1}\dot{q}(t-r/c)\,dr$ and integrate from $r = s$ to $r = \infty$, giving

$$p-p_0 = (2\pi h)^{-1}\int_s^\infty (r^2-s^2)^{-\frac{1}{2}}\,\dot{q}(t-r/c)\,dr. \tag{75}$$

This two-dimensional simple-source pressure field (75), which experimentally can be generated in a fluid confined between parallel planes a distance h apart when along a line perpendicular to them there occurs a mass outflow $q(t)$ (for example, due to an exploding wire), is seen to have a considerably more complicated form than those, (71) and (74), found in three dimensions and one dimension, respectively.

If the mass outflow $q(t)$ consists of a pulse as in figure 1, then equation (75) becomes rather simpler where s/c is very large compared with the pulse duration. Then we can approximate r^2-s^2 as $2s(r-s)$ because contributions to (75) from values of r with r/s near 1 may be shown to be much greater than contributions from larger values of r. This substitution for large s gives

$$p-p_0 = (2h)^{-1}(c/2\pi s)^{\frac{1}{2}}(d/dt)^{\frac{1}{2}}q(t-s/c), \tag{76}$$

where the derivative of $(\frac{1}{2})$th order has its usual definition:

$$(d/dt)^{\frac{1}{2}}q(t) = \int_{-\infty}^t \dot{q}(T)[\pi(t-T)]^{-\frac{1}{2}}\,dT. \tag{77}$$

Such dependence for large s on the $(\frac{1}{2})$th derivative of $q(t-s/c)$ places the two-dimensional case in a position intermediate between the one- and three-dimensional cases (74) and (71), and the pulse form (76) is seen in figure 1 to be also intermediate in character; instead of a purely positive pulse, or a once-differentiated form with deep short positive lobe followed by identical negative lobe, we observe a deep short positive lobe followed by an extended but shallower negative lobe of the same total area. Further discussion of such two-dimensional propagation is postponed to chapter 2.

In the meantime, we extend the comparisons between simple sources in one dimension and in three to questions of acoustic intensity and power output. In the one-dimensional case (74) the intensity (48) is given using (17) as

$$I = (p-p_0)u = (p-p_0)^2/\rho_0 c = c\rho_0^{-1}A^{-2}q^2(t-x/c) \tag{78}$$

consistent with a power output

$$c\rho_0^{-1} A^{-1} q^2(t) \tag{79}$$

generated at the source and transported away from it at velocity c through a tube of cross-section A.

In the three-dimensional case equation (17) is not necessarily satisfied, and we have instead (recasting (70) in terms of $q(t)$) a radial velocity

$$u_r = \partial\phi/\partial r = (\rho_0 c)^{-1} [\dot{q}(t-r/c) + (c/r) q(t-r/c)]/4\pi r. \tag{80}$$

Equation (80) is written in a form which proves that spherical waves do satisfy more and more closely the relationship

$$u_r = (\rho_0 c)^{-1}(p - p_0) \tag{81}$$

as $r \to \infty$, so that in this respect they become more and more like plane waves. In terms of the ratio ω of a typical magnitude of $\dot{q}$ to a typical magnitude of q, this approximation is seen to be good where $\omega r/c$ is *large*. For sources of sound in general (not only simple sources) with such a typical radian frequency ω, we use the term 'far field' to mean the part of the fluid whose distance r from the source is large compared with c/ω (which alternatively can be written $\lambda/2\pi$, where λ is a typical wavelength), and continue to find (section 1.6) that in the far field the plane-wave relationship (81) becomes a good approximation.

In the far field the intensity vector (54) points radially outwards from the source and has magnitude

$$I = (\rho_0 c)^{-1} \dot{q}^2(t-r/c)/16\pi^2 r^2, \tag{82}$$

just as if a power output $\quad P(t) = \dot{q}^2(t)/4\pi\rho_0 c \tag{83}$

generated at the source were transported away from it at velocity c, passing after a time delay r/c through the area $4\pi r^2$ of fluid at distance r from the source. This description of power generation by the source is in many ways acceptable, in spite of the fact that at points closer than the far field the second term in square brackets in (80) modifies (82) with an additional term

$$\rho_0^{-1} q(t-r/c)\, \dot{q}(t-r/c)/16\pi^2 r^3 = (\mathrm{d}/\mathrm{d}t)\,[\rho_0^{-1} q^2(t-r/c)/32\pi^2 r^3], \tag{84}$$

because the average value over a long period of such a time derivative of a bounded fluctuating quantity is zero. Thus the contribution to outward transport of energy made by (84) produces no net transport; in fact, we may check from equation (56) that it corresponds only to a surging to and fro of the additional kinetic energy that is required in the near field to balance the mass outflow from the source.

Thus, given fluctuations of mass transport $q(t)$ generate an acoustic power output into the far field given by (83), typical values of which differ from those for the one-dimensional case (79) by a factor

$$A\omega^2/4\pi c^2 = \pi A/\lambda^2, \tag{85}$$

where λ is a typical wavelength. This means that loudspeakers small on a scale of wavelengths, and so with areas A small compared with λ^2/π, generate acoustic power much *less* efficiently when they radiate in *three* dimensions than when their radiation is ducted one-dimensionally along a tube of cross-section A. (This statement assumes what was suggested following equation (70), that variations of mass outflow $q(t)$ spread over a region whose dimension is small compared with $c/\omega = \lambda/2\pi$ radiate like a point source; a result more fully demonstrated in section 1.6.) Examples of such superior efficiency of one-dimensional over three-dimensional sound generation are given in more quantitative detail in sections 1.11 and 1.12.

1.5   The acoustic dipole

Among those fundamental solutions of the wave equation from whose properties has been created an understanding of highly complex sources of sound, the next most important after the simple or monopole source is the dipole source. Study of the acoustic dipole in this section shows it to possess in greatly intensified form certain properties of the three-dimensional simple source noted in section 1.4: differences between far field and near field are enormously *more* pronounced, leading to the dipole's still greater inefficiency as an acoustic-power generator (a role in which simple sources in three dimensions, however inefficient relative to their one-dimensional counterparts, are found to outshine all *three*-dimensional rivals!).

We can begin to appreciate the dipole's properties by considering a solution of the wave equation (13) obtained by adding up two simple-source solutions: (i) the simple source of section 1.4 centred on the origin of coordinates (0, 0, 0), with pressure field described by equation (71) in terms of its strength $\dot{q}(t)$ and r the distance from the origin, and (ii) a simple source of equal and opposite strength $(-\dot{q})$ centred on a neighbouring point $(-l, 0, 0)$ with pressure field given by (71) with $\dot{q}$ replaced by $(-\dot{q})$ and r by r' the distance from $(-l, 0, 0)$; then

$$p - p_0 = [\dot{q}(t-r/c)/4\pi r] - [\dot{q}(t-r'/c)/4\pi r'] \tag{86}$$

represents the pressure field of the combined configuration, as illustrated

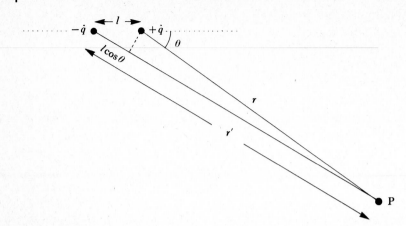

Figure 2. The sound field at P, due to a simple source of strength $\dot{q}$ at (o, o, o) and one of equal and opposite strength $-\dot{q}$ at $(-l, o, o)$, depends on the difference between the distances r and r' of P from the two sources.

in figure 2 in the case when r is large compared with l. This, in fact, is the case relevant to a dipole, which inherently involves a limiting form as l becomes small.

Expression (86) is the difference between the values at a fixed time t of expression (71) for two nearby values of r, namely r and r', whose difference cannot exceed l; figure 2 shows indeed that the difference $(r-r')$ is approximated more and more closely by $(-l\cos\theta)$ in a direction making an angle θ with the x-axis as l becomes small compared with r. The corresponding difference in the values of (71), due to differences both in the spherical attenuation factor $(4\pi r)^{-1}$ and in the time-lag (r/c), can for l small enough (see below for an estimate of how small is necessary) be expressed in terms of a derivative with respect to r:

$$p - p_0 \doteqdot (-l\cos\theta)(\partial/\partial r)[\dot{q}(t-r/c)/4\pi r]. \tag{87}$$

Figure 3 depicts another way of reaching the same result: the parallelogram indicates that, as r is distance of the point (x, y, z) from the origin so r' can be viewed as a distance from the origin of a nearby point $(x+l, y, z)$. Hence (86) is the difference between the values of (71) at (x, y, z) and at $(x+l, y, z)$ which for small enough l can be written

$$p - p_0 \doteqdot -l(\partial/\partial x)[\dot{q}(t-r/c)/4\pi r]; \tag{88}$$

this is identical with (87) since, with $r = (x^2+y^2+z^2)^{\frac{1}{2}}$,

$$\partial r/\partial x = x/r = \cos\theta. \tag{89}$$

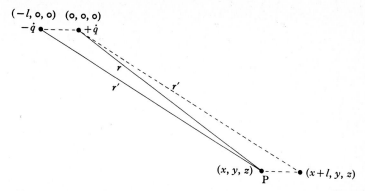

Figure 3. The distance r' of the point P with coordinates (x, y, z) from the source of strength $-\dot{q}$ at $(-l, 0, 0)$ is the same as the distance of a point with coordinates $(x+l, y, z)$ from the source of strength $\dot{q}$ at $(0, 0, 0)$.

The concept of the dipole, in acoustics as in other branches of physics, derives from the observation that expressions like (87) or (88) do not depend on l and $\dot{q}$ respectively but only on the product

$$l\dot{q}(t) = G(t), \tag{90}$$

which may be called the *strength* of the dipole. In other words, when l is small enough for equations (87) and (88) to be correct, the pressure field is the same as it would be for another pair of sources with the strength $\dot{q}$ doubled but the separation l halved.

We can answer the question of how small l must be for this to be so through the mean value theorem which states that each equation is correct if the derivative on the right-hand side is evaluated at *some intermediate value* of r between r and r'; replacing this intermediate value by r itself is only a good approximation if the derivative varies by only a small fraction of itself when r is changed by at most l. This condition is seen by inspection of the derivative (for example, when expanded out as in (92) below) to require not only r large compared with l as already assumed, but also $\omega l/c$ small, where ω is a typical frequency (ratio of a typical value of $\ddot{q}$ to a typical value of $\dot{q}$).

The latter condition for the classical dipole formula to represent a source region of finite size l like that of figure 2 is often described as the 'compactness' condition: the source region is *acoustically compact* if

$$l \ll (c/\omega) = (\lambda/2\pi). \tag{91}$$

In section 1.4 (discussion following equation (70)) we found an identical condition for mass outflow from a region of finite radius r to be represented by the corresponding simple source. Acoustical compactness is found to be generally important as the condition which allows the use of elementary solutions like simple sources and dipoles to model complex sources of sound.

It is instructive to postpone making the substitution (90) until we have expanded the derivative in (87) to give

$$p - p_0 = \cos\theta\,\{[l\dot{q}(t-r/c)/4\pi r^2] + [l\ddot{q}(t-r/c)/4\pi rc]\} \qquad (92)$$

and compared the result with the pressure fields of the individual sources. The most distinctive feature emerging from this comparison is the directional effect represented by the factor $\cos\theta$, which takes values from -1 to $+1$, but differences in the dependence upon r also repay study.

Typical values of the two terms in square brackets in (92) are in the following ratios to typical values of the pressure field (71) of the positive source alone:

$$(l/r) \quad \text{and} \quad (\omega l/c), \qquad (93)$$

where ω again signifies a typical radian frequency (ratio of a typical value of $\ddot{q}$ to a typical value of $\dot{q}$). Both ratios (93) are quantities that must be small if (92) is to be correct at all. Thus, any acoustically compact source pair produces, at distances r large compared with l, a dipole pressure field which is small compared with that due to each source separately.

Note also that of these two ratios (93) the *first* becomes by far the smaller for large enough r, and specifically in what was defined in section 1.4 as 'the far field', where

$$r \gg (c/\omega) = (\lambda/2\pi). \qquad (94)$$

The pressure excess in this far field is dominated by the second term in (92), obtained by applying the differentiation in (87) solely to the time-lag (r/c). In the near field, however, the first ratio in (93) is at least as important, becoming dominant when $\omega r/c$ is small, and this implies the importance of the first term in (92), obtained by applying the differentiation in (87) to the spherical attenuation factor $(4\pi r)^{-1}$.

Thus, the pressure field of a dipole, quite apart from its directional dependence on $\cos\theta$, has a dual structure in its dependence on r, characterised by (i) a far field (94) where excess pressure, varying as r^{-1}, results from differences in phase of signals from the two equal and opposite sources due to differences in their time of travel r/c to the point of observation, and (ii) a near field where an excess-pressure contribution proportional to r^{-2}, due to differences in the degree of spherical attenuation of the two signals, becomes dominant. This dual structure is in strong contrast to the case of a simple source, for which the pressure excess (71) varies as r^{-1} for all r, and the 'far field' is merely where the relation (81) holds, causing the radial velocity u_r also to vary in this way.

For a dipole as for a source, this simple relation between u_r and p again becomes valid in the far field. In fact the radial component of the linearised momentum equation (4), namely

$$\rho_0\, \partial u_r/\partial t = -\, \partial p/\partial r, \tag{95}$$

yields the plane-wave relation (81) in the far field, where to close approximation the $\partial/\partial r$ on the right acts only on the time-lag (r/c) in (87) (a time-lag equally present for plane waves) and not on the characteristically three-dimensional elements (the spherical attenuation factor $(4\pi r)^{-1}$ and the directional factor $\cos\theta$). It follows that the intensity vector (54) points radially outwards and has magnitude

$$I = (\cos^2\theta)(\rho_0 c)^{-1} l^2 \ddot{q}^2(t-r/c)/16\pi^2 r^2 c^2, \tag{96}$$

corresponding to an energy flux, across a sphere of radius r centred on the dipole, given by integrating (96) over the sphere's surface as $P(t-r/c)$, where the quantity

$$P(t) = l^2 \ddot{q}^2(t)/12\pi\rho_0 c^3 \tag{97}$$

can be described as the acoustic power output of the source pair.

Under the compactness condition (91) this source pair is seen by comparing (97) with (83) to be grossly inefficient, the power output being less than that of one source alone by a factor of order $(\omega l/c)^2$. This inefficiency of the dipole compared with a source is of course related to the dual structure of its pressure field, with a relatively powerful near field radiating little energy away into the much weaker far field.

Note that even with the compactness condition ($l \ll \lambda/2\pi$) the source pair generates *only where $r \gg l$* a dipole field, that itself includes a far field where $r \gg \lambda/2\pi$ and a near field. Thus, even the 'near' dipole field excludes the region with r comparable to l. If this region must be given a name then some designation like 'proximal field' is appropriate. Equation (86) must be used in the proximal field, where the two sources do not even approximate to a dipole. Outside it, the dipole approximation holds, leading to equation (92), in which the first term dominates in the near field and the second term in the far field.

An important feature of the dipole near field is that it carries a fluctuating quantity of momentum, related to the force with which the dipole acts on

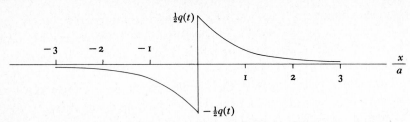

Figure 4. Mass flow across a disk, which is the cross-section of a fluid cylinder of radius a, plotted as a function of its distance x from a simple source of strength $\dot{q}(t)$ on the axis of the cylinder. The quantity plotted is also the distribution of fluid momentum per unit length along the cylinder.

the external fluid. The momentum is directed along the line joining the two sources (the *axis* of the dipole). Convergence difficulties in defining it can cause treatments careless of such matters to give incorrect answers, but the difficulties disappear if we consider momentum in a *circular cylinder* of radius a whose axis is that of the dipole, where

$$l \ll a \ll \lambda/2\pi. \tag{98}$$

The axial momentum in such a cylinder changes at a rate exactly equal to the axial force applied by the dipole because pressure forces acting on the boundary of the cylinder have no axial component.

To study the distribution of this axial momentum, we consider first that associated with just the positive source, of strength $+q(t)$. The momentum density in the near field for such a source is given by equations (69) and (70) as

$$\rho_0 \, \partial\phi/\partial x = \rho_0(x/r) \, \partial\phi/\partial r$$
$$= \rho_0(x/r) \left[m(t - r/c) + (r/c)\dot{m}(t - r/c) \right]/4\pi r^2$$
$$\doteqdot \rho_0(x/r) \left[m(t) \right]/4\pi r^2 = xq(t)/4\pi r^3, \tag{99}$$

where the expressions in square brackets are approximately equal where $\omega r/c$ is small (that is, in the near field) for reasons given following equation (70). The integral of (99) across a disk of radius a for fixed x is equal to

$$\tfrac{1}{2}xq(t) \left[|x|^{-1} - (a^2 + x^2)^{-\frac{1}{2}} \right] \tag{100}$$

and is graphed in figure 4. This represents the *momentum per unit length* of cylinder, but can alternatively be thought of as the *mass flow across the disk* in the x-direction; an interpretation that explains the discontinuous increase by $q(t)$ at $x = 0$, where new mass is emerging at a rate $q(t)$. The total momentum, obtained as the integral of expression (100) from $x = -\infty$ to $x = +\infty$ (an integral which converges at both ends) is zero, as would be expected for a simple source by symmetry.

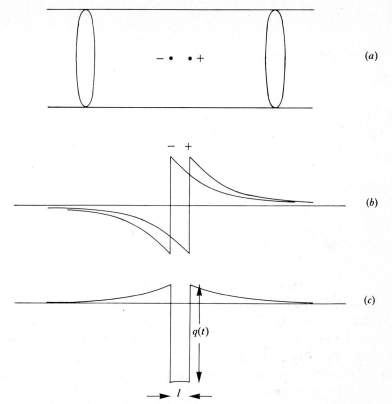

Figure 5. Distribution of momentum per unit length along the fluid cylinder (*a*), due to a pair of equal and opposite sources on its axis. By combining distributions (*b*), as given by figure 4, with opposite signs we obtain the net distribution (*c*).

For the source pair, the total axial momentum, obtained by adding up the zero values for each source separately, must also be zero. The *distribution* of the momentum is quite peculiar, however: a large negative contribution in the short region between the two sources, due to direct flow from the positive source back into the negative source, is balanced by an equal positive contribution thinly spread over the dipole near field. Figure 5 shows how such a distribution is derived as a difference between the distribution of figure 4 and the same distribution shifted so as to be centred on the negative source. This difference incorporates two discontinuities: one of $-q(t)$ at the negative source and one of $+q(t)$ at the positive source.

In the limit for small l, the negative momentum between the two sources is seen from figure 5 to be $-lq(t)$, and so the counterbalancing positive momentum spread over the dipole near field is $+lq(t)$. A dipole is of course a limiting case of a source pair as $l \to 0$, when the negative momentum in the proximal field 'gets lost' between the sources, and all the momentum in the dipole near field adds up to the positive value $+lq(t)$.

Thus, any sound-generating mechanism that can set up a dipole field must supply a momentum $lq(t)$, and so must act on the external fluid with a force equal to the rate of change, $l\dot{q}(t)$, of this momentum, which equation (90) defined as the *strength $G(t)$* of the dipole. In other words, a dipole field of strength $G(t)$ requires a force $G(t)$ acting on the fluid to set it up, the direction of the force being from the negative source towards the positive one. Conversely, we shall see that an external force acting on the fluid generates an acoustic dipole field of strength equal to that force.

In the above discussion, the force $G(t)$ is in the x-direction and the associated dipole field (88) is

$$p - p_0 = -(\partial/\partial x)[G(t-r/c)/4\pi r]; \tag{101}$$

but evidently if the force were in the y-direction or z-direction respectively we should have $\partial/\partial y$ or $\partial/\partial z$ replacing $\partial/\partial x$ in (101). When a general force $\mathbf{G} = (G_x, G_y, G_z)$ acts, the resulting dipole field is the sum of terms (101) associated with each component, each differentiated with respect to the associated coordinate, and can be written as a divergence:

$$p - p_0 = -\nabla \cdot [\mathbf{G}(t-r/c)/4\pi r]. \tag{102}$$

We call (102) the pressure field of a dipole of *vector strength* $\mathbf{G}(t)$. Its form in the far field, corresponding to the last term in (92), is

$$p - p_0 = \mathbf{r} \cdot \dot{\mathbf{G}}(t-r/c)/4\pi r^2 c \tag{103}$$

at a point with position vector $\mathbf{r} = (x, y, z)$, and the associated power output, corresponding to (97), is

$$P(t) = \dot{\mathbf{G}}^2(t)/12\pi\rho_0 c^3. \tag{104}$$

These properties of the dipole source, associated with a force generating a rate of change of momentum of the external fluid, are used in parallel with the properties of the monopole source (section 1.4), associated with a rate of change of mass outflow into the external fluid, to understand many complex acoustic phenomena.

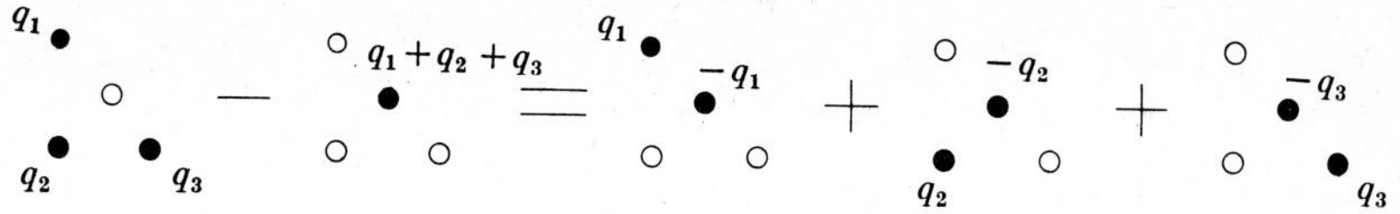

Figure 6. The difference between the fields of a small group of simple sources and of one simple source with strength equal to their total strength can be expressed as a sum of dipole fields.

1.6 Compact source regions in general

As an example of this, we may study sound generation by compact source regions in general: that is, regions with diameter small compared with $\lambda/2\pi$, where λ is a typical wavelength of the sound generated. First we consider (figure 6) a small group of *simple* sources filling such a compact region, in a case when adding the source strengths gives a total *not* much smaller than individual source strengths. We compare the pressure fields of this group of sources with that of one simple source, situated at some central point of the region and with strength equal to that total.

Figure 6 shows that the difference between the two pressure fields is simply a sum of dipole fields, each generated by a compact source pair involving a negative source at the centre and a positive source at one of the peripheral points. We inferred from the ratios (93) that the dipole field generated, at distances r much larger than the distance between such a compact pair of sources, is small compared with that due to each source separately. It follows that the whole difference field illustrated in figure 6 is small compared with that of the single source at all distances r much larger than the diameter of the source region.

Here we used the assumption that the strength of the central source (sum of the strengths of the sources in the group) is not itself small compared with individual source strengths. A group of sources satisfying this assumption generates, at distances large compared with its size (assumed compact), a pressure field close to that of a single source.

For example, a device of the 'siren' type, which allows compressed air to escape in periodic fashion from a series of holes spaced evenly round a stationary disk when an identical contiguous disk rotates so that its holes are periodically opposite those of the stationary disk, radiates sound like one simple source (equally in all directions) if it is acoustically compact. From each hole the rate of mass outflow, say $q(t)$, periodically rises sharply to a peak and falls sharply back to zero, generating an acoustic simple-source field of strength $\dot{q}(t)$. The acoustically compact group of sources located at the different hole positions then generates, at distances large compared with its size, a pressure field close to that of one simple source whose strength equals the sum of these individual source strengths.

Here we do not have to go to the 'far field' to find a pressure field close to that of a simple source; we must avoid only the 'proximal field' (see section 1.5) rather than the whole 'near field'. Certain more complicated source regions, however, generate sound close to simple-source character only in their far fields. These compact source regions (quite commonly found) involve *both* simple sources due to mass outflow *and* dipoles resulting from external forces, the near fields due to each being comparable. Section 1.5 shows that the far field of the dipoles is then small compared with that of individual sources. Hence, when the total source strength is *not* much smaller than those, the combined field of the sources dominates in the far field, where it is close (as shown in figure 6) to that of one simple source of precisely this total strength.

This is a straightforward prescription for estimating the far field of this sort of complicated, but compact, group of monopole and dipole sources. It depends only upon the sum of the monopole source strengths, which in turn is the time rate of change of the net mass outflow into the fluid that is associated with them.

Sound is often generated in fluids by the vibration of foreign bodies immersed within them. The considerations just given apply to those cases when the vibrations produce significant fluctuations in net mass outflow because bodies *are changing in volume as well as in shape*: for example, when in water containing bubbles sound is generated (as so often happens) by their pulsation.

In fact, when a foreign body within a fluid is making movements, possibly involving such changes of volume as well as of shape, their acoustic effect can be described by continuous distributions of singularities around its outside, as follows:
(i) sources, whose strengths reflect mass outflow of fluid produced by outward displacement of the body's surface;
(ii) dipoles, whose strengths reflect the force with which that surface acts on the fluid.
Such a description is easily understood from these singularities' physical interpretations given in the last two sections, while also being familiar for the motion of an incompressible fluid from texts on fluid dynamics, which derive a similar result from Green's theorem; the parallel derivation in acoustics, leading to the interpretation here given, is straightforward.

In the near field the motions due to the dipoles and due to the sources are of comparable magnitude; accordingly, it is the latter which dominate in the far field. If the body's volume, say $V(t)$, is significantly varying as a result of its surface displacements, then the different source strengths add up to a significant total, which may be written as

$$\dot{q}(t) = \rho_0 \, \dot{V}(t), \qquad (105)$$

because the net mass outflow of fluid $q(t)$ is $\rho_0 \, \dot{V}(t)$ on linear theory, where ρ_0 is the undisturbed fluid density. Thus, the far field is close to a simple-source field with this strength $\rho_0 \, \dot{V}(t)$ if the body is acoustically compact. The associated acoustic power output, by (83), is

$$(\rho_0/4\pi c) \, \ddot{V}^2(t). \qquad (106)$$

These considerations are well exemplified by sound radiation from bubbles of air or other gas in a liquid. Such bubbles commonly undergo complicated oscillations of both *shape* and *volume*, in which the shape is constrained by surface tension to oscillate about the spherical form in various modes described by the various spherical harmonics. The volume $V(t)$ does not, however (to the approximation given by linear theory), vary in any of these modes except one: that in which the bubble remains spherical and only its radius a oscillates about its undisturbed value a_0. For this one mode a restoring force more powerful than surface tension operates: that associated with the compressibility of the gas. To this restoring force the balancing *inertia* is provided almost exclusively by the density ρ_0 of the liquid, in the normal case when that of the gas is negligible by comparison.

For describing this radial mode of pulsation, the varying bubble radius a is a suitable generalised coordinate. (For the use of generalised coordinates in linearised theories of vibrations, see texts on classical dynamics.) When a makes small oscillations about its equilibrium value a_0, the radial velocity $\dot{a}$ determines local liquid velocities (by incompressible fluid dynamics) as $\dot{a} a_0^2 r^{-2}$ in the radial direction at distance r from the centre. The kinetic energy of these liquid motions (which is all the kinetic energy we take into account) is $\frac{1}{2}(4\pi\rho_0 a_0^3) \dot{a}^2$. Here, the quantity in brackets is a 'generalised inertia' for the motion.

The corresponding potential energy is calculated on the assumption, which we check later, that the bubble oscillates slowly enough for the gas density ρ_g to be effectively uniform across it, although making small oscillations with time about an undisturbed value ρ_{g0}. The relative change in ρ_g is then minus the relative change in bubble volume:

$$(\rho_g - \rho_{g0})/\rho_{g0} = -3(a - a_0)/a_0. \qquad (107)$$

By equation (50) the associated potential energy per unit volume is

$$\tfrac{1}{2}(\rho_g - \rho_{g0})^2 c_g^2 \rho_{g0}^{-1} = \tfrac{1}{2}\rho_{g0} c_g^2 [3(a - a_0)/a_0]^2, \tag{108}$$

which gives a total potential energy in the volume $\tfrac{4}{3}\pi a_0^3$ of gas equal to $\tfrac{1}{2}(12\pi\rho_{g0} c_g^2 a_0)[a - a_0]^2$. Here, the quantity in round brackets is a 'generalised stiffness' for the motion. It might be thought that changes in potential energy associated with the surface tension T (surface energy per unit area) would be important, but this is not so: the linear term in these merely alters the equilibrium point to one with gas pressure slightly in excess of fluid pressure, while only the quadratic term affects the generalised stiffness, increasing it by $8\pi T$, which is a correction of less than 1% for bubbles of radius exceeding 0.1 mm in water.

Accordingly, the radian frequency ω of the mode of bubble oscillation responsible for the far-field sound is given by general oscillation theory as

$$\omega = \left(\frac{\text{generalised stiffness}}{\text{generalised inertia}}\right)^{\frac{1}{2}} = \left(\frac{12\pi\rho_{g0} c_g^2 a_0}{4\pi\rho_0 a_0^3}\right)^{\frac{1}{2}}$$

$$= (c_g/a_0)(3\rho_{g0}/\rho_0)^{\frac{1}{2}}. \tag{109}$$

We now verify that the frequency ω is indeed small, by a factor $(3\rho_{g0}/\rho_0)^{\frac{1}{2}}$, relative to a characteristic frequency c_g/a_0 for density propagation across the bubble, so that the assumption that gas density is able to remain approximately uniform across it is plausible. The acoustical compactness condition, that ωa_0 would be small compared with the rather greater sound speed c for the liquid, is still more amply satisfied.

For air bubbles in water at $20\,^{\circ}\text{C}$, $c_g = 340\,\text{m s}^{-1}$ and $\rho_{g0}/\rho_0 = 0.0013$, giving a frequency in hertz as

$$\omega/2\pi = (680\,\text{Hz cm})/2a_0. \tag{110}$$

Thus, the sound in their far field tends to be simple-source radiation with this frequency: 680 Hz, divided by the bubble diameter in centimetres; often quite a musical note! The pulsation feels, however, a substantial damping, both by heat conduction out of the gas into the liquid, which makes the density changes depart from exact constancy of entropy, and also by energy loss due to the acoustic-power output itself.

Most common water motions that are noisy generate sound through pressure fluctuations in the flow *forcing* these bubble-pulsation modes which, as we have seen, dominate the far field even when many modes of oscillation of bubble *shape* are also present. Single bubbles make musical notes; assemblages of bubbles as in a running stream make a 'splashy' noise involving a spectrum of frequencies related to their size distribution. Hydraulic motions often involve bubbles through the special mechanism of 'cavitation' where the flow pressure becomes low enough for dissolved gas to come out of solution or for the liquid to vaporise. Cavitated flows may involve not only the sounds due to bubble pulsation noted above but also more powerful sounds due to bubble collapse on moving into higher-pressure regions.

A simple experiment to demonstrate the enhancement of sound by the presence of bubbles is to vibrate a rigid rod in water, producing a weakly audible dipole far field associated (see section 1.7 for the details) with the fluctuating force between the rod and the water; the much stronger dipole near field is not heard. However, bubbles blown through it pulsate in response to the strong pressure fluctuations in that near field and generate a much louder sound, because their *monopole* radiation (which we show how to calculate in section 1.9) has a much greater ratio of far field to near field. Quite generally, it is true that a flow with strong pressure fluctuations in the near field but weak ones in the far field, as is characteristic of radiation by dipoles and still more by quadrupoles (section 1.10), has its sound output greatly amplified by allowing bubbles to respond to its near field with volume pulsations generating a much more powerful monopole far field.

This section has shown how simple, and powerful, is the radiation from compact source regions 'in the general case' when the sum of all monopole source strengths is not specially small. This condition is satisfied when foreign bodies in the fluid are pulsating, or when volume changes occur due to other causes, such as irregular combustion. However, the flows without significant fluctuations of net mass outflow are extremely common. They produce weaker and less simple acoustic far fields that must now be studied, particularly because they become important for higher-speed motions.

1.7 Compact source regions with dipole far fields

In order to understand compact source regions without significant fluctuations of net mass outflow, we may first study, just as in figure 6, a small group of simple sources occupying a compact region, but in the particular case when adding the source strengths (rates of change of mass outflow) gives a total exactly zero. In this case the 'one simple source situated at some central point' which appears in figure 6 has zero strength, so the associated pressure field vanishes. This gives a new meaning to the basic

equation of figure 6, which now equates the whole pressure field of the
group of sources to a sum of dipole fields.

Two conclusions follow: the first, obvious from the discussion in section
1.6, is that the whole pressure field at distances much larger than the diameter
of the source system is now small compared with that of individual sources.
The more useful second conclusion is a quantitative estimate of that whole
pressure field.

To obtain this, we note that each of the component source-pair fields
in figure 6 approximates, at distances large compared with the diameter
of the source system, to a dipole field described by an equation of the form
(102). The vector **r** here represents displacement from the dipole which for
the purpose of this approximate equality can be regarded as located at either
the positive or the negative source (or, indeed, at some intermediate point);
it was taken at the 'positive' source in section 1.5, but for preference will
here be taken at the 'negative' source which for all the dipoles concerned
is the same central point. The pressure field (102) of the dipole depends
linearly on its strength **G**(t), and therefore the sum of all the different dipole
fields is equal to the pressure field (102) of a single dipole with strength
G(t) equal to the vector sum of their strengths.

This reasoning is valid provided only that the sum of the dipole strengths
is significantly different from zero, a caveat required for the same reason
as that in section 1.6 concerning sum of source strengths. Each dipole field
separately is large compared with the error involved in approximating the
corresponding source-pair field by it, but a similar result for their sum
can only be deduced when adding up the dipole strengths gives a total *not*
much smaller than individual dipole strengths.

The vector strength **G**(t) of a dipole was seen in section 1.5 to be equal
to the product of the strength $\dot{q}(t)$ of the positive source and a vector representing displacement from the negative to the positive source. In the
system now being studied, the displacement vector **r** is measured from
the negative source at the central point for each dipole, whose strength,
accordingly, is the value of $\dot{q}(t)\,\mathbf{r}$ for the positive source concerned. The
total dipole strength can therefore be written

$$\mathbf{G}(t) = \Sigma \dot{q}(t)\,\mathbf{r}, \tag{111}$$

the sum being over all the sources in the group. This vector strength (111)
of the single dipole equivalent to the group of sources is describable in
mathematical language as the *moment* of the source strengths about the
central point; note, however, that because the total source strength is zero

the value of (111) is independent of the position chosen for that central point.

A compact group of sources as in figure 6 behaves, then, *like a single source* of strength equal to the sum of their strengths if the latter is not small; but if that sum is zero it behaves *like a single dipole* of strength equal to the moment of the source strengths. In the intermediate case when the sum is small but not zero, we may require representation by a combination of a simple source at the central point *and* a dipole of strength (111), the resulting pressures being possibly comparable even in the far field. Note also that other complications, involving the idea of the 'quadrupole' source (section 1.10 below), may arise when the moment (111) is zero or small.

We saw in section 1.6 that real acoustic source regions may need to be represented by a combination of simple sources, describing local rates of change of mass outflow of fluid, and dipoles, describing external forces applied to the fluid. According to the ideas just described, the simple sources when their strengths add up to zero have a pressure field close to that of a single dipole of strength (111) located at some central point. The total pressure field at distances large compared with the diameter of the source region is then that of this central dipole, added to those of all the dipoles representing external forces.

These latter are *not* necessarily located at the central point, but there is in general only a small error in supposing them relocated there; this would mean as before that all the dipole fields could be added up to give the field of a single dipole with strength the vector sum of the strengths of *all* the separate dipoles. The error in each individual process of relocation is small when the region is acoustically compact, because the change in time-lag r/c in expression (102) then makes a small change in the associated value of **G**, while changes in the spherical attenuation factor $(4\pi r)^{-1}$ are also small. (More precise statements about the error can be made in terms of the idea of quadrupoles introduced in section 1.10.) We may then regard as small the error in using the sum of the relocated dipole fields, with the usual proviso that summing their strengths does not produce any undue degree of cancellation.

The single dipole field, to which the whole compact source region is then equivalent at distances large compared with its diameter, has a strength equal to the vector sum of (i) all the dipole strengths that represent external forces acting on the fluid, whose sum is evidently the resultant of all those forces, and (ii) a correction (111) equal to the moment of rates of change of mass outflow. The physical interpretation of dipole strength in terms of external force might lead one to imagine that the single dipole to which the complex source region is equivalent would always have strength equal to the resultant external force on the fluid; however, this correction (ii) is needed, equal to the moment of all the simple source strengths (whose sum has been supposed zero), and there are circumstances in which it can be be very important.

We saw in section 1.6 that motion of a foreign body within a fluid has acoustic effects equivalent to a continuous distribution of (i) sources related to outward displacement of the body's surface and (ii) dipoles related to the force with which that surface acts on the fluid. The total source strength is $\rho_0 \ddot{V}(t)$, where $V(t)$ is the volume of the body, and therefore the case of zero total source strength with which we are here concerned includes the very common case of a foreign body of effectively constant volume V. In this important case, however, the moment of the source distribution (i) is not in general zero. The moment of the distribution of volume displacement rate can be related to the velocity of the body's centroid (see below), and multiplication by ρ_0 and differentiation with respect to time convert this into moment of rate of change of mass outflow.

The centroid of the body is defined as the point where the mass centre would be if the density were uniform, so that its position vector $\mathbf{r}_c$ is the moment of the body's *volume* distribution divided by its total volume V. It follows that the moment of the distribution of volume displacement rate is the *rate of change* of this moment $V\mathbf{r}_c$ of volume distribution; that is, since V is constant, $V\dot{\mathbf{r}}_c$. Hence, the moment of the distribution of *mass* outflow is $\rho_0 V\dot{\mathbf{r}}_c$, and that of the source strengths is $\rho_0 V\ddot{\mathbf{r}}_c$.

We deduce the important result that *each compact body of constant volume V moving through fluid of density ρ_0 produces at distances large compared with its size a pressure field close to that of a dipole of strength*

$$\mathbf{G} = \mathbf{F} + \rho_0 V\dot{\mathbf{U}}, \qquad (112)$$

where $\mathbf{U} = \dot{\mathbf{r}}_c$ is the velocity of the centroid and $\mathbf{F}$ is the resultant force with which the body acts on the fluid; with the proviso that the approximation is degraded if the additions in (112) produce a large measure of cancellation. The theorem is valid both in the far field where the true radiated sound is found, and in the near field where it is equivalent to a theorem proved in texts on incompressible fluid dynamics (which omit, however, the elements of multiplication by ρ_0 and differentiation with respect to time introduced for acoustic purposes into the definitions of source and dipole strengths).

Note that the correction $\rho_0 V\dot{\mathbf{U}}$ in (112) can be regarded as a rate of change of momentum of 'fluid supposed to have been displaced by the body'.

We shall see that this result (112) is the key to quantitatively estimating the effectiveness of a very wide range of mechanisms of sound generation, including many that involve complicated flow fields. We illustrate its use first, however, in the simple case of a rigid body vibrating in otherwise still fluid, in rectilinear motion at a frequency such that the compactness condition (91) is satisfied.

The force $\mathbf{F}$ with which such a vibrating body acts on the fluid is commonly dominated by a term $M_v\dot{\mathbf{U}}$ proportional to the body's acceleration $\dot{\mathbf{U}}$. Here, the coefficient M_v is called the 'virtual mass' of fluid, in the sense that the body's movement with velocity $\mathbf{U}$ induces proportional movement of nearby fluid with a total momentum $M_v\mathbf{U}$, 'virtually' as if a mass M_v of fluid had to be influenced to move at the full velocity $\mathbf{U}$ of the body and the remainder were left at rest. Of course, actual fluid particles move at velocities that are fractions of $\mathbf{U}$, becoming small as distance from the body increases, and we obtain M_v by adding up the masses of such fluid particles each weighted by the fraction in question. Convergence problems can arise in this calculation but not if (as in section 1.5) we consider only fluid in a large cylinder with axis parallel to the direction of motion; furthermore, no correction to the term $M_v\dot{\mathbf{U}}$ in the force $\mathbf{F}$ can then arise from pressure forces round the boundary of that cylinder, which have zero component in the direction of motion.

If we put aside cases of greater complexity (for example, when viscous forces proportional to $\mathbf{U}$ are comparable with this $M_v\dot{\mathbf{U}}$ term, or when the body has an asymmetry with respect to the direction of motion that gives it 'anisotropic virtual inertia', allowing forces proportional to $\dot{\mathbf{U}}$ in magnitude but perpendicular in direction), we conclude that the dipole strength (112) for such a vibrating body takes the form

$$\mathbf{G} = (M_v + \rho_0 V)\dot{\mathbf{U}}. \tag{113}$$

The two terms in brackets in (113) are often comparable in magnitude: for example, it is easy to calculate (see texts on fluid dynamics) that for a circular cylinder in transverse vibration $M_v = \rho_0 V$, whereas for a *sphere* of radius a we have

$$M_v = \tfrac{1}{2}\rho_0 V, \quad \text{giving} \quad \mathbf{G} = \tfrac{3}{2}\rho_0 V\dot{\mathbf{U}} = 2\pi\rho_0 a^3\dot{\mathbf{U}}. \tag{114}$$

This value (114) for the dipole strength associated with the motion of a sphere implies when substituted in (102) a pressure field

$$p - p_0 = \tfrac{1}{2}\rho_0 a^3[\mathbf{r}\cdot\dot{\mathbf{U}}(t - r/c)/r^3] + \tfrac{1}{2}\rho_0 a^3[\mathbf{r}\cdot\ddot{\mathbf{U}}(t - r/c)/r^2 c], \tag{115}$$

which in the near field is close to the value associated with the classical potential

$$\phi = -\tfrac{1}{2}a^3[\mathbf{r}\cdot\mathbf{U}(t)/r^3] \tag{116}$$

for a sphere moving in incompressible fluid. On the other hand, the second term in (115) is dominant in the far field and alone describes the radiated sound.

When a body vibrates in air not still but moving relative to the body, both terms in (113) may be negligible compared with contributions to the dipole strength (112) *in phase with* the vibration velocity, due to aerodynamic forces fluctuating in response to fluctuations in relative velocity of the body and the air. Such forces may be of order $\rho_0 SU_0\mathbf{U}$, where U_0 is the undisturbed relative velocity and S the surface area of the body, and will exceed (113) at frequencies less than SU_0/V. A good illustration of this type of sound generation is the buzzing of insect wings.

Note that although such large aerodynamic forces may produce a corresponding 'recoil' of the whole insect's centroid, an insect with mean density ρ_m much greater than the air density ρ_0 will experience motion $\mathbf{U}$ of its centroid (here assumed coincident with its mass centre) with mass-acceleration given by $-\mathbf{F}$, the equal and opposite reaction of the wind on the body, so that the corrected dipole strength (112) becomes

$$\mathbf{F}(1-\rho_0/\rho_m), \tag{117}$$

which is not significantly different from $\mathbf{F}$. In the ocean, by contrast, creatures with mean density ρ_m very close to the density ρ_0 of the ambient water, performing rapid fin oscillations as they swim, are extremely ineffective generators of sound because the corrected dipole strength (117) is very small.

One very common source of sound arises, not from any externally caused vibration in a wind, but from the *flow instability* associated with airflow around many types of obstacles: telephone wires, tree branches, etc. Fluctuating forces $\mathbf{F}$ between the obstacle and the wind are the result of this flow instability, and generate a dipole sound field. This is almost *independent* of the extent to which the obstacle is caused to vibrate by the forces: even if it were quite free to move, the corrected dipole strength (117) would be practically equal to $\mathbf{F}$, and normally it is tethered to a pole, tree-trunk, etc. Resonance with a normal mode of vibration of a stretched wire may, however, affect the sound generated by keeping the aerodynamic forces in phase all along the wire, producing the so-called 'Aeolian tones'.

An aeroplane propeller rotating at a high angular velocity ω may fail to satisfy the compactness condition (91) if l is taken as the whole diameter of the propeller disk, and yet each 'blade element' (section of the blade at distances between s and $s+ds$ from the axis) may satisfy it because the aerofoil shape of that section is short compared with c/ω. The sound field must then be analysed as a linear superposition of dipole fields associated with different blade elements, each with strength given by (112); here, the important term in $\mathbf{F}$ is the thrust force, but the contribution to $\rho_0 V \dot{\mathbf{U}}$ due to the blade element's *centrifugal* acceleration need not be negligible.

1.8   Ripple-tank simulations

Experimental confirmation of all the essential results in sections 1.4–1.7 can be obtained if trouble is taken to use well-calibrated microphones, in conditions when they pick up only the sound generated within the source region (which normally demands working in an 'anechoic chamber'), together with equipment to measure amplitudes of any vibration of foreign bodies, as well as the aerodynamic forces that resist it. On the other hand, certain different kinds of experiment, that are not so much realisations of sound-generation phenomena as *simulations* of them, have some superior advantages as aids to understanding the principles involved, partly because they are far easier to perform but mainly because they allow the human eye to receive a clear visual representation of each phenomenon as a whole.

In these simulations, the sound waves are replaced by ripples in a glass-bottomed tank filled with water to a depth of 5 mm. The reason for the special choice of 5 mm becomes clear in chapter 3, where the close parallel between sound waves and ripples in water of this particular depth is explained: essentially, the propagation of such ripples depends on a balance between the inertia of the primarily horizontal water movements in the shallow layer and a sort of 'substitute for compressibility' in which both gravity and surface tension combine to give a pressure *defect* wherever divergence of those horizontal velocities has brought about local reduction of depth, and a pressure *excess* wherever negative divergence has caused depth increase. It is closely correct for a depth of 5 mm that these combined effects produce a ripple speed which like the speed of sound in air is independent of wavelength.

Because, however, this ripple speed is $0.22 \, \mathrm{m\,s^{-1}}$, as against the sound speed of $340 \, \mathrm{m\,s^{-1}}$ in air at $20 \, ^\circ\mathrm{C}$, the wave propagation is easy to observe clearly with a good light source by watching the moving shadows of the waves thrown on to a large screen. The fact that the propagation is two-dimensional makes it far easier to appreciate with the eye than three-dimensional propagation. To be sure, it means that the fundamental solutions studied in sections 1.4–1.7, based on the specially simple form (71) of a source in three dimensions, are not themselves being represented. It is, rather, the corresponding solutions based on the form (75) of a simple source in two dimensions (with h now the depth of the water layer), that are simulated. However, all the qualitative features of the solutions emphasised in sections 1.4–1.7 are retained: source strength is related to mass outflow and dipole strength to force; dipoles have a $\cos\theta$ directional dependence and a reduced ratio of far to near field; compact source regions radiate either like a single source or single dipole according to criteria explained in sections 1.6 and 1.7; foreign bodies have the equivalent source strength (105) or equivalent dipole strength (112) respectively in these two cases.

The best permanent record of such simulations of sound generation phenomena is a motion picture. Such motion-picture records of the moving wave shadows are available as part of a 44-minute film entitled *Aerodynamic Generation of Sound* made by the present author and Professor J. E. Ffowcs Williams under the auspices of the US National Committee for Fluid Mechanics Films. Information about purchase or rental of the film can be obtained from the distributor: Encyclopaedia Britannica Educational Corporation, 425 No. Michigan Avenue, Chicago, Illinois 60611, USA, with whose kind permission several 'stills' from the film have been included in this section. These stills give a useful picture of many of the phenomena, although readers can be recommended to take an opportunity of seeing, if possible, the motion pictures of ripple-tank phenomena and other material on sound generation in that film.

Figure 7 is a ripple-tank simulation of radiation from a simple source; the corresponding motion picture shows the concentric wave shadows moving outwards at the constant ripple speed of $0.22\,\mathrm{m\,s^{-1}}$. The wave amplitudes exhibit no directional dependence. The source is actually a foreign body moved vertically in and out of the water as indicated in figure 8; that body, being compact (much smaller than the length λ of the waves generated), is closely equivalent to a source whose strength takes the form (105) in terms of the variation in immersed volume $V(t)$.

Figure 9 shows the arrangement of two such 'plungers' required to generate a source-pair field. When they are moved up and down at the same frequency, and with the same amplitude but 180° out of phase, a dipole wave field results as in figure 10 (a motion picture again shows the wave

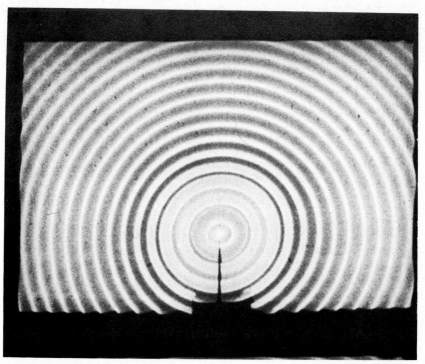

Figure 7. Ripple-tank simulation of the radiation from a simple source. [*Courtesy of the Education Development Center, Newton, Mass. U.S.A.*]

Figure 8. Mechanism for generating the ripples in figure 7. [*Courtesy of the Education Development Center, Newton, Mass. U.S.A.*]

Figure 9. Mechanism for generating the ripples simulating a dipole field. [*Courtesy of the Education Development Center, Newton, Mass. U.S.A.*]

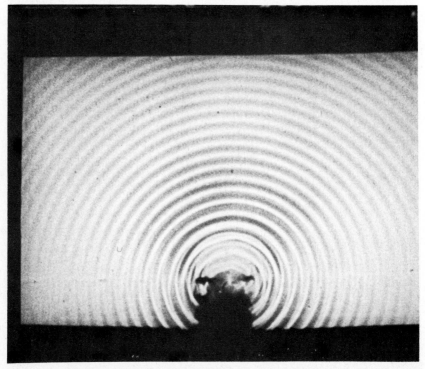

Figure 10. Ripple-tank simulation of dipole radiation. [*Courtesy of the Education Development Center, Newton, Mass. U.S.A.*]

Figure 11. Ripple-tank simulation of the field of a group of simple sources.
[*Courtesy of the Education Development Center, Newton, Mass. U.S.A.*]

shadows moving outwards). The line of the source pair is 'north–south' in figure 10, which shows the $\cos\theta$ directional distribution characteristic of the dipole; in particular, wave amplitude falls to zero in the east–west directions $\theta = \pm\frac{1}{2}\pi$, and there is a 180° phase difference between the waves where $\cos\theta$ is positive and those where $\cos\theta$ is negative.

By contrast, figure 11 shows the effect of three plungers oscillating with the same frequency but each with different amplitudes and phases, in a 'general' case when the source strengths do *not* add up to zero. Although the wave field near the plungers is exceedingly irregular in shape, the far field is close to that of a simple source. A second experiment along the same lines is carried out in the film using a single, but most irregularly shaped, plunger, with a resulting wave field very like figure 11 in appearance.

Sound is commonly generated in air by a cylindrical body moving through it, as when a cane is 'swished', and is associated with the fluctuating *sideforce* (force at right angles to the direction of motion) associated with periodic vorticity shedding from the cylinder. The observed sound is at the frequency of this sideforce fluctuation, and negligible sound is associated with the much smaller drag fluctuation at twice the frequency. The

Figure 12. 'Dipole field' generated by pulling a small coin edge-on through the ripple-tank: the maximum wave strength is found in a direction indicated by the arrow, which is perpendicular to the direction of motion of the coin. [*Courtesy of the Education Development Center, Newton, Mass. U.S.A.*]

observed sound also has a dipole-like directional maximum in the direction of that sideforce. This phenomenon is illustrated in the film when a small coin is pulled edge-on through the ripple-tank: figure 12 shows the resulting dipole wave field, with a directional maximum (indicated by the arrow) at right angles to the direction of movement.

A rather clearer picture illustrating the principle that movements of a foreign body of constant immersed volume V generate a dipole field whose strength is given by the resultant force $\mathbf{F}$ with which the body acts on the fluid, together with a correction term $\rho_0 V \dot{\mathbf{U}}$ as in equation (112), is given by figures 13 and 14. In figure 13 a dipole wave field is generated, actually, by a *pair* of such foreign bodies: cylinders with axes vertical that move with equal amplitude and phase in the north–south direction with one always remaining due east of the other. Another illustration in the film shows one such cylinder by itself generating a dipole field almost indistinguishable from figure 10. It is striking, however, that the more complex source region of figure 13, involving two such cylinders, although producing nearby

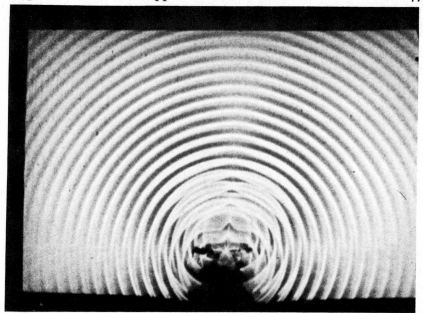

Figure 13. Dipole field generated in the ripple-tank by the vibration of two vertical cylinders in phase. [*Courtesy of the Education Development Center, Newton, Mass. U.S.A.*]

variations from a pure dipole distribution, has a far field closely equivalent to that of a single dipole.

Figure 14 is a single still from an extended piece of film illustrating the contrast between those cases when in the dipole strength (112) the force **F** between body and fluid is or is not cancelled out by the correction term $\rho_0 V \dot{\mathbf{U}}$. The point is best explained by the whole film sequence but is briefly outlined here using this still.

Figure 14 shows the last waves of a 'packet' of simple-source waves (generated by vertically oscillating a plunger for a limited period of time) as they pass two objects in the field, seen as black dots in the figure. The dot on the left marks a freely floating body, whose mass by Archimedes' principle equals the mass $\rho_0 V$ of displaced water. If the waves act on the body with a force $-\mathbf{F}$ (minus because we continue to use **F** as the equal and opposite reaction of the body on the fluid), it experiences an acceleration $\dot{\mathbf{U}}$ such that $\rho_0 V \dot{\mathbf{U}} = -\mathbf{F}$, which makes the dipole strength (112) equal to zero. This suggests that the body on the left may generate no waves, and indeed we cannot see any circular wave shadows with it as centre.

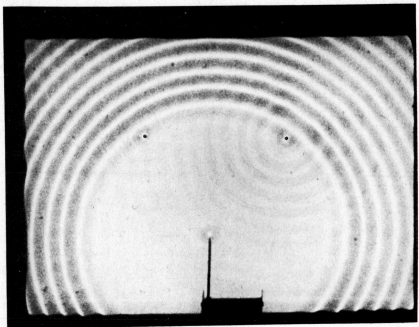

Figure 14. As the last waves in a 'packet' of simple-source ripples pass two objects marked by dots (a freely floating body on the left and a body impeded from free movement on the right) scattered waves appear only from the right-hand object. [*Courtesy of the Education Development Center, Newton, Mass. U.S.A.*]

The dot on the right represents a similar body impeded from free movement. The waves act on it with a force $-\mathbf{F}$ that does not produce accelerations $\dot{\mathbf{U}}$ that can nullify or significantly reduce the dipole strength (112). It is not surprising, therefore, that the wave packet in passing this body has caused it to generate waves, which can be seen as concentric shadows surrounding the dot on the right. These can be described as a 'scattered wave field', carrying that portion of the energy of waves incident on the body which it has succeeded in scattering. The theory of the *acoustic* scattering phenomena which this experiment simulates is given in section 1.9.

Our last ripple-tank picture looks even further ahead, towards section 1.10 on quadrupole radiation, but has also another aim: to support the procedure adopted in section 1.7 when dipoles were 'relocated'; that is, supposed moved to another position a very short distance away compared with a wavelength. Figure 15 is intended for comparison with figure 13 and, like it, was produced by oscillating two vertical rods in a 'north–south' direction about undisturbed positions with one due east of the other. The

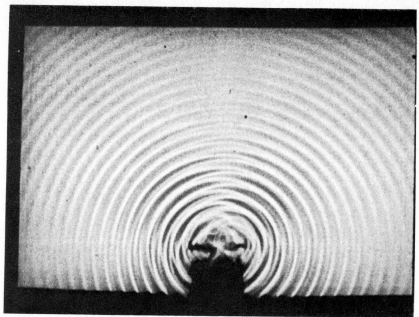

Figure 15. Quadrupole field generated in the ripple-tank by equal and opposite vibration of two vertical cylinders. [*Courtesy of the Education Development Center, Newton, Mass. U.S.A.*]

amplitudes of motion of both rods are again equal, and the same as in figure 13. However, in figure 13 their motions are in phase whereas in figure 15 they are 180° out of phase.

Thus, while figure 13 represents the *sum* of two dipole wave fields which produces, as remarked above, a far field close to that of a single dipole, in support of the argument that 'relocation' makes little difference, figure 15 gives further support to such a view by actually showing the *difference* between the wave fields of identical dipoles located at different points. From the relative strengths of the wave shadows in the two figures, we see that the far field for this *difference* (figure 15) is indeed weak compared with that for their *sum* (figure 13).

Figure 15, then, is a simulation of the considerable reduction that occurs in sound radiation from a compact source region when the *total* dipole strength is zero (as well as the total source strength). Sound radiation from *turbulent flows*, separated from any foreign bodies that can act on the air with forces, is commonly of this character with zero dipole strength. Among the special types of wave pattern characteristic of this sort of radiation is that simulated in figure 15. It is called a *quadrupole* wave pattern, essentially because two equal and opposite dipoles can be thought of as altogether four sources; notice that there are also four directions (diagonally stretching out from the source region) in which the waves are strongest. The importance of quadrupole radiation in aerodynamic sound generation is sketched in section 1.10.

1.9　Scattering by compact bodies

The ripple-tank experiments of section 1.8 included a 'scattering' experiment, illustrated in figure 14. This demonstrated no scattering of ripples as they pass a freely floating body (the one at the left-hand dot on the photograph). The explanation was that such a body, having by Archimedes' principle a mass equal to the mass $\rho_0 V$ of displaced water, moves just as that water would move if the body were absent.

The corresponding principle for sound propagation through a fluid is that there is no scattering by a body with the same mass *and the same compressibility* as those of the displaced fluid. In fact, inertia and compressibility are the only two properties of a fluid that are involved in sound propagation: a foreign body with the same values of those as the fluid it has displaced expands and contracts and oscillates to and fro exactly as that fluid would have done, and thus plays an identical part in the propagation of sound. (Note that the analogy to compressibility in ripple-tank simulations is the relation between local increase of water depth and pressure increase; that relation also is unchanged by the floating body's presence.)

In this section we study the scattering of sound waves by compact bodies with different mean compressibility and different mean density from those of the surrounding fluid. We use the principles of sections 1.6 and 1.7 to calculate the scattered sound due to these two effects as a source field and a dipole field respectively. Inequality between the body's volume changes and those of the displaced fluid produces simple-source-scattering; inequality of their momentum changes produces dipole scattering.

It might be thought impossible to apply the methods of section 1.6 and 1.7, for calculating sound generation by source regions radiating in *otherwise undisturbed fluid*, to phenomena involving an *incident sound field*. Within the framework of a linear theory, however, those methods *can* be used to calculate the scattered sound field, which is defined as the *difference* between

(I) the sound field when the body's presence disturbs the incident sound wave; and

(II) the sound field when the body is absent and the incident wave propagates through homogeneous fluid.

On linear theory such a difference of two sound fields must satisfy the equations of sound, *and* tend to zero far from the body, where (I) tends to the same incident-field value as (II). It can therefore be calculated as sound radiated into otherwise undisturbed fluid due to differences between the effects of *the body in case* (I) and those of *the fluid replacing it in case* (II).

Sections 1.6 and 1.7 tell us that this radiated sound will consist of that due to a simple source, whose strength $\dot{q}$ equals the difference between the strength values (105) for the body in case (I) and for the fluid replacing it in case (II), and that due to a dipole, whose strength $\mathbf{G}$ equals the difference between the strength values (112) in the two cases. Thus

$$\dot{q} = (\rho_0 \dot{V})_\mathrm{I} - (\rho_0 \dot{V})_\mathrm{II}, \tag{118}$$

where $V(t)$ is the volume of the body in case (I), or of the fluid replacing it in case (II), and

$$\mathbf{G} = (\mathbf{F} + \rho_0 V\dot{\mathbf{U}})_\mathrm{I} - (\mathbf{F} + \rho_0 V\dot{\mathbf{U}})_\mathrm{II} = (\mathbf{F} + \rho_0 V\dot{\mathbf{U}})_\mathrm{I}, \tag{119}$$

where $\mathbf{F}$ is the force with which the body acts on the fluid and $\mathbf{U}$ the velocity of its centroid. In (119) we have used a fact already noted in section 1.8: the dipole strength $\mathbf{F} + \rho_0 V\dot{\mathbf{U}}$ vanishes in case (II), because of the law of motion satisfied by the mass $\rho_0 V$ of fluid replacing the body in that case. We now calculate these strengths (118) and (119) for a spherical body, and then indicate briefly how the calculation can be extended to more general shapes.

We consider then a foreign body (whether solid, liquid or gas) which in its undisturbed state is a sphere of radius a_0, immersed in homogeneous fluid through which are propagated sound waves of frequency ω satisfying the compactness condition

$$\omega a_0/c \ll 1. \tag{120}$$

This implies that the 'incident wave' defined in (II) varies negligibly in phase over the spherical volume of fluid that the body displaced: if at the centre of that volume the *pressure* in the incident wave is

$$p - p_0 = p_1 e^{i\omega t}, \tag{121}$$

where in such a complex expression for a real quantity it is of course understood that the real part is to be taken, then all over the volume the same pressure variation is approximately found. The resulting variation of the total volume V of the sphere of fluid in case (II) from its undisturbed value

$$V_0 = \tfrac{4}{3}\pi a_0^3 \tag{122}$$

is
$$(V - V_0)_{\text{II}} = -V_0[(\rho - \rho_0)/\rho_0] = -V_0[(p - p_0)/\rho_0 c^2]$$
$$= -V_0(p_1/\rho_0 c^2)\,\mathrm{e}^{\mathrm{i}\omega t}, \tag{123}$$

where (37) has been used to relate changes of pressure and density at constant entropy.

Corresponding changes in volume of the body itself in case (I) are calculated by assuming that their response to uniform changes p_s at its surface is specified by a mean compressibility K for the body:

$$(V - V_0)_{\text{I}} = -V_0 K(p_s - p_0). \tag{124}$$

Here, K is the relative volume reduction per unit increase in surface pressure. For fluids, (123) shows that $K = \rho_0^{-1} c^{-2}$ so that, when our foreign body is a gas bubble like that treated in section 1.6, we must take

$$K = \rho_{\text{g0}}^{-1} c_{\text{g}}^{-2}. \tag{125}$$

For a homogeneous solid, on the other hand, K is the reciprocal bulk modulus of its material.

The surface pressure p_s in case (I) is the sum of that due to the incident wave (given by (121)) and that due to the simple source of strength $\dot{q}$ located at the centre of the sphere:

$$p_s - p_0 = p_1 \mathrm{e}^{\mathrm{i}\omega t} + \dot{q}(t)/4\pi a_0, \tag{126}$$

where in expression (71) for the excess pressure at $r = a_0$ due to a simple source the difference in retarded time $t - a_0/c$ from t can be neglected under the compactness condition (120). (Thus we *locally* use a solution (67) of Laplace's equation to represent the required solution of the wave equation.) Any effect of the dipole distribution of excess pressure can be ignored in (126) as its mean over the surface of the sphere is zero; that distribution, indeed, merely produces bulk motion of the body.

The values of $(V - V_0)_{\text{I}}$ given by (124) and (126) and of $(V - V_0)_{\text{II}}$ given by (123) are now substituted in (118) to give a *differential equation for the source strength* $\dot{q}$:

$$\dot{q} = \rho_0 V_0 K(p_1 \omega^2 \mathrm{e}^{\mathrm{i}\omega t} - \ddot{q}/4\pi a_0) - \rho_0 V_0(p_1/\rho_0 c^2)\,\omega^2 \mathrm{e}^{\mathrm{i}\omega t}, \tag{127}$$

where the two terms on the right-hand side correspond to those in (118).

Equation (127) is an equation of forced harmonic motion for a system with natural frequency

$$(4\pi a_0/\rho_0 V_0 K)^{\frac{1}{2}}; \tag{128}$$

note that, in the case of a bubble, equations (125) for K and (122) for V_0 make this resonant frequency *identical* with that calculated by a quite different method as equation (109). The solution of (127) is

$$\dot{q} = (\mathbf{1} - \rho_0 V_0 K \omega^2 / 4\pi a_0)^{-1}(K - \rho_0^{-1} c^{-2})\rho_0 V_0 p_1 \omega^2 e^{i\omega t}. \tag{129}$$

Here, the first factor re-emphasises the resonance that occurs when the frequency ω of the incident sound coincides with the natural frequency (128); as with other resonant systems the singularity at this value of ω is resolved only when we take into account damping (provided in the case of the bubble, as noted in section 1.6, partly by heat-conduction effects). The second factor in (129) confirms that no simple-source scattering occurs when the body has a mean compressibility K equal to that of the fluid.

The picture given by (129) is of a *weak* simple-source field for bodies of low compressibility K, owing to the $(\omega/c)^2$ factor present as $K \rightarrow 0$. By contrast, for more compressible bodies of foreign matter, including bubbles, there is a powerful peak in the monopole scattering effect at the resonant frequency (128).

The argument given is equally valid whether the pressure fluctuations (121) in the incident wave represent some sort of 'far field' of the source producing it, which effectively is a plane wave on the scale of the body's dimension a_0, or if those pressure fluctuations are part of a *near field*. The discussion following equation (110) emphasised how bubbles could make a resonant simple-source response to pressure fluctuations in the near field of a dipole, and enhance the sound radiated: equation (129) gives a quantitative estimate of the simple-source strength brought into play by this mechanism.

We calculate the dipole strength $\mathbf{G}$ for the scattered-sound field in a similar way to the source strength, although we do not find any possibility of resonance. We suppose that the fluid velocity $\mathbf{u}$ in the incident wave took the value

$$\mathbf{u} = \mathbf{u}_1 e^{i\omega t} \tag{130}$$

at the centre of the spherical volume of fluid that the body displaced. Note that for a travelling plane wave or, more generally, a far field, the vector $\mathbf{u}_1$ would have magnitude $p_1/\rho_0 c_0$, by (22) or (81). However, we do not confine ourselves to such cases.

By (119), we have to calculate in case (I) both the velocity $\mathbf{U}$ of the body's centroid (assumed to coincide with its mass centre) and the force $\mathbf{F}$ with which it acts on the fluid. We obtain them from two equations which they must satisfy. First, the equation of motion of the body, if it has mass $\rho_m V_0$, so that ρ_m is its mean density, is

$$\rho_m V_0 \dot{\mathbf{U}} = -\mathbf{F}. \tag{131}$$

Secondly, $\mathbf{F}$ may be related to the virtual mass of the spherical body, given by (114) as

$$M_v = \tfrac{1}{2}\rho_0 V_0. \tag{132}$$

The discussion of virtual mass preceding equation (114) related it to force for a body moving in 'otherwise still fluid'. To apply that discussion to the present case when the body is immersed in fluid moving at the velocity (130), we must work temporarily in a frame of reference moving with that velocity.

The equation of motion for any particle, in such a frame of reference in rectilinear accelerating motion, includes an inertial force

$$-Mi\omega\mathbf{u}_1 e^{i\omega t} \tag{133}$$

in addition to the other forces on the particle (supposed of mass M). This familiar feature of moving frames of reference means only that the sum of those other forces equals the particle mass times its total acceleration, which is its mass times acceleration in the moving frame of reference plus its mass times the acceleration of the frame itself. The latter term, moved with its sign changed to the other side of the equation along with the forces, becomes an effective inertial force (133). For fluid particles of density ρ_0 this force is

$$-\rho_0 i\omega\mathbf{u}_1 e^{i\omega t} \tag{134}$$

per unit volume.

Any such uniform field of force per unit volume in a fluid (gravity, for example) is balanced automatically by a pressure distribution with uniform gradient. The gradient of pressure *equals* that force per unit volume, here given by expression (134); note that this value is obtainable alternatively from the linearised momentum equation (4). Now, Archimedes' principle tells us the resultant of such a pressure distribution acting around a body of volume V_0. It is

$$\rho_0 V_0 i\omega\mathbf{u}_1 e^{i\omega t} \tag{135}$$

and has the value needed to cancel the force which would have been acting on the fluid displaced by the body in the latter's absence.

In addition to a force of minus (135) with which the body reacts on the fluid as a result of inertial forces, the body moves relative to the fluid around it (which is at rest in our frame of reference) at a velocity

$$\mathbf{U} - \mathbf{u}_1 \mathrm{e}^{\mathrm{i}\omega t} \tag{136}$$

and therefore imparts momentum to it at a rate

$$M_v(\mathrm{d}/\mathrm{d}t)(\mathbf{U} - \mathbf{u}_1 \mathrm{e}^{\mathrm{i}\omega t}). \tag{137}$$

The total force $\mathbf{F}$ with which the body acts on the fluid is therefore

$$\mathbf{F} = -\rho_0 V_0 \mathrm{i}\omega \mathbf{u}_1 \mathrm{e}^{\mathrm{i}\omega t} + M_v(\dot{\mathbf{U}} - \mathrm{i}\omega \mathbf{u}_1 \mathrm{e}^{\mathrm{i}\omega t}). \tag{138}$$

With equations (131) and (132), this gives

$$\mathbf{F} = -V_0 \frac{3\rho_0 \rho_\mathrm{m}}{\rho_0 + 2\rho_\mathrm{m}} \mathrm{i}\omega \mathbf{u}_1 \mathrm{e}^{\mathrm{i}\omega t}, \quad \dot{\mathbf{U}} = \frac{3\rho_0}{\rho_0 + 2\rho_\mathrm{m}} \mathrm{i}\omega \mathbf{u}_1 \mathrm{e}^{\mathrm{i}\omega t}, \tag{139}$$

whence by (119) the dipole strength is

$$\mathbf{G} = V_0 \frac{3\rho_0(\rho_0 - \rho_\mathrm{m})}{\rho_0 + 2\rho_\mathrm{m}} \mathrm{i}\omega \mathbf{u}_1 \mathrm{e}^{\mathrm{i}\omega t}. \tag{140}$$

The scattered sound field is that of a source of strength (129) plus that of a dipole of strength (140).

Equation (140) makes clear that the dipole strength vanishes when the mean density ρ_m of the sphere equals the fluid density ρ_0, as was obvious already from (119) and (131). Another interesting special case is that of scattering by a *fixed sphere*: this is representable by the limiting case of a sphere whose density ρ_m tends to infinity, so that by (139) its acceleration is, naturally enough, zero. The dipole strength (140) has a nonzero limit in this case.

In the above discussion of scattering by a body with mean density ρ_m different from the fluid density ρ_0, no comment was made on the effect of *gravity* when present on inducing contributions to the body's acceleration $\dot{\mathbf{U}}$ or the force $\mathbf{F}$ between it and the fluid. Those contributions, however, are steady instead of oscillatory, and so produce no scattered sound of frequency ω and indeed possess no dipole far field (103). The only effect of motion of the body under gravity on the acoustic far field is to make it that of a *moving* source and dipole. Such a field, however, is negligibly distorted when the body's velocity is small compared with c (see section 4.12 for the properties of moving sources in general).

The total power output in the scattered sound can be deduced by adding the simple-source power calculated from (83) and (129) to the dipole power calculated from (104) and (140). This power in the scattered-sound field, depending on the squares of the amplitudes $|p_1|$ and $|u_1|$, is all extracted from the incident wave.

Bodies of low compressibility produce scattered-sound fields with the unusual feature that the source and dipole components are of comparable magnitude. This is because the source strength (129) is small for such bodies. Consider, for example, the scattering of *plane sound waves*, satisfying equation (22), by a *fixed, incompressible* sphere (the limiting case $\rho_m \to \infty$, $K \to 0$). The far field, by (71) and (103), is

$$p - p_0 = V_0 p_1 (\omega/c)^2 (-1 + \tfrac{3}{2} \cos \theta) \, e^{i\omega(t - r/c)}/(4\pi r), \tag{141}$$

in a direction making an angle θ with that of the incident wave. The factor $(-1 + \tfrac{3}{2}\cos\theta)$ shows how comparable are the source and dipole components: note that they *interfere* in the forward hemisphere $\theta < \tfrac{1}{2}\pi$ but reinforce one another when $\theta > \tfrac{1}{2}\pi$ (backward scattering), having a peak numerical value for $\theta = \pi$.

The general type of scattering theory for bodies small compared with a wavelength, in which scattering of incident fields satisfying the wave equation is approximated by using properties (like virtual mass) of solutions of Laplace's equation, appears also in other branches of physics. It is known as Rayleigh scattering in electromagnetic theory, where, however, no monopole element in the scattered field occurs. Rayleigh found the same ω^2 increase in amplitude of the scattered field with frequency ω and used it to explain the blue light of the sky as due to preferential scattering by air molecules of the higher-frequency components in sunlight. From the standpoint of quantum mechanics, the theory can be regarded as a case of the Born approximation.

Extension of the calculation of dipole strength (119) to nonspherical bodies is straightforward when their shape has such symmetry about the direction of the incident-field velocity oscillations (130) that the force **F** can still be expressed in terms of a virtual mass M_v as in (138), which with (131) determines **F** and **U̇** and hence **G**. For asymmetric bodies possessing anisotropic virtual inertia, a somewhat more complicated calculation is required.

Extension of the calculation of simple-source strength involves replacing the quantity $4\pi a_0$ that occurs in equations (126)–(129) by the corresponding constant for a different body shape, that constant C being proportional to the *electrostatic capacity* of the shape in question. It has the property that a proximal-field solution close to the body exists with total source strength $\dot{q}$ and *uniform* surface-pressure excess

$$p_s - p_0 = \dot{q}/C. \tag{142}$$

Such a solution with uniform surface-pressure excess would be consistent with a change of body volume given by equation (124), so that the whole argument in equations (126)–(129) could be used with $4\pi a_0$ replaced by C.

With a general body shape the above proximal-field solution would not be equal to the actual proximal field near the body, because the *distribution* of source strength over the surface would in general be wrong, failing to satisfy the appropriate boundary conditions of continuity of normal velocity across the surface. The true proximal-field solution must be obtained by adding on a correction which is a local solution of Laplace's equation with the values of surface normal velocity needed to correct such discrepancies. This solution has zero total source strength, however, and also zero total dipole strength (since all bulk motions of the body's centroid are taken into account separately in the virtual-mass discussion) and so the associated far field can be neglected compared with the terms retained; furthermore, the associated surface-pressure distribution would normally produce body-shape distortions involving small volume changes compared with the main term (124).

1.10   Quadrupole radiation

Section 1.8 on ripple-tank simulations ended with a brief discussion of sound radiation from compact source regions when the total dipole strength as well as the total source strength is zero. Sound radiation from *turbulent flows*, separated from any foreign bodies that can act on the air with forces, was noted as being commonly of this character.

A much simpler example is that represented in figure 15; sound thus generated by two nearby equal and opposite dipoles is described as radiation from a *quadrupole*. Its far field, as comparison of figures 13 and 15 shows, is weak compared with that of the individual dipoles; essentially, for the same reason that the far field of a compact group of sources whose strengths add up to zero (section 1.7) is weak compared with that of the individual sources.

This section sketches out the connection between these two examples, and shows how a turbulent flow in a fluid generates the same sound field as a certain *distribution of quadrupoles*, with a known quadrupole strength per unit volume. Note that, while the strength of a dipole involves a direction (that of the small displacement from the negative to the positive source) and accordingly is a *vector*, the strength of a quadrupole involves *two directions* (not only that of the equal and opposite dipoles but also that of the displacement between them) and accordingly is a *tensor*. We show that the tensor quadrupole strength per unit volume takes a certain rather simple form in any turbulent flow.

The resulting quadrupole radiation may have a much weaker far field in relation to its near field than even dipole radiation has. This can mean in some cases that its far field is hardly important at all. In other cases, however, when the quadrupole strength is very large and there is no significant competition from monopole or dipole radiation, such quadrupole radiation from turbulence may produce an exceedingly disagreeable noise field which it is important to study quantitatively so as to find ways of diminishing it. Thus, studies of jet noise by methods that start from the theory of this section have led to valuable measures for noise reduction.

To obtain the quadrupole strength per unit volume, we begin by transforming the fundamental equations of motion (2) and (3) in two respects. First, we make them describe *local* rate of change ($\partial/\partial t$) of momentum or mass per unit volume; for example, the last two terms in (3) are *grouped together* as $\nabla \cdot (\rho \mathbf{u})$, which with a change of sign represents such a local rate of change of ρ. Secondly, we adopt a suffix notation which makes the arguments using tensors easier to follow: thus, the coordinates are taken as x_1, x_2, x_3 in this section and the velocity vector as (u_1, u_2, u_3), and a suffix occurring twice within any term of an equation is automatically summed from 1 to 3. Then equation (3) becomes

$$\partial \rho / \partial t + \partial(\rho u_i)/\partial x_i = 0. \tag{143}$$

Similarly, equation (2) becomes from that second transformation

$$\rho \, \partial u_i/\partial t + \rho u_j \, \partial u_i/\partial x_j + \partial p/\partial x_i = 0, \tag{144}$$

and we will achieve the goal of the first transformation, namely an equation for $\partial(\rho u_i)/\partial t$, if we add on to (144) the product of equation (143) with u_i itself. Before doing that we alter the suffix i in (143) into j (which does not alter its meaning, since *any* suffix occurring twice in a term is summed from 1 to 3) to make the product

$$u_i \, \partial \rho/\partial t + u_i \, \partial(\rho u_j)/\partial x_j = 0. \tag{145}$$

Equations (144) and (145) added together make

$$\partial(\rho u_i)/\partial t + \partial(\rho u_i u_j)/\partial x_j + \partial p/\partial x_i = 0, \tag{146}$$

which is in the form required.

These transformed equations of motion (143) and (146) have immediate physical significance and could have been derived directly, as follows. Local rate of change of mass density is equated in (143) to minus the divergence of the *mass flux vector* ρu_i. That expresses mass-conservation by equating rate of change of mass in an elementary region to minus the rate of outflow of mass from that region. Similarly, rate of change of momentum density is equated in (146) to minus the divergence of the *momentum flux tensor* $\rho u_i u_j$ plus a correction (since momentum is not conserved but changes at a rate equal to applied force) equal to the pressure force per unit volume, $-\partial p/\partial x_i$.

Actually, the last two terms in (146) can be combined to give

$$\partial(\rho u_i)/\partial t + \partial[\rho u_i u_j + (p - p_0)\delta_{ij}]/\partial x_j = 0, \tag{147}$$

where the 'Kronecker delta' δ_{ij} is equal to 1 when $i = j$ and 0 otherwise. Equation (147) expresses the idea that excess pressure $p - p_0$ creates a momentum flux $(p - p_0)\delta_{ij}$ which the δ_{ij} factor marks as being *isotropic* (that is, equal in all directions); essentially, because excess pressure acts equally in all directions. The term in square brackets in (147) is a *total momentum flux* tensor equal to rate of transport of the x_i-component of momentum in the x_j-direction due to (i) convection by the x_j-component of velocity and (ii) action of the excess pressure $p - p_0$; thus, rate of change of the x_i-component of momentum in an elementary region is equated by (147) to minus the integrated action of that total flux across the region's surface.

Transformation of the equations of motion of a fluid into the special forms (143) and (147) makes possible that estimation of how much sound is generated by turbulent fluid flows which follows from establishing a precise relationship between fluid dynamics and the linear theory of sound. In the latter theory the term in square brackets in (147) takes the value $c^2(\rho - \rho_0)\delta_{ij}$ since pressure is given a linear relationship to density based on (11) and (14) and products of small quantities like $u_i u_j$ are neglected. The equations of motion of a fluid *with only this approximation to the total momentum flux tensor* give an exactly linear wave equation for ρ, since (147) becomes

$$\partial(\rho u_i)/\partial t + c^2 \partial \rho/\partial x_i = 0 \tag{148}$$

and when we eliminate the momentum density ρu_i by subtracting $(\partial/\partial x_i)$ of (148) from $(\partial/\partial t)$ of (143) we obtain the linear wave equation

$$\partial^2 \rho/\partial t^2 - c^2 \partial^2 \rho/\partial x_i^2 = 0. \tag{149}$$

In the absence of forcing by foreign bodies immersed in the fluid, this would imply no radiation of sound.

Accordingly, the radiation of sound by a fluid flow is due entirely to *departure* of the total momentum flux (the term in square brackets in (147)) from its linear-theory approximation $c^2(\rho - \rho_0)\delta_{ij}$. We write that departure as

$$T_{ij} = \rho u_i u_j + [(p - p_0) - c^2(\rho - \rho_0)]\delta_{ij} \qquad (150)$$

and see that this *excess of momentum flux* over its value given by linear theory consists of quadratic and higher terms, due both to products of small quantities like $u_i u_j$ and to nonlinearity in the pressure–density relationship. In terms of (150) the momentum equation (147) can be written

$$\partial(\rho u_i)/\partial t + c^2 \partial \rho/\partial x_i = -\partial T_{ij}/\partial x_j, \qquad (151)$$

and when we eliminate ρu_i between this equation and the equation of continuity (143) we obtain

$$\partial^2 \rho/\partial t^2 - c^2 \partial^2 \rho/\partial x_i^2 = \partial^2 T_{ij}/\partial x_i \, \partial x_j, \qquad (152)$$

which is the linear wave equation (149) with a 'forcing term' inserted on the right-hand side.

This forcing term must be responsible for any generation of sound; that is, of fluctuations in the density ρ propagated away from the turbulent flow. Note that T_{ij} can be regarded as negligible outside the flow, where only the sound waves thus generated are present so that the approximations of linear theory are appropriate. These approximations making T_{ij} negligible may, however, be inadequate inside the flow, which constitutes therefore a bounded source region for the sound generated.

Equation (152) describes a distributed source of sound. We can relate its sound field to that of the generation mechanisms already studied (rate of change of mass outflow generating a monopole field, or force a dipole field), and at the same time consolidate understanding of those, by deriving similar equations describing distributed monopole and dipole sources in the linear theory of sound.

The hypothetical idea of distributed monopole sources involves distributed *emergence of new mass* at a rate Q per unit volume per unit time (as if a point source with mass outflow q per unit time were distributed over a finite volume to give such mass emergence per unit volume per unit time). Equation (143) expressing the rate of change of mass in an elementary region, per unit volume, would acquire a term Q on the right-hand side due to such emergence of new fluid and become

$$\partial \rho/\partial t + \partial(\rho u_i)/\partial x_i = Q. \qquad (153)$$

When we eliminate ρu_i between (153) and the linear-theory equation of momentum (148) we obtain

$$\partial^2 \rho / \partial t^2 - c^2 \partial^2 \rho / \partial x_i^2 = \partial Q / \partial t. \tag{154}$$

This is the linear wave equation (149) with the *source strength* (rate of change of mass outflow) *per unit volume* inserted on the right-hand side.

In other words, distributed monopole sources of strength $\partial Q / \partial t$ per unit volume produce a density field satisfying the linear wave equation with that source strength per unit volume on the right-hand side. The solution of this equation must be obtained as a volume distribution of the source fields given by equation (71). Thus it can be written

$$c^2[\rho(\mathbf{x}) - \rho_0] = \int [\dot{Q}(\mathbf{y}, t - r/c)/4\pi r]\,d\mathbf{y}, \tag{155}$$

where the left-hand side is the linear-theory value of $p - p_0$ at the point with position vector $\mathbf{x}$ and the right-hand side is an integral over the region of distributed sources. At a point $\mathbf{y}$ within this region the source strength is $\dot{Q}(\mathbf{y}, t)$ per unit volume at time t (where $\dot{Q}$ stands for $\partial Q / \partial t$) and so by (71) the sound generated takes the value in square brackets, again per unit volume, if

$$r = |\mathbf{x} - \mathbf{y}| \tag{156}$$

represents distance from the source at $\mathbf{y}$ to the point of reception at $\mathbf{x}$. This inference of the solution to the nonhomogeneous wave equation (154) gives the same answer (155) as is obtained by classical mathematical arguments (see texts on the wave equation), which is a check on the theories of this chapter.

Next, we consider sound that is generated without any emergence of new fluid, simply by the action of a distributed external force f_i per unit volume of fluid. Then the equation of continuity takes its usual form (143) but the equation for rate of change of momentum per unit volume acquires on its right-hand side a term f_i representing the applied force: thus, the linear-theory equation of momentum (148) becomes

$$\partial(\rho u_i)/\partial t + c^2 \partial \rho / \partial x_i = f_i \tag{157}$$

and when we eliminate ρu_i between (157) and the equation of continuity (143) we obtain

$$\partial^2 \rho / \partial t^2 - c^2 \partial^2 \rho / \partial x_i^2 = -\partial f_i / \partial x_i, \tag{158}$$

which is the linear wave equation with a forcing term $-\partial f_i / \partial x_i$ on the right-hand side.

This result gives another check on the theories of this chapter. From section 1.5, the sound field generated by such a distribution of external forces f_i per unit volume must be a distribution of dipole fields of strength f_i per unit volume. At first sight, equation (158) on comparison with (154) suggests a different conclusion: that the said sound field is a distribution of simple-source fields of strength $-\partial f_i/\partial x_i$ per unit volume. We can show, however, that those two descriptions are necessarily identical.

In fact the source field of strength $-\partial f_1/\partial x_1$ per unit volume (one of the three involved in the sum $-\partial f_i/\partial x_i$) is the limit as $\epsilon \to 0$ of a field of strength

$$\epsilon^{-1} f_1(x_1, x_2, x_3) - \epsilon^{-1} f_1(x_1 + \epsilon, x_2, x_3) \tag{159}$$

per unit volume. This places *each particular value of $\epsilon^{-1} f_1$* not only with positive sign at the corresponding (x_1, x_2, x_3) but also with negative sign at the point $(x_1 - \epsilon, x_2, x_3)$ which is where the second term in (159) takes that same numerical value. The combination of the two, by section (1.5), is a dipole of strength $(f_1, 0, 0)$, also per unit volume. Similar considerations for $i = 2$ and 3 combine with this to show that a *source distribution* of strength $-\partial f_i/\partial x_i$ per unit volume is identical, as expected, with a *dipole distribution* of strength

$$f_i = (f_1, f_2, f_3) \tag{160}$$

per unit volume.

Of these two equivalent representations of the sound field due to distributed sources, the former is relatively useless: this source distribution has total strength zero (by the divergence theorem, the volume integral of $\partial f_i/\partial x_i$ over the source region vanishes) so that its proper treatment by section 1.7 is in any case expected to involve representation by dipoles. The latter representation, by contrast, permits a convenient solution to (159) as a volume distribution of the dipole fields given by equation (102), obtained in the form

$$c^2[\rho(\mathbf{x}) - \rho_0] = -(\partial/\partial x_i) \int [f_i(\mathbf{y}, t - r/c)/4\pi r] \, d\mathbf{y} \tag{161}$$

exactly as (155) was obtained. We can furthermore from (103) write down the special form of far field associated with its dipole character as

$$c^2[\rho(\mathbf{x}) - \rho_0] = \int [r_i \dot{f}_i(\mathbf{y}, t - r/c)/4\pi cr] \, d\mathbf{y}, \tag{162}$$

where $r_i = x_i - y_i$. This form would not be at all obvious as a far-field approximation on the simple-source representation.

A further check on the work is obtained from the fact that, since the integral (155) satisfies equation (154), the *integral* on the right-hand side of (161) is the solution of the linear wave equation *with the quantity f_i on the right-hand side.* Hence expression (161), which is $-(\partial/\partial x_i)$ of that integral, satisfies the same linear wave equation with the quantity $-(\partial f_i/\partial x_i)$ on the right-hand side, which is indeed (158)! Mathematically, this deduction is valid because the linear wave operator $(\partial^2/\partial t^2 - c^2 \partial^2/\partial x_i^2)$ commutes with the operator $(\partial/\partial x_i)$.

The hypothetical distributed monopole and dipole sources just described may not be of direct utility, but the results suggest how to treat the important equation (152) for sound generated by a turbulent flow. Just as a single differentiation on the right-hand side of (158) made it necessary to recognise that the radiation was of dipole type, with strength f_i per unit volume, so the double differentiation on the right-hand side of (152) makes it necessary to identify the radiation as of *quadrupole* type, with strength T_{ij} per unit volume.

In fact, application of the dipole result shows the solution of (152) to be a dipole of strength per unit volume $-\partial T_{ij}/\partial x_j$: namely, the 'effective force' which already in the momentum equation (151) expresses how *gradients* in the total momentum flux T_{ij} across an elementary region impart change of momentum due to inequalities of flux into and out of it. Furthermore, the dipole field of strength $-\partial T_{i1}/\partial x_1$ per unit volume (one of the three involved in that sum) is the limit as $\epsilon \to 0$ of a dipole field of strength per unit volume

$$\epsilon^{-1} T_{i1}(x_1, x_2, x_3) - \epsilon^{-1} T_{i1}(x_1 + \epsilon, x_2, x_3), \tag{163}$$

which the argument following equation (159) shows equivalent to a distribution of quadrupoles, per unit volume, each defined as a dipole of vector strength $\epsilon^{-1} T_{i1}$ at (x_1, x_2, x_3) plus one of strength $-\epsilon^{-1} T_{i1}$ at $(x_1 - \epsilon, x_2, x_3)$. The other quadrupole elements. T_{ij} arise from the terms with $j = 2$ and 3 involved in the sum.

This definition shows the way in which two directions are involved in the quadrupole strength, which is accordingly a tensor. Note that one of its diagonal elements like T_{11} corresponds to an aligned configuration of dipoles (figure 16), commonly called a 'longitudinal quadrupole', while an off-diagonal element like T_{12} corresponds to dipoles separated in a perpendicular direction, called a 'lateral quadrupole'.

The work of this section indicates that a disturbance without monopole or dipole radiation, for example, a jet involving no fluctuations in mass outflow and no external forces on the fluid, may be expected to involve predominantly quadrupole radiation associated with the *excess momentum*

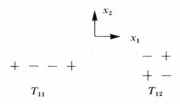

Figure 16. Illustrating how diagonal and off-diagonal elements of the tensor T_{ij} correspond respectively to a longitudinal quadrupole (aligned configuration of dipoles) and a lateral quadrupole (dipoles separated in a perpendicular direction.)

flux tensor T_{ij}. Associated with this is an extremely large ratio of near-field to far-field sound, such as is found experimentally for turbulent jets.

The corresponding solution of (152) is

$$c^2[\rho(\mathbf{x}) - \rho_0] = (\partial^2/\partial x_i\,\partial x_j) \int [T_{ij}(\mathbf{y}, t - r/c)/4\pi r]\,\mathrm{d}\mathbf{y}, \qquad (164)$$

being obtained most simply as $(\partial^2/\partial x_i\,\partial x_j)$ of the solution of the linear wave equation with T_{ij} on the right-hand side, as in the discussion following equation (162). The corresponding far field is

$$c^2[\rho(\mathbf{x}) - \rho_0] = \int [r_i r_j\,\ddot{T}_{ij}(\mathbf{y}, t - r/c)/4\pi r^3 c^2]\,\mathrm{d}\mathbf{y} \qquad (165)$$

and involves a second time-derivative from applying the two differentiations in (164) to the retarded time $t - r/c$. As the speed U of a jet increases, the fluctuations in T_{ij} defined by (150) increase like U^2 but their characteristic frequency also increases like U so the $\ddot{T}_{ij}$ in (165) increases like U^4. Thus the far-field amplitudes vary like U^4, while the intensity and power output vary like U^8. This remarkable dependence of acoustic power output on such a high power of jet speed is a direct consequence of the quadrupole character of the radiation.

The above discussion used the equation of motion of a fluid (2) with viscous stresses neglected but we find that those when included just appear as a minor additional term in T_{ij} (see section 1.13). We may also remark that the condition for acoustical compactness (91) is found to need special treatment in turbulent flows: the l which should appear in it is the effective size of *eddies* that radiate coherently, and the product ωl is typically of order the root-mean-square velocity fluctuation, which commonly leads to the compactness condition $\omega l \ll c$ being satisfied. For further discussion of these and many other matters concerning sound radiation by fluid flows, see specialised texts.

1.11   Radiation from spheres

Up to this point, emphasis in the discussion of sound generation has been on acoustically compact source regions which, however complex in other respects, may have far fields given to good approximation by quite simple rules: for example, equations (105) or (112). In an introductory chapter on the theory of sound it is appropriate to emphasise types of generation that allow such relatively simple treatment, but necessary also to convey some idea of the behaviour of noncompact source regions.

Theories of radiation from bodies of general shape lose this relative simplicity of sections 1.6 and 1.7 in the absence of the compactness condition: they are indeed extremely complex. Nevertheless some special shapes of body, including spheres, allow much simpler analysis and yet yield results rather typical of a wide range of shapes. This section on radiation from spheres is used, then, to exhibit the changes in radiation that occur at frequencies too high for the compactness condition (91) to be satisfied, as well as to suggest that at exceedingly high frequencies (satisfying an opposite condition with $\gg$ replacing $\ll$) there may appear a new simplification of a quite different type which plays an important role in the later chapters of this book.

We study first the sound generated when a spherical body pulsates radially. We suppose that its radius a makes small fluctuations around an undisturbed value a_0, as for the spherical bubble treated in section 1.6, but no longer assume that the sphere is compact.

The special simplicity of the spherical body shape in this case of *radial* pulsations comes from the obvious fact that the sound field must have spherical symmetry about the centre, and therefore can be described by a spherically symmetrical velocity potential of the form (63), where r is distance from the centre of the sphere. The arguments following that equation prove that such a velocity potential, if it involves only outward-travelling waves, may always be written in the form (69), representing a simple source at the centre of the sphere.

There is no suggestion here that such a point source with rate of volume outflow $m(t)$ is 'really present' inside the spherical body at its centre; only that the fluid in the region *outside* the sphere must move just as it would if such a point source were radiating in homogeneous fluid. The required value of the function $m(t)$ can be determined from the boundary condition that the radial velocity $\partial\phi/\partial r$ of the fluid is equal to the radial velocity $\dot{a}(t)$

of the body at its surface $r = a$. In a linear theory we write this boundary condition as

$$(\partial\phi/\partial r)_{r=a_0} = \dot{a}(t) \tag{166}$$

because any difference in the values of $\partial\phi/\partial r$ at $r = a$ and at $r = a_0$ is less than

$$|a - a_0| \operatorname{Max} |\partial^2\phi/\partial r^2| \tag{167}$$

and may be neglected as a product of small quantities.

Equation (70) for $\partial\phi/\partial r$ shows us that the boundary condition (166) can be written as a differential equation

$$m(t - a_0/c) + (a_0/c)\dot{m}(t - a_0/c) = 4\pi a_0^2 \dot{a}(t) = \dot{V}(t) \tag{168}$$

for the function $m(t)$ in terms either of the given function $a(t)$, which describes small changes of sphere radius, or of the related function $V(t)$ describing changes in sphere volume from its undisturbed value $\frac{4}{3}\pi a_0^3$. The discussion following equation (70) shows that in the acoustically compact case a good approximation to the left-hand side of (168) is $m(t)$, so that the required rate $m(t)$ of volume outflow at the source is then close to the rate of change of body volume $\dot{V}(t)$, as is of course true for any compact body (the associated source strength being $\rho_0 \dot{V}(t)$ as in (105)).

Without the compactness condition a corrected value for $m(t)$ is easily enough found, *for spheres only*, by solving (168). The solution may be obtained by use of an integrating factor $\exp(ct/a_0)$ in the form

$$m(t) = (c/a_0) \int_{-\infty}^{t} \dot{V}(T + a_0/c) \exp\left[c(T - t)/a_0\right] dT. \tag{169}$$

Another way of expressing the relationship (168) is to note that each term $V_1 \exp(i\omega t)$ in $V(t)$ generates a term

$$i\omega V_1 \exp(i\omega t) \frac{\exp(i\omega a_0/c)}{1 + (i\omega a_0/c)} \tag{170}$$

in $m(t)$.

Either approach confirms that when $V(t)$ varies very slowly $m(t)$ stays close to $\dot{V}(t)$; for example, because the fraction displayed in (170) is $1 - \frac{1}{2}(\omega a_0/c)^2 + O(\omega a_0/c)^3$ under the compactness condition ($\omega a_0/c$ small). The error becomes hardly tolerable, however, when $\omega a_0/c$ exceeds (say) 0.6.

Both approaches indicate also another interesting approximation valid when $\omega a_0/c$ is large (say, 6 or more). Then

$$m(t) \doteqdot (c/a_0) V(t + a_0/c). \tag{171}$$

We obtain this approximation by neglecting the constant term in the denominator of (170), or replacing in (169) the (relatively) very slowly varying exponential by 1.

This high-frequency approximation (171) has some interesting implications. The pressure at a point P, whose distance from the centre is r, is derived (using equations (71) and (72) for simple-source motions) as

$$p - p_0 = (\rho_0 c / a_0) \, \dot{V}[t - (r - a_0)/c]/4\pi r, \qquad (172)$$

where the *time-lag* $(r - a_0)/c$ is the time taken by a signal to reach the point P *from the nearest point of the sphere*. This implies, surprisingly but correctly, that when $\omega a_0/c$ is large the pressure fluctuations at P derive principally from signals *following the shortest path* from the body surface to P.

Furthermore, if we replace $\dot{V}(t)$ by $4\pi a_0^2 \, \dot{a}(t)$ in (172), we obtain

$$p - p_0 = (a_0/r) \rho_0 c \dot{a}[t - (r - a_0)/c], \qquad (173)$$

with some interesting echoes of formulas from one-dimensional propagation theory. At the sphere surface $r = a_0$, (173) becomes exactly the one-dimensional relationship (17) between excess pressure and the component of fluid velocity in the direction of propagation: quantities whose product is the rate of working per unit area. Hence for large $\omega a_0/c$ the sphere does work on the external fluid *at the same rate as would* a flat piston of the same surface area making the same velocity fluctuations to generate one-dimensional waves in a straight tube. In other words, the tendency noted at the end of section 1.4, for three-dimensional generation to be less efficient than one-dimensional, disappears in this high-frequency limit.

Note that, at a distance from such a piston equal to $r - a_0$ (the length of the above-noted shortest path from the body surface to P) the excess pressure in a straight tube would take a value given by (173) but without the reduction factor (a_0/r). Note also how that reduction factor in the excess pressure leads to a factor $(a_0/r)^2$ in the intensity (82) (energy flow per unit area) which may be interpreted as due to the fact that energy fed in over the surface area $4\pi a_0^2$ of the sphere is at distance r spread out over the larger area $4\pi r^2$.

For the special case of the pulsating sphere we are glimpsing remarkable high-frequency results suggesting the importance of 'rays of sound' reminiscent of the rays of light studied in geometrical optics. The pressure fluctuations at a point P are transmitted along a 'ray' defined as the shortest path from the body surface to P. If we imagine a 'ray tube' formed by a bundle of such rays extending to P and to other points near it, then in that ray tube the motions of the sphere surface generate the same pressure fluctuations as in the one-dimensional waves generated by a piston in a straight tube, but reduced for $r > a_0$ by the factor (a_0/r) needed to allow the energy flow along the ray tube to be spread over a cross-sectional area increasing like r^2.

These are principles known collectively as 'geometrical acoustics' (by analogy with geometrical optics), and later chapters are much concerned with establishing conditions for their wider validity. Here, after suggesting their possible importance through a single example, we further pursue the matter only by verifying them in an example different in one important respect.

This second example is that of radiation by a *rigid* sphere oscillating to and fro in a straight line. If the function $b(t)$ represents the small linear displacements of its centre from a fixed origin, then the boundary condition corresponding to (166) is

$$(\partial\phi/\partial r)_{r=a_0} = \dot{b}(t)\cos\theta \tag{174}$$

in spherical polar coordinates with θ the angle between the direction of those displacements and the radius vector. Thus the effective piston velocity is different at different points of the sphere, which permits a more stringent test of the hypothesis that pressure fluctuations at P depend for large $\omega a_0/c$ on piston activity at the point on the sphere surface *nearest* to P.

The previous example, where the boundary condition (166) is exactly satisfied by placing a simple source at the centre of the sphere, suggests that the boundary condition (174) might be satisfied by placing a dipole there. A dipole whose strength has magnitude $G(t)$ and direction that of the body's displacement has pressure field given by equations (92) and (90), and hence has a *radial acceleration* field

$$-\rho_0^{-1}\,\partial p/\partial r = (4\pi\rho_0)^{-1}\,[2r^{-3}G(t-r/c)$$
$$+ 2r^{-2}c^{-1}\dot{G}(t-r/c) + r^{-1}c^{-2}\ddot{G}(t-r/c)]\cos\theta. \tag{175}$$

It follows that (174) can be satisfied exactly if $G(t)$ satisfies the differential equation

$$G(t-a_0/c) + (a_0/c)\,\dot{G}(t-a_0/c) + \tfrac{1}{2}(a_0/c)^2\,\ddot{G}(t-a_0/c) = 2\pi\rho_0\,a_0^3\,\ddot{b}(t). \tag{176}$$

This equation is interesting in the compact case $\omega a_0/c$ small, since Taylor's theorem makes the left-hand side an excellent approximation, with error only $O(\omega a_0/c)^3$, to $G(t)$, checking the result (114) obtained for spheres by the approximation methods of section 1.7. More generally, we can express

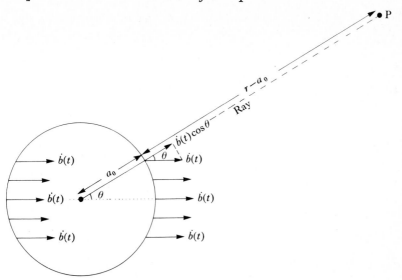

Figure 17. The pressure field at P due to an oscillating rigid sphere is transmitted according to geometrical acoustics, along a ray which is the shortest path from P to the body surface.

the relationship (176) by noting that each term $b_1 \exp(i\omega t)$ in the displacement $b(t)$ generates a term

$$2\pi\rho_0 a_0^3 (i\omega)^2 b_1 \exp(i\omega t) \frac{\exp(i\omega a_0/c)}{1 + (i\omega a_0/c) + \frac{1}{2}(i\omega a_0/c)^2} \tag{177}$$

in $G(t)$.

Our principal concern, however, is with the behaviour of $G(t)$ and of the resulting pressure field for very large values of $\omega a_0/c$. In this limit, (176) gives

$$G(t) = 4\pi\rho_0 a_0 c^2 b(t + a_0/c), \tag{178}$$

and the pressure field (92), in which when $\omega a_0/c$ is large the far-field term dominates for all $r > a_0$, becomes

$$p - p_0 = (a_0/r)\rho_0 c\dot{b}[t - (r - a_0)/c]\cos\theta. \tag{179}$$

Equation (179) verifies, for the oscillating sphere at large $\omega a_0/c$, the ideas of 'geometrical acoustics': especially, that the pressure fluctuations at a point P are transmitted along a ray which is the shortest path from P to the body surface (figure 17). This is confirmed not only by the value $(r - a_0)/c$ for the time-lag but also by the value $\dot{b}\cos\theta$ for the effective piston velocity, which as (174) shows is the radial velocity at the point on the sphere nearest to P, where (figure 17) the value of θ is necessarily the same as at P. On the other hand, the reduction factor a_0/r, which allows for the energy flow in a ray tube to be spread over a cross-sectional area increasing like r^2, appears in equation (179) exactly as in (173).

Reasons for the validity of these ideas at high frequency appear in later chapters, beginning with chapter 2 which elucidates aspects of the energy-flow arguments and chapters 3 and 4 which develop the notion of rays. We anticipate the latter discussion with one remark only: pressure fluctuations at P are dominated by sources at a minimum distance from P partly because groups of sources with a highly variable distance from P produce highly variable phase at P and tend to cancel in their total effect ('destructive interference') whereas from sources whose distance from P is more nearly stationary the phase at P is more nearly constant and their coherent fluctuations can add up to a substantial total.

Sound radiation by other motions of spheres than pulsation or rigid oscillation is not often of practical interest. We may note, however, that boundary conditions equating radial velocity to a second-order spherical harmonic can be satisfied exactly by placing a quadrupole at the centre of the sphere; these correspond to vibrations in which the instantaneous body shapes are ellipsoidal, but with volume remaining constant and the centroid at rest. General boundary conditions can be expanded in spherical harmonics, and higher-order multipoles used to deal with the higher harmonics. Each case is found to obey the above rules of geometrical acoustics in the limit of high frequency.

1.12 Radiation from plane walls

In a fluid with an internal spherical boundary making prescribed small displacements, the sound generated can be analysed fairly simply (section 1.11) and the results used both to check general theories for compact source regions and to probe the characteristics of sound radiation in the high-frequency limit. Another radiation problem amenable to fairly simple analysis at any frequency is that of fluid with a single *plane* boundary (large enough to be treated as infinite) part of which is making prescribed small displacements. It is valuable from both a practical and a theoretical standpoint to include this analysis as well as that of section 1.11 in even an introductory account of acoustics.

Thus, although sound generation by vibrating loudspeaker diaphragms in general geometrical situations is hard to calculate, a special case of some practical interest, the vibrating diaphragm flush with a plane wall, can be studied by this analysis. Other technological applications of the theory have arisen simply because radiation from plane walls is so straight-forwardly predictable that some desired pattern of sound radiation can be generated to good approximation (see below) by a well-designed array of diaphragms flush with a wall.

From the theoretical viewpoint, radiation from plane walls in the high-frequency limit has special interest as a case of a general result that ray-tracing rules need modification for sound generated from any *flat* portion of surface. Comparison with section 1.11 indicates why this is: when a sphere radiates, the pressure fluctuation at a distant point P is made up of components emanating from all over the surface, for one of which (exhibiting 'stationary phase') the value of the phase has a pronounced minimum. With a *flat* portion of surface the same is true for points P that are not too far distant, and brings about a *parallel beam* of rays according to principles suggested in section 1.11. At greater distances, however, the minimum becomes shallower and shallower and so loses its influence, while the energy in the parallel beam becomes redistributed (see below) and finally splayed out within a *narrow cone*.

Data relevant to such practical and theoretical matters may be readily obtained if we analyse sound generated in fluid filling a region $x > 0$ when part of the infinite plane wall $x = 0$ which bounds it makes prescribed small displacements. These are taken as causing a point on the wall which in the undisturbed state has coordinates $(0, Y, Z)$ to move with normal velocity $f(Y, Z, t)$ in the positive x-direction at time t; later, we suppose that f is negligible outside a finite vibrating region of the wall of linear dimension l.

A resulting sound field accurate (on linear theory) at all frequencies can in this plane-wall case be written down immediately as a linear combination of simple-source fields, such that the sources within each elementary area $dY\,dZ$ of the plane generate total mass outflow

$$q(t) = 2\rho_0 f(Y, Z, t)\,dY\,dZ. \tag{180}$$

As in section 1.11, where the sound field outside a pulsating sphere was identified as a part of the bigger sound field of a point source in homogeneous fluid, so here we identify the sound field of the vibrating plane as a part (the part $x > 0$) of the field of this distribution (180) over the plane $x = 0$ of sources radiating in homogeneous fluid.

In such unbounded fluid, by symmetry, the mass outflow must be divided equally between the part $x > 0$ and this part $x < 0$, which must receive therefore a mass outflow $\rho_0 f$ per unit area. Hence the normal velocity of fluid close to the wall in this part $x > 0$ includes

(i) a contribution $f(Y, Z, t)$ due to this mass outflow from the nearby sources; but

(ii) zero contribution from remoter sources, which produce only tangential motion in fluid close to the plane wall.

Considerations (i) and (ii) verify that the necessary boundary condition on normal velocity is satisfied by the sound field of the simple-source distribution (180); though we may note that this inference is of a type applicable only to plane walls, since it depends on consideration (ii).

The distribution (180) of simple-source fields (71) generates at the point (x, y, z) an excess pressure

$$p - p_0 = \int_{-\infty}^{\infty} \int_{-\infty}^{\infty} \frac{2\rho_0 \dot{f}(Y, Z, t - r_s/c)\, \mathrm{d}Y \mathrm{d}Z}{4\pi r_s}, \tag{181}$$

where r_s is the distance to (x, y, z) from the source at $(0, Y, Z)$:

$$r_s = [x^2 + (y - Y)^2 + (z - Z)^2]^{\frac{1}{2}}. \tag{182}$$

Although the limits of integration in (181) are $(-\infty, \infty)$ we consider only radiation from finite source regions for which f is negligible when the magnitude of Y or Z exceeds a length-scale l.

Integrals like (181) may be complicated to evaluate exactly but are greatly simplified in the *far field*, where the distance

$$r = (x^2 + y^2 + z^2)^{\frac{1}{2}} \tag{183}$$

from the origin is so large that only the leading term (proportional to r^{-1}) in an expansion of (181) in descending powers of r is significant. This then fixes the leading term (proportional to r^{-2}) in a similar expansion of the intensity and thus both the acoustic-power output and its directional distribution far from the source region.

We note first that

$$r_s = r - (y/r)\, Y - (z/r)\, Z + O(r^{-1}) \tag{184}$$

as $r \to \infty$ in a fixed direction (that is, with fixed ratios y/r and z/r). Hence (181) can be written

$$p - p_0 = \frac{\rho_0}{2\pi r} \int_{-\infty}^{\infty} \int_{-\infty}^{\infty} \dot{f}\left(Y, Z, t - \frac{r}{c} + \frac{yY + zZ}{rc}\right) \mathrm{d}Y \mathrm{d}Z + O\left(\frac{1}{r^2}\right), \tag{185}$$

where as in all far fields the variable distance r_s from the source can in the denominator of (181) be replaced by a fixed value r with error $O(r^{-2})$, but to achieve the same order of error the time-lag r_s/c must be taken to the next approximation.

In fact, for fixed r the variations in time-lag in (185) are of order l/c. Only when the compactness condition (91) is satisfied can the variation of $\dot{f}$ over a time difference of this order be neglected and all values of $\dot{f}$ taken at the same time $t-r/c$. As in the general theory of section 1.6 the whole field is then that of a single source at the origin whose strength is the combined strength of all the sources: here, twice the rate of change of mass outflow from the plane.

Note that this source strength for a compact array of vibrating diaphragms flush with a large plane wall is *twice* as much as for the same array attached to the surface of a compact body. The far-field intensity (82) is therefore four times as much, though confined to a solid angle of 2π instead of 4π; accordingly, the power output of the diaphragms is simply *doubled* by attachment to such a large plane baffle.

To investigate how a noncompact array of diaphragms radiates, we may observe that each term
$$f_1(Y, Z)\exp(i\omega t) \tag{186}$$

in the surface distribution $f(Y, Z, t)$ of normal velocity generates a term
$$(\rho_0 i\omega/2\pi r)F_1(\omega y/rc, \omega z/rc)\exp[i\omega(t-r/c)] \tag{187}$$

in the far-field pressure (185), where
$$F_1(M, N) = \int_{-\infty}^{\infty}\int_{-\infty}^{\infty} f_1(Y, Z)\exp(iMY+iNZ)\,dY\,dZ \tag{188}$$

is the two-dimensional Fourier transform of $f_1(Y, Z)$. Note that this statement includes the result for a compact array since when ω/c and hence also both M and N are small compared with $1/l$ the F_1 in (187) and (188) may be approximated by the constant $F_1(0, 0)$ giving a point-source pressure field.

In the general case, however, the far-field pressure fluctuations (187) have a *directional* dependence on y/r and z/r given by the Fourier transform of the wall velocity distribution. As in other far fields we can use equation (81) and deduce an intensity $(p-p_0)^2/\rho_0 c$, leading to a *directional distribution*
$$(\rho_0 \omega^2/8\pi^2 r^2 c)|F_1(\omega y/rc, \omega z/rc)|^2 \tag{189}$$

of the *acoustic intensity* averaged with respect to time (so that squares of sines or cosines of ωt are replaced by $\tfrac{1}{2}$).

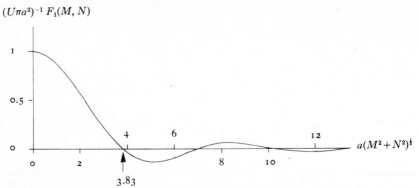

Figure 18. The function F_1 given by equation (190), specifying the far-field pressure distribution (187) of a circular diaphragm of radius a, vibrating with velocity amplitude U in a large plane wall.

These rules permit fairly easy calculations for particular diaphragm arrays: thus, a single circular diaphragm of radius a vibrating with velocity amplitude U has $f_1 = U$ or o according as $Y^2 + Z^2$ is less or greater than a^2, giving

$$F_1(M, N) = 2U\pi a(M^2 + N^2)^{-\frac{1}{2}} J_1[a(M^2 + N^2)^{\frac{1}{2}}] \qquad (190)$$

in terms of the Bessel function J_1. A nondimensional plot of $(U\pi a^2)^{-1} F_1$ as a function of $a(M^2 + N^2)^{\frac{1}{2}}$ is given in figure 18. According to (187) this specifies the directional dependence of pressure fluctuations as a function of

$$\omega a(y^2 + z^2)^{\frac{1}{2}}/cr = (\omega a/c)\sin\theta, \qquad (191)$$

where $\theta = \cos^{-1}(x/r)$ is the angle made with the x-axis.

Note that when $\omega a/c \leqslant 3.83$ (the first zero of J_1) the amplitude is a decreasing function of θ all the way to $\theta = \frac{1}{2}\pi$. For larger values of $\omega a/c$, however, the amplitude falls to zero at a particular angle

$$\theta_1 = \sin^{-1}[3.83(\omega a/c)^{-1}] \qquad (192)$$

and most of the radiation is confined within the cone $\theta < \theta_1$. Thus a single loudspeaker diaphragm can generate a narrow conical beam of sound if its radius a is large enough compared with $c/\omega = \lambda/2\pi$. However, the radiation pattern includes also 'side lobes' with $\theta > \theta_1$ with peak amplitudes 13 %, 7 %, 4 %, etc., of the central peak amplitude.

Such 'side lobes' are characteristic of most simple methods of generating narrow beams of sound. However, for technological purposes requiring some prescribed directional pattern of radiation, for example, one without side lobes, equation (187) suggests how to achieve it by using an array of diaphragms approximating a velocity-distribution function specified as the inverse Fourier transform

$$f_1(Y, Z) = (2\pi)^{-2} \int_{-\infty}^{\infty} \int_{-\infty}^{\infty} F_1(M, N) \exp(-iMY - iNZ) \, \mathrm{d}M\mathrm{d}N$$

(193)

of the desired directional distribution.

In the general case of a function $f_1(Y, Z)$ describing vibration of some array of diaphragms covering an area of length-scale l, the Fourier transform $F_1(M, N)$ must tend to zero as lM and lN become very large (see texts on Fourier analysis) and therefore equation (187) means that as $\omega l/c$ becomes very large the sound radiation must be chanelled into a narrower and narrower conical region. It is interesting to compare this general plane-wall result for $\omega l/c$ large with the ideas described in section 1.11 as geometrical acoustics.

Geometrical acoustics would predict that for $\omega l/c$ large the pressure fluctuations at each point P are transmitted along a ray which is the shortest path to P from the plane wall: that is, the normal to the wall through P. It predicts a parallel beam of rays, then, and the true conical beam deviates from this in direction by angles that do at any rate tend to zero as $\omega l/c \to \infty$.

On the other hand, the pressure fluctuation predicted by geometrical acoustics as transmitted along the ray $y = Y$, $z = Z$ from a 'piston' moving with velocity (186), namely,

$$\rho_0 c f_1(y, z) \exp[i\omega(t - x/c)],$$

(194)

bears no resemblance whatever to the far-field formula (187) even in the limit $\omega l/c$ large. Not only is there no attenuation factor r^{-1} to take into account the fact that the rays diverge conically instead of remaining parallel; the distribution across the beam is also given wrongly by the function f_1 rather than by its Fourier transform F_1.

We omit a full discussion of the general difficulty hinted at here: namely, that when geometrical acoustics predicts a parallel beam its conclusions need substantial modification in the far field. We may briefly indicate, however, the resolution of this apparent conflict between two theories.

Careful study of the accurate pressure field (181) shows that the geometrical-acoustics parallel-beam approximation (194) is closely correct in a substantial length of beam near the wall; this is followed by a region of transition to the conical-beam form (187); while at great distances from the wall the conical-beam far-field approximation becomes closely correct. Note that the transition region is where the beam width changes from order l to order $rc/\omega l$ (these are orders of magnitude of y and z within which expressions (194) and (187) respectively are significant). Those widths coincide in what turns out to be the transition region, with distance r from the wall of order $\omega l^2/c$.

In fact, geometrical acoustics is found to give a good approximation where $r \ll \omega l^2/c$; essentially, because this is the region where the phase of the received signal possesses a sufficiently pronounced minimum for destructive interference to take place away from it. Similarly the far-field approximation (187) is good where $r \gg \omega l^2/c$, while the region of redistribution of amplitude from the f_1 functional form (194) to the F_1 form (187) lies at intermediate distances r of order $\omega l^2/c$. Note that this redistribution involves no change in the total energy flow. This follows from Parseval's theorem

$$\int_{-\infty}^{\infty} \int_{-\infty}^{\infty} |f_1(y, z)|^2 \, \mathrm{d}y \, \mathrm{d}z = (2\pi)^{-2} \int_{-\infty}^{\infty} \int_{-\infty}^{\infty} |F_1(M, N)|^2 \, \mathrm{d}M \, \mathrm{d}N$$

$$(195)$$

(see texts on Fourier analysis): the energy flows in the parallel beam (194) and the conical beam (187) are $\frac{1}{2}\rho_0 c$ times the left-hand and right-hand sides respectively (as we obtain, for example, by integrating formula (189) for the intensity over the cross-sectional area of the beam).

1.13　Dissipation of acoustic energy

We end this chapter by describing mechanisms so far neglected by which acoustic energy is dissipated into heat, and ways in which these modify the linear theory of sound. We first analyse plane travelling waves, finding in common fluids that although at audible frequencies or even at ultrasonic frequencies as high as a few MHz they lose in each wavelength only an exceedingly small proportion of their energy by dissipation, this produces an exponential reduction in amplitude over distances of very many wavelengths, described as *attenuation*.

Modifications needed in several of the theories of chapter 1 to allow for dissipation can be inferred from this plane-wave result. In and around a compact source region, for example, dissipation of acoustic energy can make little difference to processes both of sound generation and of propagation out to the threshold of the far field, since this is situated only one or two wavelengths away. That far field, however, whether of simple-source or of dipole nature, has in each wavelength the characteristics of a plane wave, except in so far as an r^{-1} factor allows for acoustic energy to be gradually spread out over an area increasing like r^2. In particular the proportion of acoustic energy lost per wavelength should be much the same as for a plane wave, requiring amplitude to decrease with distance as the product of two factors: the exponential term which as in plane waves accounts for this energy loss, and the r^{-1} term permitting the energy that remains to be spread over an area increasing like r^2. Similarly, any waves treated by geometrical acoustics as propagated along ray tubes have in each wavelength the characteristics of plane waves, leading to amplitudes varying as the product of the same exponential function of distance and a geometrical factor that allows the flow of acoustic energy to be spread over a varying cross-sectional area of ray tube.

One mechanism of dissipation is *viscosity*, which in non-uniform motions allows momentum flux by the action of internal fluid stress to differ in different directions. A *stress tensor* p_{ij}, representing rate of flux of the x_i-component of momentum in the x_j-direction across unit area per unit time due to fluid stresses, expresses this anisotropy.

In hydrostatic conditions, or when viscosity is neglected, the stress tensor takes the isotropic form

$$p_{ij} = p\delta_{ij} \tag{196}$$

equal in all directions. Nonlinear theories of general motions need to take into account an anisotropic momentum flux p_{ij} due to fluid stresses, as well as a convective momentum flux $\rho u_i u_j$ as in equation (147); thus, the theory of sound generation by turbulence in section 1.10 becomes still more accurate if $p\delta_{ij}$ is replaced both in (147) and in the consequent definition (150) of T_{ij} by the true stress tensor p_{ij}. Texts on the kinetic theory show that, in perfect gases as defined following equation (23), the stress tensor can itself be viewed as an average extra momentum flux produced convectively by the large fluctuations of molecular velocities about the mean values represented by the fluid velocity u_i.

Within a *linear* theory, only the momentum flux p_{ij} due to stresses is retained since $\rho u_i u_j$ involves products of small quantities. For plane waves propagating in (say) the x_1-direction, on which we concentrate for reasons sketched out above, the one important component of p_{ij} turns out to be p_{11}.

In fact, all quantities in such a plane wave are independent of x_2 and x_3, so that a linearised momentum equation

$$\rho_0\, \partial u_1/\partial t = -\, \partial p_{11}/\partial x_1 \tag{197}$$

equates rate of change of the x_1-component of momentum to minus the gradient of its x_1-component of flux, while the linearised equation of continuity (5) becomes a relationship

$$\partial \rho/\partial t = -\rho_0\, \partial u_1/\partial x_1 \tag{198}$$

between ρ and u_1 alone. Equations (197) and (198) imply a rather simple relationship

$$\partial^2\rho/\partial t^2 = \partial^2 p_{11}/\partial x_1^2 \tag{199}$$

between ρ and p_{11}.

If in small disturbances p_{11} differed from the undisturbed pressure p_0 by a constant multiple $c^2(\rho-\rho_0)$ of the excess density, as would follow from the inviscid relationship (196) and equation (37) for reversible (constant-entropy) processes, equation (199) would be the one-dimensional wave equation. A more accurate relationship is obtained, however, when we recognise that p_{11} may vary not only with ρ but also with either $\partial\rho/\partial t$ or $\partial u_1/\partial x_1$. Since these two quantities by (198) are exactly proportional, such dependence in linearised form can be written

$$p_{11}-p_0 = c^2(\rho-\rho_0)+\delta\partial\rho/\partial t, \tag{200}$$

where the first coefficient remains c^2 because changes that are *not* too abrupt must satisfy the inviscid reversible relationships.

The coefficient δ may take into account viscosity (anisotropy of the stress tensor allowing p_{11} to exceed other components when $\partial u_1/\partial x_1$ is negative) or irreversibility due to delays in attaining thermodynamic equilibrium (with a term in $\partial\rho/\partial t$ allowing for a lag of density changes behind pressure changes), and equation (198) shows that experiments on plane waves of sound cannot distinguish between these two effects; a circumstance that has led to phrases like 'bulk viscosity', here *avoided*, being used to describe the second effect. Both contributions to δ and a less obvious contribution due to heat conduction are studied below after a brief analysis of the consequences of equation (200) for sound propagation.

Its consequences for the acoustic energy, namely W per unit volume given as the sum of (47) and (50) or for plane waves as

$$W = \tfrac{1}{2}\rho_0 u_1^2 + \tfrac{1}{2}(\rho-\rho_0)^2 c^2\rho_0^{-1}, \tag{201}$$

are by (197) and (198) that

$$\begin{aligned}
\partial W/\partial t &= -u_1\,\partial p_{11}/\partial x_1 - c^2(\rho-\rho_0)\,\partial u_1/\partial x_1 \\
&= -\partial[(p_{11}-p_0)\,u_1]/\partial x_1 + [p_{11}-p_0-c^2(\rho-\rho_0)]\,\partial u_1/\partial x_1 \\
&= -\partial I/\partial x_1 - \rho_0\delta(\partial u_1/\partial x_1)^2, \tag{202}
\end{aligned}$$

where I is an energy flux due to rate of working by the excess stress $p_{11}-p_0$ and equations (200) and (198) have been used to show that the last term is essentially negative. Equation (202) implies a rate of dissipation of acoustic energy

$$\rho_0\delta(\partial u_1/\partial x_1)^2 \tag{203}$$

per unit volume, in addition to changes due to the action of the energy flux I.

Energy dissipation proportional as in (203) to the square of a *gradient* will be found to have two consequences. For general waveforms, the pronounced maxima of dissipation rate wherever exceptionally steep gradients $\partial u_1/\partial x_1$ occur tend to diminish such large gradients. In travelling waves of sinusoidal form, furthermore, any effects of dissipation are found to depend sensitively on the *frequency*, ω. In fact the mean of the dissipation rate (203) in such waves is in the ratio $\delta(\omega/c)^2$ to the mean acoustic energy (51), implying the loss of a proportion

$$2\pi\omega\delta/c^2 \tag{204}$$

of the acoustic energy in each period $2\pi/\omega$ as the wave travels.

As mentioned above, we consider only frequencies for which this proportional energy loss (204) per period or per wavelength is very small compared with 1. Any local relationships between quantities in a sound wave, like (51) itself, can then be employed as a good first approximation in the theory.

The equation of motion derived from (199) and (200) is

$$\partial^2\rho/\partial t^2 = c^2\partial^2\rho/\partial x_1^2 + \delta\partial^3\rho/\partial x_1^2\,\partial t, \tag{205}$$

written here as an equation for ρ although it is easy to deduce from (198) and (197) that u_1 and p_{11} satisfy the same equation. A term $\rho_1(x_1)\exp(i\omega t)$ in ρ representing sinusoidal fluctuations with frequency ω must satisfy

$$(c^2+i\omega\delta)\,d^2\rho_1/dx_1^2 = -\omega^2\rho_1 \tag{206}$$

with solutions proportional to $\exp(\pm\kappa x_1)$, where

$$\kappa = i\omega(c^2+i\omega\delta)^{-\frac{1}{2}} \doteqdot (i\omega/c)(1 - \tfrac{1}{2}i\omega\delta/c^2) \tag{207}$$

to close approximation when (204) is very small. The upper and lower signs refer to waves travelling to the left and the right respectively, and each is seen to lose a fraction (204) of its energy in every wavelength $2\pi c/\omega$. For example, if a wave travels across $x_1 = 0$ into otherwise undisturbed fluid filling $x_1 > 0$ then

$$\rho_1(x_1) = \rho_1(0)\exp(-i\omega x_1/c - \tfrac{1}{2}x_1\,\omega^2\delta/c^3) \tag{208}$$

with energy distribution proportional to $\exp(-x_1\,\omega^2\delta/c^3)$.

This solution for a term $\rho_1(x_1)\exp(i\omega t)$ in $\rho(x_1, t)$ implies by Fourier analysis a solution

$$\rho(x_1, t) = (2\pi x_1\,\delta)^{-\frac{1}{2}}c^{\frac{3}{2}}\int_{-\infty}^{\infty}\rho(0, T)\exp[-(2x_1\,\delta)^{-1}c^3(T-t+x_1/c)^2]\,dT \tag{209}$$

for a wave travelling into undisturbed fluid filling $x_1 > 0$ with *arbitrary* time-variations $\rho(0, t)$ at $x_1 = 0$. This solution should be compared with the simple nondissipative plane-wave solution $\rho(0, t - x_1/c)$. Actually expression (209) is a *time mean* of $\rho(0, T)$ weighted according to a Gaussian distribution centred on $T = t - x_1/c$ with standard deviation $(x_1 \delta)^{\frac{1}{2}} c^{-\frac{3}{2}}$, so the effect of dissipation is gradually to smooth out irregularities like local regions of large gradient by averaging values over this *increasing* interval. There is a close analogy with the smoothing of distributions by diffusive effects, for which reason δ has been called the 'diffusivity of sound'.

Many fluids exhibit the *linear* dependence (204) of proportional loss of acoustic energy per wavelength upon frequency ω, at any rate when that proportion remains small as here assumed but not too small to be measured! For these fluids the observed slope of that dependence determines a value of δ, such that equation (200) can be supposed a good approximation in that range of frequencies. This way of determining δ may be more practically useful than any theory of the contributions to it from viscosity, from heat conduction and from delays in attaining thermodynamic equilibrium; particularly as effects of the delays are not easily quantified by other measurements. Nevertheless, we outline that theory for the special case of a perfect gas, mainly to show how the viscous and conductive effects can be disentangled from these 'lag' effects, and to suggest why for *certain* gases the energy dissipation per wavelength shows a more complicated dependence on ω with 'resonance' peaks.

We write
$$\delta = \delta_v + \delta_c + \delta_l, \tag{210}$$

where subscripts v, c and l indicate *viscosity*, heat *conduction*, and a *lag* in response of the density ρ to changes in

$$p_m = \tfrac{1}{3}(p_{11} + p_{22} + p_{33}), \tag{211}$$

a 'mean pressure' equal to the mean normal stress in three perpendicular directions. For a perfect gas these stresses, as already indicated, are due to momentum flux in molecular motions *relative* to the fluid velocity u_i, with (say) fluctuating velocities v_i that make p_{11} a per-unit-volume summation of Mv_1^2, where M is molecular mass. This makes $\tfrac{3}{2}(p_m/\rho)$ a per-unit-mass summation of

$$\tfrac{1}{2}M(v_1^2 + v_2^2 + v_3^2); \tag{212}$$

that is, the *translational* component of the internal energy E per unit mass. Wherever the work of compression increases E by the amount (25) in response to a density increase $d\rho$, this translational energy $\tfrac{3}{2}(p_m/\rho)$ takes slightly more than its equilibrium share of the energy increase, especially in rapid changes since other components of the energy (rotational, vibrational) need more time to adjust to the change, and the phase of p_m is accordingly ahead of that of ρ.

Although equation (200) gives no way of separating contributions to δ from viscosity and from lags in attaining thermodynamic equilibrium, this clear relation of such lags to the mean pressure p_m strongly suggests that only *departures* of the stress tensor from an isotropic form $p_m \delta_{ij}$ should be considered as true viscous effects. That viscous contribution to p_{ij} must evidently have

$$p_{11} = -\rho_0 \delta_v \, \partial u_1/\partial x_1, \quad p_{22} = p_{33} = \tfrac{1}{2}\rho_0 \delta_v \, \partial u_1/\partial x_1 \qquad (213)$$

if it is to make no contribution to p_m, since by symmetry $p_{22} = p_{33}$.

It is not obvious how these laws can be related to the usual definition of viscosity μ which in a parallel shear motion $(u_1(x_2), 0, 0)$ requires a tangential stress

$$p_{12} = p_{21} = \mu \partial u_1/\partial x_2, \qquad (214)$$

but we can construct a shear motion out of the *simple extension* $\partial u_1/\partial x_1$ by adding an equal and opposite simple compression at right angles contributing

$$p_{22} = -\rho_0 \delta_v \, \partial u_2/\partial x_2, \quad p_{33} = p_{11} = \tfrac{1}{2}\rho_0 \delta_v \, \partial u_2/\partial x_2, \qquad (215)$$

where $$\partial u_2/\partial x_2 = -\partial u_1/\partial x_1, \quad \partial u_3/\partial x_3 = 0. \qquad (216)$$

Texts on fluid dynamics show that the resulting pure straining motion in the (x_1, x_2) plane without change of density is just a parallel shear motion viewed in different axes (rotating and inclined at $45°$ to the x_1- and x_2-axes); the viscous stress system (214) becomes in these axes

$$p_{11} = -2\mu \, \partial u_1/\partial x_1, \quad p_{22} = -2\mu \, \partial u_2/\partial x_2, \quad p_{33} = 0. \qquad (217)$$

Comparison of (217) with the sum of (213) and (215) then yields

$$\delta_v = \tfrac{4}{3}\mu/\rho_0 \qquad (218)$$

as the viscous contribution to the diffusivity of sound.

Before analysing the lag contribution, we may notice a slightly less obvious contribution due to the thermal conductivity k. This is defined so that where the graph of temperature T against distance x_1 has upward curvature the rate of increase of internal energy per unit volume includes a positive term

$$k \, \partial^2 T/\partial x_1^2, \qquad (219)$$

namely the gradient of the *heat flux* $-k \, \partial T/\partial x_1$ per unit area resulting from the temperature variation. Physically this is because where $\partial^2 T/\partial x_1^2 > 0$ the

4   L W F

average molecular energy is less than the average energy of molecules in the neighbourhood, and therefore tends to be increased by diffusional processes.

To estimate from the formula (219) the resulting contribution to a slow acoustic attenuation we can use relationships between temperature, pressure, etc., for unattenuated sound waves. For given excess pressure $p - p_0$ the excess temperature by (26), (27) and (37) is

$$T - T_0 = (1 - \gamma^{-1})(p - p_0)/R\rho_0. \tag{220}$$

Hence the rate of internal energy increase (219) can be written in terms of $\partial^2 p/\partial x_1^2$ and thus by (199) of $\partial^2 \rho/\partial t^2$ as follows:

$$(1 - \gamma^{-1})(k/R\rho_0)\,\partial^2 \rho/\partial t^2. \tag{221}$$

By equation (26), this demands for given $\partial \rho/\partial t$ a contribution to $\partial p/\partial t$ so that the corresponding contribution to the rate of increase of internal energy ρE per unit volume,

$$(c_v/R)\,\partial p/\partial t, \tag{222}$$

takes this value (221). That implies in equation (200) a heat-conduction contribution

$$\delta_c = (1 - \gamma^{-1})k/\rho_0 c_v \tag{223}$$

to the diffusivity of sound.

As elements in the δ of equation (200) we have determined δ_v that allows for departure of p_{11} from the mean pressure p_{m} and δ_c that allows for departures from the no-heat-flux equation (25). Now we calculate δ_l which assumes (25) and takes $p_{ij} = p_{\mathrm{m}} \delta_{ij}$ but allows nontranslational components of the internal energy E to lag behind changes in the translational component $\frac{3}{2}(p_{\mathrm{m}}/\rho)$.

We suppose the internal energy to be found in particular modes of molecular motion numbered $n = 0, 1, 2, \ldots$, whose energy changes by $F_n\,\mathrm{d}(p_{\mathrm{m}}/\rho)$ in a small change of p_{m}/ρ but only after a time-lag τ_n. Here $n = 0$ refers to the translational energy so that $F_0 = \frac{3}{2}$ and $\tau_0 = 0$, but values of $n > 0$ refer to rotational or vibrational modes for which $\tau_n > 0$.

Now in sufficiently gradual processes the time-lags are quite negligible and we have a reversible change with $\mathrm{d}(p_{\mathrm{m}}/\rho) = R\,\mathrm{d}T$; hence

$$\mathrm{d}E = (\Sigma F_n)R\,\mathrm{d}T \quad \text{giving} \quad \Sigma F_n = c_v/R. \tag{224}$$

In changes that are more rapid, although still slow on the scale of the time-lags τ_n, we have

$$\mathrm{d}E = \Sigma F_n(1 - \tau_n\,\partial/\partial t)\,\mathrm{d}(p_{\mathrm{m}}/\rho) = (c_v/R - \Sigma F_n\tau_n\,\partial/\partial t)\,\mathrm{d}(p_{\mathrm{m}}/\rho). \tag{225}$$

To find δ_l we use equation (25) in the form $dE = p_m \rho^{-2} d\rho$ as already explained. Hence

$$p_m \rho^{-2} d\rho = (c_v/R) d(p_m/\rho) - (\Sigma F_n \tau_n)(\partial/\partial t)[(R/c_v)(p_m \rho^{-2} d\rho)], \quad (226)$$

where in the last term giving the small effect of attenuation $d(p_m/\rho)$ has been replaced by its value for unattenuated waves. Equation (226) may then be written

$$dp_m = c^2 d\rho + \delta_l(\partial/\partial t) d\rho, \quad (227)$$

where c^2 is given by (27) and

$$\delta_l = (R/c_v)^2 p_m \rho^{-1} \Sigma F_n \tau_n = (\gamma - 1)^2 (c^2/\gamma) \Sigma F_n \tau_n \quad (228)$$

is the 'lag' contribution to the diffusivity of sound.

The contributions (218), (223) and (228) to δ are all inversely proportional to density for a given temperature since μ and k and F_n are functions of temperature alone while the time-lags τ_n are multiples of a mean time between collisions, that is, inversely proportional to density. Hence the proportional energy loss per wavelength (204) is for given temperature a function of the ratio frequency to pressure, which may be measured in MHz/bar.

For monatomic gases, all the internal energy is translational and $\delta_l = 0$. With $\delta = \delta_v + \delta_c$ (two comparable contributions) the expression (204) describes adequately the results of experiments in helium and argon for frequency–pressure ratios *below* 100 MHz/bar but *above* 1 MHz/bar, conditions required for the proportional energy loss to be respectively small compared with 1 but large enough to dominate over other modifying effects such as those due to nonlinearity studied in chapter 2.

With certain diatomic gases such as nitrogen the rotational mode, say $n = 1$ with $F_1 = 1$, is the only nontranslational mode affecting attenuation; measurements of proportional acoustic energy loss per wavelength are consistent with a time-lag τ_1 of about 0.9 ns at atmospheric pressure, or six mean intervals between collisions, as if six molecular collisions are needed on the average to bring the rotational and translational energy of nitrogen molecules into equilibrium. The associated 'lag' contribution δ_l adds about 40 % to $\delta_v + \delta_c$ and then expression (204) gives good results for frequency–pressure ratios around 10 MHz/bar or less.

For significant vibrational modes of polyatomic molecules much greater time-lags are found. Even at temperatures where such a mode is scarcely excited, making the corresponding F_n so small as only slightly to affect the specific heat (224), it can dominate δ because its contribution to (228) contains the large factor τ_n. In these circumstances, however, the applicability of the theory as so far given is limited in that the approximation (225) demands small $\omega\tau_n$.

More sophisticated theories of attenuation of sound waves of higher frequency ω lead to equations (200) with frequency-dependent values of δ and even of c^2. A far better approximation to the time-lag effect in (225) replaces $(1 - \tau_n \, \partial/\partial t)$ by

$$(1 + \tau_n \, i\omega)^{-1} = (1 + \tau_n^2 \, \omega^2)^{-1}(1 - \tau_n \, \partial/\partial t), \tag{229}$$

which converts equation (228) into

$$\delta_l = (\gamma - 1)^2 (c^2/\gamma) \Sigma F_n \tau_n (1 + \tau_n^2 \, \omega^2)^{-1}. \tag{230}$$

This contribution to the diffusivity of sound is independent of frequency ω only when all the $\omega\tau_n$ are small.

Indeed, each mode makes a contribution to the proportional energy loss per wavelength (204) which rises to a maximum of

$$\pi(\gamma - 1)^2 \gamma^{-1} F_n \tag{231}$$

at $\omega = \tau_n^{-1}$ and then falls away again at higher frequencies. A plot of proportional energy loss per wavelength against frequency–pressure ratio, like that in figure 19 for CO_2, may accordingly exhibit something like a resonance peak, here as high as 0.23, at the characteristic frequency $\omega = \tau_n^{-1}$.

There is a small associated change in c^2 (the sound speed squared) from the ordinary value, when $\omega\tau_n$ is small, to a *greater* value (actually, by a factor $1 + (\gamma - 1)^2 \gamma^{-1} F_n$) when $\omega\tau_n$ is large and vibrational energy changes no longer contribute to the c_v of equation (225) after the modification (229). This greater value, appropriate to changes so rapid that the amount of vibrational energy in the gas remains constant or 'frozen' and ceases to influence the specific heat, is often called the *frozen sound speed*.

Atmospheric air is influenced by the vibrational mode in *oxygen* to exhibit a resonance peak of proportional energy loss per wavelength similar to that in figure 19, though the height of the peak is two orders of magnitude lower, being 0.0022 at 20 °C (and the associated change in sound speed is hardly appreciable). Interestingly enough, the value of the critical frequency τ_n^{-1} is critically dependent on the concentration of *water vapour*, which facilitates excitation of that mode.

Other processes involving temperature-dependent adjustments in distribution of energy in a fluid (for example, evaporation and condensation in two-phase mixtures, or ionisation and de-ionisation in high-temperature gases) can similarly lead to acoustic attenuation per wavelength rising to a

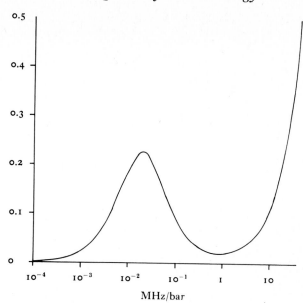

MHz/bar

Figure 19. Proportional energy loss per wavelength for sound waves in CO_2, measured as a function of the frequency–pressure ratio in MHz/bar. The results are well represented by expression (204) with (i) constant contributions to δ (generating the rapid rise beyond 1 MHz/bar) due to viscosity, heat conduction and lag in rotational energy, and (ii) a frequency-dependent contribution (230) (generating the resonance peak) due to lag in vibrational energy.

peak at a characteristic frequency for the process and then falling away again, while the sound speed changes slightly from its equilibrium value to a 'frozen' one. Note also that attenuation of a quite different kind, associated with the shear stresses (214), plays an important part wherever sound waves are propagated *tangentially to a solid wall*; for example, in sound propagation along a tube with solid walls, a case briefly analysed in chapter 2. For some further details of dissipation processes in fluids, see section 3.5.

EXERCISES ON CHAPTER 1

1. For the wave equation (13), use the one-dimensional solutions (15) and (18) to solve the following *initial-value problem*: at time $t = 0$, the fluid velocity and pressure are given by the equations

$$u = u_1(x), \quad v = w = 0, \quad p - p_0 = p_1(x);$$

at any subsequent time t, show that

$$p - p_0 = \tfrac{1}{2}[p_1(x - ct) + p_1(x + ct)] + \tfrac{1}{2}\rho_0\, c[u_1(x - ct) - u_1(x + ct)],$$

and find the fluid velocity.

2. For a fluid satisfying the *van der Waals equation of state*

$$p = RT\rho(1 - b\rho)^{-1} - a\rho^2,$$

show that the specific heat c_v is a function of temperature alone, and that the speed of sound c is given by the equation

$$c^2 = RT(1 - b\rho)^{-2}(1 + Rc_v^{-1}) - 2a\rho.$$

[The constants a and b are introduced in the van der Waals equation of state to take partial account of two effects. The $b\rho$ term reflects any increase in the momentum transport across a surface in the fluid due to a fraction $b\rho$ of the volume of space available to a molecule being effectively excluded by the short-range repulsive force-fields surrounding each of the other molecules. Conversely, the $a\rho^2$ term reflects any reduction in the pressure across a surface in the fluid due to the long-range attractive forces acting between two molecules on different sides of it.]

3. Prove that the definition (77) implies that

$$(d/dt)^{\frac{1}{2}} \exp(i\omega t) = (i\omega)^{\frac{1}{2}} \exp(i\omega t).$$

Show that a line source which extends perpendicularly between two parallel planes a distance h apart, and produces an oscillatory rate of mass outflow $q(t) = q_1 \cos \omega t$, generates a mean acoustic power output

$$\tfrac{1}{8}\rho_0^{-1} h^{-1} \omega q_1^2.$$

4. Show that, in the near field of a dipole of strength $\mathbf{G}(t) = \dot{\mathbf{H}}(t)$, the power transferred across the surface of a sphere of radius r centred on the dipole differs from its far-field value (104) by a term

$$\frac{\partial}{\partial t}\left\{\frac{1}{12\pi\rho_0 c^3}\left[\frac{3}{2}\left(\frac{c}{r}\right)\mathbf{G}^2\left(t - \frac{r}{c}\right) + 2\left(\frac{c}{r}\right)^2 \mathbf{G}\left(t - \frac{r}{c}\right)\cdot\mathbf{H}\left(t - \frac{r}{c}\right) + \left(\frac{c}{r}\right)^3\mathbf{H}^2\left(t - \frac{r}{c}\right)\right]\right\}.$$

What physical significance do you ascribe to the quantity in curly brackets?

5. For sound propagation in unbounded fluid, use the ideas of sources and dipoles to solve the most general initial-value problem; namely, given that at time $t = 0$ the fluid velocity and pressure are

$$\mathbf{u} = \mathbf{u}_I(\mathbf{x}), \quad p = p_0 + p_I(\mathbf{x}),$$

to find the pressure at any subsequent time. Note that fluid at rest with uniform pressure $p = p_0$ could be instantaneously put into the above initial state by (i) a distributed *impulse* $\rho_0 \mathbf{u}_I$ per unit volume acting at time $t = 0$ (here, the word impulse signifies, of course, a very large force acting for a very short time in such a way that the integral of the force with respect to time is equal to the impulse); and (ii) a distributed *instantaneous mass outflow* $c^{-2}p_I$ per unit volume; this is the initial value of $\rho - \rho_0$ (by instantaneous mass outflow, again, we imply a very large rate of mass outflow acting for a very short time in such a way that the integrated rate is equal to the total mass outflow). From this idea, deduce the value of $p - p_0$ at any point P at any time $t > 0$. Show that it can be written as the average value of

$$p_I + ct(\partial p_I/\partial n - \rho_0 c\nabla\cdot\mathbf{u}_I)$$

over the surface of a sphere of radius ct with P as centre (where $\partial/\partial n$ signifies a derivative along the outward normal to that surface). Verify that this agrees with the conclusion of exercise 1.

6. Show that small pulsations of a bubble of gas in a liquid lose a fraction

$$2\pi(3\rho_{g0}/\rho_0)^{\frac{1}{2}}\,(c_g/c)$$

of their energy as sound in each period of pulsation, where the undisturbed density and sound speed are ρ_0 and c for the liquid and ρ_{g0} and c_g for the gas. [For air bubbles in water at 20 °C this fraction is 0.095.]

7. The classical solution (see texts on fluid dynamics) for collapse of an empty spherical cavity in a liquid has

$$[\dot{a}(t)]^2 = \tfrac{2}{3}p_0\,\rho_0^{-1}\{a_0^3[a(t)]^{-3} - 1\}.$$

Here, $a(t)$ is the radius of the cavity at time t and a_0 is its value before motion begins, while p_0, ρ_0 are the undisturbed pressure and density of the liquid. Show that the cavity acts as an acoustic source whose strength may be written as

$$\tfrac{4}{3}\pi p_0\{a_0^3[a(t)]^{-2} - 4a(t)\},$$

so that it is negative until $a(t)$ falls to $2^{-2/3}a_0$ and thereafter is positive. [Note, however, that at those extremely small values of $a(t)/a_0$ for which this expression for the source strength tends to $+\infty$, the classical solution is inapplicable because it neglects the opposing internal pressures due to compression of the vapour inevitably present in a real cavity.]

8. Aeolian tones generated by a wire oscillating in a wind are associated with the fluctuating lift (force at right angles to the wind) due to alternate vortex shedding from the wire's upper and lower surfaces. When the wire's own oscillations keep these lift forces in phase all along the wire, with magnitude $L_1 \cos \omega t$ per unit length, find the resulting acoustic field assuming that the wire's diameter is compact but that its length $2l$ is not. With the lift forces acting in the y-direction and the wire stretching from $(0, 0, -l)$ to $(0, 0, l)$, show that at great distances

$$r = \sqrt{(x^2 + y^2 + z^2)}$$

from the wire's mid-point the pressure field takes the form

$$p - p_0 = -[L_1\,yz^{-1} \sin(\omega z l/rc)]\,(2\pi r)^{-1} \sin[\omega(t - r/c)].$$

In which direction is the amplitude greatest? In which directions does it become zero? [These latter directions bound 'side lobes' similar to those noted in section 1.12.]

9. Show that a compact sphere, whose mean compressibility K may be greater or smaller than the compressibility $\rho_0^{-1}c_0^{-2}$ of the surrounding fluid but is not greater by a *large* factor, produces more backward scattering of a plane sound wave than forward scattering if and only if the sphere's excess mean compressibility has opposite sign to its excess mean density. How does this conclusion fail if $K\rho_0\,c_0^2$ is very large?

10. It used to be common for sirens to be driven by steam. When such a fluid, of density ρ_s, emerges into the atmosphere, of different density ρ_0, at an oscillatory rate, the resulting acoustic output (for moderate flow velocities) may be estimated in two alternative ways. First, the variable *volume outflow* $m(t)$ may be regarded as pushing outwards the surrounding *air* at effectively the *same* volume flow rate, leading by the method of section 1.6 to an estimated far field equal to that of an

acoustic source of strength $\rho_0 \, \dot{m}(t)$. Alternatively, the ideas of section 1.10 could be used to view the far field as that of a simple source of strength $\rho_s \, \dot{m}(t)$ equal to the rate of change of direct mass outflow rate from the siren, plus a distribution of quadrupoles of strength (150) per unit volume. Show that for moderate flow velocities this quadrupole far field is the same as that of a simple-source distribution of strength $(-\partial^2 \rho / \partial t^2)$ per unit volume, whose total strength $(\rho_0 - \rho_s) \, \dot{m}(t)$ reconciles the conclusions of the two approaches.

11. When a rigid sphere makes small oscillations in which its centre moves always in the same straight line with displacement $b(t)$, show that the general form of the equivalent dipole strength $G(t)$ is

$$4\pi\rho_0 \, ca_0^2 \int_{-\infty}^{t} \ddot{b}(T + a_0/c) \exp\left[c(T-t)/a_0\right] \sin\left[c(t-T)/a_0\right] \mathrm{d}T.$$

Note that this may be seen to tend to the approximate form $G(t) = 2\pi\rho_0 \, a_0^3 \, \ddot{b}(t)$ when $\ddot{b}(t)$ varies slowly on a time-scale a_0/c because the distribution

$$\exp\left[c(T-t)/a_0\right] \sin\left[c(t-T)/a_0\right]$$

has centroid at $T = t - a_0/c$ and integrated value $\frac{1}{2}a_0/c$.

12. For sound radiation from a plane wall, suppose that the surface distribution (186) of normal velocity is significant only where $Y^2 + Z^2 < l^2$. Show that the resulting sound field, at a distance r from the origin large compared with l, but not necessarily large compared with $\omega l^2/c$, is well approximated by (187) provided that $F_1(M, N)$ is reinterpreted as the Fourier transform, not of $f_1(Y, Z)$ but of

$$f_1(Y, Z) \exp\left[-\tfrac{1}{2}i\omega(Y^2 + Z^2)/cr\right].$$

This agrees with the previous definition (188) if r is indeed large compared with $\omega l^2/c$.

On the other hand, where r is *small* compared with $\omega l^2/c$, the new value of $F_1(\omega y/rc, \, \omega z/rc)$ can be written

$$\int_{-\infty}^{\infty} \int_{-\infty}^{\infty} f_1(Y, Z) \exp\left[i\psi(Y, Z)\right] \mathrm{d}Y \, \mathrm{d}Z,$$

where the phase

$$\psi(Y, Z) = (\omega/cr)(yY + zZ - \tfrac{1}{2}Y^2 - \tfrac{1}{2}Z^2)$$

has a large total variation over the region $Y^2 + Z^2 < l^2$ and is stationary where $Y = y$, $Z = z$. These are circumstances when the method of stationary phase (section 3.7 below) allows the double integral to be estimated in terms of the value of $f_1(Y, Z)$ as that stationary point as

$$-2\pi i \left|(\partial^2 \psi / \partial Y^2)(\partial^2 \psi / \partial Z^2)\right|^{-\frac{1}{2}} f_1(y, z) \exp\left[i\psi(y, z)\right].$$

Show that, with this form taken by $F_1(\omega y/rc, \, \omega z/rc)$ where $r \ll \omega l^2/c$, equation (187) for $p - p_0$ agrees with the conclusions of geometrical acoustics.

2. ONE-DIMENSIONAL WAVES IN FLUIDS

2.1 Longitudinal waves in tubes and channels

The theory of sound generation and propagation based on combinations of simple-source solutions of the wave equation has been found applicable (sections 1.4–1.10) to a wide range of problems which satisfy both the *linearity* condition that disturbances are small enough for their squares to be neglected and the *compactness* condition that the source region is small compared with a wavelength. Without the compactness condition, only a much more restricted range of geometrically simple problems like those of sections 1.11 and 1.12 can be so treated. They suggest, however, another method based on ray tracing that may be applicable to a wide range of problems under the *opposite* condition that geometrical scales are large compared with a wavelength.

Although fundamental studies of why rays become significant in this limit are for chapter 4, the rule determining amplitude in geometrical acoustics from *constancy of energy flow* along a ray tube (section 1.11) is critically evaluated in this chapter as part of a general study of wave propagation in tubes and channels. Then an extended evaluation of the *linearity* condition is given and some remarkable effects of departures from it are investigated.

There is interest in identifying the circumstances under which constancy of energy flow may be disturbed by *non-uniformity* in the cross-sectional area of a ray tube (or even in the density and temperature of the fluid it traverses) allowing *reflection* of some acoustic energy in addition to dissipation as studied in section 1.13. Chapter 2 includes a general study of transmission and reflection of acoustic energy in propagation through tubes, due to variations whether gradual or abrupt in cross-sectional area or fluid state, from which those identifications in particular can be inferred (section 2.6). Chapter 2 deals, however, with propagation *not only* through abstractly defined 'ray tubes' but also through tangible tubes with solid walls...

An important example is provided by the loudspeaker 'horn'. This

[89]

enables a vibrating diaphragm so small as to be acoustically compact to generate sound not with the low efficiency it would have in three-dimensional radiation but with the much greater 'plane-wave' efficiency achieved by radiation along a solid-walled tube, whose gradual increase of cross-sectional area conserves the energy flow to deliver it from an open end large enough to radiate with good efficiency into the surrounding fluid. Further reasons for analysing sound in solid tubes containing either air or water are suggested either by problems from the study of musical wind instruments, or by the annoying phenomenon of 'water-hammer' in plumbing systems, respectively.

In fact we consider all wave propagation in fluid-filled solid tubes which is 'longitudinal', in the sense that components of fluid motion parallel to the tube axis possess far greater kinetic energy than do any components transverse to the axis. Later, the 'waveguide' theory of section 4.13 shows that a 'fundamental mode' of propagation of sound waves of any frequency is this longitudinal mode whereas propagation of other, partly transverse, modes is possible only at frequencies exceeding a certain *critical value* (for which the wavelength is comparable with a diameter). The longitudinal waves studied in chapter 2, then, are the *only* possible type of acoustic propagation below that critical frequency, as well as being interesting at all frequencies.

With only this limitation to longitudinal motions, the theory is readily generalised to allow for the tube walls to be distensible; that is, to respond to a pressure change with a proportional *change of cross-sectional area* from its local undisturbed value (though one not so great as to require transverse motions comparable with longitudinal motions). This extends the theory of water-hammer to sound waves in thin-walled elastic pipes filled with liquid; the compressibility of the liquid and the distensibility of the pipe can just be added together (equation (10) below) though the latter reduces the propagation speed (by increasing the effective compressibility) only slightly for practical plumbing systems.

Perhaps surprisingly, the theory is applicable directly to the propagation of the blood pulse in arteries. The distensibility of the arterial wall is so large that by comparison the blood's compressibility is negligibly small, and the resulting propagation velocity (section 2.2) is two orders of magnitude less than the sound speed. Nevertheless, as the pulse passes, the arterial radius expands and contracts by only a few per cent of its undisturbed value, requiring radial motions negligibly small compared with the longitudinal motions.

We see that the concept of one-dimensional waves in fluids is an elastic one! It includes waves in tubes or channels whose cross-section may have a rather general shape, together with an area that varies in response to local pressure changes. A further unexpected example is given by 'long waves' in *open* channels filled with water.

We are concerned here with channels of arbitrary cross-section filled up to a certain height. Passage of a 'long wave' (with wavelength very large compared with the depth of water) involves changes in the elevation of the water surface that produce *both* cross-sectional area changes *and* pressure changes at any fixed level that are in phase with them and therefore with each other... From this it emerges (section 2.2) that the assumptions of longitudinal-wave theory are all satisfied. To avoid misunderstanding we emphasise that these waves are not the familiar *surface waves* on water, whose treatment is postponed to chapter 3. They are phenomena of *lower frequency*, permitting the wavelength to be a large multiple of the depth; an extreme case is the tidal movement excited by the Moon with periods of half a day.

These non-acoustic examples of longitudinal waves in fluids raise as many questions as do the acoustic examples about response to cross-section gradations: for example, as tidal movements propagate up a narrowing estuary, or as the blood pulse moves through a narrowing artery. Abrupt changes can also be of interest; while there is an especial incentive to study propagation through *branching* systems, such as the cardiovascular system, or a river system with tributaries, and to investigate what proportion of the energy of a wave propagates into each branch.

The linear theory of longitudinal waves in tubes and channels is given in sections 2.1–2.7, beginning in the first two sections with cases when the properties of the tube or channel, and of the fluid, are *uniform* along it. The next four sections pursue the effects of nonuniformities in those properties, whether abrupt or gradual, and including in section 2.4 the case of branching systems. Energy dissipation, neglected in these six sections, is investigated in section 2.7: although the dissipation mechanisms of section 1.13 apply both in abstract 'ray tubes' and in solid-walled tubes, an additional, more powerful source of dissipation appears where a solid wall frictionally retards the longitudinal motions.

The whole theory is next extended to take nonlinear effects into account. These are found to yield, not merely a quantitative change of behaviour, but some qualitatively new effects of a remarkable character: especially the formation of a *discontinuous* wave (for example, the shock wave, or the hydraulic jump) out of a continuous one. Sections 2.8–2.12 give the non-linear theory for the case when the properties of the tube or channel are uniform along it, while section 2.13 indicates its extension to take into account nonuniformity of cross-section and of fluid properties, or frictional dissipation, and section 2.14 goes on to infer modifications to geometrical acoustics demanded by nonlinearity. These sections sketch in particular the principles involved in predicting (i) on which days the 'Severn bore' will form and (ii) the strength of a supersonic boom.

Theories of longitudinal waves in both abstract 'ray tubes' and solid-walled tubes, with or without nonuniformities, nonlinearity and frictional dissipation, are based on the idea that variation of the *excess pressure*

$$p_e = p - p_0 \tag{1}$$

over a cross-section is negligible. Here p_0 is the undisturbed pressure with its hydrostatic distribution, so that it is this excess pressure p_e whose gradients produce fluid accelerations. The motions forced by longitudinal gradients of p_e can be large compared with those forced by transverse gradients only if p_e varies negligibly over each cross-section.

We proceed also, for reasons given in section 1.2, to neglect variations in the density ρ over a cross-section (identified, say, by its distance x from a particular end of the tube), and take both ρ and the cross-sectional area A as varying with the local excess pressure p_e according to equations

$$\rho = \rho(p_e, S), \quad A = A(p_e, x) \tag{2}$$

specifying the compressibility of the fluid for a given entropy S and the distensibility of the tube or channel at a given position x. These general equations include, of course, simpler special cases like sound waves in rigid tubes (where A shows no variation with p_e) or long waves in open water channels (where ρ is effectively constant).

At all points of a particular cross-section the same longitudinal gradient of p_e is found, and therefore also the same fluid *acceleration* if frictional retarding forces are neglected (for their effects see section 2.7), and hence finally the same fluid velocity u, satisfying

$$\rho(\partial u/\partial t + u\,\partial u/\partial x) = -\partial p_e/\partial x, \tag{3}$$

a one-dimensional statement that density times acceleration equals the force per unit volume given by minus the gradient of excess pressure. On the other hand, the equation of continuity for longitudinal motions is influenced by area changes:

$$\partial(\rho A)/\partial t + \partial(\rho A u)/\partial x = 0 \tag{4}$$

is an equation stating that *mass per unit length* ρA changes at a rate equal to minus the gradient of the *mass flow rate* $\rho A u$.

Note that area changes constrain mass flow and so appear in (4) since no mass can pass across the tube boundary. On the other hand, area changes do not affect momentum balance or appear in (3) because the tube boundary exerts on a piece of fluid the *same momentum-changing force*, p_e per unit area, as do neighbouring pieces of fluid.

The whole of chapter 2, except for the accounts of frictional dissipation in section 2.7 and brief investigations of its effect elsewhere, is based on these general equations for longitudinal waves in fluids (2), (3) and (4) and an equation (section 2.6) of entropy conservation for a fluid particle. In sections 2.1 and 2.2, however, we calculate their implications only when properties of the fluid, including its entropy S, and of the tube or channel, are taken to be *uniform*, so that equations (2) become

$$\rho = \rho(p_e), \quad A = A(p_e), \tag{5}$$

and on the assumptions of *linear* theory: that is, we neglect squares of disturbances such as the fluid velocity u or the excess pressure p_e.

The linearised forms of the equations of momentum (3) and continuity (4) are

$$\rho_0 \, \partial u/\partial t = - \, \partial p_e/\partial x, \quad \partial(\rho A)/\partial t = -\rho_0 A_0 \, \partial u/\partial x, \tag{6}$$

where ρ_0 and A_0 are given by equations (5) for $p_e = 0$. Elimination of u from these gives

$$\partial^2 p_e/\partial x^2 = A_0^{-1} \, \partial^2(\rho A)/\partial t^2, \tag{7}$$

which in a linear theory may be written as the one-dimensional wave equation

$$\partial^2 p_e/\partial x^2 = c^{-2}\partial^2 p_e/\partial t^2 \tag{8}$$

with the wave velocity c defined by the equation

$$c^{-2} = A_0^{-1}[\mathrm{d}(\rho A)/\mathrm{d}p_e]_{p_e=0}. \tag{9}$$

Thus, when the properties of the fluid and of the tube or channel are uniform along it, linear theory leads to just the same one-dimensional wave equation as for plane waves of sound, but with the value of the wave velocity c affected by the distensibility of the tube or channel. In fact, equation (9) can be written as

$$\rho_0^{-1}c^{-2} = [(\rho A)^{-1} \, \mathrm{d}(\rho A)/\mathrm{d}p_e]_{p_e=0}$$

$$= [\rho^{-1} \, \mathrm{d}\rho/\mathrm{d}p_e + A^{-1} \, \mathrm{d}A/\mathrm{d}p_e]_{p_e=0} = K+D, \tag{10}$$

where we name the relative density increase and the relative area increase, each per unit increase of pressure, the *compressibility* K of the fluid and the *distensibility* D of the tube or channel, respectively. In other words, everything is as in the theory of plane waves of sound (where $\rho_0^{-1} c^{-2} = K$) if we accept that the *effective* compressibility of the fluid in the tube or channel is the *sum* of the true compressibility K and this distensibility D. For examples of the application of this rule, see section 2.2.

At the same time, the fluid velocity in the direction of propagation of a *travelling* wave is $(\rho_0 c)^{-1} p_e$ (with $p_e = p - p_0$) just as in section 1.1. For example, the solution

$$p_e = f(t - x/c) \tag{11}$$

of (8), representing waves travelling in the *positive* x-direction, gives on substitution in the linearised momentum equation

$$u = (\rho_0 c)^{-1} p_e \tag{12}$$

exactly as for plane sound waves.

2.2 Examples, including elastic tubes and open channels

For one important class of longitudinal waves, those in open channels, which may perhaps appear very different from sound waves, it is surprisingly easy to calculate the distensibility D, and hence to apply the whole linear theory for uniform undisturbed conditions given in section 2.1. We may consider channel cross-sections of arbitrary *shape*, which, however, is supposed uniform along the channel's length. In the undisturbed condition the fluid is at rest with a horizontal free surface, supposed of breadth b, and fills an area A_0 of each cross-section. The undisturbed pressure distribution p_0 is determined hydrostatically: in coordinates with z measured *upwards* from the undisturbed free surface (so that $-z$ is distance below it)

$$p_0 = p_a - \rho_0 g z, \tag{13}$$

where p_a is the atmospheric pressure.

At time t, with this condition *disturbed* and the pressure increased by the same excess p_e all over a particular cross-section (specified by its x-coordinate), the new altitude

$$z = \zeta(x, t) \tag{14}$$

of the free surface is determined by the condition that the pressure $p = p_0 + p_e$ must take the atmospheric value p_a at $z = \zeta$, giving

$$p_e = \rho_0 g \zeta. \tag{15}$$

Thus the whole free surface of the cross-section is lifted through the same distance $p_e/\rho_0 g$, needed to generate a pressure excess p_e below it. The corresponding increase in the cross-sectional area A of the water (if squares of ζ be neglected) is

$$A - A_0 = b\zeta = b(\rho_0 g)^{-1}p_e, \qquad (16)$$

whence the distensibility

$$D = A_0^{-1}(dA/dp_e)_{p_e=0} \qquad (17)$$

is deduced as

$$D = b(\rho_0 g A_0)^{-1}. \qquad (18)$$

Normally this distensibility (18) is so large that in (10) the compressibility K of the fluid can be neglected, giving

$$c = (gA_0)^{\frac{1}{2}} b^{-\frac{1}{2}} = (gh)^{\frac{1}{2}}, \qquad (19)$$

where the quantity h for the undisturbed cross-section is its area A_0 divided by the breadth b of its free surface: this gives a value of the *water depth averaged across the breadth*, known as the 'hydraulic mean depth' h. Note that if h takes values around 1 m to 100 m characteristic of streams, canals and rivers, the wave velocity c takes values around $3\,\mathrm{m\,s^{-1}}$ to $30\,\mathrm{m\,s^{-1}}$. The insignificance of the compressibility K in (10) is confirmed by the smallness of these values compared with the speed of sound ($1400\,\mathrm{m\,s^{-1}}$ in water).

The fundamental assumption of one-dimensional propagation, that longitudinal motions of fluid are much greater than transverse motions, can now be critically examined. The transverse motions in the plane of the cross-section are dominated by the vertical displacements (14) of the free surface. In any travelling wave those can be related to the longitudinal velocity u by equations (15) and (12) to give

$$\zeta = cu/g = hu/c, \qquad (20)$$

in which the second form is derived from the first by equation (19). Transverse motions of order $\partial\zeta/\partial t$ are then small compared with u if a typical radian frequency ω (ratio of a typical value of $\partial\zeta/\partial t$ to a typical value of ζ) satisfies

$$\omega h/c \ll 1, \qquad (21)$$

stating that the hydraulic mean depth h is 'compact' relative, not of course to the speed of sound, but to the much lower speed (19) of long waves in channels.

Alternative forms of (21) are

$$h \ll \lambda/2\pi \quad \text{and} \quad t_p \gg 2\pi(h/g)^{\frac{1}{2}}. \qquad (22)$$

The first requires that the wavelength $\lambda = 2\pi c/\omega$ is very large compared with the mean depth h; one-dimensional theory is indeed shown in chapter

3 to give errors of less than 3% in propagation velocity provided that $\lambda > 14h$. The equivalent condition on the period $t_p = 2\pi/\omega$ requires it to be large compared with a quantity that for depths ranging from 1 m to 100 m varies from 2 s to 20 s. Longitudinal waves in open channels, then, are 'long' waves, with characteristic wavelengths great compared with the mean depth, generated by forcing effects with characteristic periods long compared with these standard values of many seconds.

We indicate next how the distensibility of a thin-walled elastic tube can be estimated for use in similar wave-propagation studies. If the undisturbed tube has internal radius a_0 and thickness h, then an excess pressure p_e inside the tube must generate a *circumferential tension $a_0 p_e$ per unit length of tube*. This simple law (one of those many credited to Laplace) is understood most easily by imagining such a unit length divided into two equal parts with semicircular cross-sections, pulled apart by the internal pressure p_e acting upon a resolved area $2a_0$ with a resultant force $2a_0 p_e$ balanced by the circumferential tension acting at *both* joins between the parts.

This circumferential tension $a_0 p_e$ pulls on an area h of tube material (again per unit length), giving a circumferential *tensile stress* (force pulling on unit area normal to that force)

$$a_0 p_e/h. \tag{23}$$

In thin-walled tubes (that is, with $h \ll a$) the circumferential stress (23) greatly exceeds the magnitude of the radial stress (which varies from p_e inside to o outside). The material of the tube wall then approximates to the condition of a simple 'tensile test' performed in the circumferential direction under the stress (23).

The corresponding circumferential *strain* (relative increase in tube circumference) may be written

$$a_0 p_e/hE, \tag{24}$$

where the elastic behaviour of the material of the tube wall under small stresses is represented by a modulus E. This, if the material is isotropic, is *Young's modulus*, and otherwise is the elastic modulus in circumferential tension. The relative change $(A - A_0)/A_0$ in cross-sectional area of the fluid inside the tube is twice the relative change in circumference (24), which on the definition (17) of distensibility gives a formula

$$D = 2a_0/hE, \tag{25}$$

sufficing to bring into play the whole linear theory of propagation in uniform tubes given in section 2.1.

Before proceeding along these lines, however, we may note a consequence of the circumferential stress (23) which necessarily accompanies the circumferential strain (24): namely, strains of *opposite* sign and somewhat smaller magnitude in directions at right angles. There is no need for comment on the radial strain, which involves merely a small relative decrease of wall thickness when $p_e > 0$. Some interesting problems are raised, however, by the *longitudinal* compressive strain in the tube wall generated by such an excess fluid pressure.

We can specify its magnitude by the equation

$$\partial \xi / \partial x = - \sigma a_0 p_e / hE, \tag{26}$$

where ξ is the longitudinal displacement of a particle of the tube wall, and the strain $\partial \xi / \partial x$ represents the relative change of length of a short section of tube. The constant σ, if the material is isotropic, is *Poisson's ratio* (see texts on elasticity), while for general tubes ($-\sigma/E$) is the coefficient relating longitudinal extension to circumferential tensile stress. Typical values of σ range from 0.2 to 0.5.

Although longitudinal waves in fluid-filled elastic tubes may thus involve longitudinal displacements ξ of the thin tube walls themselves, the inertia of those displacements does not significantly influence propagation laws. We may check this by showing that the longitudinal compressive stress p_l that must be generated to overcome that inertia is small compared with (23) and does not alter the predominantly simple-tension character of the stress system in the tube wall.

In fact, the linearised momentum equation for longitudinal motions of the solid tube material, supposed of density ρ_s, is

$$\rho_s \, \partial^2 \xi / \partial t^2 = - \partial p_l / \partial x. \tag{27}$$

Equations (26) and (27) tells us that

$$\partial^2 p_l / \partial x^2 = (\rho_s \, \sigma a_0 / hE) \, \partial^2 p_e / \partial t^2 = (\rho_s \, \sigma a_0 / hE) \, c^2 \partial^2 p_e / \partial x^2, \tag{28}$$

which with (10) for c^2 and (25) for D gives

$$p_l / p_e = \tfrac{1}{2} \sigma [D/(D+K)] (\rho_s / \rho_0). \tag{29}$$

This indicates (i) that longitudinal stresses are at most of the same order of magnitude as radial stresses, and so just as small compared with the circumferential stress used in calculating the distensibility (25); and (ii) that longitudinal stresses in the tube wall are in phase with fluid pressures (whence also the velocities of solid and fluid are in phase); accordingly, solutions of the wave equation (8) with the fluid satisfying boundary conditions, either at a free open end (zero excess pressure) or at a rigidly fixed closed end (zero velocity), must satisfy also the corresponding boundary conditions for the material of the tube wall, and thus be generally acceptable.

A different type of longitudinal wave propagation in fluid-filled elastic tubes occurs, however, if the tube is 'tethered' against longitudinal displacements; that is, if its outside is held in such a way as to prevent them. Such tethering represents, indeed, the commonest condition of elastic tubes along which waves propagate. We indicate briefly the effect of this for an isotropic material: the tethering mechanism must supply a cancelling strain equal and opposite to (26) by applying forces generating in the tube wall a longitudinal *tensile* stress $\sigma a_0 p_e / h$.

This longitudinal stress is comparable with the circumferential stress (23), and produces, just as that does, strains in perpendicular directions obtained by multiplying it by $(-\sigma/E)$. Addition of one of these to the circumferential strain (24) multiplies it effectively by $(1 - \sigma^2)$. Hence finally the distensibility for *tethered* tubes is reduced to

$$D = 2a_0(1 - \sigma^2)/hE. \tag{30}$$

Wave motions with important differences in the mechanics underlying their propagation are described by the formulas (25) and (30) although the magnitudes of the distensibility contribution to equation (10) for $\rho_0^{-1} c^{-2}$ differs in the two cases to only the moderate extent implied by the $(1 - \sigma^2)$ factor.

The compressibility K of water is 5×10^{-5} bar^{-1} (where 1 bar $= 10^5$ N m^{-2} is a typical atmospheric pressure) and this is considerably greater than the distensibility of most metal pipes, whose thickness ratios $(2a_0/h)$ hardly outweigh in (25) or (30) the effects of very small compliance factors E^{-1} of order 10^{-6} bar^{-1}. Propagation through plumbing systems of 'water-hammer' (large pressure fluctuations generated by rapid closure of a valve) occurs, therefore, at speeds close to the speed of sound in water, 1400 m s^{-1}.

By contrast, the speed of longitudinal waves may fall well below this figure in piping made of more distensible materials such as plastics. A rather extreme case is presented by elastomers, such as vulcanised rubber, with compliance factors E^{-1} of order 10^{-1} bar^{-1}. Then, with distensibilities D perhaps an order of magnitude greater still, the compressibility K is completely negligible in equation (10), and the predicted wave speeds are only of order 10 m s^{-1}.

Mammalian arteries have such distensibilities of order 1 bar^{-1}, corresponding to area increases of order 10% under pressure changes of order 10^{-1} bar found in normal physiological conditions. The complicated composite structure of the arterial wall includes fibres of an elastomer, elastin, that permit distensions of this order under moderate pressure changes, together with arrangements of stiffer collagen and 'smooth-muscle' fibres that substantially reduce the increment of area produced by pressure change at the higher pressures. Propagation of large pulses may be affected by such nonlinearity of area response to pressure in ways to be studied in section 2.8; in the meantime, only a linear theory of pulse propagation is discussed.

The soft tissue within which an artery is embedded is so compliant that pressure just outside the arterial wall responds negligibly to its pulsations, as assumed in the derivation of the circumferential stress (23). That tissue is probably effective, however, for 'tethering' the arterial wall against longitudinal displacements; in which case equation (30) should be used for D, with Poisson's ratio σ in the range just under 0.5 characteristic of materials whose compliance greatly exceeds their compressibility; ratios of wall thickness to tube diameter lie in the range 0.06 to 0.10.

Measurements of arterial pulse wave velocity are best made by timing the passage of an artificially induced sharp pressure change, clearly distinguishable from *natural* pressure fluctuations that include (see section 2.4) some reflected waves as well as directly transmitted waves. Comparison with direct measures of distensibility D confirm the approximate form

$$c = (\rho_0 D)^{-\frac{1}{2}} \tag{31}$$

of equation (10), while measurements of D and E under identical conditions are broadly in accord with equation (30) though insufficiently accurate to distinguish between it and equation (25). The wave velocity may take values as low as $5\,\mathrm{m\,s^{-1}}$ in the aorta (central artery) of larger mammals, rising to values around $10\,\mathrm{m\,s^{-1}}$ in the less distensible peripheral arteries. Maximum *blood* velocities are much smaller, around $1\,\mathrm{m\,s^{-1}}$.

This section on some examples may perhaps be concluded with a simple reminder that one-dimensional waves in tubes or channels include not *only* those long waves in open channels or in elastic tubes discussed here at length, but also the extremely important case of ordinary sound waves in either abstractly defined 'ray tubes' or material tubes with *negligible* distensibility.

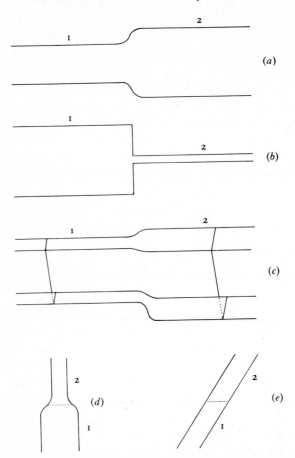

Figure 20. Illustrating a variety of junctions: (*a*) relatively abrupt widening; (*b*) discontinuous narrowing; (*c*) simultaneous change in channel breadth and depth; (*d*) junction including the horizontal interface (dotted line) between different fluids; (*e*) inclined tube with such an interface but no cross-section change.

2.3   Transmission of waves through junctions

The influence upon one-dimensional waves of nonuniformities in the properties either of the tube or channel or of the fluid is now investigated in the very simple case of a single abrupt discontinuity in properties, as an introduction to the study of various more complicated cases in the next three sections. Here, then, we analyse what happens to a travelling one-dimensional wave (11) when it reaches a junction, where the tube or channel along which it has been travelling is connected to another with different properties (and possibly containing a different fluid).

Figure 20 illustrates a variety of such junctions, connecting in each case a long tube or channel no. 1 to another long tube or channel no. 2. Case (*a*) shows a relatively abrupt *widening* of the tube or channel, producing a change in cross-sectional area from A_1 to A_2. To the question 'how abrupt must such a change be for the ensuing treatment as a discontinuity to be valid?' an answer will be found that it needs to be *compact* (spread over a region very small compared with a typical wavelength). As a contrast, case (*b*) illustrates a drastic *narrowing*, here supposed to occur quite discontinuously.

A junction between open channels of different mean depth h is shown in case (*c*): here not only the cross-sectional area but also the wave velocity (19) changes, say from c_1 to c_2. This horizontal junction is *not* a place where the density ρ of the fluid could change abruptly, because the undisturbed state before the waves arrive would not then be in hydrostatic equilibrium. However, cases (*d*) and (*e*) involving wave propagation along a vertical or inclined tube, driven either by the fluid's compressibility or the tube's distensibility or both, envisage the junction as including the horizontal interface between different fluids (or possibly the same fluid at different values of the specific entropy) with densities ρ_1 and ρ_2. The undisturbed pressure p_0 has then a straightforward hydrostatic distribution and equation (1) describes as usual any excess pressure p_e due to the presence of the waves. For the method of this section to be valid the changes in cases (*d*) and (*e*) must, again, be spread over compact regions, which for case (*e*) requires that the tube's angle of inclination be not too small.

The transmission of waves through the junction in the most general case depends on six quantities A_1, ρ_1, c_1 and A_2, ρ_2, c_2: the cross-sectional area, density and wave speed in tubes 1 and 2 respectively. Such complicated dependence on six variables (from which can be formed three independent dimensionless variables A_2/A_1, ρ_2/ρ_1, c_2/c_1) might be hard to disentangle in experiments were it not for the following theoretical arguments indicating that the essential dependence is on only one dimensionless variable $(A_2/A_1)(\rho_2/\rho_1)^{-1}(c_2/c_1)^{-1}$.

We consider the case when initially there is undisturbed fluid throughout tube no. 2 *and* in all of tube no. 1 except far to the left of the junction. Out of that region a travelling 'incident wave' given by

$$p_e = f(t - x/c_1), \quad u = (\rho_1 c_1)^{-1} f(t - x/c_1) \tag{32}$$

(compare (11) and (12)) approaches the junction, whose position is taken at $x = 0$, with $x < 0$ in tube no. 1 and $x > 0$ in tube no. 2. If $f(t) = 0$ for $t < 0$, then the wave does not reach the junction when $t < 0$ and tube no. 2 remains undisturbed while equations (32) describe fully the motion in tube no. 1.

The most general solution of the wave equation for p_e in tube no. 1 is

$$p_e = f(t - x/c_1) + g(t + x/c_1) \tag{33}$$

and in the situation after the wave has reached the junction we may expect to have to incorporate the term $g(t + x/c_1)$ representing a possible partial *reflection* of the incident wave at the junction. In the meantime a travelling wave

$$p_e = h(t - x/c_2) \tag{34}$$

may be generated in tube no. 2 at the appropriate wave speed as soon as a disturbance has reached the junction. On the other hand, we do not allow for the possibility of any wave components dependent on $t + x/c_2$ in tube no. 2 because such components would represent waves travelling in the negative x-direction from the region with large positive x, which has been assumed undisturbed.

Admittedly tube no. 2 must in practice have some finite length L, which means that any wave (34) travelling in the positive x-direction can generate when it reaches $x = L$ some reflected wave travelling in the opposite direction. In this treatment we neglect that wave reflected from the far end, *either* because we confine analysis to a time $t < 2L/c_2$ before it has had time to return and influence conditions at the junction $x = 0$ *or* because we suppose L so large that even rather small rates of wave attenuation such as may be present (sections 1.13 and 2.7) have rendered negligible at $x = 0$ the wave reflected from $x = L$. We consider then only the incident wave f, the wave reflected from the *junction* g, and the transmitted wave h.

In order to determine the unknown functions g and h from the supposedly known function f, two equations are needed, of which the first,

$$f(t) + g(t) = h(t), \tag{35}$$

expresses the need for continuity of excess pressure p_e as given by (33) for $x \leqslant 0$ and by (34) for $x \geqslant 0$. The fact that the change of properties around the junction, if not taking place discontinuously at $x = 0$, occurs in a distance l satisfying the compactness conditions

$$\omega l/c_1 \ll 1 \quad \text{and} \quad \omega l/c_2 \ll 1, \tag{36}$$

where ω is a typical frequency, means that little variation in any of the terms in (33) or (34) can be expected over distance l; this suggests that the pressure fluctuations themselves may be much the same all over the junction region, and supports the continuity condition (35) (see also (41) below).

It might be thought that the second required condition at the junction must impose continuity of the fluid velocity u, whose value

$$u = (\rho_1 c_1)^{-1} [f(t - x/c_1) - g(t + x/c_1)] \tag{37}$$

in tube no. 1 comprises for each wave component a term $(\rho_1 c_1)^{-1} p_e$ *in its direction of propagation*; while its value in tube no. 2 is, similarly,

$$u = (\rho_2 c_2)^{-1} h(t - x/c_2). \tag{38}$$

Indeed, in a case like that illustrated in figure 20(e) it *would* be correct to express continuity at $x = 0$ of the velocity u as given by (37) for $x \leqslant 0$ and by (38) for $x \geqslant 0$, since the velocity must be continuous across the interface between the two fluids.

This argument fails, however, in any case where there is a change in cross-sectional area. For example, in case (d), the velocity is indeed continuous across the *interface* but this velocity in the middle of the area contraction is *greater* than the corresponding velocity in the wider tube and *less* than the corresponding velocity in the narrower tube, where the same volume flow of fluid is forced into a smaller cross-section. In general cases, including this, it is the *volume flow Au* which is continuous through the junction, essentially because it varies in either of the fluids very little in the short length involved, while no discontinuity of volume flow occurs at the interface itself.

A rigorous test of this statement requires an estimate of the difference $A_1 u_1 - A_2 u_2$ between the volume flows into and out of the junction region. We write this as an integral over the junction region

$$A_1 u_1 - A_2 u_2 = \int_V (K+D)(\partial p_e/\partial t) \, dV, \tag{39}$$

where V is the volume of junction fluid and $(K+D)(\partial p_e/\partial t)$ is the relative rate of change of fluid volume due to compressibility and distensibility. Using (10) for $(K+D)$ and taking the orders of magnitude of $(\partial p_e/\partial t)$ and V as $\omega(\rho_0 c) u$ and Al (where ρ_0, c, u and A are intermediate values of density, sound speed, fluid velocity and cross-sectional area), we estimate the right-hand side of (39) as $(\omega l/c) Au$, which may be neglected compared with each term on the left-hand side under the compactness condition (36). Note that a similar argument from the linearised momentum equation

$$\rho_0 \, \partial u/\partial t = -\partial p_e/\partial x$$

would give

$$p_{e1} - p_{e2} = \int_{-\frac{1}{2}l}^{\frac{1}{2}l} (\rho_0 \, \partial u / \partial t) \, dx, \tag{40}$$

an integral over the length l of the junction, as of order

$$\rho_0 [\omega (\rho_0 c)^{-1} p_e] \, l = (\omega l / c) p_e, \tag{41}$$

reconfirming the correctness of the pressure continuity condition.

These two conditions reflect the special nature of *fluids*, with their capacity to transmit a *pressure* unchanged from one cross-section to another and to redirect a *volume flow* similarly into a new shape and size of cross-section. Waves in *solids* obey different laws: longitudinal waves travelling along a metal bar satisfy at an abrupt change of cross-section conditions of continuity of *force* and of *velocity*; for example, the total longitudinal forces on both sides of the small amount of material in the junction must be effectively in equilibrium as there is no other force to balance them. This argument fails for fluids, because the longitudinal force difference

$$p_{e1} A_1 - p_{e2} A_2$$

on the fluid in the junction is balanced by the resultant of the reaction on the fluid from the *wall of the tube contraction*...

Continuity at $x = 0$ of the volume flow Au as given by (37) with $A = A_1$ for $x < 0$ and by (38) with $A = A_2$ for $x > 0$ implies that

$$Y_1[f(t) - g(t)] = Y_2 h(t), \tag{42}$$

where we have introduced the notations

$$Y_1 = A_1(\rho_1 c_1)^{-1}, \quad Y_2 = A_2(\rho_2 c_2)^{-1} \tag{43}$$

to indicate those two groupings of the area, density and wave-speed variables which alone appear in this condition. Since (35) and (42) evidently determine both the reflected wave g and the transmitted wave h, we have confirmed the earlier statement that they depend on the six quantities *only* in the combination Y_2/Y_1.

The quantities Y_1 and Y_2 whose importance is shown by this analysis are called the *admittances* of the two tubes. Equations (32) make clear that for a pure travelling wave

$$Y_1 = \text{admittance} = \frac{\text{volume flow in direction of propagation}}{\text{pressure excess}} \tag{44}$$

and equations (34) and (38) imply the same physical interpretation for Y_2. The word used is chosen by analogy with the electrical engineers' nomenclature

$$\text{admittance} = \frac{\text{current flow}}{\text{potential excess}}. \tag{45}$$

In both scientific fields the quantity Y called admittance is used inter-changeably with the reciprocal quantity $Z = 1/Y$ called *impedance*; rules emerge later for determining when one is more useful than the other for waves in fluids.

Division of equation (42) by equation (35) gives

$$Y_1\frac{f(t)-g(t)}{f(t)+g(t)} = Y_2, \tag{46}$$

a simple equation for the reflected wave g whose solution is

$$g(t)/f(t) = (Y_1-Y_2)/(Y_1+Y_2). \tag{47}$$

This, by (35), implies that

$$h(t)/f(t) = 2Y_1/(Y_1+Y_2). \tag{48}$$

A first special case of these results that may be noted is that *when the admittances are exactly matched* (that is, although the cross-sections, densities and wave speeds in the two tubes may be quite different, the ratios (43) are identical giving $Y_1 = Y_2$) *there is no reflected wave*. Then the transmitted waveform $h(t)$ is identical with the incident waveform $f(t)$. This is because admittance is ratio of volume flow to pressure excess in a simple travelling wave; accordingly, simple travelling waves in two different tubes can co-exist if the tubes' admittances are equal, *reconciling* the conditions for continuity of pressure excess and of volume flow. 'Matching admittances' (also referred to as 'matching impedances' when the reci-procals $Z_1 = 1/Y_1$ and $Z_2 = 1/Y_2$ are used) is an effective method, then, of getting wave energy from one medium into another without unwanted reflections.

In the presence of a reflected wave g of the same sign as f, the ratio of volume flow to pressure excess at $x = 0$ in tube no. 1 is reduced to

$$Y_1\frac{f(t)-g(t)}{f(t)+g(t)}, \tag{49}$$

the ratio of the left-hand sides of (42) and of (35). This reduction comes about because the reflected wave increases the pressure excess at $x = 0$ while reducing volume flow by a flow $Y_1g(t)$ in the direction of its *own* propagation. Equation (46) states that to match the ratio (49) to an admit-tance $Y_2 < Y_1$ in tube no. 2 a *positive* reflection (g of the same sign as f) is required. An extreme case of this is when tube no. 2 is closed ($Y_2 = 0$), requiring $g = f$: total positive reflection.

On the other hand, when $Y_2 > Y_1$, perhaps because the tube no. 2 has larger cross-section or is more distensible or contains lighter fluid, equation (47) requires *negative* reflection with $g/f < 0$. An extreme case of this is when Y_2/Y_1 is very large: this might represent an open end (junction between a tube and the outside atmosphere) where volume flow in and out is unrestricted but contact with the atmosphere keeps pressure excess very small. (Equally it might represent the end of an open channel where it exhausts into a large reservoir.) Then $g = -f$: total negative reflection.

In any travelling wave the *energy* flow is related quite simply to admittance. The rate of transfer of energy in the direction of propagation is *velocity* times a *force* equal to cross-sectional area times pressure excess; in other words, it is volume flow times pressure excess, which in turn is the admittance times the *square* of the pressure excess. For example, in the incident wave (32) it is $Y_1 f^2(t - x/c_1)$. Similarly in the reflected wave it is $Y_1 g^2(t + x/c_1)$ but in the transmitted wave it is $Y_2 h^2(t - x/c_2)$.

Conservation of energy at the junction $x = 0$ can easily be checked: the proportion of the incident energy flow that is reflected is

$$g^2(t)/f^2(t) = (Y_1 - Y_2)^2/(Y_1 + Y_2)^2 \tag{50}$$

and the proportion transmitted is

$$Y_2 h^2(t)/Y_1 f^2(t) = 4Y_1 Y_2/(Y_1 + Y_2)^2. \tag{51}$$

These proportions are nonnegative numbers that add up to 1: all the energy that is not reflected is transmitted.

Note that if the admittances are matched only approximately (Y_2/Y_1 being close, but not equal, to 1) there is still almost perfect transmission of energy. The proportion reflected, (50), is in this case about $\frac{1}{4}(Y_2/Y_1 - 1)^2$, proportional to the *square* of the matching error; a result used further in section 2.6.

Consider finally the case when though tube no. 2 is not closed Y_2 is very small compared with Y_1, as in figure 20(b): understandably (47) in this case makes g/f near to 1 (its value for the closed tube with $Y_2 = 0$) but (48) gives the rather surprising result that h/f is almost 2. Admittedly, because Y_2 is small, the energy flow $Y_2 h^2$ in tube no. 2 is not large; in fact its ratio (51) to the incident energy flow is only about $4(Y_2/Y_1)$. Nevertheless, the transmitted pressure amplitude is nearly *twice* that in the incident wave; essentially because the nearly closed end of tube no. 1 produces a nearly total positive reflection and thus a nearly doubled pressure amplitude, to which tube no. 2 duly responds.

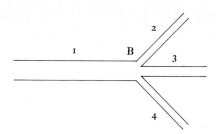

Figure 21. Illustrating a trifurcation.

2.4 Propagation through branching systems

The rules in section 2.3 for determining the reflected and transmitted waves when a single wave is incident on a single junction are the basic data from which it proves possible to calculate propagation along a tube or channel involving, say, a *sequence* of such junctions, or even (section 2.6) a continuous distribution of property changes. For *branching* systems, however, like the cardiovascular system or like a river with its tributaries, the basic data similarly needed relate not to a junction between *two* different tubes or channels but to a branch where many tubes or channels meet. It is necessary to know how such a branch (whose dimensions are, again, assumed compact) responds to an incident wave approaching it along one of those tubes or channels.

To analyse this, we number from 1 to N the tubes or channels meeting at such a branch (illustrated in figure 21 for the case of a *trifurcation*, with $N = 4$) and suppose the fluid in all of them undisturbed initially except for an incident wave (32) approaching the branch along tube no. 1. After it reaches the branch the excess pressure in that tube takes the general form (33) involving also a reflected wave. In the remaining tubes, identified by a suffix n ranging from 2 to N, the transmitted wave takes the form

$$p_e = h(t - x/c_n), \qquad (52)$$

as in (34) but with in general a different wave speed c_n in each tube. Here $x \geqslant 0$ represents distance measured from the branch $x = 0$ along each tube with $n > 1$, although we take $x \leqslant 0$ (as in section 2.3) in tube no. 1.

Note that in (52) the waveform $h(t)$ is the same in each tube because at the branch $x = 0$ the pressure must be continuous, essentially for the reasons summarised in equations (40) and (41): across a compact branch region it cannot vary significantly without forcing a local fluid flow much too large in relation to those in the adjoining tubes for mass-conservation to be preserved.

This condition of pressure continuity also imposes equation (35) as before.

The similar need for continuity of volume flow gives for a branch, however, not equation (42) but

$$Y_1[f(t) - g(t)] = \left(\sum_{n=2}^{N} Y_n \right) h(t). \tag{53}$$

The left-hand side specifies as in (42) the volume flow approaching $x = 0$ from tube no. 1, where $x \leqslant 0$ and a reflected wave is in general present. The right-hand side represents the sum of the volume flows leaving the branch (in the direction x increasing) along each of the tubes $n = 2, \ldots, N$. Here

$$Y_n = A_n (\rho_n c_n)^{-1} \tag{54}$$

(ratio of volume flow to pressure excess in a travelling wave) is the admittance of the nth tube.

Accordingly, the reflected and transmitted waveforms $g(t)$ and $h(t)$ are determined by equations (35) and (53): just the same as the equations (35) and (42) that determine them at a simple junction, except that the admittance Y_2 is replaced by the *sum* $\sum_{n=2}^{N} Y_n$ *of the admittances* of the tubes carrying the transmitted wave. It follows that all the results of section 2.3 for simple junctions can be applied to branches provided that in all of them this substitution of $\sum_{n=2}^{N} Y_n$ for Y_2 is made.

For example, the solutions (47) and (48) for $g(t)$ and $h(t)$ become

$$g(t)/f(t) = \left(Y_1 - \sum_{n=2}^{N} Y_n \right) \bigg/ \left(Y_1 + \sum_{n=2}^{N} Y_n \right) \tag{55}$$

and

$$h(t)/f(t) = 2Y_1 \bigg/ \left(Y_1 + \sum_{n=2}^{N} Y_n \right) \tag{56}$$

respectively. The same rule governs reflected and transmitted *energy*, since the energy carried away from $x = 0$ by all the transmitted waves (52) adds up to

$$\left(\sum_{n=2}^{N} Y_n \right) h^2(t), \tag{57}$$

replacing the $Y_2 h^2(t)$ of section 2.3. Therefore, equations (50) and (51) again specify the proportions of incident energy that are reflected and transmitted, if Y_2 is replaced by $\sum_{n=2}^{N} Y_n$, and again check conservation of energy.

It is a valuable feature of the *admittance* notation (54) (as opposed, say, to the impedance $Z_n = 1/Y_n$ notation) that *tubes transmitting in parallel* behave in this way exactly like a single tube of admittance equal to *the sum of their admittances.* This rule, embodying the fact that their volume flows add while their pressure excesses are the same, usefully simplifies considerations of transmission through branching systems.

A prominent branch in the cardiovascular system is the *iliac bifurcation* where the aorta (main artery issuing from the heart) divides, after passing centrally down the abdomen, into the two iliac arteries. It is instructive to consider how a pulse wave from the heart would be affected by meeting just this one bifurcation. Though such consideration ignores the real complexity of the cardiovascular system (involving large numbers of interacting branches and some gradations of arterial properties between them) its results show some similarity to observed behaviour mainly because this is such a prominent branch.

In this case with $N = 3$, the sum of the admittances Y_2 and Y_3 of the two iliac arteries turns out to be *less* than the admittance Y_1 of the aorta itself, partly because the sum of their cross-sectional areas is rather less (by around 20%, though the ratio takes widely varying values in different individuals) and partly because the aorta has greater distensibility and therefore a lower value of the wave speed c than do the iliac and other more peripheral arteries. Equation (55) implies *positive* reflection, therefore ($g/f > 0$).

Observations of pressure fluctuations in the aorta are consistent with this qualitative conclusion of a positive reflected wave. The physiological significance of the resulting enhanced pressure changes in a relatively distensible aorta is that they facilitate storage of the blood volume expelled by the heart in each stroke for delivery at a rather steady rate through the peripheral circulation.

Higher up, in the thoracic aorta, there are a number of junctions of an extremely different type, at each of which two narrow *intercostal* arteries draw away from the aorta a small quantity of blood to feed the region between two pairs of ribs. In this case with $N = 4$, the admittances Y_1 and Y_2 of the parts of the aorta proximal and distal to the branch are essentially equal and very large compared with the admittances Y_3 and Y_4 of the intercostal arteries. The sum in (56) is therefore close to Y_1 and there is almost perfect transmission of the pulse. This result, in no way surprising for the continued transmission down the aorta, implies more interestingly the same fluctuations of *pressure* propagating along the intercostal arteries as in the aorta itself.

Precisely similar considerations apply when a long wave propagating up a river estuary comes to a bifurcation or passes small tributaries. The

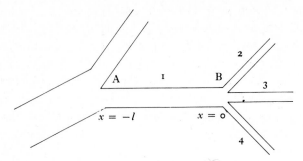

Figure 22. Interactions between two branches A (where $x = -l$) and
B (where $x = 0$).

bifurcation causes a positive reflection where the sum of the admittances of
two rivers is less than that of the estuary into which they combine. (For open
channels in general, the admittance by (19) and (54) is

$$Y = bh/\rho(gh)^{\frac{1}{2}} = bh^{\frac{1}{2}}/\rho g^{\frac{1}{2}}, \tag{58}$$

proportional to the breadth b of the water surface and to the square root
of the mean depth h.) Narrow tributaries discharging into an estuary from
the side, furthermore, transmit those fluctuations of water-surface elevation
$(p_e/\rho g)$ that are present in any long waves propagating along the estuary.

   In this introduction to transmission through branching systems we avoid
analysing complicated cases but go a step beyond the study of a single branch
to indicate *how two branches interact*, if the distance between them is not
assumed compact. This introduces a qualitatively new feature, whereas
multiple interactions between branches in a complicated system merely add
quantitatively to the volume of analysis required.

   We consider (figure 22) a branch B *exactly* like that in figure 21 except
that a length l of tube no. 1 is itself one of the tubes receiving waves transmitted through an earlier branch A. The ratio of reflected to incident wave
at B (taken at $x = 0$ as in figure 21) is still given by (55). We calculate the
effective admittance Y_1^{eff} of tube no. 1 at the branch A (where $x = -l$).
This is defined as the ratio of volume flow $A_1 u$ to pressure excess p_e at
$x = -l$, taking into account as in equations (37) and (33) the presence of
waves in *both* directions in tube no. 1 to give

$$Y_1^{\text{eff}} = Y_1 \frac{f(t+l/c_1) - g(t-l/c_1)}{f(t+l/c_1) + g(t-l/c_1)}. \tag{59}$$

If we could calculate once and for all this ratio Y_1^{eff} of volume flow to
pressure excess at branch A in tube no. 1, as influenced by the presence of

the second branch B, then the analysis of branch A would proceed in exact accordance with the rules governing a single branch with the admittance Y_1 of tube no. 1 at A replaced by Y_1^{eff} in the required sum of admittances of all the tubes that receive waves transmitted through A.

The qualitatively new feature of this problem is the time-lag l/c_1 between waveforms at A and at B which in (59) alters the f wave (travelling from A towards B) and the g wave (travelling from B towards A) in opposite senses. This is important because it prevents straightforward use of equation (55) for $g(t)/f(t)$ to calculate (59).

The way round this difficulty, universally adopted in studies of propagation through branching systems, is Fourier analysis into complex-exponential waveforms proportional to $e^{i\omega t}$. On a linear theory, whatever 'forcing effect' is exciting the waves transmitted through the system can be Fourier-analysed as a linear combination of $e^{i\omega t}$ terms for different values of ω and, if any feature of the system's response can be calculated separately for each such term, then the complete response must be the *same* linear combination of those separate responses.

For example, a *periodic* but not sinusoidal forcing effect, like the variation of excess pressure at the aortic valve (separating the heart from the aorta), is analysed as a Fourier *series* of such terms proportional to $e^{i\omega t}$ with ω taking values that are integer multiples of $2\pi/t_p$, where t_p is the period. Aperiodic forcing effects are represented by Fourier *integrals*. The pressure excess at the mouth of an estuary includes a series of $e^{i\omega t}$ terms with discrete frequencies due to the periodic tide-raising forces of astronomical origin *and* an aperiodic Fourier-integral term representing wind-generated surge effects (see the epilogue, part 1).

The fact that on linear theory a term proportional to $e^{i\omega t}$ in the forcing effect elicits responses proportional to $e^{i\omega t}$ everywhere in the system means that corresponding values of $f(t)$ and $g(t)$ in equations (55) and (59) are themselves proportional to $e^{i\omega t}$, which gives immediately

$$Y_1^{\text{eff}} = Y_1 \frac{\left(Y_1 + \sum\limits_{n=2}^{N} Y_n\right) e^{i\omega l/c_1} - \left(Y_1 - \sum\limits_{n=2}^{N} Y_n\right) e^{-i\omega l/c_1}}{\left(Y_1 + \sum\limits_{n=2}^{N} Y_n\right) e^{i\omega l/c_1} + \left(Y_1 - \sum\limits_{n=2}^{N} Y_n\right) e^{-i\omega l/c_1}}. \tag{60}$$

Reflection and transmission at branch A of waveforms proportional to $e^{i\omega t}$ can accordingly be calculated taking into account the presence of branch B provided that the effective admittance of tube no. 1 is taken as this complex number (60), dependent on ω. Such use of complex admittances dependent on frequency exactly parallels analyses of alternating-current electrical networks, where also it is common to extend results to the calculation of response to general forcing effects by Fourier synthesis.

The formula (60) for the effective complex admittance of tube no. 1 at branch A in the presence of branch B is more useful when thrown into the simplified form

$$Y_1^{\text{eff}} = Y_1 \frac{\left(\sum\limits_{n=2}^{N} Y_n\right) + iY_1 \tan(\omega l/c_1)}{Y_1 + i\left(\sum\limits_{n=2}^{N} Y_n\right)\tan(\omega l/c_1)}. \tag{61}$$

Although in section 2.7 a modification to (60) and (61) allowing for wave *attenuation* (essentially, by giving ω/c_1 an imaginary part) is noted, this simple formula (61) indicates many essential properties of the interaction when l is not too large.

If l is so small as to be compact ($\omega l/c_1 \ll 1$) equation (61) degenerates to

$$Y_1^{\text{eff}} = \sum\limits_{n=2}^{N} Y_n, \tag{62}$$

which means that the properties of the short length l of tube no. 1 become irrelevant: it is merely a region of effectively uniform pressure that acts like part of the compact branch A giving access to tubes $2, \ldots, N$. Interestingly enough, the same 'transparency' to wave propagation is demonstrated by (61) also when $\omega l/c_1$ is nearly equal to π: a tube AB whose length l is about half a wavelength 'hands on' the wave to the branch B with a phase change of π but otherwise as if the tube had not been there. Values of l that are *multiples* of half a wavelength share this 'transparency' property, though modified by attenuation effects to a rather greater extent.

Intermediate between the 'transparency' values of $\omega l/c_1$ near 0 and near π is a range of values near $\frac{1}{2}\pi$ where tube no. 1 has length about a *quarter-wavelength* and its admittance characteristics are *accentuated*, since

$$Y_1^{\text{eff}} = Y_1^2 \Big/ \left(\sum\limits_{n=2}^{N} Y_n\right). \tag{63}$$

Thus, if Y_1 is *greater* than $\sum\limits_{n=2}^{N} Y_n$ then Y_1^{eff} is greater still, while if Y_1 is *less* that $\sum\limits_{n=2}^{N} Y_n$ then Y_1^{eff} is less still.

In relation to a typical fundamental frequency of the heart beat, the aorta has length near a quarter-wavelength in normal subjects. For such a frequency, accordingly, the effective admittance of the aorta at the aortic valve is *enhanced*: a matter of physiological importance since it allows the first Fourier component in the fluctuations of volume flow through the valve to be generated with only moderate fluctuations of excess pressure produced by the heart. For some higher Fourier components, wavelengths may be low enough for the *iliac* arteries to be around a quarter-wavelength, accentuating the effect of their reduced admittance in generating positive reflection at the iliac bifurcation. General propagation analysis along lines indicated in this section but taking into account multiple interactions between branches has been found to throw considerable light on properties of the cardiovascular system.

2.5   Cavities, constrictions, resonators

The propagation in a branched system can, as we have seen, be specified conveniently by imagining any wave analysed into components proportional to $e^{i\omega t}$ and using a *complex admittance Y dependent on ω* to describe the response of any part of the system to such a component. A general formula, that takes the form (61) if wave attenuation is neglected, relates effective admittance at an earlier branch to admittances at a subsequent branch. Successive use of this formula working back from the most peripheral branches to the earliest one allows properties of the whole system to be characterised; rather as an alternating current network is analysed by combining in accordance with Kirchhoff's laws complex admittances (or impedances) dependent on frequency for 'lumped elements' of the network.

This analogy prompts the question whether one-dimensional waves in fluids can encounter any lumped elements with pure imaginary admittance similar to such common features of electrical networks as capacitances and inductances. In this section we identify close analogues to those, indicate how systems including them can be analysed, and investigate resonance conditions, in some cases analogous to those of a 'tuned circuit'.

First of all, however, we notice how the rule given in section 2.3 for calculating energy flow as admittance times the square of pressure excess has to be modified when admittances are complex. Of course any complex expression like $p_1 e^{i\omega t}$ or $Y p_1 e^{i\omega t}$ for a pressure excess or the associated volume flow implies that the *real part* is to be taken: namely,

$$\tfrac{1}{2}(p_1 e^{i\omega t} + \bar{p}_1 e^{-i\omega t}) \quad \text{or} \quad \tfrac{1}{2}(Y p_1 e^{i\omega t} + \bar{Y}\bar{p}_1 e^{-i\omega t}) \tag{64}$$

respectively, where the bars denote complex conjugates. The energy flow, or product of the two expressions in (64), includes a term

$$\tfrac{1}{4}(Y + \bar{Y}) p_1 \bar{p}_1 = (\mathscr{R}Y)\{\tfrac{1}{2}|p_1|^2\} \tag{65}$$

independent of time (where $\mathscr{R}$ stands for real part) and this is its *mean* value because the remainder just oscillates with frequency 2ω. Calculating similarly the mean square pressure excess as the quantity in curly brackets in (65), we infer the modified rule

$$\frac{\text{mean energy flow}}{\text{mean square pressure excess}} = \mathscr{R}Y; \qquad (66)$$

thus, only the *real part* $\mathscr{R}Y$ of a complex admittance Y is associated with any mean transmission of energy.

In electrical theory, capacitance C represents charge stored per unit rise in potential. A potential fluctuating as $V_1 e^{i\omega t}$ implies a fluctuating storage of charge $CV_1 e^{i\omega t}$ and an associated current flow (rate of change of charge) $Ci\omega V_1 e^{i\omega t}$ giving an admittance $Ci\omega$.

In systems transmitting one-dimensional waves, similarly, a *cavity* may be able to accept additional fluid when the pressure rises. If it is compact, so that pressure changes throughout the cavity are in phase, its capacitance C may be defined as the volume of excess fluid it accepts per unit rise in pressure. Then a pressure fluctuation $p_1 e^{i\omega t}$ implies a volume flow into the cavity $d(Cp_1 e^{i\omega t})/dt$, giving an admittance

$$Y = Ci\omega. \qquad (67)$$

Note that the real part (66) of this Y is zero: pressure excess and volume inflow are $90°$ out of phase so that there is no mean energy flow into the cavity, where mean energy remains constant, indeed, if we neglect dissipation.

If a volume V of fluid occupies the cavity in the undisturbed state, the ratio C/V represents the relative amount of extra fluid accepted per unit rise in pressure. As in equation (10) we can write this as

$$C/V = K+D, \qquad (68)$$

a sum of the fluid's compressibility K and the *cavity's* distensibility D. Only the latter increases the actual cavity volume available (by an amount DV per unit rise in pressure); but compressibility reduces the volume occupied by the existing fluid in the cavity (by an amount KV per unit rise in pressure) which equally permits acceptance of extra fluid.

An alternative form $\qquad C = V\rho_0^{-1}c^{-2} \qquad (69)$

for the capacitance of a cavity may be suggested by equation (10), which strictly has meaning, however, only for tubes. We allow our cavities to be of very general shape: then equation (69) is correct only if c is understood as the wave speed in a *tube* of the *same* distensibility D, containing fluid of the same compressibility K and undisturbed density ρ_0. In particular, (69) is true for sound waves in a rigid cavity with c the sound speed, or for long water waves in a reservoir with c^2/g equal to the mean depth h; provided that all cavities are compact. Note also that, if the cavity shape actually *is* a tube AB with the end B closed, equation (69) can be derived alternatively from (61) by taking zero admittance $\sum_{n=2}^{N} Y_n$ at B and $(\omega l/c_1)$ small, giving an effective admittance at A of

$$iY_1 \tan(\omega l/c_1) \doteqdot i(A_1/\rho_1 c_1)(\omega l/c_1) = Ci\omega, \tag{70}$$

where C takes the form (69) with V as the volume $A_1 l$ of the tube.

More generally, if in figure 22 (with AB still compact) the admittance $\sum_{n=2}^{N} Y_n$ at B is not quite zero as for a closed end but is *small enough* compared with Y_1 for the second term in the denominator of (61) to be neglected as the product of small quantities, then

$$Y_1^{\text{eff}} = \sum_{n=2}^{N} Y_n + Ci\omega; \tag{71}$$

exactly as if tube no. 1 acted like a cavity of admittance *in parallel with* tubes $2, \ldots, N$. In fact, a compact cavity of *any* shape immediately preceding a branch acts as an additional admittance $Ci\omega$ in parallel (and therefore added in) with those of the branching tubes, because fluctuations of pressure are in phase all over the cavity while the volume of extra fluid entering the cavity is the sum of that stored in the cavity and what the branches admit.

Atherosclerosis is a disease in which the aorta becomes less distensible and so able to store the blood volume expelled in each stroke only when rather high pressures are exerted by the heart. In such a case the strain on the heart may be reduced by a prosthesis that effectively adds, in parallel with the aorta's excessively low admittance, a cavity with large imaginary admittance $Ci\omega$: for example (if immunologically acceptable), a gas balloon within the thoracic aorta that makes room for considerable excess blood as it collapses under a pressure rise.

Equation (61) may be useful also in the opposite case when AB is a not necessarily compact tube terminating in a compact bulbous cavity at B $\left(\text{where therefore } \sum_{n=2}^{N} Y_r = Ci\omega \right)$ and the effective admittance at A is

$$Y_1^{\text{eff}} = iY_1 \frac{\tan(\omega l/c_1) + (C\omega/Y_1)}{1 - (C\omega/Y_1)\tan(\omega l/c_1)} = iY_1 \tan\left[\frac{\omega}{c_1}(l + l_0)\right], \tag{72}$$

where
$$l_0 = (c_1/\omega)\tan^{-1}(C\omega/Y_1). \tag{73}$$

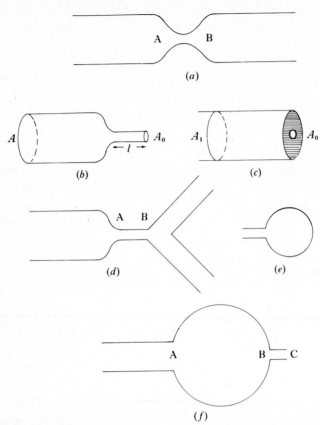

Figure 23. Illustrating a variety of constrictions: (*a*) very short constriction joining two tubes; (*b*) longer constriction joining tube to atmosphere; (*c*) small hole in solid end of tube; (*d*) constriction leading to a branch B; (*e*) the Helmholtz resonator (constriction leading to a cavity); (*f*) constriction which joins to the atmosphere a cavity at the end of a long tube.

Equation (72) equates Y_1^{eff} to the value it would have for a straight tube of increased length $l + l_0$ terminating in a *closed* end: the cavity behaves just like an extra length l_0 of tube.

The 'effective length' l_0 need *not* be compact even though the cavity itself is: for example, with ρ_0 and c in the cavity equal to their values ρ_1 and c_1 in the tube, the quantity $C\omega/Y_1$ in (73) is

$$(V\omega/\rho_1 c_1^2)/(A_1/\rho_1 c_1) = (\omega/c_1)(V/A_1), \tag{74}$$

which need not be small since a bulbous cavity of volume V can be compact without the ratio V/A_1 to the tube cross-sectional area being compact.

Effectively, the cavity makes a phase change in the reflected waveform, which by (47) becomes

$$g(t) = f(t)(Y_1 - Ci\omega)/(Y_1 + Ci\omega) \tag{75}$$

and so can be written $g(t) = f(t)\exp(-2i\omega l_0/c_1)$ (76)

with l_0 given by (73). This waveform describes a positive reflection without change in amplitude (and carrying back, therefore, all the energy flow) *delayed* by a time-lag $2l_0/c_1$: namely, the time the wave would need to travel an extra distance l_0 along the tube and back.

Electrical effects opposite in many ways to those of a capacitance result from a lumped element representing an *inductance L*, defined as magnetic flux through a circuit per unit current therein. A current fluctuating as $j_1 e^{i\omega t}$ implies a changing flux $Lj_1 e^{i\omega t}$ and so by Faraday's law an excess potential equal to the rate of change of flux $Li\omega j_1 e^{i\omega t}$, giving a pure imaginary admittance $(Li\omega)^{-1}$ of *negative* imaginary part *inversely* proportional to ω.

In one-dimensional waves, similarly, a narrow opening or *constriction* demands excess pressures to generate the accelerations needed if the volume flow through the constriction is to fluctuate. Figure 23(a) illustrates a compact constriction AB: its inductance L may be defined as the undisturbed density ρ_0 times the difference $\phi_B - \phi_A$ in velocity potential from A to B per unit volume flow from A towards B. When that volume flow fluctuates as $J_1 e^{i\omega t}$ this gives

$$\rho_0(\phi_B - \phi_A) = LJ_1 e^{i\omega t} \quad \text{and so} \quad p_A - p_B = Li\omega J_1 e^{i\omega t}, \tag{77}$$

implying the admittance $(Li\omega)^{-1}.$ (78)

The compactness of the constriction ensures continuity of volume flow across it according to equation (39) and the arguments following it, whose strength is even enhanced because in a *constricted* junction of length l the volume of fluid V is *less* than Al if A is a typical cross-sectional area outside it. There is accordingly an effectively incompressible flow, of velocity potential ϕ, within the compact constriction. The local reduction in cross-sectional area generates, however, enhanced *accelerations*, requiring significant pressure changes even though the constriction is compact. The situation is thus opposite to that of a widened compact *cavity* with pressure continuous but volume flow discontinuous across it.

A specially simple constriction is a short length l of narrow tube or channel with *uniform* reduced cross-sectional area A_0 joining some wider tube to the outside atmosphere or large reservoir (figure 23(b)). The velocity in the constriction per unit volume flow is $1/A_0$, which suggests a potential difference from A to B approximately equal to l/A_0 and so gives

$$L \doteq \rho_0 l/A_0. \tag{79}$$

This approximation can be derived also by considering expression (61) for the effective admittance at A of a tube of length l in the limiting case when the total admittance $\sum\limits_{n=2}^{N} Y_n$ at B is very large (as for a large reservoir) and $\omega l/c_1$ is small. Then

$$Y_1^{\text{eff}} \doteq Y_1[\mathrm{i}\tan(\omega l/c_1)]^{-1} \doteq (A_1/\rho_1 c_1)(\mathrm{i}\omega l/c_1)^{-1} = (L\mathrm{i}\omega)^{-1} \tag{80}$$

with $L = \rho_1 l/A_1$ in agreement with (79).

A still closer approximation allows also for how ϕ changes in the confluent motions in front of the mouth of the narrow region and beyond the farther end of the narrow region. This must be obtained by numerical integration of Laplace's equation for the actual geometry of the confluences but a correction often accurate to one significant figure is

$$L = \rho_0(l + 0.8A_0^{\frac{1}{2}})/A_0, \tag{81}$$

and this has approximate value even for constrictions of length *small* compared with a typical transverse dimension $A_0^{\frac{1}{2}}$ as in figure 23 (c).

Figure 23 (d) illustrates a uniform constriction AB that is compact but long enough for this correction (81) to be unimportant. It leads to a *branch* at B where the total admittance $\sum\limits_{n=2}^{N} Y_n$, although not infinite, is large enough compared with the admittance Y_1 of the tube AB for the second term in the numerator of (61) to be neglected as the product of small quantities. This gives

$$(Y_1^{\text{eff}})^{-1} = \left(\sum\limits_{n=2}^{N} Y_n\right)^{-1} + L\mathrm{i}\omega, \tag{82}$$

our first encounter with a law of the type which in electrical networks characterises lumped elements in series, involving added *impedance* (reciprocal admittance $Z = Y^{-1}$; here, pressure excess divided by volume flow). In fact, a compact constriction of any shape immediately preceding a branch (or a single tube as in figure 23 (a)) must satisfy this law (82), because the volume flow takes the same value at A as at B but the pressure excess at A is the sum of the pressure excess at B and the pressure difference $p_A - p_B$ across the constriction.

A special case (figure 23 (e)) is when the far end B of the constriction is a cavity with capacitance C so that equation (82) becomes

$$(Y_1^{\text{eff}})^{-1} = (Ci\omega)^{-1} + Li\omega. \tag{83}$$

Then at a certain *resonant frequency* $\omega = \omega_0$, given by the same formula

$$\omega_0 = (LC)^{-\frac{1}{2}} \tag{84}$$

as for an electrical tuned circuit, the effective impedance (83) at A drops to zero; the admittance Y_1^{eff} is infinite. This implies resonance in the sense that sinusoidal fluctuations of frequency ω_0 involving large volume flows in and out of the cavity (through the constriction) can be excited by exceedingly small pressure fluctuations outside it.

Of course, many other systems act as resonators in this sense: for example, as (63) shows, a quarter-wavelength tube with far end closed $\left(\sum_{n=2}^{N} Y_n = 0 \right)$ has infinite admittance, but it is not compact and it resonates at all frequencies that are odd multiples of the fundamental frequency. The system of figure 23 (e), called a Helmholtz resonator, has the special qualities that it effectively resonates at just one frequency ω_0 and is usefully *compact*: indeed, equations (69) and (81) make it smaller than c/ω_0 by a factor of order the square root of the ratio of cavity to constriction dimensions.

We can completely prevent waves of frequency ω_0 from entering any of the tubes beyond a certain branch, such as B in figure 21, by inserting a compact Helmholtz resonator in parallel with them, since its infinite admittance increases the sum ΣY_n in (56) to infinity and makes the transmitted wave $h(t)$ zero. We can produce the same effect in a straight tube ABC with a Helmholtz resonator opening out of it at B; treating B as a branch where wave energy can choose between entering the resonator or passing straight into BC or being reflected towards A, we see from (55) that the infinite total admittance at B brings about total negative reflection. There is no transmission into BC, the fluctuating volume flow from AB being exactly balanced by the flow in and out of the resonator.

An effect of an opposite kind is found where a tube terminates (figure 23 (f)) in a cavity AB connected through a constriction BC to the atmosphere or a very large reservoir. Equation (71) expresses the fact that a cavity's admittance $Ci\omega$ must be added to that of other elements which it precedes, giving in this case $$Y_1^{\text{eff}} = Ci\omega + (Li\omega)^{-1}. \tag{85}$$

Here, it is the *admittance* Y_1^{eff} at A which vanishes at the resonant frequency, so that the cavity behaves like a closed end! The reason why it can withstand pressure fluctuations of frequency ω_0 at the opening A opposite to the constriction without any inflow there is simply that the cavity in the absence of such an opening can resonate at that frequency: free oscillations of cavity pressure with arbitrary amplitude can occur, exactly balancing flow in and out of the constriction.

A similar behaviour is found when a uniform tube or channel AB terminates in a constriction of inductance L. Then equation (61) with $\sum\limits_{n=2}^{N} Y_n = (Li\omega)^{-1}$ gives

$$Y_1^{\text{eff}} = iY_1 \frac{\tan(\omega l/c_1) - (L\omega Y_1)^{-1}}{1 + (L\omega Y_1)^{-1}\tan(\omega l/c_1)} = iY_1 \tan\left[\frac{\omega}{c_1}(l - l_1)\right], \tag{86}$$

where
$$l_1 = (c_1/\omega)\cot^{-1}(L\omega Y_1). \tag{87}$$

In contrast to equation (72), the effective impedance of AB is that of a tube of *reduced* length $l - l_1$ terminating in a closed end! Essentially this is because AB possesses a concluding section CB of length l_1 which with the constriction at B forms a resonator opposing any oscillatory movements with frequency ω at the position C.

2.6 Linear propagation with gradually varying composition and cross-section

Sections 2.3–2.5 sketched the linear theory of longitudinal wave propagation in systems involving uniform lengths of fluid-filled tube or channel separated by compact elements: junctions, branches, cavities, constrictions. All variation in properties, either of the tube or channel or of the undisturbed fluid, occurred in the limited space of those compact elements, not in the generally more extensive lengths of tube between them.

This section, by contrast, describes (still on linear theory) effects of *gradual* variation of fluid density and compressibility, or of cross-sectional area and distensibility, along a *non-compact* length of tube or channel. Propagation through it is shown to be governed by a rule so simple that it permits immediate extension of earlier results like equation (61) to systems in which the branches separate lengths with gradually varying composition and cross-section. We explain this simple rule first through a crude line of heuristic argument from the results of section 2.3 concerning junctions, and then by a more refined analysis based on equations (3) and (4).

The heuristic argument makes the crude assumption that a length of tube involving a continuous gradation of composition and cross-section may

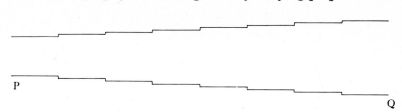

Figure 24. Gradually widening tube PQ approximated by a 'ladder'
of 'step-like' junctions.

behave as a limiting case of an approximating 'ladder' (figure 24) involving
many short uniform lengths separated by 'step-like' compact junctions at
each of which the admittance ratio Y_2/Y_1 is close to 1 and the proportion
(50) of reflected energy is therefore very small indeed. We infer, proceeding
to the limit, that no energy is reflected.

This limiting process uses the fact that expression (50) depends on the
square of the relative change of admittance: that is, of $\delta Y/Y$ or $\delta(\ln Y)$.
Indeed, this proportion (50) of reflected energy is always less than

$$\tfrac{1}{4}(\ln Y_2 - \ln Y_1)^2 \qquad (88)$$

as may easily be proved, for example, from the theorem that the arithmetic
mean exceeds the harmonic mean. If accordingly we approximate a tube
PQ, where the admittance $Y = A/\rho c$ varies (say) monotonically from Y_P
to Y_Q, by n unequal lengths of tube in each of which $\ln Y$ makes the same
small change $n^{-1}(\ln Y_Q - \ln Y_P)$, we can say that a wave travelling from P
to Q loses by reflection at each junction at most a proportion

$$\tfrac{1}{4}n^{-2}(\ln Y_Q - \ln Y_P)^2 \qquad (89)$$

of its energy flow. This in turn is not more than the same proportion (89)
of the *initial* energy flow that enters the tube at P. The ratio of energy flow
at Q to that at P is therefore at least

$$1 - \tfrac{1}{4}n^{-1}(\ln Y_Q - \ln Y_P)^2, \qquad (90)$$

since the proportion lost by reflection at n junctions is at most n times (89)
and since any further reflection of the reflected energy can only increase
the ratio.

This 'ladder' of uniform lengths of tube may be regarded as approximating
more and more closely to the actual continuous gradation of values of
$\ln Y$ as $n \to \infty$, when the ratio (90) tends to 1. In this limit, we infer that the
energy flow at P is transmitted with negligible reflection all the way to Q.
In a general tube with gradually varying properties we make this inference

of perfect transmission of wave-energy flow along every stretch with monotonically varying Y and may then combine the results to infer it for the tube as a whole.

These crude heuristic arguments lead, then, to a picture of a wave travelling from P to Q (say, in the positive x-direction) with velocity c that varies gradually with x but with *constant* energy flow: that is, with the product $Y p_e^2$ of the admittance and the square of the pressure excess constant. Such a wave would satisfy

$$p_e = [Y(x)/Y(0)]^{-\frac{1}{2}} f\left(t - \int_0^x c^{-1} \, dx\right), \qquad (91)$$

where $f(t)$ is the waveform at $x = 0$ and $\int_0^x c^{-1} \, dx$ is the time taken for the wave to travel from 0 to x. Equation (91) generalises equation (11) for a uniform tube or channel. The corresponding expression for a wave travelling in the negative x-direction would be

$$p_e = [Y(x)/Y(0)]^{-\frac{1}{2}} g\left(t + \int_0^x c^{-1} \, dx\right). \qquad (92)$$

In either case the volume flow *in the direction* of propagation would be $Y(x)p_e$.

We momentarily postpone the more refined arguments for the close accuracy of equation (91) in cases with properties varying gradually enough (in a sense to be made more precise) while we sketch its implications. For long water waves in channels it makes the free-surface elevation $p_e/\rho g$ vary as $Y^{-\frac{1}{2}}$ where, for a cross-section of breadth b and mean depth h, as in (58),

$$Y = bh^{\frac{1}{2}}/\rho g^{\frac{1}{2}}. \qquad (93)$$

Thus, for propagation along a channel with gradually varying b and h,

$$\text{wave elevation} \propto h^{-\frac{1}{4}} b^{-\frac{1}{2}}: \qquad (94)$$

a result over a century old known as Green's law.

Another special case of the simple rule (91) is one of the principles of geometrical acoustics: the determination of amplitude variation along a ray tube of gradually varying cross-section by assuming constancy of energy flow. Such amplitude variation in sound radiation from a sphere of large enough radius a_0 in uniform undisturbed fluid, for example, was found in section 1.11 to contain the factor a_0/r, explained as due to a constant energy flow along the ray tube spread over a cross-sectional area increasing like r^2. With different rays (associated, for example, with a different body shape) the energy flow at a distance x along a ray tube would be spread over an area $A(x)$ following a different law of variation, making p_e vary like

$$[A(x)/A(\text{o})]^{-\frac{1}{2}},$$

in agreement with (91) for uniform undisturbed fluid.

When properties of the fluid exhibit gradual variation the rays may again be different: for example, variations in the wave speed c are shown in chapter 4 to cause refraction (bending) of the rays just as in geometrical optics, essentially because they change the condition of stationary phase. At the same time the energy flow along a ray tube, namely the product of its cross-sectional area A with the acoustic intensity $I = p_e^2/\rho_0 c$, remains constant if and only if p_e varies as $(A/\rho_0 c)^{-\frac{1}{2}} = Y^{-\frac{1}{2}}$, again just as in (91). These considerations enhance the importance of (91) as the appropriate rule in all problems of geometrical acoustics for determining amplitude distributions once the rays have been found.

In the cardiovascular system, equation (91) governs propagation along arteries of gradually varying cross-sectional area A and distensibility D. Along the aorta, for example, both A and D gradually *decrease* with increasing distance from the heart. It follows, therefore, from equation (31), that $Y = A/\rho_0 c$ takes a form $A(D/\rho_0)^{\frac{1}{2}}$ which similarly decreases, and hence the amplitude of the wave (91) *increases*: this time in proportion to $A^{-\frac{1}{2}}D^{-\frac{1}{4}}$. Such a slow increase in amplitude of the pressure fluctuations along the aorta is actually observed by means of a catheter (thin tube attached to a pressure gauge, inserted into a peripheral artery and pushed a known distance into the aorta).

In all *branching* systems equation (61), for the effective admittance at A (where $x = -l$) of a tube AB taking into account reflections from a branch B (where $x = \text{o}$), can be readily extended to cases when the properties of the tube AB vary gradually along it. This is especially because the same factor $[Y(x)/Y(\text{o})]^{-\frac{1}{2}}$ appears in the direct wave (91) and in the reflected wave (92), cancelling therefore in an expression like (59). In fact, the ratio Y_1^{eff} of the volume flow at A to the pressure excess, with the latter taken as the sum of (91) and (92), becomes

$$Y_1^{\text{eff}} = Y_1^{\text{A}} \frac{f(t+l/c_{\text{AB}})-g(t-l/c_{\text{AB}})}{f(t+l/c_{\text{AB}})+g(t-l/c_{\text{AB}})}, \tag{95}$$

where Y_1^{A} is the admittance $Y(-l)$ of the tube AB itself at A and c_{AB} is the mean wave speed along AB defined by

$$l/c_{\text{AB}} = \int_{-l}^{0} c^{-1}\,\mathrm{d}x. \tag{96}$$

On the other hand, equation (55) determining the ratio of reflected waveform $g(t)$ to incident waveform $f(t)$ at $x = 0$ (the branch B) applies with Y_1 taken as the admittance $Y_1^B = Y(0)$ of the tube AB at B. Accordingly, the arguments leading to equation (61) now give the slightly modified equation

$$Y_1^{\text{eff}} = Y_1^A \frac{\left(\sum\limits_{n=2}^{N} Y_n\right) + i Y_1^B \tan(\omega l/c_{\text{AB}})}{Y_1^B + i\left(\sum\limits_{n=2}^{N} Y_n\right) \tan(\omega l/c_{\text{AB}})}, \tag{97}$$

which can be used for calculating properties of branching systems of tubes with gradually varying properties just as readily as (61) can for systems of uniform tubes. For example, there is a quarter-wavelength resonance condition, now defined by $\omega l/c_{\text{AB}} = \frac{1}{2}\pi$, and the resonant value, replacing (63), of the effective admittance at A is $Y_1^A Y_1^B \Big/ \left(\sum\limits_{n=2}^{N} Y_n\right)$.

The question remains: just how gradually must the cross-section and the fluid properties vary for the constant-energy-flow rule (91) to be a good approximation? All the preceding examples of the rule's utility illustrate the need to answer this question. The crude line of argument used gives little help in doing so, however, beyond suggesting that changes in admittance must be gradual enough to be regarded as a succession of very small changes: presumably, spread over a noncompact region since a substantial change of admittance within a compact region was seen in section 2.3 to imply continuity of volume flow and of pressure excess rather than of wave energy flow.

We must go back to the full equations (2), (3) and (4) for longitudinal waves to obtain a more quantitative answer. Equation (2) reminds us that with nonuniform undisturbed properties both ρ and A depend on another variable (S and x respectively) besides p_e. Equation (10) can still be used to define a local wave speed c, however, if it is interpreted as meaning

$$\rho_0^{-1} c^{-2} = K + D = [\rho^{-1}(\partial\rho/\partial p_e) + A^{-1}(\partial A/\partial p_e)]_{p_e=0}, \tag{98}$$

where the two partial derivatives keep constant respectively S and x. The quantity $\rho_0(x)$ is the value of the density ρ when both $p_e = 0$ and S takes its local undisturbed value.

Now since longitudinal waves without dissipation satisfy

$$\partial S/\partial t + u\,\partial S/\partial x = 0, \tag{99}$$

expressing entropy conservation for a fluid particle, the definition (98) of K gives

$$\rho^{-1}(\partial\rho/\partial t + u\,\partial\rho/\partial x) = K\,\partial p_e/\partial t, \tag{100}$$

where in linear theory a term $Ku\partial p_e/\partial x$ on the right-hand side may be neglected as the *product* of disturbances from the equilibrium state $u = 0$, $p_e = 0$. Differentiating the area A of a cross-section at constant x, however, we obtain

$$A^{-1}\partial A/\partial t = D\partial p_e/\partial t. \tag{101}$$

Hence the equation of continuity (4), which may be written

$$\rho^{-1}(\partial\rho/\partial t + u\partial\rho/\partial x) + A^{-1}[\partial A/\partial t + \partial(Au)/\partial x] = 0, \tag{102}$$

becomes on linear theory

$$(K+D)\,\partial p_e/\partial t + A_0^{-1}\,\partial(A_0 u)/\partial x = 0, \tag{103}$$

giving by (98)

$$\partial p_e/\partial t = -\rho_0 c^2 A_0^{-1}\,\partial(A_0 u)/\partial x = -cY^{-1}\partial J/\partial x, \tag{104}$$

where $J = A_0 u$ is the linearised volume flow and $Y(x) = A_0/\rho_0 c$ is the admittance. But the momentum equation (3) in linearised form can be written

$$\partial J/\partial t = -(A_0/\rho_0)\,\partial p_e/\partial x = -cY\partial p_e/\partial x. \tag{105}$$

Eliminating J from (104) and (105) we obtain

$$\partial^2 p_e/\partial t^2 = cY^{-1}(\partial/\partial x)(cY\partial p_e/\partial x) \tag{106}$$

as the modified form of equation (8) when c and Y are functions of x.

This exact linear equation (106) for longitudinal waves possesses no simple exact solution, but the accuracy of a simple expression like (91) can be estimated through the degree of approximation to which it satisfies (106). The result of substituting (91) into the right-hand side of (106) turns out to be

$$[Y(x)/Y(0)]^{-\frac{1}{2}}\left\{\ddot{f}\left(t - \int_0^x c^{-1}\,dx\right) - \tfrac{1}{2}cY^{-\frac{1}{2}}(cY^{-\frac{1}{2}}Y')'f\left(t - \int_0^x c^{-1}\,dx\right)\right\}, \tag{107}$$

while evidently its substitution into the left-hand side gives the same expression with only the first term in curly brackets. If ω^2 is the ratio between typical values of $\ddot{f}$ and f, the relative error (ratio of the second term to this first term) can be estimated, on expanding the derivative $(cY^{-\frac{1}{2}}Y')'$, as

$$\frac{1}{2}\left[\frac{c'}{c} - \frac{1}{2}\left(\frac{Y'}{Y}\right) + \frac{Y''}{Y'}\right]\frac{Y'}{Y}\bigg/\left(\frac{\omega}{c}\right)^2. \tag{108}$$

If those relative rates of change of c and of Y *and also of Y'*, which appear here in square brackets, *all* have small ratios to $\omega/c = 2\pi/\lambda$, then this error is small of order the square of those ratios. This condition for good accuracy is of the type expected from discussions in chapter 1 of the geometrical-acoustics approximation which is one case of the constant-energy-flow rule. It is important to note, however, that gradual variation of cross-section and composition (that is, of A_0, ρ_0, c and hence of $Y = A_0/\rho_0 c$) is not enough: their variation must also be *smooth*, in the sense that the gradient Y' itself varies gradually. On the other hand, any function of x other than $Y^{-\frac{1}{2}}$ as a factor in (91) would have produced a term in $\dot{f}$ in (107) and therefore, in general, a larger order of error.

Whenever p_e is given by (91), equation (105) can be solved exactly for J giving

$$J = [Y(x)/Y(\text{o})]^{-\frac{1}{2}}\left[Y(x)f\left(t - \int_0^x c^{-1}\,dx\right) + \tfrac{1}{2}c\,Y'(x)f_1\left(t - \int_0^x c^{-1}\,dx\right)\right],$$

(109)

where $df_1/dt = f(t)$. The correction involving f_1 to the crude approximation $J = Yp_e$ may be generally significant since its relative order is larger than the theory's error term (108). It effectively gives a frequency-dependent imaginary part to the local admittance since when $f(t)$ is proportional to $e^{i\omega t}$ the ratio J/p_e becomes

$$Y(x) - i[c\,Y'(x)/2\omega].$$

(110)

The constancy of energy flow is not affected, however, as equation (66) shows, by that imaginary part of the local effective admittance (110).

Thus under the conditions for good accuracy of (91), which include gradual variation of Y', the wave energy flow is transmitted with good accuracy; as is required in, say, a loudspeaker horn. By contrast, a *discontinuity* in the gradient $Y'(x)$ implies a discontinuity in the effective admittance (110), allowing a *reflected* signal which might be calculated through a formula such as (47) from the theory of junctions.

In the other limit of *compact* changes of cross-section and composition with relative gradients very large compared with ω/c, the time-derivatives on the left-hand sides of equations (104) and (105) become negligible, in support of the proposition that both the pressure excess p_e and the volume flow J show little spatial variation. The relative error in taking p_e independent of x over a length l may be estimated from (106) as of order at most

$$\omega^2 l^2 \operatorname{Max}(c^{-1}Y)/\operatorname{Min}(cY),$$

(111)

which is close to the square $(\omega l/c)^2$ of the compactness ratio except in regions like constrictions (section 2.5) with large ratios of maximum to minimum cross-sectional area.

Intermediate behaviour when $\omega l/c$ is neither small nor large can be illustrated by an exact solution of (106) for special distributions of $c(x)$ and $Y(x)$, supplementing similar calculations of special cases in section 1.11. The *exponential* loudspeaker horn satisfies

$$c = \text{constant}, \quad Y = Y_1 e^{\alpha x}, \tag{112}$$

and is used to achieve an area change as rapidly as possible without violating the constant-energy-flow condition that (108) be small. For this horn, equation (106) becomes exactly

$$\partial^2 p_e/\partial t^2 = c^2(\partial^2 p_e/\partial x^2 + \alpha \partial p_e/\partial x) \tag{113}$$

for which solutions proportional to $e^{i\omega t}$ are readily found by solving an ordinary differential equation in x with constant coefficients. This leads to

$$p_e = p_1 \exp\left[i\omega t - i(\omega^2 c^{-2} - \tfrac{1}{4}\alpha^2)^{\frac{1}{2}} x - \tfrac{1}{2}\alpha x\right] \tag{114}$$

for propagation in the direction x increasing, while equation (105) enables the corresponding solution for J to be found, giving

$$J/p_e = Y(x)\left[(1 - c^2\alpha^2/4\omega^2)^{\frac{1}{2}} - i(c\alpha/2\omega)\right]. \tag{115}$$

Note that these solutions change character when $\omega < \tfrac{1}{2}\alpha c$ and the square root becomes imaginary.

Equation (114) shows at high frequency that expected small departure of the order of magnitude $c^2\alpha^2/\omega^2$, suggested by (108), from the proportionality to $Y^{-\frac{1}{2}} \exp[i\omega(t - x/c)]$ indicated in (91). Note, indeed, that the *amplitude* in this horn shows no departure at all from proportionality to $Y^{-\frac{1}{2}}$, provided only that

$$\omega > \tfrac{1}{2}\alpha c; \quad \text{that is,} \quad \alpha < 2\omega/c = 4\pi/\lambda. \tag{116}$$

The departure from (91), rather, is a slight increase in the speed of travel of the wave crests from c to $c(1 - c^2\alpha^2/4\omega^2)^{-\frac{1}{2}}$; we do not dwell here on the consequences of this though we return to it in chapter 3. Of more immediate significance is the change in the effective admittance (115): the imaginary part agrees exactly with the value (110) suggested by general theory but it is interesting that a less substantial *reduction in the real part* also occurs. The horn is effective, then, only if the condition (116) is satisfied with a substantial margin to spare: the limiting value $\omega = \tfrac{1}{2}\alpha c$ is indeed a 'cut-off frequency' where the real part of the effective admittance (115) falls to zero whence by (66) the horn ceases to transmit any energy.

In the other limit, when a short stretch of exponential area increase with $c\alpha/\omega$ large forms a compact *junction*, one of the two possible real forms of

$-\mathrm{i}(\omega^2 c^{-2} - \tfrac{1}{4}\alpha^2)^{\frac{1}{2}} x$ in (114) becomes close to $+\tfrac{1}{2}\alpha x$ corresponding to changes in the junction with p_e practically independent of x, while the other is close to $-\tfrac{1}{2}\alpha x$ corresponding to changes with J practically independent of x. Thus this exponential-horn example illustrates the low-frequency behaviour characteristic of compact junctions as well as exhibiting at frequencies above the cut-off frequency the constant-energy-flow behaviour chiefly studied in this section.

2.7 Frictional attenuation

Like the introduction to the linear theory of sound (chapter 1), the foregoing introduction to the linear theory of one-dimensional waves in fluids ends with a discussion of wave energy dissipation and its effects: here including wave attenuation again (the gradual exponential decrease of energy flow in a travelling wave), and some consequences of it for branching and resonant systems. Admittedly the mechanisms of energy dissipation described in section 1.13 can be fully effective for one-dimensional waves in fluids: indeed, when that idea is used to represent propagation along an abstract ray tube they again may be the only mechanisms involved. However, a much greater rate of energy dissipation, and therefore of wave attenuation, can be additionally produced in a *solid-walled* tube or channel by friction.

Texts on the dynamics of fluids explain that since 'irrotational' flows (without vorticity) necessarily involve a nonzero tangential velocity of fluid at any solid boundary they can approximately represent real fluid motions only if thin viscous 'boundary layers' intervene between them and solid boundaries. Across such a layer the tangential velocity falls from the value demanded by the external irrotational flow to the zero value demanded of fluid in immediate contact with the solid boundary. The vorticity Ω is nonzero in the boundary layer, whose thickness is determined by the fact that vorticity generated at the solid surface, besides being convected by the largely tangential motions in the layer, is also diffused with a diffusivity

$$\nu = \mu/\rho \qquad (117)$$

equal to the ratio (called 'kinematic viscosity') of viscosity to density. Viscous tangential stresses (section 1.13) are the mechanism of this diffusion: also, they dissipate energy in the layer, and generate a *frictional force* between the solid boundary and the fluid.

Frictional attenuation of longitudinal waves in solid-walled tubes or channels should result, therefore, from viscous retardation and associated energy dissipation in boundary layers that separate the irrotational flows studied in sections 2.1–2.6 from solid boundaries. The phenomena lend themselves to ready estimation if, as assumed in this section, the diffusivity (117) is small enough to make the boundary layers thin compared with cross-sectional dimensions.

A good general idea of such a boundary layer's thickness and other properties in motion oscillating sinusoidally with frequency ω is obtained if we consider a large volume of fluid in a region $z > 0$ bounded by the solid *plane* wall $z = 0$ and subjected to a pressure gradient $\partial p_e/\partial x$ in the x-direction which is *spatially uniform* but oscillates sinusoidally in time. We here neglect any variability in the density ρ and viscosity μ due to the supposedly small variations in pressure or temperature.

The fluid flow determined by these conditions is a simple shearing motion in which the y- and z-components of fluid velocity are everywhere zero and the x-component $u(t, z)$ varies only with time and the distance z from the solid wall. The viscous tangential stress

$$p_{xz} = -\mu \partial u/\partial z \qquad (118)$$

represents (section 1.13) the flux of x-momentum in the z-direction per unit area per unit time. The one other nonzero component of tangential stress is p_{zx}, which has the same value (118) but no effect on the motions: only flux in the z-direction can produce change of momentum because that is the only direction in which fluxes vary. In fact the viscous force per unit volume is

$$\mu \partial^2 u/\partial z^2 \qquad (119)$$

in the x-direction; namely, minus the gradient of (118), representing rate of change of x-momentum due to difference between flux into and flux out of a fluid element.

The reasons why the simple shearing motion is produced are that it satisfies the equation of continuity $\nabla \cdot \mathbf{u} = 0$ for a constant-density flow and that the total force per unit volume on each fluid element,

$$-\partial p_e/\partial x + \mu \partial^2 u/\partial z^2, \qquad (120)$$

being purely in the x-direction and independent of x and y, can generate only x-momentum of magnitude independent of x and y. We deduce by equating this force (120) to density times acceleration

$$\rho \partial u/\partial t \qquad (121)$$

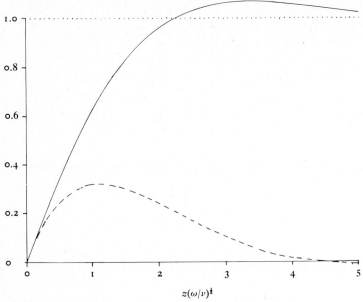

$$z(\omega/\nu)^{\frac{1}{2}}$$

Figure 25. The Stokes boundary layer: an oscillatory motion parallel to a solid wall is modified at small distances z from the wall by the factor given in curly brackets in equation (124), whose real and imaginary parts are the plain line (with asymptote 1, the dotted line) and the broken line, respectively.

(where any convective acceleration is identically zero because u is independent of x), a second-order equation of 'diffusion' type with diffusivity (117), suitable for matching the condition

$$u = 0 \quad \text{when} \quad z = 0 \tag{122}$$

at the solid boundary to a state uninfluenced by it for large positive z.

Although the equation so obtained is easily solved for any variations of $\partial p_e/\partial x$ with time about a zero mean, the conclusions (like others in this chapter) can be expressed most clearly through the form they take when $\partial p_e/\partial x$ is proportional to $\exp(i\omega t)$ as suggested at the outset; other variations can then be treated by Fourier synthesis. In these circumstances the u derived from equating (120) to (121) also varies as $\exp(i\omega t)$ and satisfies

$$\mu \partial^2 u/\partial z^2 - \rho i \omega u = \partial p_e/\partial x, \tag{123}$$

an ordinary differential equation with constant coefficients. Its solution satisfying the boundary condition (122) is

$$u = (\rho i \omega)^{-1}(-\partial p_e/\partial x)\{1 - \exp[-z(i\omega/\nu)^{\frac{1}{2}}]\}, \tag{124}$$

where the exponential of *minus* $z(i\omega/\nu)^{\frac{1}{2}}$ can appear since for large positive z this term becomes negligible and (124) tends therefore to the external irrotational solution

$$u_{\mathrm{ex}} = (\rho i\omega)^{-1}(-\partial p_e/\partial x), \qquad (125)$$

whereas no term involving the exponential of *plus* this quantity $z(i\omega/\nu)^{\frac{1}{2}}$ with positive real part can be present because its magnitude would increase without limit as $z \to \infty$.

The factor in curly brackets in (124), by which Stokes found that such an oscillatory irrotational solution (125) is modified within the boundary layer, is plotted as a function of $z(\omega/\nu)^{\frac{1}{2}}$ in figure 25. Its real part (the plain line) describes velocity oscillations in phase with those external motions: note that it is less than 1 in the inner portion of the boundary layer with

$$z(\omega/\nu)^{\frac{1}{2}} < 2.2$$

but slightly exceeds 1 (though by at most 7%) in an outer portion. A boundary-layer thickness might be arbitrarily defined as the thickness

$$5(\nu/\omega)^{\frac{1}{2}} \qquad (126)$$

of the layer beyond which this real part is within 3% of 1; but the factor 5 in (126) would be reduced to about 2 if a limit of 7% (excluding the outer layer) were used.

Nevertheless, any thickness, however defined, varies as $(\nu/\omega)^{\frac{1}{2}}$, as equation (124) makes plain, and as might be expected from the idea that the solid boundary generates vorticity that diffuses away from it (see (117)) with diffusivity ν over a distance of order $(\nu/\omega)^{\frac{1}{2}}$ in a time of order $1/\omega$ related to the period of fluctuation of the vorticity source strength. Actually, it may be shown that the value of this vorticity source strength on a plane wall is $(-\partial p_e/\partial x)$, whose phase leads by 90° that of the external irrotational flow (125) and therefore that of the plain line in figure 25. The vorticity as represented by the *gradient* of that plain line is, accordingly, largest near the wall where it has been generated with positive sign for half a period, but decreases to small negative values in the region $z(\omega/\nu)^{\frac{1}{2}} > 3.3$ where some effects of generation with the opposite sign during the previous half-period still predominate.

The broken line in figure 25, being the imaginary part of the factor in curly brackets, representing a part of the velocity response (125) in the boundary layer in phase with the pressure force $(-\partial p_e/\partial x)$. Such a velocity component in phase with force can occur where force is balanced by viscous resistance, which depends on instantaneous velocity, rather than by that

inertial *reactance* (depending on acceleration) which produces the 90°
phase-lag in (125). The corresponding vorticity (gradient of the broken
line), in phase with the vorticity source strength, is positive very near the
wall, where it has been generated with positive sign for a quarter-period,
but effects of generation with negative sign during the preceding half-period
bring about a stronger and closer region of negative vorticity than in the case
of the plain line.

We have noted in earlier sections the importance of volume-flow response
to fluctuations of excess pressure. The Stokes boundary layer (124) produces
a defect of volume flow past unit breadth of wall, relative to the volume flow
in irrotational motion, of

$$\int_0^\infty (u_{ex} - u)\, dz = (\rho i \omega)^{-1} (-\partial p_e/\partial x)(\nu/i\omega)^{\frac{1}{2}}, \tag{127}$$

where the integration is taken from o to ∞ but the integrand vanishes of
course outside the boundary layer. The last factor in (127) is such (since
$i^{-\frac{1}{2}} = (1-i)\, 2^{-\frac{1}{2}}$) that numerically equal contributions to volume-flow defect
are made by its real and imaginary parts, corresponding to the plain and
broken lines in figure 25: the defect in phase with u_{ex} (the net area by which
the plain line falls below the dotted line) is equal to the defect that lags by
90° behind u_{ex} (the area beneath the broken line).

The quantity (127) on multiplication by ρ becomes the defect of fluid
momentum per unit area of wall, whose rate of change

$$(-\partial p_e/\partial x)(\nu/i\omega)^{\frac{1}{2}} \tag{128}$$

must be the viscous retarding force applied by unit area of solid boundary
in opposition to the pressure force $(-\partial p_e/\partial x)$ acting on unit volume of fluid.
Thus, (128) is the component $(-p_{xz})_{z=0}$ of *viscous tangential stress*, which
can also be calculated from (118) and (124) to give the same result. The
fact that this viscous retarding force (128) involves equal contributions *in
phase* with the pressure force and *lagging by* 90° is reflected in figure 25
by the fact that the plain and broken lines have the same slope at the
origin.

The pressure force $(-\partial p_e/\partial x)$ per unit volume can do *work*, however,
only because the fluid velocity in the boundary layer includes a component
in phase with it (broken line in figure 25). The integral of that component
with respect to z is obtained from (127) as

$$(\rho\omega)^{-1}(-\partial p_e/\partial x)(\nu/2\omega)^{\frac{1}{2}} \tag{129}$$

since the real part of $-i^{-\frac{3}{2}}$ is $2^{-\frac{1}{2}}$. Accordingly the mean rate of working on the fluid per unit area of solid boundary is

$$\tfrac{1}{2}(\rho\omega)^{-1}|\partial p_e/\partial x|^2(\nu/2\omega)^{\frac{1}{2}}, \tag{130}$$

which must equal the mean rate of viscous dissipation of energy, per unit area of boundary layer, since the average energy of the fluid does not change. The formula (130) so derived for mean energy dissipation per unit time per unit area can alternatively be obtained from (124) by integrating the mean energy dissipation rate per unit volume,

$$\tfrac{1}{2}\mu|\partial u/\partial z|^2 = \tfrac{1}{2}(\rho\omega)^{-1}|\partial p_e/\partial x|^2\exp[-z(2\omega/\nu)^{\frac{1}{2}}], \tag{131}$$

across the boundary layer.

Results like equations (127) for volume-flow defect and (130) for energy dissipation which have here been derived exactly under certain simplifying assumptions (especially, a stationary plane wall and a spatially uniform pressure gradient) can be used to good approximation for rather general oscillatory motions in tubes and channels. Provided that the solid boundaries of cross-sections have radii of curvature large compared with the thickness of the predicted boundary layer, its properties should be similar to those for a plane wall (see texts on boundary layers for a detailed discussion; note that axial nonuniformity of pressure gradient on a scale of wavelengths should be even less influential). Approximately, the above equations can be used with z as *normal distance* from the solid boundary, even when the latter pulsates.

In particular, the energy dissipation rate is approximately (130) per unit *area* of solid boundary, or

$$\tfrac{1}{2}s(\rho\omega)^{-1}|\partial p_e/\partial x|^2(\nu/2\omega)^{\frac{1}{2}} \tag{132}$$

per unit length of tube, where s is the solid perimeter of a cross-section (perimeter excluding any *free* surface). Similarly, the linearised volume flow is

$$J = (\rho i\omega)^{-1}(-\partial p_e/\partial x)[A_0 - s(\nu/i\omega)^{\frac{1}{2}}], \tag{133}$$

where in the square brackets the A_0 resulting from the irrotational motion (125) integrated over the cross-sectional area is reduced by a 'displacement area' term resulting from multiplying the perimeter s by the volume-flow defect (127) per unit breadth.

For motions of frequency ω, equation (133) can be rewritten as a modified form of equation (105), as follows:

$$\partial J/\partial t = -cY\partial p_e/\partial x[\mathrm{1} - (\nu/i\omega)^{\frac{1}{2}}(s/A_0)]. \tag{134}$$

The real part of the factor in square brackets represents an increased effective inertia due to volume-flow defect, while the imaginary part represents the resistive action of wall friction.

On the other hand, longitudinal waves with sheared velocity distributions still satisfy a linearised equation of continuity in the form

$$\partial p_e / \partial t = -c Y^{-1} \partial J / \partial x, \tag{135}$$

as in (104), even though it can no longer be deduced by a method starting from equation (4) since u varies over a cross-section. The argument why p_e varies negligibly over a cross-section (equation (1)) still holds, however. Furthermore, the relative rate of change of volume of a fluid element due to pulsation of the cross-section A and variability of the sectional volume flux J can be written

$$A^{-1}(\partial A / \partial t + \partial J / \partial x) \tag{136}$$

and equated to minus the relative change of density averaged over the element. This can be written, using a bar to denote such a cross-sectional average, as

$$-\rho^{-1}\overline{(\partial \rho / \partial t + u \partial \rho / \partial x)} = -K \partial p_e / \partial t, \tag{137}$$

where equations (99) and (100) have been used for each particle of fluid separately. Equations (136) and (137) with (101) now give in linearised form

$$(K+D)\, \partial p_e / \partial t + A_0^{-1}\, \partial J / \partial x = 0, \tag{138}$$

which with (98) yields (135) as before.

Under the conditions studied in section 2.6, when cross-section and composition vary *gradually* (if at all) equations (104) and (105) were found to have a good approximate solution (91) for p_e. This fact makes a solution to (134) and (135) easy to obtain since they are precisely in the form of (104) and (105) with c and Y replaced by

$$c[1 - (\nu/i\omega)^{\frac{1}{2}}(s/A_0)]^{\frac{1}{2}} \quad \text{and} \quad Y[1 - (\nu/i\omega)^{\frac{1}{2}}(s/A_0)]^{\frac{1}{2}} \tag{139}$$

respectively; note that this substitution changes cY into the coefficient of $\partial p_e / \partial x$ on the right of (134) while leaving cY^{-1} unaltered.

The chief importance of these changes, by factors assumed to differ only slightly from 1, lies in the predicted change in c^{-1}, since that occurs inside the *exponent* in (91) with $f(t) = \exp(i\omega t)$. On a first binomial approximation to the effect (139) of the viscous boundary layer, c^{-1} changes to

$$c^{-1}[1 + \tfrac{1}{2}(\nu/i\omega)^{\frac{1}{2}}(s/A_0)]$$

$$= c^{-1}[1 + \tfrac{1}{2}(\nu/2\omega)^{\frac{1}{2}}(s/A_0)] - \tfrac{1}{2}ic^{-1}(\nu/2\omega)^{\frac{1}{2}}(s/A_0), \tag{140}$$

where the change in real part signifies a minor wave-speed reduction due to the increased effective inertia while the appearance of an imaginary part in c^{-1} describes wave attenuation due to frictional resistance.

For example, in a uniform tube or channel, equation (140) implies a travelling wave of the form

$$p_e = p_1 \exp\{i\omega t - i\omega x c^{-1}[1 + \tfrac{1}{2}(\nu/2\omega)^{\frac{1}{2}}(s/A_0)] - \tfrac{1}{2}\omega x c^{-1}(\nu/2\omega)^{\frac{1}{2}}(s/A_0)\}$$

(141)

with an associated mean energy flow

$$\tfrac{1}{2}Yp_1^2 \exp\{-\omega x c^{-1}(\nu/2\omega)^{\frac{1}{2}}(s/A_0)\}.$$

(142)

Note that this exponential decay of energy flow agrees with equation (132) for rate of energy dissipation per unit length of tube, which with $\partial p_e/\partial x$ replaced approximately by $(-i\omega/c)p_e$ becomes

$$\tfrac{1}{2}s\rho^{-1}\omega c^{-2}|p_e|^2(\nu/2\omega)^{\frac{1}{2}} = \tfrac{1}{2}Yp_1^2\omega c^{-1}(\nu/2\omega)^{\frac{1}{2}}(s/A_0)$$

(143)

at $x = 0$: exactly the rate of reduction of (142) per unit length of tube.

Equation (143) implies, incidentally, that the proportional energy loss *per wavelength* is approximately equal to the ratio of the boundary-layer thickness (126) to the section dimension A_0/s. By contrast, the energy dissipation rate in the absence of solid boundaries studied in section 1.13 included a viscous contribution that is negligibly small compared with this value (143), because longitudinal waves in a solid-walled tube or channel demand viscous dissipation associated with velocity variations over such a boundary-layer thickness instead of over a wavelength.

All the results of sections 2.4–2.6 for branching and resonant systems are immediately extensible to take frictional attenuation into account if the simple substitution of c^{-1} by (140) is made in all formulas. Then equation (96) gives $\omega l/c_{AB}$ *a negative imaginary part*, which modifies the conclusions that can be drawn from the fundamental branching formula (97). Essentially, this frictional damping effect, as might be expected, makes all resonance peaks finite by limiting the closeness with which the complex number $\omega l/c_{AB}$ can approach a pole of $\tan(\omega l/c_{AB})$ such as $\tfrac{1}{2}\pi$.

Additionally, the substitutions (139) requires us to make slight modifications to the admittance values on the right-hand side of (97). These modifications involve in a first binomial approximation the replacement of each Y by

$$Y[1 - \tfrac{1}{2}(\nu/i\omega)^{\frac{1}{2}}(s/A_0)].$$

(144)

The effect is often negligibly small, but can disturb an otherwise good admittance match.

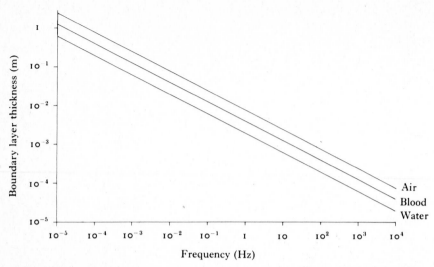

Figure 26. Boundary-layer thicknesses (126) for water, blood and air (with kinematic viscosities taken as 1, 4 and 15 mm² s⁻¹ respectively) plotted against the frequency in hertz, $\omega/2\pi$.

The above sketch of how the linear theories of one-dimensional waves in fluids are modified to take into account frictional attenuation can be filled in easily enough at most points, but on two matters omitted altogether a reference to specialised texts may be required. One is the treatment needed if the predicted boundary-layer thickness (126), whose values are plotted in figure 26 for applications involving water, blood and air (with ν around 1, 4 and 15 mm² s⁻¹ respectively), is comparable with cross-section dimensions. Methods effective in such circumstances include one for circular cross-sections which has been applied successfully to the blood pulse in smaller arteries.

A more radically different approach is required if the motions are on such a scale that *turbulent* friction predominates. This applies to tidal motions, for example: although a viscous boundary layer based on a semi-diurnal tidal period would, as figure 26 shows, be reasonably thin (about 0.5 m), a far stronger diffusion due to turbulence is found (see texts on boundary layers) in motions at such a high Reynolds number. For approaches to the estimation of its effects on propagation, it often suffices to use, as exemplified above, knowledge (in this case, experimentally derived) of the turbulent boundary layer's volume-flow defect as a function of pressure gradient and frequency.

2.8   Nonlinear theory of plane waves

Ramifications of the linear theory of one-dimensional waves in fluids were traced in considerable variety in the first half of chapter 2, after the introduction in chapter 1 to the linear theory of sound waves in three dimensions. The second half of chapter 2 is concerned with *nonlinear* effects, which have been neglected completely up to this point: it seeks to analyse the consequences of wave amplitudes *not* so small that their squares and products can be neglected. The information obtained is valuable for simply establishing under what circumstances nonlinear terms *are* effectively negligible. It is also of much wider interest because it brings to light remarkable phenomena of some importance, qualitatively different from any that could be accounted for by linear considerations.

This widening of scope to embrace the extra complexities of nonlinear phenomena needs, however, to be accompanied initially by strict limitations in other respects. We concentrate in sections 2.8–2.11 on *plane waves* of sound, although noting at several points that the results are applicable also to general longitudinal waves in a *uniform* tube or channel with friction neglected, and explicitly returning to the case of long water waves in a uniform open channel in section 2.12. By discarding in all these five sections any complications due to nonuniform composition or cross-section or due to attenuation or three-dimensional effects, we are enabled to focus attention directly upon the special features which nonlinear terms in equations of motion introduce into even those very simple properties of plane sound waves that are fully set out already on linear theory in section 1.1.

A brilliant mathematical discovery made by one of the greatest mathematicians of the mid-nineteenth century (Riemann) laid the foundations of all subsequent work on the nonlinear theory of plane waves of sound. This discovery, amounting to a transformation of the equations of motion into a form remarkably tractable for waves of any amplitude, led in due course to an excellent level of understanding of the subject.

On the other hand the scintillating brilliance of the mathematics involved in making this initial transformation had in the longer term a hypnotic effect upon acousticians, leading to a certain stagnation in the nonlinear theory of sound, associated with general belief that the whole success in understanding the subject had depended on that initial mathematically brilliant transformation. For many decades this precluded extension of the results to any other conditions of propagation, including the important case of one-dimensional propagation in tubes and channels with gradually varying composition and cross-section, because no transformation with similar properties could be found in those cases.

In this section, to avoid inducing such hypnosis, we precede an account of Riemann's mathematical transformation itself by a chain of simple physical arguments leading to exactly the same result. These may, perhaps produce a clearer physical understanding of what Riemann's theory and the conclusions from it mean, and lay a good foundation for accounts later in this book of the modern developments in nonlinear wave theory; beginning with an account in section 2.13 of nonlinear propagation in tubes and channels of gradually varying composition and cross-section, which in section 2.14 is applied to nonlinear propagation along the 'ray tubes' of geometrical acoustics. Further developments of nonlinear theory are then postponed until the epilogue.

We consider, then, a wave of arbitrary amplitude and waveform that is 'plane' in the sense that all motion of the fluid is in, say, the x-direction with the x-velocity u, the pressure p and the density ρ dependent only upon x and the time t; and we assume dissipative processes to be negligible so that the entropy per unit mass S, which is taken to be uniform initially, remains uniform and constant. Then we ask if the development of such a plane nondissipative wave of arbitrary amplitude can be predicted by simple physical arguments presupposing only a knowledge of linear acoustics.

It is reasonable to attempt an answer by noting that if we consider any particular position $x = x_1$ and time $t = t_1$ then there must be *some* space-interval around $x = x_1$ and *some* time-interval around $t = t_1$, both so small that if x and t respectively remain in those intervals then disturbances of u and p from their values u_1 and p_1 at (x_1, t_1) remain small enough for *linear* theory to describe their behaviour reliably. Admittedly the x-velocity u exhibits small disturbances from a nonzero value u_1 rather than from zero as assumed in the linear theory developed in chapters 1 and 2. If, however, we analyse the disturbances in those intervals in a special frame of reference moving with uniform velocity u_1, with position in that frame of reference specified by a new space-coordinate.

$$x - u_1 t, \tag{145}$$

then the velocity in *that* frame is $u - u_1$, which remains everywhere small (just as in linear theory) in the intervals in question.

At the same time, we must assign to the sound speed c in the interval a particular 'local' value c_1 for these small perturbations of the pressure p about its value p_1 at (x_1, t_1). At the given constant entropy S, a particular density $\rho = \rho_1$ corresponds to $p = p_1$, and c_1^2 is the derivative $\partial p/\partial \rho$ for changes at constant S (section 1.2) *taken at* $\rho = \rho_1$, $p = p_1$. It tends to be largest wherever p_1 (and hence also, at constant S, the *temperature*) is greatest.

Now the linear theory of sound gives a general plane-wave solution (section 1.1) for the small pressure perturbation $p - p_1$, and also for the fluid velocity $u - u_1$ in the frame of reference for which $x - u_1 t$ is the space-coordinate, as

$$p - p_1 = f(x - u_1 t - c_1 t) + g(x - u_1 t + c_1 t), \qquad (146)$$

$$u - u_1 = (\rho_1 c_1)^{-1} f(x - u_1 t - c_1 t) - (\rho_1 c_1)^{-1} g(x - u_1 t + c_1 t), \qquad (147)$$

in the intervals where these quantities, which we shall write as δp and δu respectively, remain very small. It follows that

$$\delta u + (\rho_1 c_1)^{-1} \delta p \text{ is a function of } x - (u_1 + c_1) t \text{ alone}, \qquad (148)$$

and similarly that

$$\delta u - (\rho_1 c_1)^{-1} \delta p \text{ is a function of } x - (u_1 - c_1) t \text{ alone}, \qquad (149)$$

in these small intervals.

A method of increasing greatly the value of these results makes use of an integral whose form they suggest; namely,

$$\int_{p_0}^{p} (\rho c)^{-1} \, \mathrm{d}p = P(p), \quad \text{say}, \qquad (150)$$

where p_0 is some reference pressure such as the initial pressure of the fluid before the plane waves appeared. Evidently, in a very small change δp about $p = p_1$,

$$\delta P = P'(p_1) \delta p = (\rho_1 c_1)^{-1} \delta p. \qquad (151)$$

The quantity on the left of equation (148) can accordingly be written as the very small change $\delta(u + P)$ which the quantity $u + P(p)$ makes from its value at (x_1, t_1) in the small intervals around that point. Within those small intervals it states that

$$\delta(u + P) = 0 \quad \text{when} \quad \delta x - (u_1 + c_1) \delta t = 0, \qquad (152)$$

where δx and δt are the small changes in x and t from their values x_1 and t_1; this is because $\delta(u + P)$ is a function of $x - (u_1 + c_1) t$ alone which is zero at (x_1, t_1) and hence also for all (x, t) with the same value of $x - (u_1 + c_1) t$ as at (x_1, t_1).

Equation (152) suggests that we consider *a curve in space–time C_+* at all points of which the differential relation

$$\mathrm{d}x = (u + c) \, \mathrm{d}t \quad \text{along} \quad C_+ \qquad (153)$$

is satisfied; evidently, C_+ is the locus of a point which always moves forwards at the local wave speed c relative to the local fluid velocity u. Equation (152) states that along such a C_+ curve through (x_1, t_1) the quantity $u + P$ is *stationary* at (x_1, t_1) itself: a remarkable result because, as (150) shows, the quantity $u + P(p)$ is defined without involving in any way the point (x_1, t_1); the entire argument that we have given must therefore apply also to a small neighbourhood of any *other* point on the C_+ curve! Since a function that is stationary at *all* points of a curve C_+ must take a constant value all along C_+ we deduce Riemann's first result:

$$u + P = \text{constant along a } C_+ \text{ curve } dx = (u + c)\,dt. \tag{154}$$

An identical argument starting from (149) and again using (151) yields Riemann's second result:

$$u - P = \text{constant along a } C_- \text{ curve } dx = (u - c)\,dt. \tag{155}$$

Evidently a C_- curve is the locus of a point which always moves backwards (that is, in the *negative* x-direction) at the local wave speed c relative to the local fluid velocity u.

Note that the reasoning leading to these important results in the non-linear theory of plane waves of sound is equally valid for other kinds of longitudinal waves of arbitrary amplitude, in tubes or channels with *uniform* cross-section and composition, because according to equation (12) these waves exhibit the same local relationships between expressions for pressure excess and fluid velocity as in (146) and (147), and an integral exactly like (150) can be similarly defined. On the other hand, the reasoning does demand uniform composition and in particular uniform *entropy* S; otherwise the integrand of (150) is not simply a function of p alone but varies also with S, which is not in general constant along a C_+ or a C_- curve but tends rather to remain constant along a fluid particle path $dx = u\,dt$. Cross-sectional properties varying as a function of x (and therefore equally failing to remain constant along C_+ or C_-) would equally nullify the reasoning.

We now confirm the validity of (154) and (155) by deducing them directly from the full nonlinear equations of motion, using Riemann's original method extended to general one-dimensional waves in *uniform* tubes or channels. We define c for waves at arbitrary excess pressure p_e in this general case by extending equation (9) in the form

$$c^{-2} = A^{-1}d(\rho A)/dp_e, \tag{156}$$

where equations (2) with uniform entropy S and cross-sectional properties make both ρ and A, and hence also c, functions of p_e alone; and we define P as

$$P = \int_0^{p_e} (\rho c)^{-1} \mathrm{d}p_e. \tag{157}$$

Then the momentum equation (3) divided by ρ becomes

$$\partial u/\partial t + u\partial u/\partial x + c\partial P/\partial x = 0, \tag{158}$$

while the equation of continuity

$$\partial(\rho A)/\partial t + u\partial(\rho A)/\partial x + \rho A\partial u/\partial x = 0 \tag{159}$$

(see (4)) divided by $\rho A/c$ becomes by (156) and (157)

$$\partial P/\partial t + u\partial P/\partial x + c\partial u/\partial x = 0. \tag{160}$$

Adding (158) and (160), we obtain

$$\partial(u+P)/\partial t + (u+c)\,\partial(u+P)/\partial x = 0, \tag{161}$$

which is just another way of writing (154)! Subtracting (160) from (158) we obtain

$$\partial(u-P)/\partial t + (u-c)\,\partial(u-P)/\partial x = 0, \tag{162}$$

which similarly proves (155). The arguments are fully as compelling as before though, perhaps, giving no explanation of why the integral P should be written down at all.

As a first indication of the uses of (154) and (155), prior to further examples in section 2.9, we consider an *initial-value problem*: if at an initial instant $t = 0$ all the fluid except that in a certain slab (finite interval of values of x) is undisturbed (that is, with zero values of both u and $p_e = p - p_0$, and hence also zero P) while the disturbance within the slab may be large, but of course with constant entropy, then we ask how disturbances propagate ahead of and behind the slab in the subsequent motion.

To attack this problem by the Riemann method, we draw a space–time diagram: the (x, t) plane of figure 27. The disturbance is at $t = 0$ confined to the slab between the points B (for back) and F (for front). To analyse the subsequent development of the flow we imagine a considerable number of C_+ curves (such that $\mathrm{d}x/\mathrm{d}t = u+c$) drawn on the diagram, including those we designate C_+^B and C_+^F which originate when $t = 0$ at the points B and F; and similarly with the C_- curves (such that $\mathrm{d}x/\mathrm{d}t = u-c$).

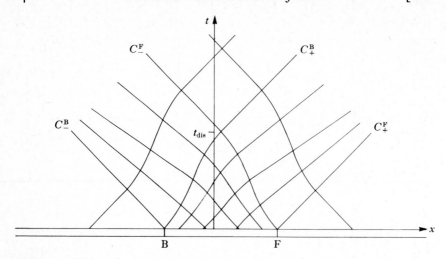

Figure 27. Initial-value problem treated by the Riemann method. The disturbance initially confined to BF becomes disentangled, when $t = t_{\text{dis}}$, into two simple waves (that between C_+^{B} and C_+^{F} travelling to the right, and that between C_-^{B} and C_-^{F} travelling to the left) with an undisturbed region in between.

The shape of these curves is not known in advance, since it depends on how u and c vary, but certain useful facts about them can be inferred, as follows.

Equation (155) tells us that along each C_- curve $u - P$ takes a constant value: different in general, of course, for different C_- curves. However, C_-^{F} and all curves ahead of it originate from the region ahead of F where $u = P = 0$: the constant value of $u - P$ on each of *these* therefore can only be zero. We conclude that

$$u = P(p_e) = \int_0^{p_e} (\rho c)^{-1} dp_e \quad \text{ahead of} \quad C_-^{\text{F}}. \tag{163}$$

Throughout this region ahead of C_-^{F}, which grows progressively *wider* than 'the region ahead of the slab' where we asked how disturbances would propagate, an important simplifying relationship (163), a sort of extension of equation (12), has already been established.

Identical arguments imply, evidently, that

$$u = P \quad \text{behind} \quad C_-^{\text{B}}, \tag{164}$$

while similar arguments for the C_+ imply that

$$u = -P \quad \text{behind} \quad C_+^{\text{B}} \tag{165}$$

and that

$$u = -P \quad \text{ahead of} \quad C_+^{\text{F}}. \tag{166}$$

Actually, the definitions in (154) and (155) imply that C_+^F must lie ahead of C_-^F, so that in the whole region of (166) equation (163) also holds, which can only mean that $u = P = 0$: evidently, it is the region of space–time that remains undisturbed because it has not yet been reached by a wave. Similar remarks apply to the region of (164), included *within* the region of (165): that region behind C_-^B is, then, also undisturbed, with $u = P = 0$.

More surprisingly there is, after a certain time $t = t_{dis}$ (figure 27) when C_-^F intersects C_+^B, a region lying *between* those curves where (163) and (165) both apply, leading once more to $u = P = 0$. During the time $0 < t < t_{dis}$ the disturbances become *disentangled* and thereafter propagate as two 'simple waves' (one forwards and one backwards) with an undisturbed region in between.

To understand the special properties of these 'simple waves' consider the forward simple wave, in the region ahead of C_-^F between C_+^B and C_+^F. Equation (163) satisfied in this region implies that the quantity $u + P$, which is constant by (154) along each C_+, is simply $2u$, so that u itself is constant along each C_+. Therefore, by (163), p_e is constant and so also c is constant. Any C_+ curve (154) in this simple-wave region therefore has

$$u + c = \text{constant} \quad \text{along} \quad C_+, \tag{167}$$

so that it is a straight line in the (x, t) plane as indicated in figure 27.

Simple waves in general, which are studied in section 2.9, may travel either *forwards* relative to the fluid, when they necessarily involve such C_+ 'curves' that are all straight lines on each of which the flow velocity u and excess pressure p_e take constant values related by equation (163); or *backwards* (like the other simple wave in figure 27) when they involve straight C_- curves with u and p_e constants related by (165) along each. The remarkable conclusion from Riemann's analysis is not so much that simple waves exist as that any finite disturbance must in a finite time t_{dis} disentangle itself into a pair of simple waves travelling in opposite directions and separated by an undisturbed region. This conclusion shows the power of the analysis by means of the curves C_+ and C_-: for a wider mathematical perspective in which those curves are viewed as special cases of the characteristic curves of hyperbolic systems of partial differential equations, see texts on differential-equation theory.

2.9 Simple waves

The expression 'simple wave' is used, appropriately enough, to signify a rather straightforward extension of the linear-theory concept of a travelling plane wave (section 1.1) to disturbances of arbitrary amplitude. In section 2.9 we investigate mechanisms of generation of simple waves, and also analyse their propagation laws. Much of the interest *lies* in how simply the propagation of these nonlinear phenomena can be calculated, but a deeper interest progressively emerges. An enigma perhaps hinted at in figure 27 – in brief, what happens when two C_+ curves intersect? – is defined gradually in section 2.9, although answers to this important question are postponed to later sections.

We take the positive x-direction as the direction of travel of the simple wave. Such a wave is then defined as a region where the constant value (155) taken by $u - P$ along each C_- curve is in every case zero as in (163), giving

$$u = P(p_e) = \int_0^{p_e} (\rho c)^{-1} dp_e. \tag{168}$$

Its C_+ curves, by contrast, are straight lines along each of which u takes a constant value, generally different for each; as do p_e, c and (equation (167)) the value $u + c$ of dx/dt along the C_+ curve. Note that this brief description applies identically to the backward-moving simple wave in figure 27, since the changes of sign in x and in u needed to make *its* direction of travel the positive x-direction force equation (165) which it satisfies to take this form (168) and force the value $u - c$ of dx/dt on *its* straight line along which u and p_e are constant to become $u + c$.

Simple waves differ from the travelling plane waves of linear theory only in that (i) the admittance for unit cross-section, $(\rho c)^{-1}$, is, as (168) shows, a *differential* admittance depending on the excess pressure p_e and representing increased fluid velocity u per unit increase in p_e; and (ii) different excess pressures travel at the different wave speeds c appropriate to each, *relative* to the associated fluid velocity, so that absolutely they travel at speed $u + c$.

Simple waves are generated not only as end-products from a rather general initial-value problem as in section 2.8; they arise also from certain types of boundary-value problem. In particular, when a uniform tube or channel filled with initially undisturbed fluid begins to be excited *at one end* the immediately resulting disturbance in the tube takes the form of a simple wave (one only this time). Figure 28 indicates this for three types of excitation: in each case, with x increasing along the tube away from the

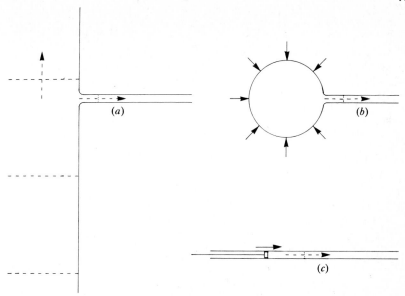

Figure 28. Three types of excitation of simple waves: (*a*) by pressure changes occurring at an end opening into a large reservoir; (*b*) by volume flow out of a compact cavity; (*c*) by displacement of a piston.

excitation position, the entire tube is filled with C_- curves which originate in undisturbed fluid, so that the constant value taken by $u - P$ along each must be zero as for a simple wave.

Pressure excitation is illustrated in figure 28(*a*): the wave is excited by pressure changes occurring at an open end with characteristic frequencies for which the tube or channel cross-section is compact. Those pressure changes, which might be generated, for example, by long-wave propagation in a large external reservoir, are nearly but (as we shall see) not exactly continuous across the junction to the adjoining channel, whose compactness, on the other hand, prevents any significant reaction on the reservoir motions from the volume flow through the orifice (compare figure 20(*b*) analysed on linear theory at the end of section 2.3).

Volume-flow excitation is illustrated in figure 28(*b*): the end opens this time into a cavity which is itself compact and whose volume changes excite the wave by generating a defined volume flow through the end of the tube. This excitation is like the forcing of blood into the aorta by contractions of the heart, but highly simplified because in figure 28(*b*) we suppose the tube to be uniform and the fluid initially undisturbed.

6

Displacement excitation is illustrated in figure 28 (c): the simple wave is excited this time by displacement of a *closed* end, which may be thought of as effectively a *piston* whose movements in the tube (like those of a loud-speaker diaphragm) force into motion the initially undisturbed fluid. This figure illustrates a boundary-value problem involving a *moving* boundary: the piston surface.

With each of these methods of excitation the resulting simple wave is fully specified in terms of the variations of pressure excess or volume flow at the open end $x = 0$, or in terms of the displacement $x = H(t)$ of the piston. For example, if pressure excess p_e varies as $\tilde{p}_e(t)$ at $x = 0$, then corresponding variations $c = \tilde{c}(t)$ and $u = \tilde{u}(t)$ can be inferred from (156) and (168) respectively, so that the straight C_+ curve emitted from $x = 0$ at time $t = \tau$ is fully specified by the equations

$$p_e = \tilde{p}_e(\tau), \quad u = \tilde{u}(\tau), \quad c = \tilde{c}(\tau), \quad x = [\tilde{u}(\tau) + \tilde{c}(\tau)](t - \tau). \tag{169}$$

Note that the speed $\tilde{u}(\tau) + \tilde{c}(\tau)$ of successive signals increases with τ in a *compressive* wave (when the pressure excess $\tilde{p}_e(\tau)$ carried by successive signals is rising) leading to an enigma whose resolution we defer about what happens when later signals overtake earlier ones.

Pressure excitation at a compact end open to a reservoir involves continuity, not of the excess pressure itself (as the *linearised* momentum equation has been shown in (40) and (41) to imply), but of a more complicated quantity

$$\tfrac{1}{2}|\mathbf{u}|^2 + \int_0^{p_e} \rho^{-1}\, dp_e. \tag{170}$$

This is derived from the *unlinearised* momentum equation for irrotational flow with uniform entropy, shown in texts on fluid dynamics to imply that expression (170) has gradient $-\partial\mathbf{u}/\partial t$. It follows that the change in (170) across a compact junction is negligible: reservoir motions *with a given value of* (170) at each time t determine the pressure $\tilde{p}_e(t)$ at the end of the tube so that the associated value of (170) (with $|\mathbf{u}|^2 = [\tilde{u}(t)]^2$ related to $\tilde{p}_e(t)$ according to (168)) is identical. The appearance of the $\tfrac{1}{2}|\mathbf{u}|^2$ term does not, however, make this condition in practice greatly different from one of continuity of pressure.

Similarly, in volume-flow excitation at a compact end, the quantity assured as continuous in nonlinear theory is strictly the *mass flow* $\rho A u$, which by equation (4) has gradient $-\partial(\rho A)/\partial t$. Specification of $\rho A u$, again, with ρA a function of p_e and (168) making u a function of p_e, determines $\tilde{p}_e(t)$ and leads to the same solution (169).

Excitation by a piston displacement $x = H(t)$ specifies the fluid velocity u at the piston as $\tilde{u}(t) = \dot{H}(t)$, from which $\tilde{p}_e(t)$ is inferred by (168) and $\tilde{c}(t)$ by (156). Then the C_+ signal emitted from $x = H(\tau)$ at time $t = \tau$, carrying excess pressure $\tilde{p}_e(\tau)$ and travelling at constant velocity $\tilde{u}(\tau) + \tilde{c}(\tau)$ satisfies the equation

$$x = H(\tau) + [\tilde{u}(\tau) + \tilde{c}(\tau)] (t - \tau). \tag{171}$$

For such a tube which being closed by a piston cannot admit new fluid it is of interest to check that our solution satisfies mass-conservation. At a fixed time t the mass of fluid between successive C_+ signals (171) emitted at times τ and $\tau + \mathrm{d}\tau$ is

$$\tilde{\rho}(\tau) \, \tilde{A}(\tau) \, [x(\tau) - x(\tau + \mathrm{d}\tau)], \tag{172}$$

where $\tilde{\rho}(\tau)$ and $\tilde{A}(\tau)$ are the fluid density and cross-sectional area associated with the excess pressure $\tilde{p}_e(\tau)$, and the distance $x(\tau) - x(\tau + \mathrm{d}\tau)$ between the signals can be inferred by differentiating (171) with respect to τ. This, with $\dot{H}(\tau) = \tilde{u}(\tau)$, gives the differential mass (172) as

$$\tilde{\rho}(\tau) \, \tilde{A}(\tau) \, \tilde{c}(\tau) \, \mathrm{d}\tau - \tilde{\rho}(\tau) \, \tilde{A}(\tau) \, [\mathrm{d}\tilde{u}(\tau) + \mathrm{d}\tilde{c}(\tau)] (t - \tau)$$
$$= - \mathrm{d}\{\tilde{\rho}(\tau) \, \tilde{A}(\tau) \, \tilde{c}(\tau) \, (t - \tau)\} \tag{173}$$

since by (168) and (156)

$$\tilde{\rho}\tilde{A}\mathrm{d}\tilde{u} = (\tilde{c})^{-1} \tilde{A}\mathrm{d}\tilde{p}_e = \tilde{c}\mathrm{d}(\tilde{\rho}\tilde{A}). \tag{174}$$

Integrating (173) from τ to t determines $\tilde{\rho}(\tau) \, \tilde{A}(\tau) \, \tilde{c}(\tau) \, (t - \tau)$ as the mass between the signal emitted at time τ and the piston, which is emitting its current signal at time t. Mass-conservation is nicely demonstrated by this expression which is the total mass traversed in time $(t - \tau)$ by a C_+ signal moving through a mass $\tilde{\rho}(\tau) \, \tilde{A}(\tau)$ of fluid per unit distance at the *relative* velocity $\tilde{c}(\tau)$. Putting $\tau = 0$ (time of emission of a signal carrying zero disturbance) gives the total mass of fluid in the wave at time t as $\rho_0 A_0 c_0 t$ by either method. The corresponding calculation in the open-end case (169) requires removal of the term $H(\tau)$ in (171), leading in (172) to an extra term $\tilde{\rho}(\tau) \, \tilde{A}(\tau) \, \tilde{u}(\tau) \, \mathrm{d}\tau$ whose integral from 0 to t is the additional mass introduced at the open end.

Our examples show the importance of all those relationships that link different variables in a simple wave. In particular, the relationship (168) between u and p_e determines the pressures that resist a given motion of a piston (or a given volume-flow injection from a contracting cavity) and, conversely, determines the inflow accompanying pressure changes at an open end. Corresponding values of c, again, govern the propagation of the signals. We now calculate these relationships explicitly in two cases of practical importance.

The first is a case of plane waves of sound in a perfect gas. Such a gas (one of very low density compared with the same substance in a condensed phase) was seen in section 1.2 to satisfy the equation

$$c^2 = dp/d\rho = \gamma p/\rho \tag{175}$$

in changes at constant entropy, where $\gamma = c_p/c_v$ is the ratio of the specific heats. In conditions where γ remains approximately constant (for example, in monatomic gases $\gamma = \frac{5}{3}$, while in atmospheric air $\gamma = 1.40$ to two decimal places for temperatures between 100 K and about 1000 K) equation (175) gives

$$(p/p_0) = (\rho/\rho_0)^\gamma, \quad (c/c_0) = (\rho/\rho_0)^{\frac{1}{2}(\gamma-1)} \tag{176}$$

if $\rho = \rho_0$ and $c = c_0$ when $p = p_0$. It follows that

$$P = \int_{p_0}^{p} (\rho c)^{-1} dp = \int_{\rho_0}^{\rho} \rho^{-1} c \, d\rho = 2(c - c_0)/(\gamma - 1). \tag{177}$$

This simple linear relationship between P and c for a perfect gas with γ constant means that equation (168) for a simple wave becomes

$$c = c_0 + \tfrac{1}{2}(\gamma - 1) u. \tag{178}$$

For air with $\gamma = 1.40$ this implies an increase in the speed c of wave propagation relative to the fluid by one-fifth of the fluid velocity u in the direction of propagation. The absolute speed $u + c$ of a signal is thus increased above c_0 by

$$u + c - c_0 = \tfrac{1}{2}(\gamma + 1) u, \tag{179}$$

which for $\gamma = 1.40$ is five-sixths due to convection by the fluid motion and one-sixth due to the increased propagation speed.

The corresponding density and pressure values are obtained from (176) as

$$\rho/\rho_0 = [1 + \tfrac{1}{2}(\gamma - 1)(u/c_0)]^{2/(\gamma-1)}, \quad p/p_0 = [1 + \tfrac{1}{2}(\gamma - 1)(u/c_0)]^{2\gamma/(\gamma-1)}. \tag{180}$$

Figure 29, which plots these ratios, shows how the excess pressure resisting motions of a piston into undisturbed air increases much more than proportionally to the piston velocity u for larger values of u/c_0.

Secondly, we consider long waves in open channels, with cross-section uniform and arbitrary except that the *breadth* b of the free surface remains constant for all values of the free-surface level $z = \zeta$ occurring in the wave. This makes equation (16) exact, although equation (19) for the wave speed relative to the fluid becomes

$$c = (gA)^{\frac{1}{2}} b^{-\frac{1}{2}} \tag{181}$$

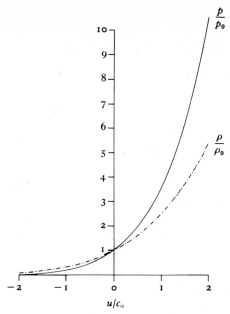

Figure 29. The pressure p and the density ρ in a simple wave, plotted against the fluid velocity u in the direction of propagation, for a perfect gas with ratio of specific heats 1.40 and undisturbed pressure, density and sound speed p_0, ρ_0 and c_0.

when the water's cross-sectional area changes from A_0 to A. Accordingly,

$$P = \int_0^{p_e} (\rho_0 c)^{-1} \, dp_e = g^{\frac{1}{2}} b^{-\frac{1}{2}} \int_{A_0}^{A} A^{-\frac{1}{2}} \, dA = 2(c - c_0), \qquad (182)$$

another linear relationship between P and c.

Simple waves in open channels of uniform cross-section with the breadth b of the free surface constant therefore satisfy equations closely similar to (178), (179) and (180):

$$c = c_0 + \tfrac{1}{2} u; \quad u + c - c_0 = \tfrac{3}{2} u; \quad A/A_0 = (1 + \tfrac{1}{2} u/c_0)^2. \qquad (183)$$

Equations (183) have often been described as stating that such simple waves behave exactly 'like those in a perfect gas with $\gamma = 2$'; not that any such gas exists! The analogy may actually be useful, in the limited context of simple waves, for inferring open-channel results from acoustic results with general γ. Note that the excess speed $u + c - c_0$ of a signal is here two-thirds due to convection by the fluid motion and one-third due to increased propagation speed.

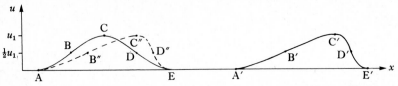

Figure 30. Illustrating waveform distortion for a perfect gas with specific heats in the constant ratio 1.40. Plain lines: distortion of the waveform ABCDE into A′B′C′D′E′ after time t, when each value of u is found a distance $c_0 t + \frac{1}{2}(\gamma + 1)ut$ to the right (thus, the distance CC′ is $c_0 t + \frac{1}{2}(\gamma + 1)u_1 t$, whereas BB′ and DD′ are $c_0 t + \frac{1}{4}(\gamma + 1)u_1 t$ and AA′ and EE′ are only $c_0 t$). Broken line: all these distances are reduced by $c_0 t$ when plotted against $X = x - c_0 t$ as AB″C″D″E, exhibiting the distortion more conveniently.

All cases of simple waves exhibit this increase of excess signal speed with u: a property which progressively *distorts the waveform*. Instead of the graph of u as a function of x having at a later time the same shape as at an earlier time (section 1.1) the signals carrying larger values of u are shifted forwards relative to low-amplitude signals by an amount increasing with u. For cases satisfying the linear relationship (179), which includes (183) as a special case $\gamma = 2$, the waveform suffers a *simple shearing distortion* (figure 30); after a time t each value of u is found a distance $c_0 t + \frac{1}{2}(\gamma + 1)ut$ to the right (plain line) depending linearly on u.

Distortion of the waveform with time may more conveniently be depicted in a frame of reference which moves at the wave speed c_0 for undisturbed fluid. In this frame, a signal moves at the speed

$$v = u + c - c_0 \tag{184}$$

and the space-coordinate is $\quad X = x - c_0 t. \tag{185}$

Where the excess signal speed v is a simple multiple of u, as in (179), it shears the graph of u with respect to X in time t by the amount $\frac{1}{2}(\gamma + 1)ut$ (broken line in figure 30) proportional to distance above the horizontal axis. This causes sections of the waveform with positive slope to become less steep while sections with negative slope become steeper. For the latter sections, which are *compressive* in that p_e increases with time, the previous enigma arises in a new form: what happens when that negative slope becomes infinite?

The distortion of quite general simple waves, which need not satisfy (179), becomes a uniform shearing if we choose to redefine 'waveform' as the graph of v as a function of X. Here, v is constant (because u and c separately are) along C_+ curves satisfying $dX/dt = v$. Thus, in any simple wave viewed in

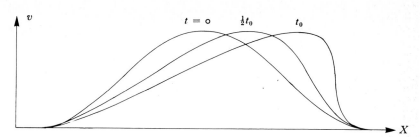

Figure 31. Plots of excess signal velocity v against $X = x - c_0 t$ show the distortion of a general simple wave. After time t (here taking values 0, $\frac{1}{2}t_0$ and t_0) each value of v is found at a point displaced a distance vt to the right, so that the waveform is subjected to unit rate of shear.

this frame of reference moving with velocity c_0, the variable v is one satisfying the remarkable equation

$$\partial v/\partial t + v\,\partial v/\partial X = 0, \tag{186}$$

which states that signals carrying a given value of v travel at velocity v.

Figure 31 shows that this gives any waveform a *unit rate of shear*: after a time t a given value of v is found at a point displaced a distance vt to the right, while a greater value $(v + \delta v)$ is found at a point $(v + \delta v)t$ to the right, whose distance δX to the right of the previous point has therefore *increased* by $(\delta v)t$. This implies that the *reciprocal slope* $\delta X/\delta v$ of the waveform for given v *increases at a unit rate* with respect to time: an important result that can be arrived at alternatively by differentiating (186) with respect to X and deducing that

$$(\partial/\partial t + v\,\partial/\partial X)[(\partial v/\partial X)^{-1}] = 1. \tag{187}$$

The result implies, perplexingly enough, that any *negative* value of a reciprocal slope must in some finite time increase to zero, corresponding to *infinite slope*. This happens first for a given waveform after a time

$$t_0 = \text{Min}\,[-(\partial v/\partial X)^{-1}], \tag{188}$$

and the enigma of what may happen after that time is, again, merely noted.

Waveform distortion is analysed above as a change in a *spatial* distribution with time: a description especially suitable to an initial-value problem like that of section 2.8. For certain boundary-value problems, like those of figures 28(a) and 28(b), distortion may more conveniently be viewed as a change in a *temporal* distribution (like that which is specified at an open end) with distance x along the tube. This requires us to use not a moving spatial reference (185) but a *time* reference varying with position, at a rate

equal to a *reciprocal* signal velocity. Indeed the theory takes the same form as illustrated in figure 31 if a waveform is regarded as the dependence on a modified time-coordinate
$$X_1 = c_0^{-1} x - t \qquad (189)$$

of, not an excess signal velocity v but a *defect in reciprocal signal velocity*:
$$v_1 = c_0^{-1} - (u+c)^{-1}. \qquad (190)$$

Evidently, this quantity v_1 is constant along a C_+ curve for which, *with x replaced by* t_1, equation (189) gives
$$dX_1/dt_1 = c_0^{-1} - (dt/dx) = c_0^{-1} - (u+c)^{-1} = v_1, \qquad (191)$$

so that the distortion rules of figure 31 and equations (186), (187) and (188) all apply with every variable given suffix 1. In particular, equation (188) then gives the *distance* along the tube at which the enigma of a temporal waveform with infinite slope first appears.

With displacement excitation (boundary conditions applied neither at a fixed time nor at a fixed location but at a moving piston) we may think of how the waveform once produced is distorted from either of the points of view just described. Alternatively, we may use equation (173) for the differential mass between two neighbouring C_+ curves, which implies that two such curves first run together, leading to an infinite slope of the waveform, at the instant when (173) first vanishes: namely,
$$t = \text{Min}\,\{d[\tilde{\rho}(\tau)\,\tilde{A}(\tau)\,\tilde{c}(\tau)\,\tau]/d[\tilde{\rho}(\tau)\,\tilde{A}(\tau)\,\tilde{c}(\tau)]\}. \qquad (192)$$

In all these problems, the analysis up to such a moment when C_+ curves first run together is seen to be relatively tractable, but the conceptual difficulties of extending it beyond that moment appear substantial, and must be investigated in detail in later sections.

2.10   Shock waves

Deeper considerations, necessary to resolve the enigma noted in section 2.9, are offered initially for plane waves of sound excited by a particularly simple movement of a piston. Extensions of the analysis to more general methods of excitation, and more general types of longitudinal waves in fluids, follow in sections 2.11 and 2.12 respectively.

Figure 28(c) depicts how a piston whose acceleration has initially a *positive* x-component (towards the fluid) generates sound waves in which later signals, carrying higher excess pressures, tend to overtake the earlier ones. The corresponding distortion of the spatial waveform is illustrated

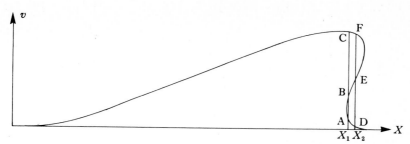

Figure 32. The waveform obtained by continuing the construction of figure 31 to the later time $t = \frac{3}{2}t_0$: its physical impossibility is implied by its indication of three different values A, B and C for v at $X = X_1$ or three values D, E and F for v at $X = X_2$.

in figure 31 up to the moment (specified by equation (188)) when signals first run together and an infinite gradient results. The same theory of wave-form shearing would, however, predict at subsequent instants an *impossible* waveform (figure 32) involving three different values of the excess signal velocity at any position in a certain spatial interval (see points A, B, C on the curve at $X = X_1$ or points D, E, F at $X = X_2$). Section 2.11 analyses what really happens at those subsequent times, using inferences from study of a simpler case in this section.

Similar difficulties are absent only for a piston whose x-component of acceleration is always $\leqslant 0$ (figure 33). Later signals are then slower than earlier ones so that no overtaking can occur. An interesting special case is *impulsive* motion away from the fluid, where the piston has zero acceleration except during an extremely brief interval when its velocity in the negative x-direction increases very fast indeed (figure 34). During that brief interval it sends out a fan of signals with rapidly decreasing velocities all from essentially the same point. This 'centred simple wave' is illustrated in figure 34 for air with $\gamma = 1.40$, the corresponding spatial waveforms being plotted in the manner of figure 30 at times o and t. The initially discontinuous waveform has become much less steep at time t: the rarefaction signal has then travelled to the right absolutely (plain line) although in a frame moving at the undisturbed sound speed (broken line) the shearing through a distance $\frac{1}{2}(\gamma + 1)ut$ has moved signals carrying negative values of u to the left.

A sharp contrast to the easing of steep parts of the waveform in expansive waves (figures 33 and 34) is the predicted steepening for compressive waves shown in figure 31, leading after a finite time to the impossible waveform of figure 32. In the special case directly analogous to figure 34 the impossible situation is predicted to occur immediately (figure 35): impulsive piston

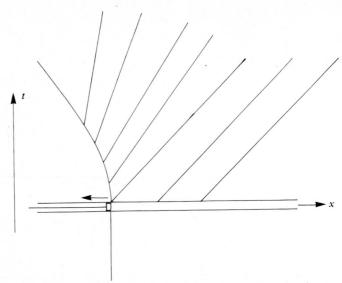

Figure 33. Simple wave generated by piston acceleration that is negative (*away* from the adjoining fluid).

motion into the fluid sends out, once more from essentially the same point, a fan of signals with rapidly *increasing* velocities. At once they run into each other; and at once the shearing produces an impossible waveform (figure 35); essentially because what immediately precedes that for general wave-forms, namely the infinite slope shown in figure 31, is generated *initially* by such an impulsive piston motion.

We now consider what features in real plane waves of sound prevent these impossible deformations of waveforms in general, and then analyse, taking those features into account, the real wave generated by *impulsive* motion of a piston into fluid. Note that the only features which can modify our conclusions on simple-wave propagation are dissipative processes, since the Riemann theory (section 2.8) underlying our conclusions is exact for waves subject only to nondissipative processes. Out of the various dis-sipative processes considered in sections 1.13 and 2.7, we must find therefore whether any can produce effects big enough and fast enough to annul a powerful and rapid tendency for transformations of waveform like those between figures 31 and 32.

At first sight this seems unlikely, as dissipative processes were found to produce in typical cases only gradual changes of waveform: by attenuation at a moderate rate due to wall friction (section 2.7) and at a usually far

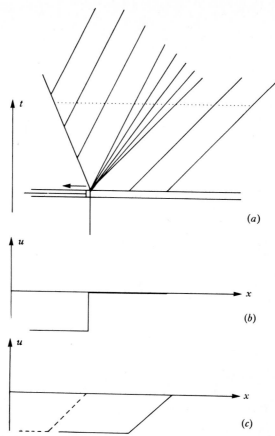

Figure 34. Centred simple wave generated by impulsive motion of a piston away from the adjoining fluid: (a) C_+ curves; (b) initial waveform; (c) waveform at the time t indicated by a dotted line. Broken line: waveform at time t expressed alternatively as a function of $X = x - c_0 t$.

slower rate due to the diffusivity of sound δ (section 1.13). We need to know, however, whether any such process can act powerfully and rapidly as soon as waveform slopes become extremely large. In that special situation, we find that wall friction continues to be too weak but, interestingly enough, diffusivity comes into its own and becomes exceedingly effective.

This is because the diffusivity of sound δ as defined in section 1.13 makes an addition

$$\delta \partial \rho / \partial t \tag{193}$$

to the longitudinal compressive stress p_{11} *above* the value which it would have in a nondissipative process. For normal rates of change of density

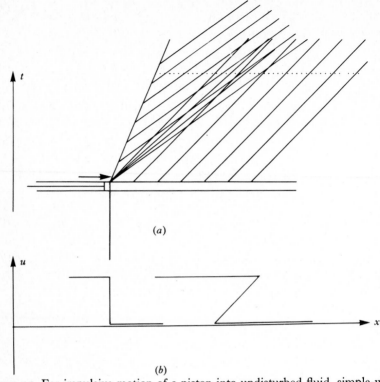

Figure 35. For impulsive motion of a piston into undisturbed fluid, simple-wave theory predicts immediate impossibilities: (*a*) intersections of C_+ curves; (*b*) formation of a three-valued waveform.

$\partial \rho / \partial t$, this extra compressive stress (193) is quite small and slow in its effect (section 1.13). By contrast, where the compressive part of a waveform is being *steepened*, so that the local rate of increase of density is tending to *infinity* as a certain instant is approached, the additional compressive stress (193) rises in proportion till it reaches such an enormous local value as to bring the whole steepening process to a halt: essentially, by retarding signals behind it and accelerating signals in front of it. It is then possible for an enormously steep compressive wave called a 'shock wave' to continue to propagate at a speed intermediate between the signal speeds behind it and in front of it by means of a balance between steepening and dissipation: the local steepness of the waveform adjusts itself so that the powerful and rapid tendency to further steepening is cancelled by the large effects of the local stresses (193).

In discussing this process here for one type of excitation and in section

2.11 for more general types, we do not emphasise questions of the exact value of δ. Actually, doubling the diffusivity halves the required steepness of waveform in the shock wave and so doubles its thickness; however, the shock wave remains so extremely thin compared with other characteristic dimensions as to behave with or without the doubling like essentially the same discontinuous change in density. Accordingly, any attempt to improve on the linearised approximation (193) to the added compressive stress, or to take into account effective frequency-dependence in the 'lag' contribution to δ (section 1.13), is here omitted.

The fact that energy dissipation by the mechanisms of section 1.13 can produce significant effects in changes so extremely abrupt that the extra compressive stress (193) is substantial is related to the fact that those mechanisms reduce the energy of a *sinusoidal* wave in a single period by a fraction proportional to $\omega\delta$, which can be substantial if $\omega\delta$ is large enough. High rates of change *reduce*, on the other hand, the effectiveness of the frictional dissipation mechanism of section 2.7, with a proportional energy loss per period decreasing like $\omega^{-\frac{1}{2}}$ as ω increases (equation (142)). Thus the convection-driven shock wave, though it can interact with a frictional boundary layer, is essentially a diffusion-resisted process.

Impulsive piston motion into the fluid, on which we now concentrate, produces *immediately* a discontinuous waveform (figure 35), which has signals behind it travelling faster than the undisturbed speed c_0 of the signals ahead of it, leading at once in the absence of dissipation to impossible consequences. We now analyse the possibility that the wave may almost instantaneously become stabilised in an almost discontinuous form: the 'shock wave'; that is, a transitional region with an extremely small thickness determined so that the extra compressive stress (193) prevents further steepening, by retarding signals behind it and accelerating signals ahead of it so that the shock wave itself propagates at some intermediate speed into the undisturbed fluid (figure 36).

We develop the theory in two forms: the first, which in studies of shock waves for practical purposes is quite sufficient, treats the shock wave as effectively a discontinuity (but a dissipative one) where the fluid velocity u changes instantaneously from zero to the constant value u_1 which it takes between the piston and the shock wave, and where we seek to determine how other variables such as pressure and density may change by using conservation of mass and other physical principles. The second form treats the actual shock-wave structure and thickness and

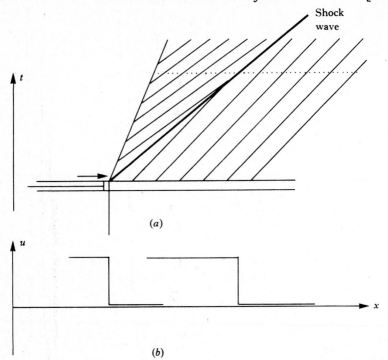

Figure 36. Generation of a shock wave by the impulsive motion of a piston into undisturbed fluid: (a) the extra compressive stress (193) in the almost discontinuous shock wave is large enough to slow down signals (propagated along C_+ curves) behind it and accelerate signals ahead of it, so that (b) the shock wave can propagate with an unchanging waveform, shown here both initially and at the later time indicated in (a) by the dotted line.

confirms that the first form gives a correct impression of the nature of the wave when it has propagated over distances much larger than that thickness.

The first treatment is of interest as showing how the momentum and mass-conservation principles, and associated thermodynamics, given quantitative expression for continuous motions in sections 1.1 and 1.2 respectively, can also be applied quite differently to analyse the possibility of *discontinuous* motions of fluid. Such analysis may determine, in a case like that of figure 36, the speed U of propagation of the discontinuous wave (which can as mentioned above be expected to *exceed* c_0) as well as the pressure p_1 and density ρ_1 in the region behind it, given that the fluid velocity u_1 in that region is equal to the piston velocity.

Thus, the mass of undisturbed fluid with density ρ_0 that is traversed at speed U by unit area of shock wave per unit time is

$$\rho_0 U. \tag{194}$$

Conservation of mass requires this fluid to emerge at the same rate behind the shock wave, at *relative* velocity $U - u_1$ and density ρ_1, giving

$$\rho_1(U - u_1) = \rho_0 U; \tag{195}$$

while its rate of acquisition of momentum, namely the mass rate (194) times the x-velocity u_1 it acquires, can be equated to the net force $p_1 - p_0$ in the x-direction on fluid being traversed by unit area of shock wave to give

$$\rho_0 U u_1 = p_1 - p_0. \tag{196}$$

It is of some interest to solve equations (195) and (196) for U and u_1 in terms of the pressures and densities only as

$$U^2 = \left(\frac{\rho_1}{\rho_0}\right)\frac{p_1 - p_0}{\rho_1 - \rho_0}, \quad u_1^2 = \frac{(p_1 - p_0)(\rho_1 - \rho_0)}{\rho_0 \rho_1}. \tag{197}$$

Here, the 'finite-difference' equation for the shock-wave speed U generalises to discontinuities of arbitrary amplitude the differential relationship $c^2 = \mathrm{d}p/\mathrm{d}\rho$ to which it reduces for small disturbances, while the equation for u_1 similarly generalises the equation $u = (p - p_0)/\rho_0 c$. In fact, the momentum and mass-conservation laws yield the same type of information as they do in linear theory (section 1.1) but no more, and thermodynamic considerations are needed as in section 1.2 to complete the analysis.

Since dissipative processes are important inside a shock wave, we cannot expect entropy to remain constant for fluid traversed by it. Entropy, by the second law of thermodynamics, must *increase* through dissipation of mechanical energy into heat. We can, however, make a budget of *total energy*, including internal energy of the fluid as well as kinetic energy. If the internal energy per unit mass is changed from E_0 to E_1 on being traversed by the shock wave, while the kinetic energy increases from 0 to $\frac{1}{2}u_1^2$, the rate of change of total energy of fluid being traversed by unit area of shock wave at the mass rate (194) is

$$\rho_0 U(\tfrac{1}{2}u_1^2 + E_1 - E_0) = p_1 u_1. \tag{198}$$

The right-hand side of (198) expresses the fact that such fluid is subjected by adjacent fluid to forces p_1 behind and p_0 in front (see (196)) of which, however, only the fluid behind is in motion (with velocity u_1) and therefore doing work (at the rate $p_1 u_1$).

Substitution in (198) of the expressions (197) for U and u_1 gives an

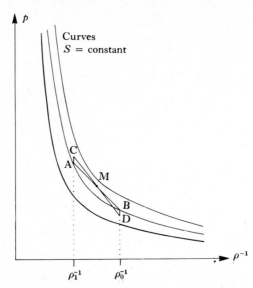

Figure 37. In a pressure–volume diagram, the increase of internal energy at a shock wave is the area under the straight line joining points representing conditions ahead of and behind it. Two points A and B on the same constant-entropy curve cannot satisfy this condition, but by rotating the line AB about its mid-point M we obtain points C and D which do.

important general law, due to Hugoniot, relating the internal energy change across a shock wave to the values of the pressure and density ahead of and behind it:

$$E_1 - E_0 = \tfrac{1}{2}(p_1 + p_0)(\rho_0^{-1} - \rho_1^{-1}). \tag{199}$$

This equation is best interpreted by means of a pressure–volume diagram (figure 37) on which graphs of pressure p against specific volume ρ^{-1} at constant entropy S are shown. It equates the internal energy change to the *area under the straight line* joining the points (ρ_0^{-1}, p_0) and (ρ_1^{-1}, p_1) representing the fluid states ahead of and behind the shock wave.

Note that these could not be points like A and B on the *same* constant-entropy curve because along such a curve $dE = -p\,d\rho^{-1}$ giving $E_1 - E_0$ as the area $\displaystyle\int_A^B p\,d\rho^{-1}$ under the *curve* between A and B. This characteristically lies below the straight line AB, which means that equation (199) cannot be satisfied: conservation of entropy across a shock wave is impossible.

In order to find points C and D that do satisfy equation (199), we begin to rotate AB clockwise about its mid-point M, which does not alter the right-hand side of (199) but rapidly increases the left-hand side as E_1

increases by the energy change at constant volume associated with the increase of entropy between A and C, while E_0 decreases by the amount associated with the decrease of entropy between B and D. Rotating in this way thus increases $E_1 - E_0$ steadily until the previous discrepancy (area between the curve AB and the straight line AB) is annulled.

This construction for obtaining solutions of (199) confirms that the entropy per unit mass S is greater on the high-pressure side of any shock wave than on the low-pressure side. In other words, discontinuous waves involving a *rise* in pressure are allowed by the second law of thermodynamics, and may be named 'shock waves'. By contrast, a discontinuous wave involving a fall in pressure, say from C to D, although equally satisfying the Hugoniot equation (199), could never occur because it would imply decreasing entropy. Its impossibility causes no embarrassment, however, since we know that only *compressive* waves (with p increasing) show any tendencies to steepening, and hence to shock-wave formation! Any momentary expansive discontinuity instantly eases into a continuous form (figure 34).

Actually the assumption used in proving this, namely that the signal speed $u + c$ in a simple wave *increases with* p, is not a necessary consequence of thermodynamic laws but seems to be generally true in fluids. Furthermore, it is *exactly* equivalent to the assumption used above in analysing figure 37: that the constant-entropy (ρ^{-1}, p) curve is concave upwards $(\mathrm{d}^2\rho^{-1}/\mathrm{d}p^2 > 0)$. This equivalence follows from the equations $\mathrm{d}u = \mathrm{d}p/\rho c$ and $\mathrm{d}p = c^2\mathrm{d}\rho$ in a simple wave which can be shown to imply that

$$\mathrm{d}(u+c)/\mathrm{d}p = \tfrac{1}{2}\rho^2 c^3 \mathrm{d}^2\rho^{-1}/\mathrm{d}p^2. \qquad (200)$$

It means that the fluids in which only compressive waves can steepen to form discontinuities *are* the fluids in which only compressive discontinuities can satisfy the second law of thermodynamics.

For the important special case of a perfect gas with constant specific heat c_v (and therefore with

$$\gamma = (R+c_v)/c_v \qquad (201)$$

constant) the Hugoniot equation (199) can be solved explicitly. In a perfect gas (section 1.2) the change $E_1 - E_0$ in internal energy depends only on the change in the temperature $p/R\rho$, giving

$$E_1 - E_0 = c_v[(p_1/R\rho_1)-(p_0/R\rho_0)] = (\gamma-1)^{-1}(p_1\rho_1^{-1}-p_0\rho_0^{-1}) \qquad (202)$$

for constant $c_v = (\gamma-1)^{-1}R$. In terms of the *shock-wave strength*, defined as the relative pressure rise

$$\beta = (p_1-p_0)/p_0, \qquad (203)$$

equations (199) and (202) may be solved for the density ratio ρ_1/ρ_0 to give

$$\rho_1/\rho_0 = [2\gamma + (\gamma + 1)\beta]/[2\gamma + (\gamma - 1)\beta], \tag{204}$$

a simple relationship between points like C and D in figure 37. Then equations (197) give

$$\frac{U}{c_0} = \left(1 + \frac{\gamma + 1}{2\gamma}\beta\right)^{\frac{1}{2}}, \quad \frac{u_1}{c_0} = \frac{\beta}{\gamma}\left(1 + \frac{\gamma + 1}{2\gamma}\beta\right)^{-\frac{1}{2}} \tag{205}$$

for the shock-wave speed U (which, as expected, comes out always *greater* than the undisturbed sound speed c_0) and the fluid velocity u_1 behind the shock wave. The sound speed c_1 behind the shock wave satisfies

$$\left(\frac{c_1}{c_0}\right)^2 = \frac{p_1/\rho_1}{p_0/\rho_0} = (1 + \beta)\frac{2\gamma + (\gamma - 1)\beta}{2\gamma + (\gamma + 1)\beta}, \quad \frac{u_1}{c_1} = \frac{\beta}{\gamma}\left(1 + \frac{\gamma - 1}{2\gamma}\beta\right)^{-\frac{1}{2}}(1 + \beta)^{-\frac{1}{2}}, \tag{206}$$

where the first expression is of interest also as the *temperature ratio* T_1/T_0 across the shock wave and the second may be described as the *local Mach number* of the fluid motion behind the shock wave.

We see that shock waves exist for any positive strength β and that all the above ratios are increasing functions of β. As $\beta \to \infty$ the density ratio ρ_1/ρ_0 and the Mach number u_1/c_1 tend to asymptotic values $(\gamma + 1)/(\gamma - 1)$ and $\{2/\gamma(\gamma - 1)\}^{\frac{1}{2}}$ respectively, while the other ratios increase without limit. All the ratios are plotted in figure 38 for atmospheric air with $\gamma = 1.40$; these plots are reasonable approximations provided that the temperatures T_0 and T_1 ahead of and behind the shock wave *both* lie in the region from 100 K to about 1000 K.

Figure 38 under these conditions, or the corresponding diagram constructed from Hugoniot's law (199) for other conditions or other fluids, gives an only slightly simplified specification of the entire fluid motion due to impulsive motion of a piston with velocity u_1 into undisturbed fluid in the state specified by subscript zero. It allows the shock-wave strength $\beta = (p_1 - p_0)/p_0$ to be immediately read off for the given value of u_1/c_0, after which the shock-wave speed U and the values of density and temperature between piston and shock wave can be immediately inferred. Figure 38 shows, in particular, that impulsive piston motions at speeds comparable to c_0 or more are resisted by extremely large excess pressures and generate very high air temperatures.

The main over-simplification in this picture is the lack of any estimate of shock-wave thickness. We conclude this section with an account of how the thickness of, at least, a *weak* shock wave can be estimated.

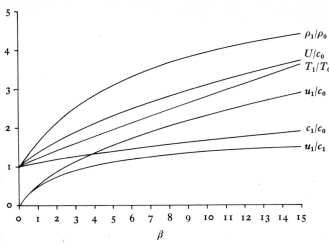

Figure 38. For a shock wave of strength β moving into a previously undisturbed perfect gas with specific heats in the ratio 1.40, nondimensional forms of its speed U and of the changes across it (of fluid velocity from o to u_1, of density from ρ_0 to ρ_1, of temperature from T_0 to T_1 and of sound speed from c_0 to c_1) are plotted for o $< \beta <$ 15.

An important clue is given by the Hugoniot law (199) that emphasises the straight line joining the initial and final states in the pressure–volume diagram (figure 37). Further consideration shows that line to have still greater significance: as a graph of longitudinal compressive stress p_{11} against specific volume ρ^{-1} through the thickness of the shock wave. To see this we consider that moving plane within the shock wave on which the density continues to take a particular value ρ and let the longitudinal stress on that plane be p_{11} and the fluid velocity u. Then the mass-conservation and momentum arguments for fluid traversed at speed U by the shock wave lead to the same equations (195) and (196) as before for the fluid emerging through that plane if ρ, p_{11} and u replace ρ_1, p_1 and u_1, and as before they give the first of equations (197) which in the form

$$p_{11}-p_0 = \rho_0^2 \, U^2(\rho_0^{-1}-\rho^{-1}) \qquad (207)$$

is the stated linear relation between p_{11} and ρ^{-1} inside the shock wave.

The longitudinal stress in a shock wave, then, follows a straight line when plotted against ρ^{-1}, although (figure 37) the similar plots of pressure against ρ^{-1} in constant-entropy processes are all curved upwards. Hence the *additional* longitudinal stress p_{11}, over and above its value in such a process, given in (193) as $\delta\partial\rho/\partial t$, rises to a peak in the middle of a shock wave and

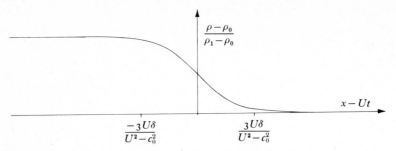

Figure 39. The density distribution (209) within a weak shock wave: note that 90 % of the density rise occurs between the points marked, requiring for given x only the extremely short density rise time (210).

falls away at the sides. This determines the *rate* at which the density must complete its change across the shock wave.

We give the details for the case of small relative change in density across the shock wave, when the small-disturbance arguments (section 1.13) leading to (193) should be most reliable and when we can to good approximation take the straight line (207) in figure 37 as AB: essentially, because the entropy change at the shock wave (that alters it to CD) is proportional to the area between AB and the curve which becomes small of order the *cube* of the overall density change as A approaches B whereas the additional stress (193), or height of AB above the curve, remains bigger (of order the *square*). The equation for ρ which thus retains only the squares of disturbances is

$$\delta \partial \rho / \partial t = (U^2 - c_0^2)(\rho_1 - \rho)(\rho - \rho_0)/(\rho_1 - \rho_0), \tag{208}$$

where the right-hand side is the distance between the straight line (207) and a parabolic approximation to that constant-entropy curve which intersects it at $\rho = \rho_0$ and $\rho = \rho_1$ and has $dp/d\rho = c_0^2$ at $\rho = \rho_0$. The solution of equation (208) for a shock wave centred on $x = Ut$, namely

$$\rho - \rho_0 = (\rho_1 - \rho_0)\{1 + \exp[(U^2 - c_0^2)(x - Ut)/U\delta]\}^{-1}, \tag{209}$$

is plotted in figure 39.

This result demonstrates the extraordinary rapidity of changes in a shock wave: the density rise time, defined as the time within which the central 90 % of the density rise occurs, is

$$6\delta/(U^2 - c_0^2), \tag{210}$$

which under atmospheric conditions is of order 10^{-9} seconds divided by the relative density rise $(\rho_1 - \rho_0)/\rho_0$. Normally this process is so rapid that only the viscous and heat-conduction contributions δ_v and δ_c can make their full contribution to δ (section 1.13). More important, the density rise is completed in so short a time, and over so short a distance (the wave speed times the rise time), that treatment of a shock wave as essentially a discontinuity satisfying equations (203)–(206) is valid for all practical purposes.

2.11 Theory of simple waves incorporating weak shock waves

The view of a shock wave as essentially a discontinuity, that propagates into still air at a wave speed U rather greater than the undisturbed sound speed c_0, has been established in section 2.10 for the very special case of a shock wave reaching its full strength *immediately* as a piston's impulsive motion generates a discontinuous waveform: the shock wave then remains essentially discontinuous through the balance between convection and diffusion, and retains its strength as long as the piston continues to move at the velocity u_1 compatible with uniformity of conditions behind the shock wave. We now investigate the possibility of analysing a much more general type of propagation, in which distortion of a simple wave (section 2.9) proceeds *continuously* up to the moment at which the waveform slope is tending to infinity, when diffusion can become effective and help make a shock wave: essentially a *discontinuity* again, but one whose strength can grow to a maximum as waveform shearing proceeds further, and subsequently *decay* after the piston ceases to do work on the fluid.

For this analysis we need to be able to infer from the relevant physical laws motions of fluid incorporating regions with continuously varying conditions *separated* by a moving discontinuity. Fortunately, in such fluid motions the relationships satisfied at the discontinuity remain essentially those derived in section 2.10: even though the shock-wave strength and speed may vary with time, the laws governing rates of change of mass, momentum and energy for the fluid being traversed by unit area of shock wave at any one instant still take the form of equations (195), (196) and (198) provided that the subscripts o and 1 in these equations are replaced by *subscripts a and b* denoting conditions immediately *ahead of and behind* the shock wave, and provided that the velocities U and u_b of the shock wave and the fluid immediately behind it are taken to be velocities in that frame of reference in which the fluid velocity u_a immediately ahead of the shock wave is zero; thus, U and u_b are velocities *relative* to the fluid motion immediately ahead of the shock wave. From the resulting equations we can again derive the Hugoniot law (199), which after these modifications becomes

$$E_b - E_a = \tfrac{1}{2}(p_b + p_a)(\rho_a^{-1} - \rho_b^{-1}). \tag{211}$$

It might be thought that the relationships governing the continuously varying part of the fluid motion must similarly take the form of the simple-wave equations derived in sections 2.8 and 2.9. Such a statement cannot be exactly true, however, in the region behind the shock wave, because the work of those sections is all founded on the assumption that the entropy per unit mass S remains uniform and constant. This is a good assumption in fluid where there has been no dissipation (for example, ahead of the shock wave) but equation (211) implies that all fluid traversed by the shock wave suffers an increase in S whose magnitude varies with the shock-wave strength $\beta = (p_b - p_a)/p_a$. Accordingly, a shock wave of changing strength β must leave behind it a fluid motion with nonuniform entropy S, where simple-wave relationships cannot hold exactly.

Fortunately there is a large class of important wave motions incorporating shock waves where those relationships do nevertheless hold to extremely close approximation: namely, those where any shock wave remains relatively *weak*. We noted at the end of section 2.10 that the Hugoniot law makes the entropy change at the shock wave small of order β^3 when the shock-wave strength β is small: essentially, because the area in figure 37 between the straight line AB and the curve becomes small of order the cube of the distance AB as that decreases. A *weak* shock wave, therefore, may involve negligibly small entropy change.

For example, in a perfect gas with constant specific heat c_v the entropy (section 1.2) satisfies

$$T\mathrm{d}S = \mathrm{d}E + p\mathrm{d}\rho^{-1} = c_v\mathrm{d}T + p\mathrm{d}\rho^{-1}, \quad \text{where} \quad T = p/R\rho; \tag{212}$$

equations which with (201) imply that

$$S = c_v \ln(p/\rho^\gamma) + \text{constant}, \tag{213}$$

in agreement with equation (176) which makes p/ρ^γ constant when S is constant. The entropy change $S_b - S_a$ at a shock wave can therefore be written, using (204) for the density ratio ρ_b/ρ_a, as

$$(S_b - S_a)/c_v = \ln(1 + \beta) - \gamma \ln\{[2\gamma + (\gamma + 1)\beta]/[2\gamma + (\gamma - 1)\beta]\}. \tag{214}$$

Its expected proportionality to β^3 as $\beta \to 0$ becomes obvious from the expression

$$(\gamma^2 - 1)\beta^2(1 + \beta)^{-1}[2\gamma + (\gamma + 1)\beta]^{-1}[2\gamma + (\gamma - 1)\beta]^{-1} \tag{215}$$

for the derivative of the right-hand side of (214) with respect to β, which evidently behaves like $(\gamma^2 - 1)\beta^2/4\gamma^2$ as $\beta \to 0$ although growing more slowly than that for larger β. We infer that the entropy change itself satisfies

$$(S_b - S_a)/c_v \sim (\gamma^2 - 1)\beta^3/12\gamma^2 \quad \text{as} \quad \beta \to 0, \tag{216}$$

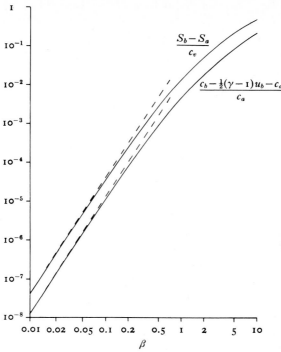

Figure 40. The nondimensional forms (214) and (217) of the changes in entropy S and in the quantity $c - \frac{1}{2}(\gamma - 1) u$ at a shock wave in a perfect gas with $\gamma = 1.40$, logarithmically plotted against its strength β. Broken lines: limiting forms as $\beta \to 0$.

in proportion to the cube of the shock-wave strength, while growing more slowly than this for larger β.

Figure 40 confirms this behaviour for atmospheric air with $\gamma = 1.40$, by plotting both the exact expression (214) for $(S_b - S_a)/c_v$ as a function of β and also the limiting form (216) which becomes $0.041\beta^3$ as $\beta \to 0$. The agreement is close for small β but when $\beta = 0.5$ the value of $(S_b - S_a)/c_v$ has increased only to 0.0027; about half the value indicated by the limiting form. This means, by equation (213), that the error in taking the pressure for given density as given by the constant-entropy equation (176) would still be only 0.27% for this shock-wave strength $\beta = 0.5$, although becoming far greater for larger strengths. We conclude that for flows incorporating shock waves that are 'weak' in the sense that their strength β is less than about 0.5, the errors involved in assuming constant entropy throughout the flow may be extremely small, even though the small entropy change plotted in figure 40 results from processes of energy dissipation within the shock wave that are essential (section 2.10) to its very existence.

A plane wave entering still air, therefore, as long as any shock waves in it have strengths less than about 0.5, should satisfy to close approximation in every *continuous* part of the wave the relationships that Riemann derived on the assumption of constant entropy: namely, the simple-wave relationships including equation (178). We can make an independent check on this conclusion since it implies that to close approximation the quantity $c - \frac{1}{2}(\gamma - 1)u$ should have the same value c_0 on both sides of such a shock wave. Equations (205) and (206), on the other hand, specify exactly the relative change in this quantity across a shock wave as

$$\frac{c_b - \frac{1}{2}(\gamma - 1)\, u_b - c_a}{c_a} = \left(1 + \frac{\gamma + 1}{2\gamma}\, \beta\right)^{-\frac{1}{2}} \left[(1 + \beta)^{\frac{1}{2}} \left(1 + \frac{\gamma - 1}{2\gamma}\, \beta\right)^{\frac{1}{2}} - \frac{\gamma - 1}{2\gamma}\, \beta\right] - 1$$

$$(217)$$

in a frame of reference in which the velocity u_a immediately ahead of the shock wave is zero. Figure 40 shows how exceedingly small this is (less than 0.001) for $\beta < 0.5$.

This confirms the conclusion which so facilitates the study of simple waves: that, even after the appearance of a discontinuity in the form of a weak shock wave, all continuous parts of the wave satisfy the same propagation laws as were established in section 2.9. Application of those laws to calculate waveform distortion beyond the moment at which they predict an infinite slope is now seen to be valuable because even though the whole waveform so calculated is of the type described (figure 32) as 'impossible', *each continuous portion* of the real waveform must be a portion of the waveform so calculated.

That real waveform as a whole, therefore, must incorporate a discontinuity somewhere in the spatial interval where the calculated simple wave is three-valued: in figure 32, for example, the discontinuity might be from A to C, separating continuous portions to the right of A and to the left of C. Alternatively, it might be from D to F, separating continuous portions to the right of D and to the left of F. Experimental evidence, that steepening does culminate in a waveform incorporating some such discontinuity, is shown in figure 41.

The choice between different possible positions of the discontinuity within this interval at each instant during propagation of the wave can be made on the basis of the shock-wave equations that we have derived by applying physical laws of mass-conservation, momentum and energy to the

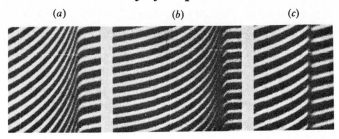

Figure 41. Formation of a shock wave in air. Interferogram of waveform at successive instants (*a*), (*b*) and (*c*). Each dark line is essentially a graph of minus the density (on a suitable scale) against distance. In (*a*) they have moderate slope, in (*b*) they have become steeper, and in (*c*) discontinuous.

fluid being traversed by the discontinuity. A natural but tedious procedure for this, formerly thought to be the only one possible, is to use an expression such as

$$(\rho_b/\rho_a)^{\frac{1}{2}}(p_b-p_a)^{\frac{1}{2}}(\rho_b-\rho_a)^{-\frac{1}{2}} \tag{218}$$

for the shock-wave velocity relative to the fluid immediately ahead of it in order to determine the shock wave's displacement between successive instants and so compute its position by small steps forward in time from when it first appeared. A much more convenient procedure discovered later makes use of the mass-conservation principle directly to determine the position of the shock wave at any one instant independently of any calculations of its earlier positions.

This improved procedure starts from the knowledge that any simple-wave solution, including even an 'impossible' one like that of figure 32, is an exact solution to equations of motion that include the mass-conservation equation or equation of continuity. In other words, it satisfies a principle of conservation of the total mass of fluid per unit cross-sectional area $\int \rho dx$. The density ρ in a simple wave is uniquely related to the fluid velocity u or to the excess signal speed v (section 2.9) and so, when any waveform like that of figure 32 has been determined from the rules of distortion by simple shearing explained in figure 31, it is straightforward to derive the associated distribution of ρ as a function of x.

Figure 42 shows such a density distribution derived from the 'impossible' waveform of figure 32 and yet necessarily satisfying the conditions that the integral $\int \rho dx$ (area between the curve and the x-axis) accounts for an unchanging total mass of fluid per unit cross-sectional area. On the other hand, the *real* waveform incorporating the discontinuity must satisfy the same mass-conservation conditions. This fixes the position of the dis-

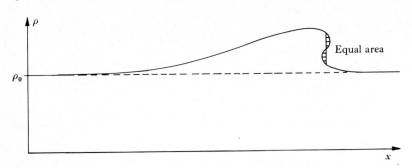

Figure 42. Because conservation of mass must be satisfied both by the density distribution derived from simple-wave theory and by the true density distribution incorporating a shock wave, that discontinuity must be placed where it does not alter the area between the graph of ρ and the x-axis.

continuity conveniently and uniquely as that which makes no change to the integral $\int \rho \mathrm{d}x$. Figure 42 shows how easy it is to draw in by eye the vertical chord which leaves unchanged the area between the curve and the x-axis because it cuts off *lobes of equal area* on the two sides of the line between it and the curve.

Additional simplification in the practical use of this procedure derives from the fact that the limits of density variation within a weak shock wave are such that the simple-wave relationship between the density ρ and the excess signal speed v is to sufficient approximation linear between those limits. This implies that the construction carried out on the graph of ρ against x (figure 42) can be applied directly to the graph of v against X (figure 32) with, to close approximation, an identical result (leading, actually, to the choice of DEF as the correct position of the discontinuity in figure 32); since a *linear* transformation of the ordinate (from ρ to v) cannot alter the equality of area to the two lobes in figure 42.

Although a full analysis of why this further approximation degrades only slightly the theory's accuracy is omitted, we may note here that a weak shock wave even at the upper limit of strength 0.5 induces in atmospheric air a change of $0.359 c_a$ in excess signal speed v and that the density ratio takes a value $\rho_b/\rho_a = 1.333$, only 2.6 % greater than is given by assuming a linear dependence on v; this is a maximum percentage error for $\beta \leqslant 0.5$. Also, the idea that as a result the construction involving lobes of equal area may be applied directly to the 'impossible' waveform of v against X can be checked in the special case (figure 35) of *impulsive* motion of a piston into fluid. Evidently, it places the discontinuity at exactly the centre of the 'Z'

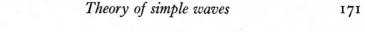
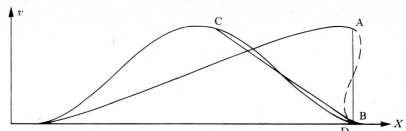

Figure 43. General construction for a relatively weak shock wave AB formed in the course of waveform distortion (Whitham's law). Broken lines: discarded equal-area portions of the waveform after shearing for a time t. During this time the line CD of reciprocal slope $-t$, cutting off lobes of equal area on the initial waveform, turns into the vertical line AB.

shape in figure 35, leading to the conclusion that the excess shock-wave speed $U - c_0$ is one-half of the excess signal speed $u_1 + c_1 - c_0 = v_1$ behind the shock wave. Actual values of $(U - c_0)/(u_1 + c_1 - c_0)$ calculated from equation (205) and (206) lie between 0.500 and 0.543 for $0 < \beta < 0.5$, confirming the conclusions of 'weak-shock-wave' theory to reasonable approximation in this range of strengths.

A very straightforward procedure for determining waveform distortion in simple waves incorporating weak shock waves results from the above considerations: with the waveform regarded as the graph of excess signal speed v against $X = x - c_0 t$, we can describe distortion as taking place (figure 31) at unit rate of shear for all time provided that a vertical discontinuity cutting off lobes of equal area on each side of it appears wherever necessary to keep the waveform one-valued. Figure 43 shows the distortion during time t calculated in this way: as in figure 31 each value of v is found at a point displaced a distance vt to the right, but those lobes (of equal area) which are discarded in favour of the discontinuity AB are shown as broken lines.

It is instructive to ask what line CD drawn on the initial waveform would be distorted (at unit rate of shear) into this vertical straight line AB after time t. The answer is in two parts: CD must be a line of reciprocal slope $-t$ because unit shear, which increases the reciprocal slope of any line at unit rate, must raise it to zero after time t; and the areas of lobes cut off between CD and the curve on both sides of CD must be equal since both areas are unchanged by the shearing process which distorts them into the equal lobes between AB and the distorted waveform. The existence of such a chord CD on the initial waveform satisfying these two conditions (having reciprocal slope $-t$ and cutting off lobes of equal area) is a guarantee that after time t

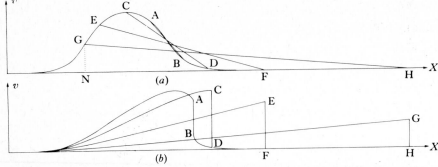

Figure 44. Whitham's law applied to a waveform consisting of a single compression pulse: on the initial waveform (*a*), the four lines of different reciprocal slope $-t$ cutting off lobes of equal area specify the discontinuous changes in v across the shock wave at those four different times t, when the distorted waveforms are as in (*b*).

the waveform will include a shock wave where v increases discontinuity from its value at D to its value at C.

This yields Whitham's law: that a certain simple construction carried out on an initial waveform determines the whole future history of the formation, growth and decay of a weak shock within it as the wave progresses. We simply draw, for each of a succession of values of t greater than the shock-wave formation time (188), a chord of reciprocal slope $-t$ cutting off lobes of equal area on both sides between it and the curve, in order to indicate the successive times t at which the shock wave involves discontinuous changes between pairs of values of v given by the end-points of those successive chords.

Figure 44(*a*) shows such a construction carried out on an initial waveform that consists of a single compression pulse (where both excess pressure and excess signal speed rise to a positive peak and then fall back to zero), such as is generated by a piston moving into the fluid a certain distance and then coming to rest again. Each of the diagonal lines of negative slope, cutting off lobes of equal area on both sides of it, indicates what discontinuous increase in v is to be found after a time equal to minus its reciprocal slope. Note how the shock wave, although formed with zero strength at the time t_0 when the pulse has the form with infinite slope shown in figure 31, soon becomes a substantial discontinuity where v changes between its values at A and at B. It reaches maximum strength (involving a change from C to D) at a later time equal to minus the reciprocal slope of CD, and then begins to decay to a weaker discontinuity EF, while an asymptotic behaviour after a large time t is given by the chord GH with very small reciprocal slope. For these four chords the corresponding waveforms are shown in figure 44(*b*): note how the shock wave initially formed inside the compression pulse begins to move to the front, through its wave speed exceeding the signal speed ahead of it, until it becomes a *head shock wave* EF moving into undisturbed fluid ahead of the rest of the pulse.

The asymptotic behaviour of this head shock wave is given by equating in figure 44(*a*) the area under the *curve* GH (which asymptotically is the total area Q under the initial waveform) to the area under the *chord* GH which is exactly

$$\tfrac{1}{2}(\text{NH})(\text{NG}) = \tfrac{1}{2}t(\text{NG})^2 \tag{219}$$

since the reciprocal slope of GH is $-t$. The excess signal speed $v_1 = \text{NG}$ behind the head shock wave therefore satisfies

$$v_1 \sim (2Q/t)^{\frac{1}{2}}: \tag{220}$$

an asymptotic *inverse-square-root law* of decay of shock-wave strength with time. The asymptotic shape of the wave is a shallow *triangle* (figure 44(*b*)) of constant area Q as required by mass-conservation, the *pulse length* NH being asymptotically

$$(2Qt)^{\frac{1}{2}}. \tag{221}$$

Although mass is thus conserved, the total wave energy in the pulse is proportional to the length (221) times the *square* of the amplitude (220) and so tends to zero in proportion to $Q^{\frac{3}{2}}t^{-\frac{1}{2}}$, its rate of decrease with time being therefore proportional to $(Q/t)^{\frac{3}{2}}$. Of course all dissipation of mechanical energy into heat takes place *inside* the shock wave, at a rate specified by the entropy rise, proportional to shock-wave strength cubed. This by (220) is proportional to $(Q/t)^{\frac{3}{2}}$, and detailed evaluation of coefficients (actually carried out in equations (263)–(269) below for a considerably more general case) shows a perfect balance between rate of change of total wave energy in the pulse and rate of dissipation within the shock wave in this limit as $t \to \infty$, as well as at earlier times.

Study of the many interesting cases of waveform distortion that involve the appearance of more than one shock wave is here limited to just one case: an initial waveform consisting of a compression followed by an expansion as in figure 45. This involves formation of two shock waves at positions and times corresponding to the two inflexion points with negative waveform slope. Whitham's law shows that both shock waves grow to a maximum strength and then start to decay; they become a head shock wave just as in figure 44 and a *tail shock wave* raising the reduced pressure of the air at the

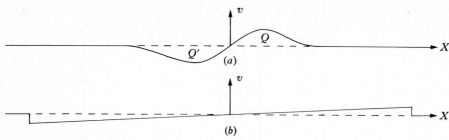

Figure 45. For an initial waveform (*a*) consisting of a compression followed by an expansion, the asymptotic waveform as $t \to \infty$ is an N-wave (*b*).

tail of the pulse up to the undisturbed value. The tail shock wave, at a wave speed intermediate between the undisturbed value c_0 behind it and the reduced signal speed ahead of it, moves gradually *backwards* in the frame of reference specified by the coordinate $X = x - c_0 t$. The asymptotic behaviour is a so-called N-wave, incorporating a triangular pulse satisfying (220) and (221) as in figure 44, where Q is the area of the part of the initial waveform above the X-axis, followed by an *inverted* triangle of length $(2Q't)^{\frac{1}{2}}$ (where Q' is the area of the part of the initial waveform below the X-axis) terminating in a tail shock wave where v changes by an amount $(2Q'/t)^{\frac{1}{2}}$.

We see that nonlinear effects can bring about drastic transformations of acoustic waveforms, in which compressive portions change into discontinuities while the slope of expansive portions becomes asymptotically very small like $1/t$. These transformations obliterate much of the information concerning an initial waveform: in the example just given, the asymptotic behaviour depended *only* on the areas Q and Q' of its positive and negative portions.

While these remarkable features have been established for the temporal distortion of a spatial waveform, we conclude this section by noting that a practically identical theory exists for the way in which a *temporal* waveform varies with *position x*. Defining waveform in this case (see (189) and (190)) as the dependence of the defect in reciprocal signal velocity, v_1, on a *modified time-coordinate*, which we write as X_1 while replacing x by t_1, we can specify propagation of all continuous parts of the wave by the rule (equation (191)) that each signal carrying a particular value of v_1 moves along a path satisfying $dX_1/dt_1 = v_1$. Again, insertion of any discontinuity must comply with mass-conservation, which for temporal waveforms requires conservation of the total *mass flux* $\int \rho u \, dt$ across different positions x. For weak shock waves we can show as above that the relationship between the local mass flux ρu and the defect in reciprocal signal velocity v_1 is to sufficient approximation linear. Hence any discontinuity can be inserted by means of a procedure involving conservation of the waveform area $\int v_1 \, dX_1$ and the whole theory of this section can be used to analyse the spatial distortion of temporal waveforms provided that v, X and t are replaced by v_1, X_1 and t_1 throughout.

2.12   Hydraulic jumps

The theory of simple waves, developed in section 2.9 for *general* longitudinal propagation in uniform tubes or channels, leads to an enigma that has been resolved in sections 2.10 and 2.11 for only the special case of plane waves of sound (through the theory of the shock wave). Although in an introductory account of nonlinear wave theories we do not wish to analyse exhaustively the possibilities for discontinuous wave propagation, we now indicate the form it takes in another special case: that of long waves in open channels of constant breadth, for which explicit relations between the variables in a simple wave are given in equations (181)–(183). Various special properties, and certain common features, that characterise discontinuous waves where the cross-sectional area of fluid may change, are exemplified here, then, by the theory of the 'hydraulic jump': a phenomenon of common occurrence in natural streams as well as of practical significance in hydraulic engineering projects.

We survey this theory in an order rather different from that of section 2.10: before discussing mechanisms that may act to resist further distortion of a waveform wherever something close to a discontinuity has appeared, we enumerate the conditions that a longitudinal motion in a channel of constant breadth must satisfy at any discontinuous wave. As in section 2.10 we consider in the first place a discontinuous wave propagating into undisturbed fluid, such as might be generated by a piston moving into it impulsively with velocity u_1, although we anticipate from the discussion in section 2.11 that the same equations, properly interpreted, will be applicable to discontinuities appearing within continuous wave motions.

A discontinuous wave moving at speed U into undisturbed water of cross-sectional area A_0 and density ρ_0 traverses a mass of fluid

$$\rho_0 A_0 U \tag{222}$$

per unit time. Conservation of mass requires this fluid to emerge at the same rate behind the discontinuous wave, at *relative* velocity $U - u_1$ and with a new cross-sectional area A_1, although with unchanged density ρ_0 since in

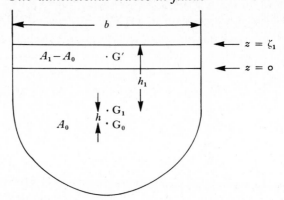

Figure 46. Ahead of and behind a hydraulic jump, moving along a channel of constant breadth b, the water cross-sections are (i) A_0, bounded by the undisturbed water level $z = 0$, and having centroid G_0; (ii) A_1, bounded by the elevated water level $z = \zeta_1$ and having centroid G_1. The difference in area, $A_1 - A_0$, is equal to $\zeta_1 b$ and has centroid G' at $z = \frac{1}{2}\zeta_1$.

open-channel flows (section 2.2) compressibility is negligible compared with distensibility. This gives

$$\rho_0 A_1(U - u_1) = \rho_0 A_0 U, \tag{223}$$

a close parallel to the shock-wave equation (195). The form of the momentum equation, however, is more complicated than (196), at least on the right-hand side, because in open channels the pressures ahead of and behind the discontinuity act on different cross-sectional *areas* as well as varying hydrostatically over those areas.

Immediately behind the discontinuity the excess pressure (15) takes the value $\rho_0 g \zeta_1$, where ζ_1 is the altitude of the free surface above its undisturbed level. The net force in the x-direction on fluid being traversed by the discontinuity can be divided into two parts (figure 46): (i) a force $\rho_0 g \zeta_1 A_0$ on fluid *below* the undisturbed water level due to such an excess pressure acting over the cross-sectional area A_0 of that fluid; (ii) a force

$$\int_0^{\zeta_1} \rho_0 g(\zeta_1 - z)\, b\, dz = \tfrac{1}{2}\rho_0 g \zeta_1^2 b \tag{224}$$

on fluid *above* the undisturbed water level, due to the integrated action, over an area $b\, dz$ at each altitude z above that level, of a pressure difference $\rho_0 g(\zeta_1 - z)$ between the fluid *behind* the discontinuity and the fluid subjected to opposing atmospheric pressure *within* the discontinuity. Equating this combined force, after a substitution

$$\zeta_1 b = A_1 - A_0 \tag{225}$$

as in (16), to the rate of acquisition of fluid momentum (namely, the mass rate (222) times the x-velocity u_1 which the fluid acquires) gives

$$\rho_0 A_0 U u_1 = \tfrac{1}{2}(A_0 + A_1)\rho_0 g \zeta_1 = \tfrac{1}{2}(A_0 + A_1)(\rho_0 g/b)(A_1 - A_0). \quad (226)$$

Equations (223) and (226) are readily solved for U and u_1 in terms of only A_0 and A_1 (apart from the constants g and b) as

$$U^2 = \frac{gA_0}{b}\frac{A_1(A_1+A_0)}{2A_0^2}, \quad u_1^2 = \frac{g(A_1-A_0)^2}{bA_0}\frac{A_1+A_0}{2A_1}. \quad (227)$$

On the right-hand side of each of these equations the *first* factor is the value given by linear theory for a wave of small amplitude moving into the undisturbed fluid; we may note, as expected, that the speed U of a discontinuous wave exceeds the speed $(gA_0/b)^{\frac{1}{2}}$ of signals of small amplitude by a correction factor which increases with the area-ratio A_1/A_0.

This dependence of equations (227) on only a single ratio A_1/A_0 is in striking contrast with the corresponding shock-wave equations (197) which depend upon two ratios, the pressure-ratio p_1/p_0 and the density-ratio ρ_1/ρ_0, whose interrelationship is to be determined by the Hugoniot law derived from an *energy* equation. For long waves in open channels, one and the same ratio A_1/A_0 determines *both* the rate of generation of momentum (like p_1/p_0) *and* the mass-flow balance (like ρ_1/ρ_0). Any information that we might get later about dissipation of mechanical energy into heat at the discontinuity would be irrelevant to the determination of U and u_1 in terms of A_1/A_0 by mass and momentum considerations, which are quite uninfluenced by the resulting negligibly small temperature rise.

A discontinuous wave moving along an open channel in accordance with equations (227) is commonly called a 'hydraulic jump' of strength

$$\beta = (A_1 - A_0)/A_0,$$

in terms of which those equations become

$$U/c_0 = (1 + \tfrac{1}{2}\beta)^{\frac{1}{2}}(1+\beta)^{\frac{1}{2}}, \quad u_1/c_0 = \beta(1+\tfrac{1}{2}\beta)^{\frac{1}{2}}(1+\beta)^{-\frac{1}{2}}, \quad (228)$$

where c_0 is the undisturbed wave speed $(gA_0/b)^{\frac{1}{2}}$. These ratios, together with

$$c_1/c_0 = (1+\beta)^{\frac{1}{2}}, \quad (229)$$

are plotted as functions of β in figure 47. Note that a hydraulic jump of strength exceeding 2.21 accelerates the water to a 'supercritical' velocity u_1 (one greater than the local speed c_1 of disturbances of small amplitude). Figure 47 enables the strength β of the jump generated by any impulsive

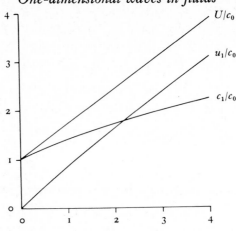

Figure 47. For a hydraulic jump of strength β moving into an undisturbed water channel of constant breadth, nondimensional forms of its speed U and of the changes across it (of fluid velocity from o to u_1 and of wave speed from c_0 to c_1) are plotted for o < β < 4.

piston motion to be inferred from the ratio u_1/c_0 of piston velocity to the wave speed in the undisturbed fluid.

We now calculate the rate at which the mass (222) of fluid traversed by the hydraulic jump in unit time loses mechanical energy as

$$F_1 u_1 - \rho_0 A_0 U(\tfrac{1}{2} u_1^2 + gh). \tag{230}$$

Here $F_1 u_1$ is the rate at which work is done on that fluid by the fluid behind, pushing it at velocity u_1, and $\tfrac{1}{2} u_1^2 + gh$ is the sum of the kinetic energy and extra potential energy which it acquires per unit mass (figure 46). These statements require that h is the vertical displacement of the centroid of the fluid cross-section at the jump, and that F_1 is the integral over the area A_1 of the quantity $\rho g(\zeta_1 - z)$ by which the water pressure exceeds the atmospheric pressure: by considering only rate of working by pressure *differences* from atmospheric pressure we avoid the need to subtract any rate of working against the atmosphere's pressure by rising fluid.

Evidently this integrated pressure difference F_1 over the area A_1 takes the form

$$A_1 \rho_0 g h_1, \tag{231}$$

where h_1 is the *average* depth of a point in A_1 below the free surface; that is, h_1 is the depth of the centroid G_1 of the area A_1. Now since the area A_1 includes (figure 46) a portion possessing area A_0 with its centroid G_0 a distance h below G_1, and a portion of area $(A_1 - A_0)$ with its centroid

G' at a depth $\frac{1}{2}\zeta_1$ below the surface, we can write the difference $h_1 - \frac{1}{2}\zeta_1$ as the distance $A_0 h/(A_1 - A_0)$ of G_1 below G', giving

$$F_1 = A_1\rho_0 g[\tfrac{1}{2}\zeta_1 + A_0 h/(A_1 - A_0)]. \tag{232}$$

As a check on this important formula for integrated pressure force behind the jump, we may write down the integrated pressure force over the cross-section A_0 ahead of the jump by a similar argument as

$$F_0 = A_0\rho_0 gh_0 = A_0\rho_0 g[-\tfrac{1}{2}\zeta_1 + A_1 h/(A_1 - A_0)], \tag{233}$$

where h_0 is the depth of the centroid G_0 of the area A_0 below *its* free surface so that $h_0 + \frac{1}{2}\zeta_1$ is the depth of G_0 below G'; and then note that the difference

$$F_1 - F_0 = \tfrac{1}{2}(A_1 + A_0)\rho_0 g\zeta_1 \tag{234}$$

agrees with the net force calculated earlier by a different method as the right-hand side of (226).

The rate of loss of mechanical energy at the jump (230) is the amount by which the rate of working on the fluid at the jump exceeds its rate of gain of kinetic and potential energy. The rate of working $F_1 u_1$, with F_1 given by (232) and u_1 and ζ_1 substituted from (223) and (225) as $U(A_1 - A_0)/A_1$ and $(A_1 - A_0)/b$ respectively, is

$$F_1 u_1 = \tfrac{1}{2}\rho_0 Ug(A_1 - A_0)^2/b + \rho_0 A_0 Ugh. \tag{235}$$

Hence when we calculate the difference (230) the two terms $\rho_0 A_0 Ugh$ cancel, but the other terms, with u_1^2 substituted from (227), yield a residual rate of loss of mechanical energy

$$\rho_0 Ug(A_1 - A_0)^3/4bA_1. \tag{236}$$

A somewhat complicated argument leads, then, to a rather simple expression for rate of mechanical energy loss, with a dependence on the *cube* of the area change which is still more explicitly evident than in the corresponding rules for shock waves. With hydraulic jumps, on the other hand, there is a greater diversity in the means by which the required loss may be effected.

The reason for this is that, although only dissipative effects are neglected in the theory of plane waves of sound, so that they only can provide (as discussed in section 2.10) the balance against convective effects that shock-wave formation requires, some other effects also are neglected in the theory of long waves in open channels. These are effects associated with any departure from the long-wave assumption, that the longitudinal scale of variations is much larger than the mean depth h. Evidently, wherever the steepness of a waveform is locally becoming large, we can expect some departure from the accuracy of this assumption. Its nature will be probed deeply in later studies, beginning with chapter 3, but we anticipate those later discussions here with an inevitably oversimplified summary of the general conclusions.

The discussion of long waves in section 2.1 suggests that if a distorted waveform contains, in the sense of Fourier analysis, significant sinusoidal components with wavelengths *not* large compared with the depth then those components need not travel at the long-wave velocity c relative to the fluid velocity u. In fact, we show in chapter 3 that they travel at a *lower* speed. This suggests a method of getting rid of energy at the necessary rate (236) as follows: shearing distortions, by generating an exceptionally steep portion of a waveform produce more and more substantial sinusoidal components of short wavelength that lag behind the propagation of the rest of the wave, until some equilibrium waveform that is nearly a discontinuity is ultimately reached with energy slipping backwards from it at the necessary rate (236) as a train of relatively short waves behind the jump.

Several features make the true picture more complicated than this. Some are a consequence of the at first surprising result in chapter 3 that waves whose 'phase velocity' (velocity of propagation of a given phase of a sinusoidal wave such as the crest) *decreases with decreasing wavelength* have a velocity of energy propagation *less* than the phase velocity. In equilibrium propagation of a jump, all waves generated have at the jump itself the same phase; it follows that their phase velocity equals the velocity of the jump. They have, then, a phase velocity $U - u_1$ relative to the fluid behind the jump, and this value, being less than the long-wave speed c_1 as figure 47 shows, determines the *length* of those waves that are generated: they are waves of such a length that each crest can keep a constant distance behind the jump. Their *energy* travels more slowly, however, and so leaks backwards relative to the jump, and their amplitude must adjust itself so that this *rate* of loss takes the necessary value (236).

Beyond these aspects to be explained fully in chapter 3 there are some further complications (epilogue, part 2): the amplitude of any waves of the kind just described cannot exceed a certain limit without substantial energy being dissipated in foaming at the crests (as in the marine phenomenon known as 'white horses'). Experimentally a true 'undular jump', without significant foaming and losing most of the requisite energy through a train of waves behind it, is found only for strengths less than around 0.3. At greater strengths foaming and associated turbulence at the very first crest

Figure 48. Undular bore on the river Severn.

dissipate most of the requisite energy locally: little undulation is found behind the jump at these strengths.

In England the spectacular distortion of the tidal waveform as it travels up the gradually narrowing Severn estuary generates at high spring tides a hydraulic jump locally called the 'bore' and this name is used widely as an alternative to 'hydraulic jump'. A crude practical rule is that *undular bores* appear at strengths less than around 0.3: the Severn bore under most conditions is of this type (figure 48). Stronger, *turbulent bores* with a steeper, violently foaming front and at most a weak wave train behind are generated, however, in certain conditions (figure 49). Two additional complexities may be noted which are brought finally in the epilogue through refined nonlinear theories to a relatively complete resolution: although even the undulations behind the bore are subject to significant nonlinear influences, and although some fraction of the requisite energy loss occurs by turbulent dissipation in every bore and this fraction depends on other variables besides the strength, nevertheless a certain rather simple mathematical approach treats the waves nonlinearly but succeeds in evaluating their amplitude as a function of the bore strength and that energy loss.

Figure 49. Turbulent bore on another part of the river Severn. The bores in figures 48 and 49 were both photographed by Dr D. H. Peregrine and are reproduced here by his kind permission.

In spite of all the complexities associated with undulation and turbulence as methods of energy removal from a hydraulic jump, it remains valuable for many purposes of approximate analysis to treat the jump as effectively a discontinuity satisfying equations (228) and (229). The stronger, turbulent bores are the ones most closely resembling a discontinuity. For undular bores, just such a discontinuity in the *mean* water level, defined as the water level averaged over the wavelength of the undulations, occurs. With this interpretation, we may consider whether a theory of propagation of simple waves incorporating weak bores as discontinuities on the lines of section 2.11 is possible.

This requires that the simple-wave relationships be satisfied at all points behind or ahead of the bore: a hypothesis which can be checked by calculating the value of $(c - c_0 - \frac{1}{2}u)/c_0$ (zero on simple-wave theory) behind that bore which satisfies (228) and (229) as

$$(c_1 - c_0 - \tfrac{1}{2}u_1)/c_0 = (1 + \beta)^{\frac{1}{2}} - 1 - \tfrac{1}{2}\beta(1 + \tfrac{1}{2}\beta)^{\frac{1}{2}}(1 + \beta)^{-\frac{1}{2}}. \qquad (237)$$

Expression (237) varies between o and -0.0035 for $0 < \beta < 0.5$. This suggests that a weak-bore theory identical to the weak-shock-wave theory of section 2.11 may once more be applicable in this range of strengths. Such a theory would apply the mass-conservation condition exactly as in figure 42 but to a graph of $\rho_0 A$ (the mass per unit length) against x. We again find that the relationship of this quantity to excess signal speed v is close enough to the linearised relationship (within 3.5 % in the range $0 < \beta < 0.5$) for the construction to be applied directly to the waveform specified by the dependence of v on X.

The method so inferred for analysing propagation of simple waves including weak discontinuities is exactly that of section 2.11: specifically, this waveform for v is distorted at unit rate of shear, while discontinuities appear in such positions that the waveform remains one-valued and of constant area. As all consequences are identical they need not be repeated. We conclude this section, however, by checking the predictions from this method in the one case of a discontinuous wave calculated accurately above. In that case described by equations (228) and (229) the weak-bore theory would, just as in figure 35, give a value $\frac{1}{2}$ for the ratio $(U - c_0)/(u_1 + c_1 - c_0)$ instead of the accurate ratio

$$[(1 + \tfrac{1}{2}\beta)^{\frac{1}{2}}(1+\beta)^{\frac{1}{2}} - 1]/[\beta(1 + \tfrac{1}{2}\beta)^{\frac{1}{2}}(1+\beta)^{-\frac{1}{2}} + (1+\beta)^{\frac{1}{2}} - 1]; \quad (238)$$

but expression (238) varies only between 0.500 and 0.542 for hydraulic jumps of strengths between o and 0.5, again suggesting that this is the range of strengths within which weak-bore theory may give results of reasonable accuracy.

2.13 Nonlinear propagation with gradually varying composition and cross-section

The nonlinear theory of simple-wave propagation is developed in the preceding sections 2.8–2.12 for any fluid of *uniform* undisturbed composition confined within a tube or channel of uniform undisturbed cross-section. Under these conditions the basic properties of a simple wave while it remains continuous are readily established, for initial-value problems in equations (156)–(163) and for boundary-value problems in equations (168)–(171), while the associated shearing of the waveform proceeds according to equations (184)–(191). Although *discontinuity* formation is analysed above in two cases only (plane sound waves and long waves in open channels), those do suggest that any simple-wave propagation generating only weak discontinuities may be treated to good approximation by modifying the uniformly sheared continuous waveform with such *area-conserving* discontinuities as are required to keep it one-valued.

We now ask the question whether this nonlinear theory of simple waves can be extended so as to apply to cases with nonuniform undisturbed composition and cross-section, like those analysed on linear theory in sections 2.3–2.6. To this question, for cases like those of sections 2.3–2.5 involving discontinuities of admittance, the answer must be negative: a fundamental feature of the simple wave, with its identically zero value of $u - P$ along *each* C_- characteristic (equation (168)), is that no propagation whatever in the negative x-direction accompanies the wave's progress in the positive x-direction; evidently this feature excludes the possibility of reflected waves such as are necessarily produced (section 2.3) at any discontinuity of admittance. This line of argument does not, however, exclude cases of gradually varying admittance, which have been shown in section 2.6 for variation gradual enough on a scale of wavelengths to involve essentially *no* reflected wave, at least according to linear theory.

In this section, then, we use a combination of ideas from sections 2.6 and 2.9 to seek a nonlinear theory of the simple-wave propagation of a pulse through some tube or channel whose undisturbed cross-section, together perhaps with the undisturbed composition of the fluid it contains, exhibits spatial variation gradual on a scale of the pulse length. We confine ourselves, however, to relatively weak pulses in the sense of section 2.11, partly to keep the analysis simple and partly to allow extension to cases when the predicted waveform is many-valued by adding area-conserving discontinuities.

We seek, then, to calculate nonlinear effects on propagation of a relatively weak pulse along a tube or channel whose undisturbed cross-section $A_0(x)$, together with the undisturbed wave speed $c_0(x)$ and density $\rho_0(x)$ of the fluid it contains, exhibits gradual spatial variation on a scale of the pulse length. It would be rather formidable to approach this problem from the full equations of motion (2), (3), (4) and (99). An easier approach is suggested by the simple physical arguments used at the beginning of section 2.8 to infer Riemann's results (154) and (155).

The idea of this approach is that in a local frame of reference moving at the fluid velocity u the variables are locally changing in accordance with *linear* theory: an idea used in section 2.8 to derive equations (146) and (147) for the local behaviour. However, in cases where the undisturbed composition and cross-section exhibit gradual spatial variation, the appropriate linear-theory behaviour is as indicated in section 2.6: a general expression for p_e on linear theory, for example, is the sum of (91) and (92). With a similar expression for u (equal, in fact, to (91) minus (92) divided by the product of the density and the sound speed) this can give results regarding ratios of the changes in different physical quantities along curves

$$C_+ : dx = (u+c)\,dt, \quad \text{and} \quad C_- : dx = (u-c)\,dt, \tag{239}$$

generalising the conclusions (152)–(155).

Some of the complication that might be expected to appear in those generalisations owing to the variability with x in the undisturbed admittance $A_0/\rho_0 c_0$ is greatly reduced, however, because of the assumption that the whole pulse is short compared with a distance in which significant changes in A_0, ρ_0 and c_0 occur. If the pulse is travelling in the positive x-direction then *that part* of any C_- curve which lies within the pulse remains entirely in a region where A_0, ρ_0 and c_0 are effectively constant. This means that the conditions along such a part of a C_- curve can effectively be given the form (155) appropriate to uniform cross-section and composition and that the constant can be taken as zero as in (163) because the C_- curve originates in the undisturbed region ahead of the pulse.

The relationship between the fluid velocity u and the excess pressure p_e takes the same form as in (163), then, except that the dependence of ρ and c on p_e may vary with position:

$$u(x,t) = \int_0^{p_e} \frac{dp_e}{\rho(p_e, x)\,c(p_e, x)}. \tag{240}$$

The value of x on the right-hand side of (240) is taken constant because, when we integrate the equation $du = dp_e/\rho c$ along the C_- curve from (x, t) to the region $p_e = 0$ ahead of the pulse, x changes too little to affect the values of ρ and c. Locally, then, we can use simple-wave relationships appropriate to uniform cross-section and composition between the fluid velocity u and the excess pressure p_e and hence also the wave speed $c(p_e, x)$.

We have established already the fact that essential features of simple-wave propagation are retained when composition and cross-section vary gradually on a scale of the pulse length: propagation is in one direction only, in the sense that a zero signal is propagated along the C_- curves, and local relationships between different variables in a simple wave are unchanged. Note that these include the expression for the *signal velocity* $u+c$; that is, the value of dx/dt along one of those C_+ curves that carry the pulse itself. For example, in a perfect gas with constant specific heats confined in a rigid tube, with spatial variation of tube cross-section and perhaps of the undisturbed temperature (so that the undisturbed sound speed $c(0, x) = c_0(x)$ is a function of position), we can deduce from (240) the result (178) in the only slightly modified form

$$c = c_0(x) + \tfrac{1}{2}(\gamma - 1)u, \quad \text{giving} \quad u + c = c_0(x) + \tfrac{1}{2}(\gamma + 1)u. \quad (241)$$

Just the same equations but with $\gamma = 2$ hold, as in (183), for an open channel with breadth $b(x)$ gradually varying but independent of water elevation if $c_0(x)$ is the undisturbed wave speed $[gA_0(x)/b(x)]^{\frac{1}{2}}$ and $A_0(x)$ the gradually varying undisturbed water cross-section.

Results of the analysis up to this point, based on consideration of changes along C_- curves, are valid for pulses of arbitrary *amplitude*, provided that cross-section and composition vary gradually on the scale of the pulse length. Consideration of changes along C_+ curves, however, requires restriction to relatively weak pulses if a treatment simple enough for an introductory account of nonlinear effects is to result. Note that a C_+ curve, unlike any C_- curve, remains within the pulse over long distances so that changes in cross-section and composition *cannot* be neglected. Note also that, although a relatively weak pulse involves only slight departures (equation (241)) of the signal velocity $u + c$ from the linear-theory wave speed $c_0(x)$, the distorting effects of such small departures over long distances of propagation have been shown in section 2.9 to be very significant indeed. We analyse, then, a *competition* between the waveform-changing influences of gradually varying composition and cross-section and of relatively weak nonlinear effects.

The idea that along a C_+ curve (following a signal travelling relative to the local fluid velocity u at the wave speed c) variables should locally change in accordance with linear theory is now applied taking into account that we have already established propagation in one direction only (the positive x-direction). The appropriate variation in excess pressure on linear theory is given by equation (91): the quantity propagated forwards unchanged is not p_e itself but $p_e Y^{\frac{1}{2}}$, where the admittance Y varies gradually with x.

For relatively weak waves such use of the approximate linear-theory conclusions on the influence of gradual changes of cross-section and composition is particularly plausible. It also lacks ambiguity, because for small amplitudes we can take the variation in admittance Y along the tube or channel as given by its *undisturbed* form $[A_0(x)/\rho_0(x)c_0(x)]$. The distribution of excess pressure, then, follows the rule

$$p_e[A_0(x)/\rho_0(x)c_0(x)]^{\frac{1}{2}} = \text{constant along } dx = (u + c)\,dt, \quad (242)$$

with u related to p_e by equation (240); this makes the distribution suffer, over long distances, significant changes due both to waveform-shearing effects and to admittance-change effects, whereas any modifications, due to attempted incorporation of a 'disturbed' value for Y, produce only small changes by comparison. Use of the rule (242), which these arguments render plausible, is further subjected to checks of consistency in the next section (equations (263)–(269)), but can be justified mathematically only as a limiting case of a far more intricate theory, not here reproduced.

Study of how the waveform shearing, analysed in section 2.9 under conditions of uniform cross-section and composition, is modified by the effect of the term in square brackets in equation (242) is quite straight-forward. First we must decide whether to describe distortion of the *spatial* waveform with time, achieved in section 2.9 through use of the variables v (equation (184)) and $X = x - c_0 t$, or to describe distortion of the *temporal* waveform with distance as was done using v_1 (equation (190)) and

$$X_1 = c_0^{-1} x - t.$$

Actually, the latter must be chosen: although the *nonlinear* effects can be described in either manner, the cross-section and composition vary with *distance* and produce waveform changes with distance rather than with time; furthermore, on linear theory these were described in terms of a variable $\int c_0^{-1} dx - t$ (equation (91)) which is a natural generalisation of X_1.

In terms of v_1 (the 'defect in reciprocal signal velocity') the reciprocal signal velocity $(u+c)^{-1}$ is $c_0^{-1} - v_1$, which means that equation (242) can be written

$$[(c_0^{-1} - v_1)\, \partial/\partial t + \partial/\partial x]\, [v_1/V_0(x)] = 0, \tag{243}$$

where we define the function $V_0(x)$ so that in the limit of weak disturbances

$$p_e[A_0(x)/\rho_0(x)\, c_0(x)]^{\frac{1}{2}} = [v_1/V_0(x)] \times \text{constant}. \tag{244}$$

For example, in either case leading to (241) (a rigidly confined perfect gas with constant specific heats, or water in an open channel with breadth independent of water elevation) we have for weak disturbances

$$v_1 = c_0^{-1} - (u+c)^{-1} \sim \tfrac{1}{2}(\gamma+1)u/c_0^2 \sim \tfrac{1}{2}(\gamma+1)p_e/\rho_0 c_0^3 \tag{245}$$

so that we may take $V_0(x) = A_0^{-\frac{1}{2}}(x)\rho_0^{-\frac{1}{2}}(x) c_0^{-\frac{5}{2}}(x). \tag{246}$

Equation (243) is solved by a transformation of variables already in part suggested. We take

$$X_1 = \int_0^x c_0^{-1} dx - t, \tag{247}$$

in generalisation of equation (189), and use

$$V_1 = v_1/V_0(x) \qquad (248)$$

as a new dependent variable. Then it pays to define

$$T_1 = \int_0^x V_0(x)\,\mathrm{d}x \qquad (249)$$

as a new time-like variable (integral with respect to distance of a reciprocal-velocity function: the scale function $V_0(x)$ for the defect v_1 in reciprocal signal velocity). Use of the new independent variables (247) and (249) changes the partial derivatives in (243) into

$$\partial/\partial t = -\,\partial/\partial X_1 \quad \text{and} \quad \partial/\partial x = c_0^{-1}\,\partial/\partial X_1 + V_0\,\partial/\partial T_1, \qquad (250)$$

giving for the first square bracket in (243)

$$[(c_0^{-1} - v_1)\,\partial/\partial t + \partial/\partial x] = v_1\,\partial/\partial X_1 + V_0\,\partial/\partial T_1. \qquad (251)$$

Finally, with the change (248) of dependent variable, (243) becomes

$$\partial V_1/\partial T_1 + V_1\,\partial V_1/\partial X_1 = 0. \qquad (252)$$

Surprisingly, this approach has transformed the rather general and quite complicated problem of this section to the same equation (186) shown in section 2.9 to describe uniform waveform shearing! All the results, while the waveform remains continuous, then follow exactly as illustrated in figure 31 but with the variables t, x, v replaced by T_1, X_1, V_1. In particular, the reciprocal slope of the waveform, $(\partial V_1/\partial X_1)^{-1}$, increases (for a given value of V_1) at a unit rate with respect to change in T_1. Given the waveform at $x = 0$ (that is, at $T_1 = 0$) this means that infinite waveform slope appears first for a value of T_1 given by the equation

$$T_1 = \mathrm{Min}\,[-(\partial V_1/\partial X_1)^{-1}]_{T_1=0} \qquad (253)$$

as in (188). By equations (247)–(249) this gives

$$\int_0^x [V_0(x)/V_0(0)]\,\mathrm{d}x = \mathrm{Min}\,[(\partial v_1/\partial t)^{-1}]_{x=0} \qquad (254)$$

as an equation determining the distance x within which infinite waveform slope must appear (necessitating formation of a discontinuity), given the waveform at $x = 0$.

Equation (254) shows that discontinuity formation is *delayed if $V_0(x)$ is a decreasing function*, when evidently the value of x for which the left-hand side takes a given value (the minimum reciprocal waveform slope with respect to time at $x = 0$) is *raised*. For example, in the cases when $V_0(x)$ is as in (246), *increasing cross-sectional area* has this effect: by attenuating the amplitude of the signal and hence also the excess signal speed, it retards waveform steepening.

In a loudspeaker horn with area distribution $A_0(x)$, the left-hand side of (253) is $\int_0^x [A_0(0)/A_0(x)]^{\frac{1}{2}} \, dx$, which may easily grow too slowly for any shock-wave formation within the length of the horn to be possible. Indeed, in the exponential horn (section 2.6) with $A_0(x) = A_0(0)\, e^{\alpha x}$ this integral can never exceed

$$\int_0^\infty e^{-\frac{1}{2}\alpha x} \, dx = 2\alpha^{-1}, \tag{255}$$

however long the horn may be, so that by (245) and (254) a shock wave can form *only* if

$$\text{Max} \, (\partial p_e/\partial t)_{x=0} > \rho_0 c_0^3 \alpha/(\gamma+1): \tag{256}$$

a very restrictive condition.

By contrast, when $V_0(x)$ is *increasing*, the distance x at which the condition (254) for discontinuity formation must be satisfied is *reduced*. This happens, for example, as the influence of the tides propagates up a *gradually* narrowing estuary (though not one with abrupt changes of cross-section, which as shown in section 2.3 reflect much of the energy). Even though the tidal period is so great that the time rate of change on the right-hand side of (254) must be quite small, the left-hand side can be written

$$\int_0^x [b(0)/b(x)]^{\frac{1}{2}} \, [h(0)/h(x)]^{\frac{1}{4}} \, dx \tag{257}$$

in terms of the mean depth $h(x) = A_0(x)/b(x)$, and this integral (257) can rise, after substantial reductions in $b(x)$ and $h(x)$, to values large enough for the condition for discontinuity formation to be amply satisfied.

These arguments make it easy to appreciate how a bore tends to form as the tide propagates up a gradually narrowing estuary, like the Severn. Actually, they somewhat overestimate the likelihood of bore formation because they neglect any dissipation of wave-energy flow by turbulence. The left-hand side of (242), which represents the square root of the energy flow on a linear-theory approximation, should be modified by an attenuation factor representing a slow exponential decay with distance (the turbulent-dissipation analogue of a similar attenuation factor (142) due to viscous dissipation); by (244) the same factor must be incorporated into $V_0(x)$, where it limits the growth of the left-hand side of (254). This method of analysis has proved successful in explaining why it is only at the higher spring tides that the Severn bore forms.

It is not only while a wave remains continuous that the transformation in equations (247)–(249) reduces each problem to the case of uniform composition and cross-section (equation (252)). We can easily adapt the argument at the end of section 2.11 to show that when, in the variables T_1, X_1, V_1, the uniform shearing process has produced a many-valued continuous waveform it is an *area-conserving* discontinuity that must be inserted to render it one-valued. At each position x (that is, for each given T_1), the total mass flux $\int \rho Au \, dt$ in the pulse must take the same value whether for the real motion or for the exact continuous (but many-valued) calculated motion. Because at each position x, for weak pulses, the relationship between ρAu and v_1 is linear, the conservation of $\int \rho Au \, dt$ implies the *conservation of the waveform area* $\int V_1 \, dX_1$. It follows that all the conclusions of section 2.11, as well as of section 2.9, can be applied without change if the variables T_1, X_1 and V_1 are used.

In particular, the method illustrated in figure 44, for calculating from a single initial waveform all discontinuities possible in subsequent waveforms, can in these variables be immediately applied. *Asymptotically* this gives, from a single compression pulse at $x = 0$ (that is, at $T_1 = 0$) with area

$$Q = (\textstyle\int V_1 \, dX_1)_{T_1=0} = (\textstyle\int v_1 \, dt)_{x=0}/V_0(0), \tag{258}$$

a waveform for large x preceded by a head shock wave where V_1 rises from 0 to $(2Q/T_1)^{\frac{1}{2}}$ as in (220), making v_1 rise from 0 to

$$V_0(x) \left[2\left(\textstyle\int v_1 \, dt \right)_{x=0} \Big/ V_0(0) \int_0^x V_0(x) \, dx \right]^{\frac{1}{2}}. \tag{259}$$

This shock wave is followed by a *triangular* pulse (signal falling to zero at a constant rate) whose duration (that is, in terms of X_1, whose length) increases as in (221) like

$$(2Q T_1)^{\frac{1}{2}} = \left\{ 2\left(\textstyle\int v_1 \, dt \right)_{x=0} \int_0^x [V_0(x)/V_0(0)] \, dx \right\}^{\frac{1}{2}}. \tag{260}$$

These remarkably general asymptotic rules (259) and (260) for nonlinear pulse propagation, of which many interesting special cases are given in the next section, illustrate the power of the transformation method for treating problems with gradually varying composition and cross-section.

2.14 Nonlinear geometrical acoustics

Theories of one-dimensional waves in fluids are not limited in their application to fluids confined within material tubes or channels. They have another major role, as foreshadowed in section 2.1: propagation of sound in three-dimensional systems with geometrical scales greatly exceeding a typical wavelength can be approximated by regarding signals as travelling one-dimensionally along abstract *tubes of rays*, defined so that on any ray the time for a signal to reach a distant reception point is a minimum. This condition picks out the main signal reaching that point for a reason to be analysed in detail in chapters 3 and 4: briefly, it is that different signals near to the minimum-time signal have phases nearly constant, so that their *coherent* fluctuations combine into a total much more significant than can result from the *destructive* interference of signals highly variable in phase.

In sound propagation through homogeneous fluid, discussed in chapter 1, such minimum-time paths are shortest-distance paths; in other words, rays are straight lines. By contrast, propagation through fluid whose undisturbed speed of sound exhibits spatial variation, gradual on a scale of wavelengths, follows *curved* minimum-time paths. Curved rays can be calculated as in geometrical optics, and described as obeying laws of *refraction* to be discussed further in chapter 4.

The variation of acoustic amplitude along a ray tube was investigated on linear theory in section 2.6: energy flow remains constant as the wave passes along the tube (which, having no material existence, cannot produce any of the *frictional* attenuation of energy flow studied in section 2.7). It follows (equation (91)) that

$$p_e Y^{\frac{1}{2}} = p_e[A_0(x)/\rho_0(x) c_0(x)]^{\frac{1}{2}} \tag{261}$$

remains constant under changes of x and t satisfying $dx = c_0(x) dt$: that is, under changes which keep $t - \int c_0^{-1} dx$ constant. In the transformed variables T_1, X_1 and V_1 defined in equations (244)–(249) this means that the *waveform* (graph of V_1 as a function of X_1) remains unchanged on linear theory as T_1 changes.

A modification of this result, taking nonlinear effects into account for relatively weak disturbances, was suggested in section 2.13: essentially, the signal carrying an unchanged value of (261) along the tube travels at a modified speed $dx/dt = u + c$ which is greater for larger values of p_e and less for smaller ones. This distorts the pulse in such a way that in the transformed variables T_1, X_1, V_1 a continuous initial waveform, instead of remaining unchanged, is subjected to a uniform shearing at a unit rate with respect to T_1, until when T_1 takes the value (253) this gives it infinite slope. Thereafter, it must be further modified by inserting *shock waves*, such as keep it one-valued while conserving the area of the waveform. Asymptotically, any initial compression pulse tends to a *triangular* waveform specified by equations (259) and (260).

These results are directly applicable to estimating nonlinear effects on three-dimensional propagation of relatively weak disturbances. Note that we use for the same reasons as before the basic idea of geometrical acoustics that signals follow rays: that is, minimum-time paths. We also once more use the ray paths calculated from linear theory: that is, from the spatial distribution of the *undisturbed* sound speed c_0. Any attempt, for relatively weak disturbances, to use the true signal speed would make only small changes to these ray paths, and hence to the distribution of the cross-sectional area $A_0(x)$ in (261) with distance x along a ray tube; changes unimportant compared with large variations in the square bracket in (261) that are found along any ray tube on linear theory. A consistent approach, then, is to use the true signal speed where its relatively small excess produces *cumulatively* a big effect (by waveform shearing) but not where its effects through ray deformation) remain relatively small.

In this approach to nonlinear geometrical acoustics, it is interesting to study the energy flow along a ray tube beyond that point where the formation of a shock wave initiates dissipation. We make such calculations here, however, only for a perfect gas with constant specific heats and in the asymptotic limit when equations (259) and (260) hold. In those, the relation (245) between v_1 and u allows us to write

$$\left(\int v_1 \, dt\right)_{x=0} = \tfrac{1}{2}(\gamma+1)c_0^{-2}(0)H, \quad \text{where} \quad H = \left(\int u \, dt\right)_{x=0} \tag{262}$$

is a measure of the total forward displacement of the fluid in the pulse at the ray tube's initial point $x = 0$. Using (245) again, with (205), we deduce that the relatively weak shock wave's asymptotic strength is $2\gamma c_0(x)/(\gamma+1)$ times (259); this gives

$$\beta \sim 2\gamma[c_0(x)/c_0(0)]\,[V_0(x)/V_0(0)]\left\{H \middle/ (\gamma+1)\int_0^x [V_0(x)/V_0(0)]\,dx\right\}^{\frac{1}{2}} \tag{263}$$

as an equation specifying the asymptotic decay of the head shock-wave's strength along a ray tube in terms of H and of the function (246). The duration (260) of the triangular pulse headed by this shock wave is similarly

$$t_p \sim \left\{(\gamma+1)H\int_0^x [V_0(x)/V_0(0)]\,dx\right\}^{\frac{1}{2}} \middle/ c_0(0). \tag{264}$$

Note that at a fixed position x the excess pressure p_e in the asymptotic triangular pulse varies linearly from $p_0(x)\beta$ to 0 during the time t_p, while at each instant the wave energy flowing past that position is the square of

(261). Its integral with respect to time for the duration t_p (that is, the total wave energy which the pulse carries past the position x) is accordingly

$$\tfrac{1}{3} t_p \, p_0^2(x) \, \beta^2 [A_0(x)/\rho_0(x) \, c_0(x)], \tag{265}$$

which by (263) and (264) and (246), with $p_0(x) = \gamma^{-1}\rho_0(x) \, c_0^2(x)$, is

$$\tfrac{4}{3}(\gamma+1)^{-\frac{1}{2}} \rho_0(0) \, c_0^2(0) \, A_0(0) \left\{ H^3 \Big/ \int_0^x [V_0(x)/V_0(0)] \, \mathrm{d}x \right\}^{\frac{1}{2}}. \tag{266}$$

This wave energy passing the position x is seen to be a monotonically decreasing function of x. Its spatial rate of decrease

$$\tfrac{2}{3}(\gamma+1)^{-\frac{1}{2}} \rho_0(0) \, c_0^2(0) \, A_0(0) \, [V_0(x)/V_0(0)] \left\{ H \Big/ \int_0^x [V_0(x)/V_0(0)] \, \mathrm{d}x \right\}^{\frac{3}{2}} \tag{267}$$

must represent wave energy dissipated per unit length of ray tube.

Such dissipation can occur only *inside* the shock wave, where the entropy rise is given for weak shock waves by equation (216) as

$$c_v(\gamma^2 - 1) \, \beta^3 / 12\gamma^2 \tag{268}$$

per unit mass. Multiplying (268) by the undisturbed temperature $T_0(x)$, where $c_v \, T_0(x)$ equals $c_0^2(x)/\gamma(\gamma-1)$, gives the energy dissipation by the shock wave per unit mass of fluid, and multiplying that by the mass $\rho_0(x) \, A_0(x)$ per unit length of ray tube gives

$$(\gamma+1)\rho_0(x) \, c_0^2(x) \, A_0(x) \, \beta^3 / 12\gamma^3 \tag{269}$$

as the dissipation of wave energy per unit length. Satisfactorily, the use of (263) for β and (246) for $V_0(x)$ checks exactly the identity between (267) and (269) as expressions for this dissipation per unit length arrived at by two *independent* routes.

In an isothermal atmosphere the undisturbed speed of sound is constant, which simplifies the above formulas not only because $c_0(x)$ can be replaced by a constant c_0 but because all rays are straight lines. With *straight* rays, the area distribution for a thin ray tube necessarily follows a simple quadratic law

$$\frac{A_0(x)}{A_0(0)} = \frac{\partial(y+x\eta,\, z+x\zeta)}{\partial(y,z)} = \begin{vmatrix} 1+x\partial\eta/\partial y & x\partial\eta/\partial z \\ x\partial\zeta/\partial y & 1+x\partial\zeta/\partial z \end{vmatrix} = 1 + \Delta x + jx^2, \tag{270}$$

where (η, ζ) is the two-dimensional vector field of projections onto the cross-section $x = 0$ of unit vectors along ray directions, while Δ and j are the values of its divergence $\partial\eta/\partial y + \partial\zeta/\partial z$ and Jacobian $\partial(\eta, \zeta)/\partial(y, z)$ at the origin. The asymptotic form of a compressive pulse generated by forward displacement H in such a ray tube is a triangular pulse whose duration, by (246), (264) and (270), is

$$t_{\mathrm{p}} \sim \left\{ (\gamma+1) H \int_0^x [\rho_0(x)/\rho_0(0)]^{-\frac{1}{2}} (1 + \Delta x + jx^2)^{-\frac{1}{2}} \, dx \right\}^{\frac{1}{2}} \Big/ c_0, \qquad (271)$$

headed by a shock wave whose strength, by (263), is

$$\beta \sim 2\gamma [\rho_0(x)/\rho_0(0)]^{-\frac{1}{2}} (1 + \Delta x + jx^2)^{-\frac{1}{2}} (H/c_0 t_{\mathrm{p}}). \qquad (272)$$

We note further simplifications where the undisturbed density ρ_0 is effectively constant. The most extreme simplification is found when, in addition, $j = 0$, giving a linear law of expansion of ray-tube area (270). This is characteristic of *cylindrical* pulse propagation; for example, such as may be generated by an exploding wire. The duration t_{p} of the asymptotic triangular pulse becomes

$$t_{\mathrm{p}} \sim \{2(\gamma+1) H \Delta^{-1} [(1+\Delta x)^{\frac{1}{2}} - 1]\}^{\frac{1}{2}}/c_0 \sim [2(\gamma+1) H]^{\frac{1}{2}} (x/\Delta)^{\frac{1}{4}}/c_0 \quad (273)$$

as $x \to \infty$, and in the same limit

$$\beta \sim \gamma [2H/(\gamma+1)]^{\frac{1}{2}} \Delta^{-\frac{1}{4}} x^{-\frac{3}{4}}; \qquad (274)$$

an interesting *inverse-three-quarters* power law of asymptotic decay of cylindrical shock waves. (Note that we do not consider any special case with $\Delta = j = 0$ – locally parallel rays – having concluded in section 1.12 that geometrical acoustics *cannot* be applied asymptotically if the ray-tube area is not increasing.)

In the more general case $j > 0$ when the ray-tube area increases quadratically with distance, as in spherical pulse propagation, for example, the integral in (271) increases only *logarithmically* for large x and the pulse duration t_{p} (proportional to the *square root* of that integral) shows therefore a very slow increase: nonlinear waveform distortions accumulate only slowly because the rate of distortion is continually and substantially reduced by spherical attenuation. Similarly, according to (272) the shock wave decays faster than signals do on a linear theory by a factor $(H/c_0 t_{\mathrm{p}})$ falling away only very slowly (like the inverse square root of a logarithm).

This section has so far been concerned with systems throughout which geometrical scales greatly exceed a typical wavelength: the basic assumption of geometrical acoustics. However, the approach can be used also in certain systems to only part of which that assumption applies.

Any *compact* source region satisfies, of course, the opposite condition: its scale is small compared with a wavelength. Section 1.6 shows on linear theory that its *far field*, under certain fairly general circumstances, can be approximated as that of *one simple source* of strength $\dot{q}(t)$ equal to the rate of change of total mass outflow from the source region. Section 1.4 establishes, in the *far field* of any simple source, relationships later described (section 1.11) as those of geometrical acoustics but, as emphasised in figure 1, the signal propagated along each ray tube in the far field is proportional to $\dot{q}$; by contrast, it is proportional to q when the conditions of geometrical acoustics apply everywhere (as for the radiation from large spheres studied in section 1.11).

We have seen that, with relatively weak disturbances, several wavelengths of propagation are needed for nonlinear effects to produce significant waveform distortion. This suggests that the conversion of near-field mass outflow q into far-field signals should be described adequately by linear theory, while the further propagation over long distances of those far-field signals (proportional to $\dot{q}$) may be treated by geometrical acoustics, in the version of this section taking nonlinear effects into account. Note, however, that these signals cannot normally consist of a single compression pulse; for example, when the total mass outflow from the sources rises to a positive maximum q_{max} and then falls back to zero, the far field $\dot{q}$ consists (figure 1) of a compression pulse followed by an expansion pulse of equal area, which nonlinear effects convert (figure 45) into an N-wave.

Thus, in an atmosphere with uniform undisturbed density and sound speed, a simple-source far-field velocity distribution

$$u = \dot{q}(t - r/c_0)/(4\pi r \rho_0 c_0) \tag{275}$$

generates at a radius r_0, which we take as a ray tube's initial cross-section $x = 0$ (so that $x = r - r_0$) for application of the above analysis, a displacement

$$H = \left(\int u \, dt \right)_{x=0} = q_{max}/(4\pi r_0 \rho_0 c_0) \tag{276}$$

for the 'compression' part of the waveform, and an equal and opposite displacement for the 'expansion' part. The ray-tube area ratio (270) is $(r_0 + x)^2/r_0^2$, and so the asymptotic form of the pulse is an N-wave consisting of a front triangular pulse whose duration by (271) is

$$t_p \sim \left\{ (\gamma + 1) H \int_0^x r_0 (r_0 + x)^{-1} dx \right\}^{\frac{1}{2}} \Big/ c_0$$

$$= \{ (\gamma + 1) q_{max} [\ln (r/r_0)]/4\pi \rho_0 c_0^3 \}^{\frac{1}{2}} \tag{277}$$

and an equal and opposite rear triangular pulse. The strength (272) of both shock waves is asymptotically

$$\beta \sim 2\gamma r_0 (r_0 + x)^{-1} H / c_0 t_p$$

$$= 2\gamma [q_{max}/4\pi(\gamma + 1)\rho_0 c_0]^{\frac{1}{2}}/r[\ln(r/r_0)]^{\frac{1}{2}}. \tag{278}$$

These approximations are only slightly sensitive to the value chosen for the radius r_0 where nonlinear distortion of the far field (275) is assumed to begin.

Similar considerations apply to two-dimensional propagation of a cylindrical pulse from a long uniform line of compact sources: linear theory may be used to estimate the generation of the far-field signal, here proportional (figure 1) to the *derivative of* $(\frac{1}{2})th$ *order* of $q(t)$, while nonlinear geometrical acoustics describes its subsequent development. Once more an *N-wave* results, since even a purely positive mass outflow (for example, from an exploding wire) is seen in figure 1 to generate a far field with both compression and expansion phases.

Although from most source regions rays emerge in all directions, there are certain remarkable exceptions to that rule: any source region *travelling at a speed U greater than the undisturbed sound speed c_0* emits rays only at the acute angle

$$\theta = \cos^{-1}(c_0/U) \tag{279}$$

to its direction of motion; rays carrying the famous 'supersonic boom'. We explore this result mainly for an isothermal atmosphere in which c_0 is a constant.

In such an atmosphere, the sounds emitted by a source whose distance from an observer is *decreasing at a rate exceeding c_0* are heard by him *in reverse* ('pap pep pip pop pup' is heard as 'pup pop pip pep pap'!) because sounds emitted at earlier times are left behind by the faster approach of the source. In between this rather unfamiliar situation, and the ordinary case when the sounds are heard in the right order because the source's distance from the observer decreases *slower* than c_0, lies a critical condition. Whenever the distance decreases at *exactly* the rate c_0 all the sounds (vowels and consonants!) are heard together as one single 'boom'.

This condition is satisfied, with a source travelling along a straight path at a constant speed $U > c_0$ and a stationary observer *not* on that path, when the angle θ between the path and the straight ray joining the source to the observer takes the value (279): the source's velocity component $U\cos\theta$ *towards* the observer is then c_0. This makes it possible for a 'boom' to travel along the ray in question.

The law defining rays as straight lines making the angle (279) with the path of the source is really just a special case of the general view of rays as minimum-time paths: the instant when emitted signals reach the observer becomes steadily earlier while the source's distance from him is decreasing at a rate exceeding c_0, but starts to become steadily later for signals emitted after that rate of decrease becomes less than c_0. When it is exactly c_0 the signals emitted (at the angle (279)) reinforce one another because at the position of the observer they satisfy the stationary-phase condition.

Note that the rays emitted at each instant fill a forward-facing *cone* of semi-angle (279) whose axis is the path of the source. The rays in *all* of these cones (each emitted at different instants) fill all space: wherever an observer is, he must be on one such ray. That ray and all those close to it form a 'ray tube' whose cross-sectional area A_0 increases linearly, in direct proportion to the distance from the source path. This means that $j = 0$ in equation (270); it follows that in a *uniform* atmosphere the strength of the supersonic boom falls off as the inverse-three-quarters power of distance as in equation (274).

For the estimations of supersonic boom strength at ground level needed during the design of supersonic aircraft, however, it is essential to take into account the large variation in undisturbed density ρ_0 between its relatively high values near the ground and its far lower values at those altitudes where flight at speeds greater than the sound speed can be economic. This variation greatly reduces the boom strength at ground level, as the square bracket in (272) already suggests.

We sketch how the effect can be estimated quantitatively for steady flight along a straight path at altitude $x = h$. On rays in a plane through that path making an angle ψ with the downward vertical, the altitude at a distance r along a ray is

$$z = h - r \sin\theta \cos\psi, \tag{280}$$

where θ is the angle (279) between the ray and the path. An atmosphere with c_0 constant satisfies

$$-g\rho_0 = \mathrm{d}p_0/\mathrm{d}z = \gamma^{-1}c_0^2\,\mathrm{d}\rho_0/\mathrm{d}z \tag{281}$$

giving

$$\rho_0 \exp(\gamma g c_0^{-2} z) = \text{constant.} \tag{282}$$

Thus by (280)

$$\rho_0(r)/\rho_0(r_0) = \exp[\gamma g c_0^{-2}(r - r_0)\sin\theta\cos\psi]. \tag{283}$$

If r_0 here is an initial value of r in the far field of the source region, so that geometrical acoustics can be applied for $r > r_0$, then with the ray-tube area proportional to r we can replace equation (270) by

$$A_0(r)/A_0(r_0) = r/r_0. \tag{284}$$

After these substitutions equation (271) with $x = r - r_0$ gives for the asymptotic duration of the compression phase of the pulse generated by the aircraft

$$t_p \sim \left\{ (\gamma + 1) H_1 \int_{r_0}^{r} \exp\left[-\tfrac{1}{2} \gamma g c_0^{-2} (r - r_0) \sin\theta \cos\psi \right] r^{-\frac{1}{2}} dr \right\}^{\frac{1}{2}} \bigg/ c_0.$$

(285)

Here,
$$H_1 = r_0^{\frac{1}{2}} H = r_0^{\frac{1}{2}} \left(\int u \, dt \right)_{r = r_0},$$

(286)

which in the far field on linear theory is a quantity independent of the precise choice of r_0: in fact, $r_0^{\frac{1}{2}} u$ follows a curve, as in figure 1, proportional to the $(\tfrac{1}{2})$th derivative of the mass outflow (which here is that due to air being pushed aside by the forward part of the aircraft and sucked in again at the rear), and H_1 is the area of the positive lobe of that curve.

The value of (285) at ground level, where the exponential is typically reduced to around 0.3, is close to the value obtained by replacing the upper limit of integration by ∞; note also that r_0 can be replaced by 0 in this integral with little loss of accuracy because $g c_0^{-2} r_0$ is small compared with 1. These two simplifications make

$$t_p \sim [(\gamma + 1) H_1 / c_0]^{\frac{1}{2}} [2\pi / (\gamma g \sin\theta \cos\psi)]^{\frac{1}{4}}.$$

(287)

We make the second approximation also in (272), giving at ground level (where $r \sin\theta \cos\psi = h$)

$$\beta \sim 2\gamma \exp\left(-\tfrac{1}{2} \gamma g h / c_0^2 \right) [H_1 / (\gamma + 1) c_0 h]^{\frac{1}{2}} [\gamma g (\sin^3\theta \cos^3\psi) / 2\pi]^{\frac{1}{4}}$$ (288)

as the asymptotic strength of the shock wave which heads the compression pulse. This compression pulse is followed by an expansion pulse of approximately the same area, and the two together form an N-wave involving two shock waves of strength (288) separated by an interval of steady fall of pressure of duration $2t_p$.

The exponential term in (288) helps to keep typical values of β, for transport aircraft at a representative supersonic cruising altitude such as 17 km, below 0.001. Note that the boom strength is a maximum vertically under the flight path (where $\psi = 0$) and falls away to half that maximum where $\psi = 70°$: at about 40 km to either side of the flight path. The duration $2t_p$ varies less markedly and is typically around 0.5 s.

Various improvements to the above approximate method of calculation can be made in attempts to achieve greater accuracy. The linear-theory far field can be improved upon: the simple-source distribution, associated with mass outflow from each point of the flight path positive when the forward part of the aircraft is passing it and negative afterwards, is combined with a distribution of vertical dipoles associated with the lift force between the aircraft and the air; their far field has a $(\cos \psi)$ directional distribution which (unfortunately) increases the signal most in the vertically *downward* direction $\psi = 0$. On the other hand, variations in the undisturbed sound speed c_0 (which typically falls from $340 \, \mathrm{m \, s^{-1}}$ at the ground to around $300 \, \mathrm{m \, s^{-1}}$ where the aircraft flies) make useful reductions in the boom strength at the ground by increasing $V_0(x)/V_0(0)$ in (263). This results from the $c_0^{-\frac{5}{2}}$ effect in (246), modified by any changes in $A_0^{-\frac{1}{2}}$ as rays are refracted so as to splay out more with the angles to the horizontal reduced (especially for lower angles).

EXERCISES ON CHAPTER 2

1. We have seen that there is no need, in analysing one-dimensional wave propagation, to calculate transverse motions, which have too little energy to affect the propagation. Nevertheless, we invite the reader to attempt such a calculation.

For the linear theory of long-wave propagation along open channels (section 2.2), show that the longitudinal velocity u satisfies

$$\partial u/\partial x = -b\dot{\zeta}/A_0,$$

where b is the breadth of the free surface and A_0 the undisturbed cross-sectional area of the water. Infer from this a first approximation to the transverse velocities v and w as follows: $v = \partial\phi/\partial y$ and $w = \partial\phi/\partial z$, where $\partial^2\phi/\partial y^2 + \partial^2\phi/\partial z^2$ takes a value $b\dot{\zeta}/A_0$ independent of y and z, while the normal derivative of ϕ vanishes on the cross-section's solid boundary and takes the value $\partial\phi/\partial z = \dot{\zeta}$, independent of y, on the free surface $z = 0$. Note that these values are compatible by the two-dimensional divergence theorem. Show that

$$\Phi = \phi - \dot{\zeta}z - (b\dot{\zeta}/4A_0)(y^2 + z^2)$$

is an even function of z satisfying the two-dimensional Laplace equation

$$\partial^2\Phi/\partial y^2 + \partial^2\Phi/\partial z^2 = 0,$$

so that standard methods of solving Laplace's equation in two dimensions are available for finding the transverse motions.

For a semicircular cross-section, use polar coordinates with

$$y = r \cos \theta, \quad z = -r \sin \theta \quad (0 < r < \tfrac{1}{2}b, \, 0 < \theta < \pi)$$

to seek a solution in the form

$$\Phi = \sum_{n=1}^{\infty} a_n \, r^{2n} \cos 2n\theta.$$

Obtain the boundary condition for Φ on $r = \tfrac{1}{2}b$ as

$$\partial\Phi/\partial r = \dot{\zeta}(\sin \theta - 2\pi^{-1}) \quad (0 < \theta < \pi),$$

and deduce that

$$a_n = -2\dot{\zeta}(\tfrac{1}{2}b)^{1-2n}/[\pi n(4n^2 - 1)].$$

Finally, use the fact that the first term dominates the series for Φ to obtain a very rough sketch of the streamlines of the transverse flow when $\zeta > 0$.

2. At a branch B, a uniform tube AB of length l_1 and cross-sectional area A_1 is joined onto *two* uniform tubes: one of cross-sectional area A_2 and length l_2, and one of cross-sectional area A_3 and length l_3, each terminating in a closed end. For sound waves in air of uniform density ρ_0 and sound speed c, find the effective admittance Y_1^{eff} of the system at A. Deduce that resonant frequencies ω satisfy the equation

$$\sum_{n=1}^{3} A_n \tan (\omega l_n/c) = 0,$$

when A is a closed end. Give also the condition for resonance when A is an open end.

3. Show that, for sufficiently long waves in water of density ρ_0, a reservoir whose horizontal water surface has area S behaves like a capacitance $S/\rho_0 g$.

Such a reservoir is connected by a uniform canal of breadth b and depth h to the open sea. Resonance at the tidal frequency ω is observed although the canal length is only $\frac{1}{4}\pi\omega^{-1}(gh)^{\frac{1}{2}}$, Deduce that S is close to the value $b\omega^{-1}(gh)^{\frac{1}{2}}$.

4. Stenosis, which means any constriction of an artery, may (for example) result from external pressure by an enlarged organ. Consider a sinusoidal component of the blood pulse, of frequency ω and wave speed c, incident upon a constriction of length l with $\omega l/c$ small. Estimate the proportion of wave energy which is reflected as

$$\frac{\overline{(A_1/A)}^2 \, (\omega l/c)^2}{4 + \overline{(A_1/A)}^2 \, (\omega l/c)^2},$$

where A_1 is the area of the unconstricted tube and $\overline{A_1/A}$ signifies an average value of A_1/A within the constriction.

5. For a crude estimate of tides, the western end of the Mediterranean might be represented as a constriction (the Straits of Gibraltar) with effective length 50 km and cross-sectional area 4 km², followed to the east by a long channel of width 160 km and cross-sectional area 160 km², which still further to the east widens gradually enough for reflections of tidal energy to be neglected. Show that, on this model, the amplitude of tidal rise and fall with period $12\frac{1}{2}$ hours would be about one-third as much, and the time of high water a little over 2 hours later, on the Mediterranean side of the Straits of Gibraltar as on the Atlantic side. Show that a further halving of tidal amplitude would be expected where the channel width has increased to 500 km and the cross-sectional area to 800 km² (around Majorca).

[Note: this crude model gives a first rough idea of why the tides in the Mediterranean are so modest, and gives a good estimate (5×10^6 m³ s⁻¹) of the amplitude of volume-flow fluctuations in the Straits, but needs to be improved by allowing for some reflection effects. Actual tidal amplitudes are rather less, and the time-lags considerably smaller.]

6. The distribution of cross-sectional area A along an exponential horn is

$$A = A_1 \exp (\alpha x) \quad (x \geqslant 0).$$

Show that for frequencies well below its cut-off frequency the horn behaves like an inductance $L = \rho_0 \, \alpha^{-1} A_1^{-1}$, so that its reponse to a wave incident from $x < 0$ is the same as that of a constriction of length α^{-1} and uniform cross-section A_1.

7. Show that there is no cut-off frequency for a conical horn.

A sound wave $p_e = f(t-x/c)$ travels along a tube whose cross-sectional area A takes a constant value A_1 for $x < x_1$ but is equal to $A_1 x_1^{-2} x^2$ for $x > x_1$. Show that the transmitted wave in the conical horn $x > x_1$ has

$$p_e = x_1 x^{-1} f(t-x/c) - \tfrac{1}{2}cx^{-1} \int_{-\infty}^{t-x/c} f(T) \exp\left[\tfrac{1}{2}cx_1^{-1}(T-t+x/c)\right] \, \mathrm{d}T,$$

and find the reflected wave.

8. Air at $20°$ C in a very long tube of uniform cross-section is initially at rest, bounded on the left (at $x = 0$) by the surface of a closely fitting piston. The piston is then accelerated rapidly (away from the air) to a velocity $u = -U$ which it reaches at time τ when the piston surface is at $x = -l$. Thereafter, the piston moves at constant velocity $u = -U$.

At any time $t > \tau$ find which stretch of the tube is filled with air moving at velocity $u = -U$. Show that this coincides with the stretch $x \leqslant -l$ if U is about 280 m s^{-1}. What then is the air temperature in that stretch?

9. The cross-sectional area of water in a channel takes the uniform value A_0 in the undisturbed state and the value $A_0 + b\zeta$ (where b is constant) wherever the water level is raised through a distance ζ. Show that in a simple wave travelling along the channel the Bernoulli constant (170) takes the value $c_0 u + \tfrac{3}{4}u^2$, where $c_0 = (gA_0)^{\frac{1}{2}} b^{-\frac{1}{2}}$ is the undisturbed wave speed and u is the fluid velocity.

The channel opens into a large reservoir of uniform depth h_R and long-wave speed $c_R = (gh_R)^{\frac{1}{2}}$. Show that a wave propagating along a straight side of the reservoir past the channel opening generates a simple wave in the channel. Find an equation relating the fluid velocity $\tilde{u}(t)$ just inside the open end of the channel to the fluid velocity $u_R(t)$ produced just outside it by the wave in the reservoir.

10. Show that viscous attenuation of longitudinal waves in tubes or channels, as studied according to linear theory in section 2.6, produces effects which can be expressed in terms of the coordinate $X = x - c_0 t$ introduced later, in (185), as follows. The sinusoidal waveform $p_e = p_1 \exp(-ikX)$ decays as

$$p_e = p_1 \exp\left[-ikX - \epsilon(ik)^{\frac{1}{2}} t\right],$$

where

$$\epsilon = \tfrac{1}{2}sA_0^{-1}(\nu c)^{\frac{1}{2}}$$

is proportional to the length s of the solid part of the cross-section's perimeter. Deduce that, when X and t are used as independent variables, then the attenuation of a pulse of *arbitrary* waveform is represented by the equation

$$\partial p_e/\partial t = -\epsilon[\partial/\partial(-X)]^{\frac{1}{2}} p_e.$$

Infer from this that a theory allowing *both* for weak nonlinear distortions of the pulse waveform *and* for gradual attenuation in viscous boundary layers on solid walls should be based on the replacement of equation (186) by

$$\partial v/\partial t + v \, \partial v/\partial X = -\epsilon[\partial/\partial(-X)]^{\frac{1}{2}} v;$$

that is, by

$$\left[\frac{\partial}{\partial t} + v(X, t)\frac{\partial}{\partial X}\right] v(X, t) = \epsilon \int_X^\infty \frac{\partial}{\partial \xi} v(\xi, t) \frac{\mathrm{d}\xi}{[\pi(\xi-X)]^{\frac{1}{2}}}.$$

11. Shock waves are often generated for experimental purposes by means of the *shock tube*, a device whose principles are as follows. A thin diaphragm at $x = 0$

separates the *working* fluid, initially at rest in $x > 0$ at pressure p_0 and density ρ_0, from a *driver* fluid initially at rest in $x < 0$ at a much greater pressure p_2 and density ρ_2. The diaphragm is ruptured instantaneously.

Assuming that the interface between the two fluids is instantaneously jerked into motion at a constant velocity u_1, show that this generates a shock wave only in the working fluid. What does it generate in the driver fluid?

If the working fluid is a perfect gas with constant specific heats in the ratio γ_0, show that the pressure p_1 at the interface is related to u_1 by the equation

$$u_1 = (p_1 - p_0)\, \{\tfrac{1}{2}\rho_0[(\gamma_0 + 1)\, p_1 + (\gamma_0 - 1)\, p_0]\}^{-\frac{1}{2}}.$$

By considering the dynamics of the driver fluid, taken as a perfect gas with constant specific heats in the ratio γ_2, show that

$$u_1 = 2(\gamma_2 - 1)^{-1}\, (\gamma_2\, p_2/\rho_2)^{\frac{1}{2}}\, [1 - (p_1/p_2)^{(\gamma_2 - 1)/2\gamma_2}].$$

Prove that these two equations uniquely determine the strength $\beta = (p_1 - p_0)/p_0$ of the resulting shock wave.

[The assumption that the interface velocity u_1 is attained *instantaneously* is retrospectively justified by the conclusion that all the dynamical equations are satisfied, and that the fluid velocity and pressure are continuous across the interface between the two fluids. Alternatively, we might have anticipated this result from dimensional analysis: the data of the problem can define quantities with the dimensions of *velocity* such as $(p_2/\rho_2)^{\frac{1}{2}}$ and $(p_1/\rho_1)^{\frac{1}{2}}$, but no quantities with the dimensions of *time*.]

12. Use Whitham's law to discuss the development of the following simple wave. At time $t = 0$, a single compression pulse, with area Q_1 under the curve of excess signal speed v against distance X, is moving forward behind a weaker compression pulse, with area $Q_2 < Q_1$ under its corresponding curve. The distance l between the *rear ends* of the compression pulses is great enough for the pulses to be separated by an undisturbed region. Show graphically how Whitham's law yields a construction for determining the process of *union* in which, after the rear shock wave catches up the forward shock wave, they then unite into a single shock wave.

Show also that, for sufficiently large l, the union occurs when the time t satisfies

$$l = (2Q_1\, t)^{\frac{1}{2}} - (2Q_2\, t)^{\frac{1}{2}},$$

and that the jump in excess signal speed v at the united shock wave is thereafter

$$[2(Q_1 + Q_2)/t]^{\frac{1}{2}}.$$

What are the jumps in excess signal speed at the two shock waves just before union?

13. Hydraulic jumps frequently appear as *stationary* phenomena in the steady flow of a stream, and can be crudely analysed as follows for a stream of uniform breadth b and bottom slope α. At a place where the stream's depth is h, the mean flow velocity u over a cross-section is taken as J_0/bh, where J_0 stands for the stream's constant volume flow. Then we put $u\, du/dx = g(\alpha - dh/dx) - fu^2/h$, equating a crude estimate $u\, du/dx$ for the mean fluid acceleration to the component of gravitational acceleration parallel to the free surface, minus a crude correction for turbulent frictional resistance per unit mass. Show that these equations may be combined into an equation

$$dh/dx = (\alpha h^3 - f h_c^3)/(h^3 - h_c^3),$$

where h_c is the *critical depth* at which the flow velocity is equal to the long-wave speed.

This differential equation poses an enigma similar to those analysed in sections 2.9–2.12; namely, the gradient dh/dx becomes infinite at $h = h_c$. On the other hand, this gradient becomes zero where h takes the steady-state value

$$h_c(f/\alpha)^{\frac{1}{3}} = h_s.$$

The reader is invited to sketch the solution curves (i) when $\alpha < f$, so that $h_s > h_c$, and (ii) when $\alpha > f$, so that $h_s < h_c$. Note that case (i) allows a smooth transition between steady flow at $h = h_s$ and an asymptotic solution $dh/dx \to \alpha$ with a horizontal free surface (corresponding to flow into a reservoir). Show also that no such smooth transition is possible in case (ii).

In case (ii), note that the steady flow with $h = h_s$ is supercritical (the flow velocity exceeds the long-wave speed). This means that a hydraulic jump propagating *upstream* can be held stationary by that opposing steady flow. Show that the depth h_1 behind such a stationary hydraulic jump satisfies the equation

$$h_1 h_s(h_1 + h_s) = 2h_c^3,$$

of which the only positive solution exceeds h_c. Show how this allows a unique transition between the solution $h = h_s$ and a specified solution curve which tends for large x to a particular uniform level of horizontal free surface.

14. An illustration of nonlinear geometrical acoustics with quadratic area dependence of ray-tube area A_0 on distance (section 2.14) is given by the sound field of a *decelerating supersonic projectile*. We invite the reader to calculate this for radiation into air of effectively uniform density ρ_0.

Consider the ray cones emitted by such a projectile at times t and $t + \delta t$ when its velocity is U and $U - f \delta t$ respectively; here, f is its rate of deceleration. Show that the area $A_0(r)$ between the cones varies with the distance from the point of emission in proportion to

$$r + (r^2/l),$$

where

$$l = (U^2 - c_0^2) U(c_0 f)^{-1}.$$

Deduce that, if $r_0 \ll l$ is an initial value of r in the far field of the source region, so that geometrical acoustics can be applied for $r > r_0$, then for large r/r_0 the duration $2t_p$ of the N-wave generated by the projectile is given by

$$t_p = \{(\gamma + 1) H_1 l^{\frac{1}{2}} \ln [1 + 2(r/l + r^2/l^2)^{\frac{1}{2}} + 2(r/l)]\}^{\frac{1}{2}}/c_0,$$

where H_1 is given by equation (286). Note that this expression makes a *transition*, between behaviour where $r \ll l$ with

$$t_p \doteqdot \{2(\gamma + 1) H_1 r^{\frac{1}{2}}\}^{\frac{1}{2}}/c_0,$$

rather as in the case (273) of a cylindrical pulse, to behaviour where

$$t_p \doteqdot \{(\gamma + 1) H_1 l^{\frac{1}{2}} \ln (4r/l)\}^{\frac{1}{2}}/c_0,$$

rather as in the case (277) of a spherical pulse. Interpret such a transition in terms of the geometry of the ray cones.

3. WATER WAVES

3.1 Surface gravity waves

Even an introductory treatment of waves in fluids must survey *dispersion* effects: a complicated group of phenomena found in a large majority of wave-propagation systems in fluids; namely, those in which the wave speed *c varies with the wavelength*, and possibly also with the direction of propagation. Thus, even while amplitudes remain very small so that a linear theory can be used, the propagation velocity takes different values for different sinusoidal components of a general disturbance.

Chapters 3 and 4 are primarily concerned with giving a general view of the linear theory of dispersion effects. The particular wave systems described in these chapters (mainly gravity waves), although of substantial importance in their own right, are used above all to exemplify dispersion phenomena; all nonlinear effects are omitted.

By contrast, chapter 2 deals with nondispersive systems, in which there is a fixed wave speed *c*, independent of wavelength, for all disturbances of very small amplitude; but develops nonlinear theories allowing for the dependence of signal velocity on the magnitude of the signal. That nonlinear analysis of nondispersive systems, with signal velocity independent of wavelength but dependent on signal magnitude, is approximately of the same order of difficulty as the following *linear* analysis of *dispersive* systems, with signal magnitudes too small to affect a signal velocity that varies, however, with wavelength (and possibly also with the direction of propagation). Accordingly, both analyses are appropriate to an introductory account, but the far harder nonlinear analysis of dispersive waves (involving a 'competition' between the variation of signal velocity with wavelength and with signal magnitude) requires the use of much more advanced methods, and is postponed to the epilogue.

Isotropic dispersion, with the wave speed *c* independent of the direction of propagation, though varying with wavelength, is investigated in chapter 3. After wave systems of this kind have been analysed in sections 3.1–3.5, the next three sections give the theory of a remarkable and important feature of such systems: that a speed known as the *group velocity*, quite different from the wave speed c at which individual crests and troughs move, plays a fundamental role in their propagation. That theory is applied to particular systems in sections 3.9 and 3.10.

This chapter is followed by chapter 4, giving a natural extension of the theory to *anisotropic* dispersive waves with speed c dependent on the direction of propagation as well as on the wavelength. Again, an essential role is played by a group velocity, which now is different in *direction* as well as in magnitude from the velocity of crests and troughs. This vector group velocity turns out to be the key to a generalised analysis of rays and their properties, which both extends to general *linear* systems the results in the geometrical-acoustics limit given in chapters 1 and 2 and puts them on a much firmer foundation.

The principal examples of dispersive waves used in both chapters 3 and 4 are *gravity waves*, driven by a balance between a fluid's inertia and its tendency, under gravity, to return to a state of stable equilibrium with heavier fluid underlying lighter. Chapter 4 is concerned with such waves inside a fluid which in that undisturbed equilibrium state has a *continuous* decrease of density with altitude: the so-called internal gravity waves. Meteorologists have shown that the stratification of density within parts of the atmosphere is such that internal gravity waves appear, and significantly affect certain observed processes, while oceanographers have shown the importance of internal gravity waves within regions of the ocean with substantial density stratification. Since gravity as a restoring force is necessarily always in one particular direction, there is no reason for isotropy ('all directions being equal') in gravity-wave propagation, and internal gravity waves are found to be markedly anisotropic.

The existence of fluids both in liquid and in gaseous phases allows also a quite different type of stable equilibrium with heavier fluid underlying lighter: one with homogeneous liquid (for example, water) separated by a horizontal surface from a homogeneous gas above it (for example, air). The density variation is then discontinuous, and is confined to a *surface*: 'the water surface' (or in general the interface between the liquid and the gas). Disturbances to this state of equilibrium take the form of *surface gravity waves*, that cannot move far from the surface: as we shall see, not more than one wavelength away. Accordingly, their propagation over distances of many wavelengths is in horizontal directions only. Since the existence of a vertical restoring force due to gravity can make no distinction between different *horizontal* directions, the waves are isotropic (all propagation directions are equal). The effective fluid inertia, however, associated with a depth of penetration of disturbances depending upon wavelength, causes *dispersion*: a dependence of wave speed upon wavelength.

We describe surface gravity waves disturbing a flat water–air interface (though the same analysis applies to other liquid–gas interfaces). The waves on a flat water surface studied in chapter 2 are exclusively 'long' waves, with the water depth a very small fraction of a wavelength. In those, disturbances can fill the whole depth without contravening the rule against penetration to more than one wavelength: section 2.2 indicates, indeed, that the excess pressure is approximately uniform over a cross-section. The effective fluid inertia, for small disturbances of the water surface, is then independent of wavelength, and the waves are nondispersive. In section 3.3 we rederive this nondispersive behaviour as one limiting case of a linear theory of surface gravity waves that in all *other* cases predicts dispersion.

In this general linear theory of gravity waves on a homogeneous body of water, compressibility proves to be negligible so that the water density may be taken as a constant ρ_0. This is a conclusion reached already for long waves in section 2.2, where the distensibility of the water cross-section is shown to exceed by many orders of magnitude the compressibility of water: a result equivalent to the smallness of the wave speed c compared with the speed of sound. We furthermore check in section 3.3 that for water of given depth h the greatest possible wave speed c is the long-wave speed $(gh)^{\frac{1}{2}}$, which is small compared with the speed of sound c_s provided that

$$h \ll c_s^2 g^{-1} = (1400 \text{ m s}^{-1})^2 (9.8 \text{ m s}^{-2})^{-1} = 200 \text{ km}. \tag{1}$$

All known water masses satisfy this condition which ensures, as verified below in equation (5), that compressibility can be neglected. In the present section we neglect also viscosity and other dissipative effects, postponing their study to a special section 3.5 which (rather like sections 1.13 and 2.7 in previous chapters) analyses water-wave *attenuation*.

We use once more the notation of section 2.2: in coordinates with z measured *upwards* from the free surface, the undisturbed pressure distribution is hydrostatically determined as

$$p_0 = p_a - \rho_0 g z, \tag{2}$$

where p_a is the atmospheric pressure. The excess pressure due to a disturbance is defined as

$$p_e = p - p_0. \tag{3}$$

The linearised momentum equation may be written as

$$\rho_0 \, \partial \mathbf{u}/\partial t = -\nabla p_e, \tag{4}$$

where, as at the outset of chapter 1, the nonlinear term in the acceleration, $\mathbf{u} \cdot \nabla \mathbf{u}$, has been neglected and we deduce by 'taking the curl' that on linear theory the vorticity field is independent of time: 'vorticity stays put, however much other quantities may be propagated'. The 'rotational part' of the velocity field, induced by this stationary vorticity field, is independent of time, has $p_e = 0$ (by equation (4)) and therefore does not disturb the flatness of the water surface. The remaining part of the velocity field is irrotational and so can be written as the gradient $\nabla \phi$ of a velocity potential ϕ: only this part disturbs the water surface or exhibits the fluctuations associated with wave propagation.

The equation of continuity for an incompressible fluid is $\nabla \cdot \mathbf{u} = 0$, giving for this irrotational, propagating part of the velocity field Laplace's equation

$$\nabla^2 \phi = 0. \tag{5}$$

This may be contrasted with the wave equation for sound waves (section 1.1) where a term $c_s^{-2} \, \partial^2 \phi/\partial t^2$ appears on the right-hand side. However, for water waves propagating horizontally with wave speed c, an individual term on the left-hand side (in fact, a second derivative in the direction of propagation) is of order $c^{-2} \partial^2 \phi/\partial t^2$, enormously greater than $c_s^{-2} \partial^2 \phi/\partial t^2$ under the condition (1) that ensures $c \ll c_s$: Laplace's equation (5) is therefore an excellent approximation for describing water waves.

Although Laplace's equation can never describe the propagation of waves in any fluid wholly bounded by stationary surfaces, it can do so in combination with the conditions satisfied at a *free* water surface. As found in section 2.2, these require that

$$p_e = \rho_0 g \zeta \tag{6}$$

on the disturbed water surface

$$z = \zeta(x, y, t), \tag{7}$$

in order that equations (2) and (3) will permit the pressure $p = p_0 + p_e$ to take the atmospheric value p_a on that surface. In applying this condition, we are taking gravity as the *sole* force restoring the flatness of the free surface: thus, we ignore any pressure discontinuity across the air–water interface due to surface tension. This is later (section 3.4) found to be negligible for all waves save 'ripples' (very short ones).

With the irrotational, propagating part $\mathbf{u} = \nabla \phi$ of the velocity field, equation (4) associates an excess pressure field

$$p_e = -\rho_0 \, \partial \phi/\partial t, \tag{8}$$

as in the theory of sound. Equations (6) and (7) then tell us that

$$[\partial\phi/\partial t]_{z=\zeta} = -g\zeta, \tag{9}$$

a boundary condition complicated by the shape $z = \zeta$ of the free surface not being known in advance. In a linear theory, however, this complication disappears and equation (9) can be written

$$[\partial\phi/\partial t]_{z=0} = -g\zeta, \tag{10}$$

because the difference between the left-hand sides of (9) and (10) is, by the mean value theorem, equal to the product of the small disturbance ζ to the free surface and another small-disturbance term: the value of the second derivative $\partial^2\phi/\partial z\partial t$ at a point intermediate between the undisturbed and disturbed surfaces $z = 0$ and $z = \zeta$.

A second boundary condition connects the irrotational, propagating part $\mathbf{u} = \nabla\phi$ of the velocity field to that vertical displacement ζ of the free surface with which it is associated. The rate of change of ζ, following a particle of fluid, is equal to the vertical component of $\nabla\phi$ at the surface:

$$\partial\zeta/\partial t + \mathbf{u}\cdot\nabla\zeta = [\partial\phi/\partial z]_{z=\zeta}. \tag{11}$$

In a linear theory equation (11) can be simplified both by neglecting on the left-hand side the convective rate of change $\mathbf{u}\cdot\nabla\zeta$ as the product of small disturbances and by replacing the value at $z = \zeta$ on the right-hand side by a value at $z = 0$ as in (10). This gives

$$\partial\zeta/\partial t = [\partial\phi/\partial z]_{z=0}. \tag{12}$$

Equations (10) and (12) are linear boundary conditions, to be applied at the fixed boundary $z = 0$ of the region $z \leqslant 0$ in which Laplace's equation (5) must be satisfied. Differentiating equation (10) with respect to t eliminates ζ between them to give

$$\partial^2\phi/\partial t^2 = -g\partial\phi/\partial z \quad \text{on} \quad z = 0 \tag{13}$$

as the boundary condition for the velocity potential ϕ itself. Solutions of (5) in $z \leqslant 0$, subject to this boundary condition (13), represent surface gravity waves; the surface displacement ζ being deducible from the velocity potential ϕ by either (10) or (12).

3.2 Sinusoidal waves on deep water

After thus establishing that a linear theory of surface gravity waves demands merely the solution of the familiar Laplace equation (5) for irrotational incompressible flow, subject to a special boundary condition (13) at the undisturbed position $z = 0$ of the free surface, we first of all study solutions that describe travelling sinusoidal waves. It is these solutions which demonstrate the dispersive property: dependence of wave speed on wavelength. Conversely, we shall see that the effect of dispersion is often to take a general (nonsinusoidal) disturbance and cause different sinusoidal components of it to be found, at some subsequent instant, *in different places*; evidently, this is a mechanism through which a water surface may locally develop an approximately sinusoidal shape (though section 3.6 brings some surprises regarding details of how the mechanism works).

These considerations suggest the value of seeking sinusoidal-wave solutions of Laplace's equation in a water mass $z \leqslant 0$ subject to the condition (13) at its upper boundary $z = 0$. We may also need to apply, at any stationary lower boundary of the water mass, the appropriate condition: for irrotational flow this is the vanishing of the normal component of fluid velocity, which is the normal *derivative* of the velocity potential ϕ. Any irrotational-flow solution so obtained involves, of course, a nonzero tangential velocity at the boundary, which in a viscous fluid has to be reconciled with the exact boundary condition of zero fluid velocity at a stationary solid surface through the intervention of a thin dissipative boundary layer (section 2.7) between the irrotational flow and the surface.

Although in section 3.3 sinusoidal-wave solutions of Laplace's equation are thus constructed to satisfy not only an upper but also a lower boundary condition, and in section 3.5 the associated viscous boundary layers are studied, we treat first a much simpler phenomenon: waves on water so deep that the exact boundary condition is automatically satisfied on the bottom because motions associated with surface waves cannot penetrate so far down. Surface waves under these circumstances are described as 'waves on deep water'; the condition is satisfied (as stated in section 3.1) for any water mass whose depth everywhere exceeds the wavelength.

This statement is easily verified: a velocity potential that represents a sinusoidal wave propagating in the positive x-direction with wave speed c is

$$\phi = \Phi(z) \exp\left[i\omega(t - x/c)\right] = \Phi(z) \exp\left[i(\omega t - kx)\right], \tag{14}$$

where ω and k, the radian frequency and wavenumber, satisfy equations

$$\omega = 2\pi/t_{\mathrm{p}}, \quad k = \omega/c = 2\pi/\lambda \tag{15}$$

in terms of the period t_{p} and the wavelength $\lambda = ct_{\mathrm{p}}$. The function $\Phi(z)$ represents the dependence of a (possibly complex) amplitude of the motions on the distance $(-z)$ below the surface. Expression (14) (of which, of

8   LWF

course, it is understood that the *real part* represents the velocity potential ϕ) satisfies Laplace's equation (5) if $\Phi(z)$ satisfies

$$\Phi''(z) - k^2\Phi(z) = 0. \tag{16}$$

The general solution of this ordinary differential equation is a linear combination of a solution e^{kz} and another solution e^{-kz}: the *first* of these satisfies a deep-water boundary condition, in that the motions it represents become smaller and smaller at positions far below the surface where z is large and negative; indeed, on a lower boundary where $z < -\lambda$, it has fallen to less than $e^{-2\pi} = 0.00187$ and become quite negligible.

For waves on deep water, then, we must take

$$\Phi(z) = \Phi_0 e^{kz}, \tag{17}$$

where Φ_0 is a constant (the value of Φ at $z = 0$). We must avoid including any term proportional to e^{-kz}, because that would increase exponentially where z is large and negative (exceeding $e^{2\pi} = 535$ at positions more than a wavelength below the surface).

Evidently, equations (14) and (17) imply that $\partial^2\phi/\partial t^2$ is $-\omega^2\phi$ while $\partial\phi/\partial z$ is $k\phi$, so that the boundary condition (13) on $z = 0$ yields

$$\omega^2 = gk \tag{18}$$

as the relationship between frequency and wavenumber for gravity waves on deep water. This *dispersion relationship* can also be written in terms of the wave velocity c, by (15), as

$$c = \omega/k = (g/k)^{\frac{1}{2}} = (g\lambda/2\pi)^{\frac{1}{2}}. \tag{19}$$

Such dependence of the wave velocity on the square root of the wavelength implies a very substantial variation over the range of wavelengths of interest. The coefficient $(g/2\pi)^{\frac{1}{2}}$ is $1.25\ \mathrm{m}^{\frac{1}{2}}\mathrm{s}^{-1}$, so that *when λ is measured in metres*, c in $\mathrm{m\,s}^{-1}$ and the period $t_p = \lambda/c$ in seconds, equation (19) gives

$$c = 1.25\lambda^{\frac{1}{2}}, \quad t_p = 0.80\lambda^{\frac{1}{2}}. \tag{20}$$

For the range of typical surface gravity waves with wavelengths λ from 1 m to 100 m, then, the wave velocity c varies from $1.25\ \mathrm{m\,s}^{-1}$ to $12.5\ \mathrm{m\,s}^{-1}$ and the period t_p from 0.8 s to 8.0 s. Furthermore, surface waves with λ taking rather lower values down to 0.1 m (with $c = 0.4\ \mathrm{m\,s}^{-1}$, $t_p = 0.25$ s) are found in section 3.4 to be still almost pure gravity waves (in the sense that the effects of surface tension on them remain very small), and waves with λ as high as 1000 m (with $c = 40\ \mathrm{m\,s}^{-1}$, $t_p = 25$ s) are still effectively waves on deep water in oceanic regions with depths of several kilometres. There is interest, then, in sinusoidal waves on deep water with a wide range of velocities and periods.

Sitting on a beach, one commonly observes an interval around (say) 8 s between arrivals of successive wave crests. If one assumes those crests to be the successive effects, modified by propagation through shallow water near the beach, of successive crests of sinusoidal waves that have approached the coastline through deep water, one infers from (20) for those latter waves a wavelength 100 m (giving $t_p = 8$ s). We shall see in sections 3.3 and 3.6 that this inference is basically correct; the reduction in wavelength from such a large deep-water value to values several times smaller near the beach itself results from shallow-water effects.

One major contrast between waves on deep water and 'long' waves (that is, waves long compared with their depth) is in the relative magnitudes of the horizontal and vertical motions: though vertical motions in long waves are much smaller than horizontal motions (section 2.2), both are of equal amplitude in waves on deep water. In fact, equations (14) and (17) give for the velocity components in the x-direction (horizontal) and z-direction (vertical) the values

$$\partial\phi/\partial x = -\mathrm{i}k\Phi_0\,\mathrm{e}^{kz}\exp\left[\mathrm{i}(\omega t - kx)\right], \quad \partial\phi/\partial z = k\Phi_0\,\mathrm{e}^{kz}\exp\left[\mathrm{i}(\omega t - kx)\right], \quad (21)$$

of which both vary sinusoidally with time *with the same amplitude* $k\Phi_0\,\mathrm{e}^{kz}$. That amplitude depends on position, of course; as previously noted, it decreases exponentially with distance $(-z)$ below the surface. At any *fixed* position, however, the oscillating velocity components (21) differ only in their phase; that of the horizontal component $\partial\phi/\partial x$ *lags* by 90° (represented by the $-\mathrm{i}$ factor) behind that of the vertical component $\partial\phi/\partial z$. This means that the *velocity vector rotates in a clockwise sense, keeping always the same magnitude* $k\Phi_0\,\mathrm{e}^{kz}$ as its horizontal and vertical components oscillate in quadrature.

The linear theory of sinusoidal waves on deep water, then, predicts that at any fixed point the fluid *speed* remains constant, while the *direction* of fluid motion rotates with angular velocity ω. The velocity of a particle of fluid, which may suffer small oscillating displacements from that fixed point, may be taken as satisfying the same law, since on linear theory the difference between the small fluid velocities found at the fixed point and at another point displaced from it by a small amount can be neglected as the product of small quantities. Thus, a particle of fluid displaced by the waves from the position (x, y, z) moves with constant speed $k\Phi_0\,\mathrm{e}^{kz}$ and with direction

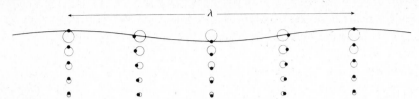

Figure 50. Motion of fluid particles (on linear theory) in a sinusoidal wave of length λ travelling from left to right on deep water. The maximum surface elevation is 0.02λ, and particles on the surface describe circles of this radius. The particles shown at mean depths 0.05λ, 0.1λ, 0.15λ and 0.20λ describe circles of radius 0.0146λ, 0.0106λ, 0.0078λ and 0.0057λ respectively. In each case the particle's instantaneous position on its circular path is shown.

rotating with angular velocity ω; in other words, it describes a circle of radius

$$\omega^{-1}k\Phi_0\,e^{kz}. \tag{22}$$

Note that the phase of the motions (21) is independent of z, since the factor e^{kz} is everywhere real and positive. Accordingly, in any vertical plane $x = $ constant those motions cause all particles of fluid to rotate in phase, though with radii depending as in (22) on distance $(-z)$ below the surface, while of course as x increases the phase of the motions decreases at the rate $k = 2\pi/\lambda$. Figure 50 illustrates this linear-theory approximation to the motion of fluid particles in sinusoidal waves on deep water.

The excess energy in water waves (like that in sound waves) is, on linear theory, divided equally between (i) kinetic energy and (ii) energy associated with the restoring force – which for gravity waves is the gravitational potential energy ρgz per unit volume.* Evidently, the corresponding value of the potential energy per unit horizontal area, at a point where the local depth of water is h, is obtained by integrating $\rho gz\,dz$ from the bottom $z = -h$ to the free surface $z = \zeta$, giving

$$\int_{-h}^{\zeta} \rho gz\,dz = \tfrac{1}{2}\rho g(\zeta^2 - h^2). \tag{23}$$

At that point, then, the excess *potential energy per unit horizontal area* over the value for an undisturbed free surface $(\zeta = 0)$ is

$$\tfrac{1}{2}\rho g\zeta^2; \tag{24}$$

proportional, as expected on a linear theory, to the *square* of a displacement from a position of stable equilibrium. Note that while a *raised* free surface

* Because variations of density ρ have been shown in section 2.1 to be negligible for water waves, the suffix zero is dropped from ρ_0 throughout the remainder of chapter 3.

adds potential energy by insertion of new fluid *above* the position $z = 0$ (that is, with positive potential energy), a depressed free surface adds potential energy by *removal* of fluid from *below* the level $z = 0$ (that is, with negative potential energy).

In the language of the general linear theory of vibrations (see texts on dynamics) the form (24) of the excess potential energy per unit horizontal area suggests that, locally, a suitable *generalised coordinate* describing the propagation of gravity waves would be the displacement ζ of the free surface, provided that the *kinetic* energy per unit horizontal area can correspondingly be written as a multiple of $(\partial\zeta/\partial t)^2$. Such a form is, indeed, readily derivable for sinusoidal waves.

A general expression for the kinetic energy in fluid flow satisfying Laplace's equation (5) is

$$\tfrac{1}{2}\rho \int (\phi \,\partial\phi/\partial n)\,\mathrm{d}S: \tag{25}$$

a surface integral over the boundary of the fluid, with $\partial/\partial n$ representing differentiation along the outward normal. This statement is proved in texts on fluid dynamics by using the divergence theorem to express (25) as a volume integral of $\tfrac{1}{2}\rho\nabla\cdot(\phi\nabla\phi)$ which, when (5) is satisfied, is the kinetic energy per unit volume $\tfrac{1}{2}\rho(\nabla\phi)^2$. In water waves there is no contribution to (25) from the bottom, where $\partial\phi/\partial n = 0$, and the contribution to (25) from the free surface can on linear theory be approximated by the same integral over the undisturbed free surface $z = 0$. This gives, with an error appropriate to linear theory of order the cubes of disturbances, a kinetic energy

$$\tfrac{1}{2}\rho[\phi\,\partial\phi/\partial z]_{z=0} \tag{26}$$

per unit horizontal area.

Now, condition (12) on $z = 0$ equates $\partial\phi/\partial z$ to $\partial\zeta/\partial t$, while for sinusoidal waves on deep water equations (14) and (17) give $\phi = k^{-1}\partial\phi/\partial z$. Their *kinetic energy per unit horizontal area* is therefore

$$\tfrac{1}{2}\rho k^{-1}(\partial\zeta/\partial t)^2. \tag{27}$$

Here, the proportionality to k^{-1} corresponds to the fact that the surface motions $\partial\zeta/\partial t$ carry with them a layer of fluid (figure 50) penetrating a distance proportional to (though rather less than) a wavelength $2\pi/k$ below the surface.

Expressions (24) and (27) can be regarded as specifying, for sinusoidal waves on deep water, a generalised stiffness ρg (coefficient of $\tfrac{1}{2}\zeta^2$ in the potential energy) and generalised inertia ρk^{-1} (coefficient of $\tfrac{1}{2}(\partial\zeta/\partial t)^2$ in the kinetic energy) per unit area of water surface. Equation (18) can then be interpreted in terms of the general principle that ω^2, the square of the frequency of a vibration, is equal (so that the kinetic and potential energies will have equal *average* values) to the ratio of a generalised stiffness to a generalised inertia; in this case, gk.

When the elevation of the free surface takes the form $\zeta = a \cos(\omega t - kx)$, with amplitude a and ω related to k as in (18), the total wave energy W per unit horizontal area (sum of (24) and (27)) takes the constant value

$$W = \tfrac{1}{2}\rho g a^2. \tag{28}$$

(This wave of amplitude a is one for which $\Phi_0 = i\omega a/k$.) For surface waves, which can only propagate horizontally, such a wave energy W per unit area is the analogue of the wave energy W per unit *volume* in sound waves (section 1.3).

We may conclude this section by asking how accurate is the dispersion relationship (18) for waves that are not exactly of the sinusoidal form (14). We note that (14), being independent of the y-coordinate, implies purely two-dimensional motions, identical in every plane (like that of figure 50) parallel to the xz-plane: wave crests, for example, must stretch indefinitely at right angles to such a plane. However, for sinusoidal waves to satisfy $\omega^2 = gk$ to good approximation, it suffices merely that their dependence on y is so gradual that the $\partial^2\phi/\partial y^2$ term in Laplace's equation (5) is very much smaller than the other two terms (the terms taken to balance each other exactly in deriving (16), (17) and (18)). Broadly speaking, this means 'long-crested' waves: waves whose crests extend coherently in the y-direction for distances of many wavelengths. While mentioning this, however, we postpone to section 3.6 any first detailed discussion of *departures* from the behaviour of pure sinusoidal waves.

3.3 Sinusoidal waves on water of arbitrary, but uniform, depth

Using linear theories, we have established (i) in section 3.2 for waves on 'deep' water (in the sense that its depth everywhere exceeds a wavelength λ) the *dispersive* property that the wave speed depends on λ, varying like $(g\lambda/2\pi)^{\frac{1}{2}}$; and (ii) in section 2.2 for 'long' waves (in the sense that λ is a large multiple of the depth) the *nondispersive* property that the wave speed is independent of λ, taking the value $(gh)^{\frac{1}{2}}$ in water of uniform depth h. In this section we study the transition between those extremes by determining the wave speed on water of *arbitrary*, but uniform, depth h, applying at the solid bottom $z = -h$ the boundary condition of zero normal velocity

$$\partial\phi/\partial z = 0 \quad \text{on} \quad z = -h \tag{29}$$

appropriate to irrotational flow. We defer the consideration of associated viscous boundary layers to section 3.5, where (as in section 2.7) we shall find their effects on the waves to be primarily attenuative.

The boundary condition (29) modifies the calculations of section 3.2 only by specifying differently the required solution of the ordinary differential equation (16) for the amplitude $\Phi(z)$ of the general expression (14) for sinusoidal waves. Instead of specifying a solution (17) which becomes zero as z becomes large and negative, it specifies a solution satisfying $\Phi'(-h) = 0$. This means that, in the general linear combination

$$\Phi_1 e^{kz} + \Phi_2 e^{-kz} \tag{30}$$

of elementary solutions of (16), we have

$$\Phi_1 e^{-kh} = \Phi_2 e^{kh}, \tag{31}$$

and putting both sides of (31) equal to $\frac{1}{2}\Phi_0$ we can write the solution as

$$\Phi(z) = \Phi_0 \cosh [k(z+h)]. \tag{32}$$

Here and elsewhere we use the classical notation

$$\cosh x = \tfrac{1}{2}(e^x + e^{-x}), \quad \sinh x = \mathrm{d}(\cosh x)/\mathrm{d}x = \tfrac{1}{2}(e^x - e^{-x}) = \cosh x \tanh x \tag{33}$$

for the hyperbolic functions. They are plotted in figure 51; note the zero slope of the graph of $\cosh x$ at $x = 0$ confirming that (32) satisfies $\Phi'(-h) = 0$.

Whereas equation (14) requires (just as in section 3.2) that $\partial^2 \phi / \partial t^2$ is $-\omega^2 \phi$, it implies with (32) that on $z = 0$

$$\partial \phi / \partial z = [\Phi'(0)/\Phi(0)] \phi = (k \tanh kh) \phi, \tag{34}$$

so that the free-surface boundary condition (13) yields

$$\omega^2 = gk \tanh kh \tag{35}$$

as the relationship between frequency ω and wavenumber k for gravity waves on water of arbitrary, but uniform, depth h. This dispersion relationship can be written in terms of the wave speed c as

$$c = \omega/k = (gk^{-1} \tanh kh)^{\frac{1}{2}}: \tag{36}$$

a result agreeing with *both* the limiting forms noted above, $(g/k)^{\frac{1}{2}}$ in the deep-water limit (kh large) and $(gh)^{\frac{1}{2}}$ in the long-wave limit (kh small), because in those limits $\tanh kh$ asymptotes to 1 and to kh respectively (figure 51).

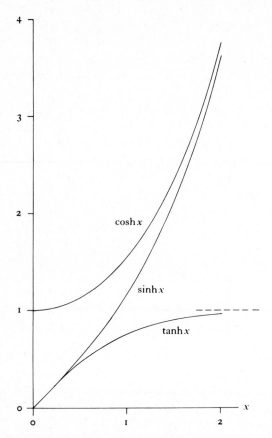

Figure 51. Full lines: graphs of the hyperbolic functions cosh x, sinh x and tanh x. Broken line: asymptotic value, 1, taken by tanh x for large x.

Numerically, $(\tanh kh)^{\frac{1}{2}}$ lies between 0.97 and 1 for $kh > 1.75$. This means, in terms of $\lambda = 2\pi/k$, that expression (36) for c falls short of the deep-water value $(g\lambda/2\pi)^{\frac{1}{2}}$ by at most 3 % provided that

$$h > 0.28\lambda. \tag{37}$$

This condition for the dispersion to be predicted to good accuracy by deep-water theory may appear surprising since the motions predicted by that theory are only reduced to 17 % of their surface amplitude in a distance 0.28λ below the surface; more significantly, however (as we shall see), their contributions to the kinetic energy (27) (and hence to an inertia coefficient) are in the same distance reduced to as little as 3 %.

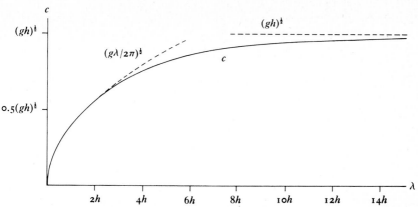

Figure 52. The wave speed c given by linear theory for waves of varying length λ on water of uniform depth h. Note the transition between the deep-water value $(g\lambda/2\pi)^{\frac{1}{2}}$ and the long-wave value $(gh)^{\frac{1}{2}}$.

At the other end of the scale, $(\tanh kh)^{\frac{1}{2}}/(kh)^{\frac{1}{2}}$ lies between 0.97 and 1 for $kh < 0.44$. This means, in terms of $\lambda = 2\pi/k$, that expression (36) for c falls short of the long-wave value $(gh)^{\frac{1}{2}}$ by at most 3 % provided that

$$h < 0.07\lambda. \tag{38}$$

As foreshadowed in section 2.2, the depth h needs to be a very small proportion (under 7%) of the wavelength for the wave speed to be given to good accuracy by the expression $(gh)^{\frac{1}{2}}$ independent of λ derived on the assumption of effectively one-dimensional motions.

It is, then, only in the range of depths *between* 0.07λ and 0.28λ (corresponding to wavelengths between 14h and 3.5h) that the dependence of c upon λ, given by (36) as

$$c = [(g\lambda/2\pi)\tanh(2\pi h/\lambda)]^{\frac{1}{2}}, \tag{39}$$

departs significantly from *both* its limiting forms. Figure 52 exhibits this by plotting c against λ for fixed h according to (39) and showing how it makes a smooth transition between the 'parabolic' limiting form $(g\lambda/2\pi)^{\frac{1}{2}}$ for $\lambda < 3.5h$ and the constant asymptote $(gh)^{\frac{1}{2}}$ for $\lambda > 14h$.

It is interesting also to plot the wave speed c against the depth h for fixed *frequency* ω; especially (see section 3.8) because as sinusoidal waves on deep water approach a coastline, passing through water of gradually less and less depth, the frequency tends to remain constant (so that the number of crests reaching the beach per unit time is equal to the number approaching

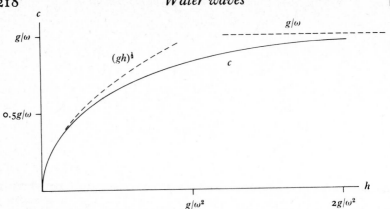

Figure 53. The wave speed c for waves of frequency ω on water of gradually varying depth h. Note the transition between the long-wave value $(gh)^{\frac{1}{2}}$ and the deep-water value g/ω.

the coastline). Now, equation (35) with k replaced by ω/c can be written

$$c\omega/g = \tanh(\omega h/c) \tag{40}$$

and used to determine how c varies for fixed frequency ω as h gradually becomes less and less. Evidently, a transition is made between the deep-water limit when the right-hand side of (40) is 1, giving $c = g/\omega$, and the long-wave limit when it is $(\omega h/c)$, giving $c = (gh)^{\frac{1}{2}}$; figure 53 exhibits this transition.

For example, a wave of period 8 s was shown in section 3.2 to have, in deep water, $c = 12.5\,\mathrm{m\,s^{-1}}$ and $\lambda = 100\,\mathrm{m}$. Both the wave speed and the wavelength are *reduced*, however, with passage into gradually shallower water, *down the curve* in figure 53. Near the beach, at a depth of (say) 1 m, c is reduced (for this 8-second wave) to $3.1\,\mathrm{m\,s^{-1}}$ and so λ is reduced to 25 m: a fourfold reduction.

One normally observes waves approaching a beach with their crests nearly parallel to it, even though they may have originated as wave crests on deep water moving towards the coastline at a substantial angle thereto. The alignment of the crests to the direction of the depth contours near the beach is the result of the reduction in wave speed as the depth becomes less: along any crest approaching the coastline obliquely those parts which first enter the shallow water are slowed down while parts still in deep water continue to move forward fast and the crest swings round as a result (figure 54).

Having (for water of arbitrary depth) noted certain consequences of the dispersion relationship, we analyse now (as we did for deep water in section

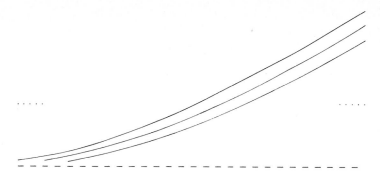

Figure 54. Alignment of crests approaching a beach of uniform slope. The broken line indicates the shore. Dotted lines: position where the depth is $\lambda/2\pi$ (so that $kh = 1$).

3.2) the water motions associated with the waves. Equations (14) and (32) give for the velocity components in the horizontal and vertical directions the values

$$\partial\phi/\partial x = -\mathrm{i}\{k\Phi_0 \cosh\left[k(z+h)\right]\}\exp\left[\mathrm{i}(\omega t - kx)\right], \tag{41}$$

and

$$\partial\phi/\partial z = \{k\Phi_0 \sinh\left[k(z+h)\right]\}\exp\left[\mathrm{i}(\omega t - kx)\right], \tag{42}$$

exhibiting *different* amplitudes of sinusoidal variation, represented by the expressions in curly brackets. The graphs of the cosh and sinh functions (figure 51) indicate the extent of the difference: the two amplitudes become closely equal only where $k(z+h)$ is large; that is, in the deep-water limit (*kh* large) and with the exception of water close to the bottom $z = -h$ (where, in any case, the motions in that limit are relatively very small). In the other limiting case of long waves (*kh* small), the amplitude of vertical movement remains small compared with the amplitude of horizontal movement (as assumed in the one-dimensional theory of chapter 2) by a factor $\tanh\left[k(z+h)\right]$ which for $-h < z < 0$ varies between 0 and $\tanh kh$; that factor, for example, is at most 0.41 in those waves with $h = 0.07\lambda$ for which one-dimensional theory gives the wave speed with a 3 % error, while for longer waves it is less still.

    At any fixed position, just as for waves on deep water, the oscillations of the velocity components (41) and (42) differ by 90° in their phase: that of the horizontal component $\partial\phi/\partial x$ lags by 90° (represented by the $-\mathrm{i}$ factor) behind that of the vertical component $\partial\phi/\partial z$. On linear theory the *same* expressions (41) and (42) describe the velocity components of a particle of fluid which may suffer small oscillating displacements from that fixed

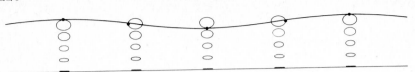

Figure 55. Paths of fluid particles in a sinusoidal wave of length λ travelling from left to right on water of depth $h = 0.16\lambda$. As in figure 50, the maximum surface elevation is 0.02λ. A particle's instantaneous position on its elliptical path is here shown only for those in the top row, but the motion of every particle in the same vertical line is (once more) in phase.

point. From this we would deduce, as in section 3.2, *if both velocity amplitudes were* $k\Phi_0 \cosh[k(z+h)]$, that the particle must describe a circle of radius $\omega^{-1}k\Phi_0 \cosh[k(z+h)]$; however, the true motions, with the amplitude of the vertical velocities reduced by a factor $\tanh[k(z+h)]$, follow a circular path foreshortened in the vertical direction by just that factor; in other words, an ellipse with major and minor semi-axes

$$\omega^{-1}k\Phi_0 \cosh[k(z+h)] \quad \text{and} \quad \omega^{-1}k\Phi_0 \sinh[k(z+h)]. \tag{43}$$

Figure 55 illustrates this linear-theory approximation to the motions of fluid particles in waves of length λ on water of a particular depth $h = 0.16\lambda$, *intermediate* between those values 0.07λ and 0.28λ where the long-wave and deep-water limits give reasonable approximations. The horizontal motions are 54% greater at the surface than on the bottom, contradicting *both* the one-dimensional assumption of approximately uniform distribution *and* the deep-water assumption of relatively small motion near the bottom. (For the effects of viscous boundary layers between the solid bottom and any nearby horizontal motions, see section 3.5.) The vertical motions at the surface are as much as 18% greater than the horizontal motions near the bottom, while the ratio of vertical to horizontal motions varies from 0 at the bottom to 0.76 at the surface, neither approaching close to the deep-water value 1 nor remaining small as on one-dimensional theory.

In water substantially deeper, the ellipses of figure 55 would become almost circular, as in figure 50, except near the bottom where they would be flattened but relatively small. In much *shallower* water, they would all be greatly flattened and exhibit almost uniform lengths of major axis.

The excess potential energy takes the value (24), per unit horizontal area, for waves on water of any depth, as its derivation (23) shows. Similarly, the kinetic energy per unit horizontal area takes always the value (26), but the relationship between ϕ and $\partial\phi/\partial z$ on $z = 0$ is modified as in (34) for water of arbitrary depth h. This relationship, with condition (12) equating $\partial\phi/\partial z$ to $\partial\zeta/\partial t$, gives, then,

$$\tfrac{1}{2}\rho(k\tanh kh)^{-1}(\partial\zeta/\partial t)^2 \tag{44}$$

as the kinetic energy of the waves per unit horizontal area. Thus, the generalised inertia per unit area of water surface is *increased* above its deep-water value ρk^{-1} to

$$\rho(k\tanh kh)^{-1}. \tag{45}$$

This new value, with the unchanged generalised stiffness ρg, allows the dispersion relationship (35) to be interpreted as setting ω^2, the frequency squared, equal to the ratio of stiffness to inertia. As usual, this gives the kinetic and excess potential energies equal *average* values; and the total wave energy W per unit area takes the same form (28) as before.

For waves on deep water the generalised inertia ρk^{-1} per unit area is interpreted in section 3.2 as due to surface motions $\partial\zeta/\partial t$ carrying with them comparable movements in a layer of fluid of thickness proportional to k^{-1}. Accordingly, it may appear surprising that a solid bottom, curtailing the depth of fluid able to move, should *increase* the generalised inertia and so reduce the frequency. The explanation is that $\partial\zeta/\partial t$ represents the *vertical* component of surface motion: as kh is reduced, the increase in the ratio of horizontal to vertical motions produces a greater change in kinetic energy per unit area for given $\partial\zeta/\partial t$ than does any decrease in the available volume of fluid.

For example, in the long-wave limit the amplitude of the horizontal motions is $(kh)^{-1}$ times that of the surface vertical motions, making their kinetic energies per unit mass in the ratio $(kh)^{-2}$, which with a mass ρh per unit area given an inertia coefficient $\rho h(kh)^{-2}$, in agreement with the long-wave limit of the exact value (45). Of course, if the horizontal displacement of surface fluid were taken as generalised coordinate, the generalised inertia would in the long-wave limit take the value ρh independent of wavelength; but then the generalised stiffness would be modified to $\rho g(kh)^2$, once more making ω^2 equal to ghk^2 to correspond with nondispersive waves of speed $(gh)^{\frac{1}{2}}$. For further important information on energy relationships in waves on water of arbitrary depth h, see section 3.8.

3.4 Ripples

Surface waves are analysed in sections 3.1–3.3 with gravity taken as the sole force restoring the flatness of the air–water interface $z = \zeta$; across that interface, the atmospheric pressure p_a is assumed continuous with the water pressure so that the latter's excess p_e over its undisturbed value $p_0 = p_a - \rho g z$ takes the boundary value $\rho g \zeta$ (equation (6)). In this section, the theory is modified by allowing for an additional flatness-restoring force: *surface tension*, which generates a discontinuity proportional to the interface's *curvature* between the air and water pressures. The resulting correction to the surface value of p_e, proportional to interface curvature, bears, in comparison with the uncorrected term $\rho g \zeta$, proportional to interface displacement, a ratio varying as the inverse square of the wavelength. Accordingly, it is important only for those rather short waves commonly called *ripples*: in practice, waves of length less than around 0.1 m, for which, however, it markedly alters, as we shall see, the character of the dispersion relationship.

The elementary definition of the surface tension, T, is in terms of the idea that two adjacent pieces of water surface with a straight frontier between them pull on each other with equal and opposite forces T per unit length of that frontier at right angles to it; here, T is measured in newtons per metre, taking the value 0.074 N m^{-1} for water. Before sketching that idea's relationship to the general physics of liquids, we show how easily the resulting modifications to the theory of surface waves can be inferred from it.

In long-crested waves propagating in the x-direction the force T across unit length of a frontier parallel to the wave crests (that is, in the y-direction) has a *vertical component* which on linear theory can be written

$$T \, \partial \zeta / \partial x, \tag{46}$$

representing the small z-component of a force T acting in a surface inclined upwards with a slope $\partial \zeta / \partial x$. It follows that, on a strip of water surface bounded by two such frontiers a distance δx apart, there acts a net vertical force

$$T(\partial^2 \zeta / \partial x^2) \, \delta x \tag{47}$$

representing the difference between the values of the vertical component (46) of tension at (say) $x + \delta x$ and of an opposing tension at x. Since the area of such a strip of water surface, of width δx and unit length, is δx, the upward *force per unit area* acting on it must be $T \partial^2 \zeta / \partial x^2$. This, evidently, must be balanced by an excess of air pressure over water pressure at the surface; no mass-acceleration term can appear in such a balance because in a thin layer of surface water the mass per unit area tends to zero with the layer's thickness.

The boundary value of the water pressure $p_0 + p_e$ is therefore *less* than the atmospheric pressure p_a by an amount $T \partial^2 \zeta / \partial x^2$, proportional to a linear approximation to the surface's curvature. This implies, with the hydrostatic relationship $p_0 = p_a - \rho g z$, that the excess pressure is

$$p_e = \rho g \zeta - T \, \partial^2 \zeta / \partial x^2 \tag{48}$$

on $z = \zeta$. In sinusoidal waves of wavenumber $k = 2\pi/\lambda$, the boundary value (48) takes the simpler form

$$p_e = (\rho g + Tk^2)\, \zeta. \tag{49}$$

At this point it would be straightforward to go through all the arguments given in equations (6)–(13) for surface gravity waves, but starting from this modified boundary value (49) in place of (6); the dispersion relationship would then be deduced as in the steps leading up to (18) for the deep-water case or to (35) for water of arbitrary, but uniform, depth. This whole procedure would, however, be a waste of time!

It is much quicker to recognise that equation (49) is *exactly the same* as equation (6) with

$$g \quad \text{replaced by} \quad g + \rho^{-1}Tk^2. \tag{50}$$

This means that, for sinusoidal waves of wavenumber k, the effect of surface tension is exactly the same as the effect of such an increase (50) in g, the acceleration due to gravity. All the properties of the waves must, accordingly, be specified correctly if we just make this substitution in the conclusions of the preceding sections.

In particular, the dispersion relationship (18) for surface waves on deep water becomes, with surface tension thus taken into account as well as gravity through the substitution (50),

$$\omega^2 = (g + \rho^{-1}Tk^2)\, k \tag{51}$$

in terms of frequency; while the wave speed (19) becomes

$$c = [(g + \rho^{-1}Tk^2)/k]^{\frac{1}{2}}. \tag{52}$$

Note that in such formulas the effect of the substitution (50) is negligible wherever the correction term $\rho^{-1}Tk^2$, proportional as foreshadowed earlier to the inverse square of the wavelength $\lambda = 2\pi/k$, is very small compared with g itself; that is, if k is small compared with

$$k_m = (\rho g/T)^{\frac{1}{2}}. \tag{53}$$

This condition for the waves to be effectively pure gravity waves, in the sense that surface tension is negligible by comparison, is equivalent to requiring that the wavelength λ be large compared with

$$\lambda_m = (2\pi/k_m) = 2\pi(T/\rho g)^{\frac{1}{2}}. \tag{54}$$

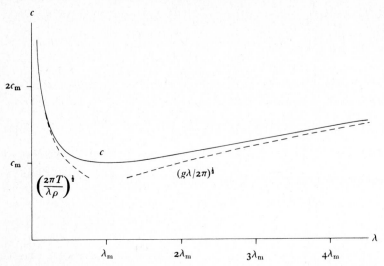

Figure 56. The wave speed c for ripples on deep water. Note the transition between the capillary-wave value $(2\pi T/\lambda\rho)^{\frac{1}{2}}$ and the gravity-wave value $(g\lambda/2\pi)^{\frac{1}{2}}$. This occurs around $\lambda = \lambda_m$, the wavelength for minimum wave velocity given by equation (55).

For example, water with $T = 0.074 \, \text{N m}^{-1}$ and $\rho = 1000 \, \text{kg m}^{-3}$ has

$$k_m = 360 \, \text{m}^{-1} \quad \text{and} \quad \lambda_m = 0.017 \, \text{m}, \tag{55}$$

supporting the earlier statement that waves of length exceeding $0.1 \, \text{m}$ are pure gravity waves. Indeed, the error in the gravity-wave formula $(g/k)^{\frac{1}{2}}$ is less than 3% if $\lambda > 0.07 \, \text{m}$. The word 'ripples' is a convenient one to use for those waves for which surface-tension effects *are* significant, with length less than $7 \, \text{cm}$.

The reason for the employment of a suffix m is that the expression (52) for the wave speed takes a *minimum* value,

$$c = c_m = (2g/k_m)^{\frac{1}{2}} = 0.23 \, \text{m s}^{-1} \text{ for water,} \tag{56}$$

at $k = k_m$ (or $\lambda = \lambda_m$). Such a minimum appears because both surface tension and gravity produce effects on the wave-speed squared, c^2, which become large as the wavenumber k becomes, respectively, large or small; the effects are just equal, and the combined effect least, for $k = k_m$.

Figure 56 illustrates the dispersion relationship (52) for waves on deep water by plotting c against λ to show (i) the minimum wave speed c_m reached at $\lambda = \lambda_m$, (ii) the running together with the gravity-wave curve for $\lambda > 4\lambda_m$, and (iii) the tendency for wave speeds to rise again when

λ/λ_m is very small. It is when $\lambda < \frac{1}{4}\lambda_m$ that the simplified forms of (51) and (52) with gravity neglected,

$$\omega^2 = \rho^{-1}Tk^3 \quad \text{and} \quad c = (\rho^{-1}Tk)^{\frac{1}{2}}, \tag{57}$$

become accurate to within 3 %. Such waves in which the only significant restoring force is surface tension, the force responsible for 'capillary attraction', are often called 'capillary waves'.

Capillary waves on water are waves with $\lambda < 4\,\text{mm}$. Their frequency $\omega/2\pi$ takes values exceeding 70 Hz, well into the acoustic range, so that it is easy to excite them by striking a tuning fork and placing the tines in the water. For example, the tuning fork for the violinist's A string, with frequency $\omega/2\pi = 440\,\text{Hz}$, generates by (57) waves of length $\lambda = 1.3\,\text{mm}$, with velocity $0.57\,\text{m s}^{-1}$. Such waves are, in fact, observed, but only rather close to the tines; for reasons discussed in section 3.5, waves of such high frequency suffer extremely rapid attenuation and, in consequence, cannot be seen beyond a radius of a few centimetres.

Note that the nature of the *paths of fluid particles* in waves on deep water is unaltered from that illustrated in figure 50 when we make the substitution (50) to allow for the effect of surface tension. Again, the limiting value that the depth must exceed for the deep-water results to be accurate remains the same: about a wavelength, λ, if boundary conditions are to be satisfied to extreme accuracy, or as little as 0.28λ if the test is taken to be 3 % accuracy in the dispersion relationship (condition (37)).

Effectively, this means that in waves of any 'ordinary' depth, exceeding a few centimetres, all *ripples* (whose wavelengths cannot, as we have seen, exceed 7 cm) may be regarded as deep-water phenomena. Another way of putting this is to say that effects of finite depth (section 3.3) and of surface tension (section 3.4) modify the theory of gravity waves on deep water (section 3.2) in *disjoint* ranges of wavelength: large and small respectively.

Nevertheless, there is a special reason why it is desirable to study the dispersion relationship for ripples on water of arbitrary, but uniform, depth; namely, the possibility of choosing a much *lower* than 'ordinary' depth of water in a 'ripple tank' so that (section 1.7) the ripples will mimic sound waves by being as closely as possible nondispersive. The idea is to select the depth so as to produce cancellation between two opposing divergences from the asymptotic long-wave speed: a decrease (figure 52) as λ falls to values comparable with the depth and an increase (figure 56) as λ falls to values where the effective value of g is augmented by surface-tension effects.

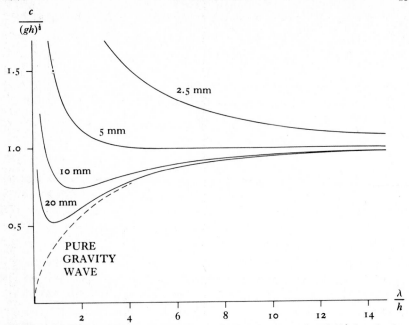

Figure 57. Ratio of the wave speed c to the long-wave value $(gh)^{\frac{1}{2}}$ for ripples on shallow water of depths h equal to 2.5 mm, 5 mm, 10 mm and 20 mm. The pure gravity-wave curve (figure 52) is given for comparison as a broken line.

Indeed, when we apply the substitution (50) in equation (36), we obtain the dispersion relationship

$$c^2 = (g + \rho^{-1}Tk^2)\,k^{-1}\tanh kh \tag{58}$$

for ripples on water of arbitrary, but uniform, depth h. Expanding the right-hand side in ascending powers of k (corresponding to λ decreasing from large values) we obtain

$$gh + (-\tfrac{1}{3}gh^3 + \rho^{-1}Th)\,k^2 + O(k^4), \tag{59}$$

where the two terms in the coefficient of k^2 represent, respectively, the two opposing tendencies just mentioned. Evidently, they cancel for the particular water depth

$$h = (3T/\rho g)^{\frac{1}{2}} = 5\text{ mm for water}, \tag{60}$$

which was that mentioned in section 1.7 as used in the ripple-tank simulation of sound waves.

Figure 57, which plots c against $\lambda = 2\pi/k$ according to the exact dispersion relationship (58) for various water depths, confirms that $h = 5$ mm is the value that closely maintains constancy of c down to wavelengths as low as possible; an accuracy of 3 % being achieved whenever λ exceeds about 2 cm (just 4 times the depth, instead of 14 times as condition (38) gives for pure gravity waves). The corresponding wave speed, about 22 cm s^{-1}, allows a visual representation of acoustic phenomena enormously slowed down (section 1.7).

We conclude this section with a discussion of the energy carried by ripples, coupled with a sketch of the relationships between the idea of surface tension and the general physics of liquids. Contributions to the internal energy of a liquid are made both by the average kinetic energy associated with the thermal motions of molecules and by the average potential energy associated with the forces of interaction between molecules. These contributions are of opposite sign because, except in statistically rare cases of molecules that have approached quite unusually close to other molecules, the forces of interaction are *attractive*, producing negative potential energies. Indeed, the internal energy per unit mass, called E here as in section 1.2, is far smaller (for example, by three orders of magnitude for water) than the kinetic energy per unit mass, whose huge value is nearly cancelled by a numerically nearly equal *negative* potential energy per unit mass.

Statistically uniform contributions to the liquid's negative potential energy are made at all positions in the liquid, with one exception. Molecules *extremely* close to the free surface can interact with only a *smaller than average number* of molecules; accordingly, the contribution to the liquid's negative potential energy made by such molecules extremely close to the free surface is numerically only a fraction of that of molecules in general. This affects the total internal energy of the liquid by adding to the contribution which is statistically uniform, and therefore directly proportional to mass, a *positive* correction proportional to the area of the free surface and associated with the deficiency in negative contribution from molecules within about 1 molecular diameter of the free surface. Quite a significant contribution may be made from such an exceedingly thin layer because the reduced negative potential energy per molecule is far larger numerically than the *net* internal energy per molecule.

Accordingly, the internal energy of a mass M of liquid whose free surface has area A can be written

$$ME + AT \tag{61}$$

in terms of E, the internal energy per unit mass of bulk liquid, and T, the positive correction required per unit area of free surface. It follows that a small change of shape in a fixed mass M of liquid produces an energy change $T\delta A$, proportional to any small change δA in its free-surface area.

In other words, the work that must be done to stretch the free surface takes exactly the value $T\delta A$ indicated by the idea that the stretching is resisted by a *surface-tension* force T per unit length of free-surface frontier. We see that the quantity T can be viewed *either* as an effective tensile force per unit length of frontier acting within the free surface and measured in $\mathrm{N\,m^{-1}}$ (newtons per metre), *or* as an additional potential energy per unit area of free surface, measured in $\mathrm{J\,m^{-2}}$ (joules per metre squared). Although the quantity's significance is made physically much clearer from the latter standpoint, the former, more elementary idea must lead to identical conclusions.

For simplicity of exposition the above discussion has been tacitly based on an approximation, to the effect that the change in *internal energy* when the free surface is stretched gives the work done; in terminating it we note that the *free energy* (internal energy minus the product of temperature and entropy) is the quantity which from the standpoint of accurate thermodynamics changes by an amount equal to the work input (with any isothermal heat input excluded). It is strictly correct, therefore, to amend (61) so as to make T the change in *free energy* per unit change in area for a given mass of liquid, and to identify as the surface's contribution to the potential energy of a wave the excess free energy that it carries.

Any sinusoidal waves on a water surface increase its area by a proportion which, averaged over a wavelength, may be written as the average of

$$[1 + (\partial\zeta/\partial x)^2]^{\frac{1}{2}} - 1; \tag{62}$$

on a small-amplitude approximation, this is the average of $\frac{1}{2}(\partial\zeta/\partial x)^2$, which for waves of wavenumber k can be expressed as

$$\tfrac{1}{2}k^2\overline{\zeta^2} \tag{63}$$

in terms of the mean square vertical displacement $\overline{\zeta^2}$. This additional surface area (63) per unit area, when multiplied by the surface tension T, becomes the additional potential energy per unit area which, with the gravitational contribution $\frac{1}{2}\rho g\overline{\zeta^2}$, makes a total potential energy per unit area

$$\tfrac{1}{2}(\rho g + Tk^2)\,\overline{\zeta^2}. \tag{64}$$

We arrive thus at the conclusion that the effect of surface tension on wave energy, just as on every *other* aspect of sinusoidal waves on water, is represented accurately if we make the substitution (50) in the pure gravity-wave value. The generalised stiffness of the free surface is increased by a factor $1 + (\rho g)^{-1}Tk^2$, while the frequency squared (51) for the associated vibrations increases (as expected on general grounds) by the same factor. Finally, making the substitution (50) in (28) gives the total wave energy, W.

3.5 Attenuation

Those sinusoidal waves on a water surface that are described in sections 3.1–3.4 suffer attenuation through three main processes of energy dissipation, each of which is somewhat analogous to one of the attenuative mechanisms studied already in sections 1.13 and 2.7. *Bottom friction* (analysed first in this section) is the most important wherever the water depth is substantially less than a wavelength so that the waves induce significant horizontal motions near the bottom (figure 55); the associated energy dissipation takes place in a boundary layer between them and the solid bottom, like that described for long waves in section 2.7, with a thickness δ given (when it is laminar) by the graph marked 'water' in figure 26.

Waves on deep water, on the other hand, involve no movements, and therefore no frictional dissipation, near the bottom. Their attenuation proceeds at a relatively slower rate such as is associated with the mechanism which we next study: *internal dissipation* by viscous stresses acting throughout the wave. This, rather like the viscous contribution to sound-wave attenuation analysed in section 1.13, becomes substantial only at quite low values of the wavelength.

Thirdly, we describe *surface dissipation*: a source of attenuation associated with departures of the surface tension T from the value that it takes in conditions of equilibrium. This mechanism, reminiscent of the 'lag' contribution to acoustic attentuation (section 1.13) may be especially important for water covered with a thin film of surface contaminant.

First, then, we estimate wave attenuation by bottom friction, making the assumption that the boundary layer is laminar, though our *method* of estimation is found appropriate also to motions on such a scale that turbulent friction predominates. We recall from section 2.7 that a sinusoidally oscillating pressure gradient $(-\partial p_e/\partial x)$ near a plane wall produces within a thin laminar boundary layer velocity components u varying as shown in figure 25 *with the distance from the wall* (taken there as z, corresponding to $z+h$ in this chapter). The broken line represents a part in phase with the forces $(-\partial p_e/\partial x)$, which directly counteract the viscous resistance to those motions; while the plain line represents motions which lag in phase by 90° behind the forces because they react *inertially*, as does the irrotational motion

$$u_{\text{ex}} = (\rho i\omega)^{-1}(-\partial p_e/\partial x) \tag{65}$$

external to the boundary layer. Within the boundary-layer thickness δ there is a net defect of volume flow past unit breadth of wall, relative to the volume flow in irrotational motion, of

$$\int_{-h}^{-h+\delta} (u_{ex} - u)\, dz = (\rho i\omega)^{-1}(-\partial p_e/\partial x)(v/i\omega)^{\frac{1}{2}}. \tag{66}$$

The real and imaginary parts of (66) are of equal magnitude: the volume-flow defect in phase with u_{ex} (the net area by which the plain line in figure 25 falls below the dotted line) is equal to the defect that lags by 90° behind u_{ex} (the area beneath the broken line).

The effect of such a boundary layer on waves in water, of a depth h assumed much larger than the boundary-layer thickness, is made easy to estimate through the mathematical concept of the *displacement thickness* δ_1 defined by the equation

$$\int_{-h}^{-h+\delta} (u_{ex} - u)\, dz = u_{ex}\delta_1, \tag{67}$$

where from (65) and (66) we have

$$\delta_1 = (v/i\omega)^{\frac{1}{2}} = (v/2\omega)^{\frac{1}{2}}(1-i). \tag{68}$$

This concept states that *irrotational* motions outside the boundary layer attached to a wall $z = -h$ are just those which would be calculated neglecting viscosity in a region with boundary

$$z = -h + \delta_1, \tag{69}$$

where a condition $\partial\phi/\partial z = 0$ of zero flow through the boundary is applied. This is because equation (67) implies a volume flow

$$\int_{-h}^{-h+\delta} u\, dz = u_{ex}(\delta - \delta_1) \tag{70}$$

in the boundary layer, exactly as would be the case if the velocity took the irrotational-flow value u_{ex} throughout it but the lower boundary were shifted to $-h+\delta_1$. Because this position (69) of boundary, for calculations of the irrotational flow neglecting viscous effects, yields the correct distribution of volume flow parallel to the boundary, a condition of zero flow through it can appropriately be applied.

Texts on fluid dynamics develop at length the idea of displacement thickness along lines briefly sketched above, at least for *steady-flow* boundary layers; then δ_1 is a real number and the idea has the simple physical interpretation that the boundary layer's effect on the external irrotational flow is just as if, with viscous effects absent, the solid boundary were displaced a distance δ_1 into the fluid. It may at first seem confusing that for an *oscillatory flow* boundary layer δ_1 is complex so that this simple physical interpretation cannot be used. Nevertheless, the mathematical idea of the identity between the flow external to the boundary layer and a solution of the irrotational-flow equations with (69) as boundary remains equally valid, and is just another illustration of the effectiveness of complex-number treatments of oscillations and waves.

For sinusoidal waves of wavenumber k the frequency obtained by solving Laplace's equation (5) for purely irrotational flow in water of depth h is

$$\omega_{\mathrm{irr}} = [(g + \rho^{-1}Tk^2)\,k\tanh kh]^{\frac{1}{2}}. \tag{71}$$

It follows that the frequency ω in the presence of a viscous boundary layer takes the same value (71) but with the depth h given a small change to $h - \delta_1$. This changes ω to

$$\omega = \omega_{\mathrm{irr}}[1 - \delta_1\,\partial(\ln\omega_{\mathrm{irr}})/\partial h]$$

$$= \omega_{\mathrm{irr}}[1 - (\nu/2\omega)^{\frac{1}{2}}(1-\mathrm{i})\,k/(\sinh 2kh)]. \tag{72}$$

The small negative real part of the correction factor in square brackets is relatively unimportant; it implies only a small proportional increase in the oscillation period above $2\pi/\omega_{\mathrm{irr}}$, due to viscous action. By contrast, the small positive imaginary part of the quantity in square brackets corresponds, by equation (14), to an exponential attenuation of amplitude with time. Although in each period $2\pi/\omega_{\mathrm{irr}}$ there is only a small proportional loss of amplitude

$$2\pi(\nu/2\omega)^{\frac{1}{2}}\,k/(\sinh 2kh), \tag{73}$$

the resulting attenuation when accumulated over many periods can be very great.

The proportional *energy* loss per period, which is of course double the value in (73), can be written

$$[2\pi(\nu/2\omega)^{\frac{1}{2}}\,h^{-1}]\{(2kh)/(\sinh 2kh)\}. \tag{74}$$

The square bracket in expression (74) represents the proportional loss of wave energy in each period due to bottom friction as calculated in section 2.7 in the long-wave limit; this, to a close enough approximation, is the ratio of the boundary-layer thickness (graph marked 'water' in figure 26) to the depth. Expression (74) reconfirms this limiting value because as $kh \to 0$ the modifying factor in curly brackets tends to 1. Figure 58 shows how this modifying factor falls below 1, as λ/h decreases, till it enormously

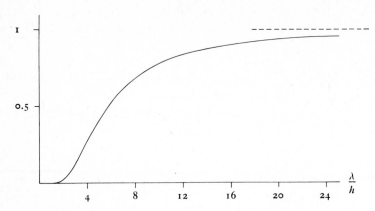

Figure 58. Ratio of energy loss by bottom friction to its value in the long-wave limit, plotted against the wavelength-depth ratio λ/h. (This is a graph of the modifying factor in curly brackets in (74).)

reduces the proportional energy loss per period below its long-wave value as less and less of the wave energy is located where bottom friction can help to dissipate it.

For waves on *deep* water, with (say) $h > 0.5\lambda$, figure 58 confirms the expected lack of significant attenuation by bottom friction. These waves do *not* induce near the solid bottom any motions which can lead, in an intervening boundary layer, to specially enhanced shearing involving substantial viscous dissipation of energy; consequentially, the dominant cause of attenuation may be internal dissipation in shearing motions within the main body of the wave.

Just as a special argument involving the displacement thickness δ_1 simplifies the analysis of water-wave attenuation by viscous dissipation within the boundary layer (and the results, just as in section 2.7, can be checked by a direct calculation of the dissipation rate) so a special simplifying argument suggested by Stokes readily determines, in water of arbitrary depth, the attenuation due to viscous dissipation *outside* any boundary layer. The associated rate of wave energy loss may simply be added to that resulting from dissipation within the boundary layer, although in the commonest circumstances either the one or the other is dominant.

Sinusoidal water waves propagating in the x-direction involve, as we have seen, water motions in only the x- and z-directions, subject of course to the equation of continuity

$$\partial u/\partial x + \partial w/\partial z = 0. \tag{75}$$

In such motions the x- and z-components of the stress tensor take the form

$$p_{xx} = p - 2\mu\partial u/\partial x, \quad p_{xz} = -\mu(\partial u/\partial z + \partial w/\partial x), \\ p_{zx} = -\mu(\partial u/\partial z + \partial w/\partial x), \quad p_{zz} = p - 2\mu\partial w/\partial z, \tag{76}$$

in terms of deviations (see section 1.13) from a 'mean pressure' p (there called p_m) resulting from rates of shear $\partial u/\partial z$ and $\partial w/\partial x$, or from that combination of equal and opposite simple extensions $\partial u/\partial x$ and $\partial w/\partial z$ which is equivalent to a shearing motion in different axes.

We recall also from section 1.13 the definition of the components (76) which makes, for example, p_{xx} and p_{xz} the x- and z-components of the rate of flux of x-momentum across unit area per unit time. The divergence of that flux, which by (75) and (76) is

$$\partial p_{xx}/\partial x + \partial p_{xz}/\partial z = \partial p/\partial x - \mu\nabla^2 u, \tag{77}$$

gives the net rate of loss of x-momentum from unit volume of fluid; thus, the viscous stresses modify the x-component of the linearised momentum equation (4) to

$$\rho_0 \partial u/\partial t = -\partial p_e/\partial x + \mu\nabla^2 u \tag{78}$$

while, similarly, the z-component becomes

$$\rho_0 \partial w/\partial t = -\partial p_e/\partial z + \mu\nabla^2 w. \tag{79}$$

Stokes pointed out that these equations of motion for a viscous fluid are satisfied exactly when we represent water waves by a solution of Laplace's equation (5) for irrotational flow, since by differentiating (5) with respect to x and z respectively we deduce that the terms carrying the coefficient μ in (78) and (79) are identically zero. Accordingly, it is *only the boundary conditions* proper to a viscous fluid which such a solution fails to satisfy through, for example, inducing nonzero horizontal motions at the bottom; we have explained how to remedy this failure by insertion of a boundary layer.

On the other hand, we can *imagine* an experiment where no boundary layer would appear because an extensible solid bottom was caused to make, in its own plane, just those horizontal back-and-forth movements that particles of fluid are supposed to make near the bottom (figure 55) according to irrotational-flow theory. In such an imagined motion *without* a bottom boundary layer any energy dissipation must be that *internal* dissipation which we are now seeking to calculate. Note that the motions of the bottom in its own plane would do no work on the fluid because in irrotational flow, by (76), the tangential stress is

$$p_{xz} = -2\mu\,\partial^2\phi/\partial x\partial z, \tag{80}$$

a quantity vanishing on $z = -h$, where $\partial\phi/\partial z = 0$ for all x.

In this imagined experiment, however, the boundary condition at the *surface* would still not be satisfied exactly by the irrotational flow. For example, the tangential stress (80) does *not* vanish at the surface. It represents an x-component of force per unit area with which the water is acting on a thin surface layer, equilibrium of which would therefore be possible only if an equal and opposite external x-component of force were applied to that layer. Again, our boundary condition for the irrotational motion (namely, (48) when surface tension is taken into account) is deduced by assuming that the z-component of force per unit area with which the water acts on a thin surface layer (counteracting the other forces $p_0 - T\partial^2\zeta/\partial x^2$) is the pressure $p = p_0 + p_e$. Equation (75) shows, however, that the true z-component of force per unit area exceeds p by an amount

$$-2\mu\,\partial w/\partial z = -2\mu\,\partial^2\phi/\partial z^2, \tag{81}$$

and we infer again that the motion could be sustained only if an equal and opposite external z-component of force were applied to the surface layer.

Maintenance of the irrotational motions of the linear theory of surface waves would require, then, not only those bottom movements required to maintain the no-slip condition but also an external force

$$(2\mu\,\partial^2\phi/\partial x\partial z,\, 0,\, 2\mu\,\partial^2\phi/\partial z^2) \tag{82}$$

per unit area applied to the free surface. Unlike the bottom movements, this force would do *work* on the fluid at a rate

$$2\mu[(\partial\phi/\partial x)\,\partial^2\phi/\partial x\partial z + (\partial\phi/\partial z)\,\partial^2\phi/\partial z^2]_{z=0} \tag{83}$$

per unit area, where on linear theory the free-surface value can be replaced by the value on $z = 0$ because the difference is of the order the cube of the disturbances.

Stokes' ingenious idea was to recognise that the average value of this rate of working (83) required to maintain the unattenuated irrotational motions of sinusoidal waves must exactly balance the rate at which the same waves when propagating freely would lose energy by internal dissipation! Furthermore, the average of (83) for sinusoidal waves of wavenumber k is the average of

$$2\mu[k^2\phi\,\partial\phi/\partial z + k^2\phi\,\partial\phi/\partial z]_{z=0}, \tag{84}$$

where the two terms in (83) have been reduced to identical forms, respectively by writing the mean product of the x-derivatives of two sinusoidal quantities as k^2 times the mean product of the quantities themselves and by making the substitution $\partial^2\phi/\partial z^2 = k^2\phi$ (equation (16)).

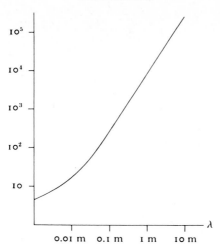

Figure 59. Number of periods required for the energy of sinusoidal waves of length λ on deep water to be reduced by a factor e through internal viscous dissipation.

Surface waves lose energy by internal dissipation, then, at a rate per unit area which is the average of (84). According to (26), this is $8\nu k^2$ times the average kinetic energy per unit area, or $4\nu k^2$ times the total wave energy (of which exactly half is kinetic). Thus, internal dissipation produces a *proportional* loss of wave energy $4\nu k^2$ per unit time, or

$$8\pi\nu k^2\omega^{-1} \tag{85}$$

per period. This value may be added directly to the contribution (74) from bottom friction, which, however, is absent for waves on deep water.

Figure 59 shows the number of periods required for the energy of sinusoidal waves on deep water to be reduced by a factor of e through internal viscous dissipation; thus, the reciprocal of (85) is plotted as a function of wavelength. Typical gravity waves are seen to be very lightly damped: the time needed for the energy in waves of length 1 m and 10 m to be reduced by a factor of e is 8000 periods and 250000 periods respectively. Even quite short gravity waves with $\lambda = 0.1$ m still require 250 periods; waves with $\lambda = 0.01$ m under the combined influence of gravity and surface tension, however, decay far more rapidly, needing only 16 periods, and the energy of pure capillary waves at the very low wavelength 0.001 m is reduced by a factor of e in 4 periods. Those results can be summed up in the comment made already in section 3.4: among waves on deep water, it is the *ripples* which are pre-eminently susceptible to viscous damping.

All the above results are stated in terms of a decay time, in periods, for a sinusoidal wave train infinite in extent. We have not, as yet, studied the properties of wave trains with spatial gradation of amplitude; these are analysed in the immediately following sections. It would be natural to suppose that waves generated by a fixed source oscillating at a fixed frequency should lose energy with distance from it at the same proportional rate *per wavelength* as the infinite sinusoidal train loses *per period* (that is, (74) due to bottom friction and (85) due to internal dissipation). This supposition, however, would be mistaken: it tacitly assumes that energy travels at the wave velocity c instead of at a speed calculated below (sections 3.6 and 3.8) as varying from $\frac{1}{2}c$ for gravity waves to $\frac{3}{2}c$ for capillary waves. In these two cases, respectively, the quantity plotted in figure 59 must be decreased or increased by 50 % to give the number of *wavelengths* in which the energy carried by the propagating wave signals falls by a factor of e. Even so, for $\lambda > 30$ m this distance can be halfway round the globe and it indeed proves possible to observe gravity waves in that range of wavelengths travelling over such an oceanic distance of 2×10^7 m, even though some additional *nonlinear* effects of interaction with other wave components may over such great distances act to extract energy from them.

The foregoing discussion, based on an imaginary experiment with the unattenuated irrotational motions of a sinusoidal wave maintained by boundary forcing, including the application of a force (82) per unit area of free surface, suggests the question: in what way *besides* attenuation may wave motions under ordinary conditions (with no external force applied) deviate from those irrotational motions? For example, how does the tangential stress p_{xz} make the transition from its irrotational-flow value (80) to zero at the surface?

The answer, easy to calculate using the fact that derivatives of u and w appearing in p_{xz} satisfy equations of exactly the same form as the equations (78) and (79) for the velocity, is that p_{xz} falls from its external irrotational-flow value to zero at the surface according to the laws of the Stokes boundary layer represented graphically in figure 25. Boundary layers at the surface, where shearing stress falls to zero, and at the bottom, where tangential velocity falls to zero, have the same structure and the same small thickness δ given by the graph marked 'water' in figure 26.

Wave attenuation by internal dissipation is *not* significantly affected by the circumstance that the dissipation associated with shearing stress falls to zero within such a very thin surface boundary layer from its 'internal' level within the wave. We conclude this section, however, by mentioning circumstances when greatly *enhanced* dissipation can occur within a surface boundary layer.

It is departures of the surface tension T from its equilibrium value that can result in such surface dissipation. In a fluid such that small wave motions generate small variations in T, a net x-component of force

$$(\partial T/\partial x)\,\delta x \qquad\qquad (86)$$

must act on a strip of surface of width δx with frontiers of unit length parallel to the y-axis, even though on linear theory the same small variations make no change to the z-component (47). In the surface boundary layer, therefore, the tangential stress changes from (80) not to zero but to the value

$$p_{xz} = -\,\partial T/\partial x \qquad\qquad (87)$$

needed to balance the x-component $(\partial T/\partial x)$ of surface force per unit area.

There are conditions when in the surface boundary layer the tangential stresses increase in magnitude so enormously from the internal value (80) to the surface value (87) that the resulting *surface dissipation* (extra viscous dissipation due to *enhanced* shearing stresses within the surface boundary layer) greatly exceeds the rate of internal dissipation. This is the mechanism responsible for the proverbial calming effect of 'oil on troubled waters'.

Although the physics of surface-tension variability is highly complicated, a qualitative indication of why T may vary as waves stretch and unstretch a water surface is suggested by the physical principle that the free energy is a *minimum* in thermodynamic equilibrium; accordingly, when an area A of water surface is *rapidly* stretched to a value $A+\delta A$ the increase in free energy can significantly exceed the value that it would take in a slower, reversible change through a sequence of equilibrium states. Quantitatively, the resulting enhancement of tension in a stretched surface is found to be particularly great for water covered by an exceedingly thin layer of a contaminant such as oil.

3.6 Introduction to group velocity

Water waves of sinusoidal shape are analysed at length in sections 3.2–3.5. Detailed water movements in waves of small amplitude are derived (figures 50 and 55), together with their energy content and, also, estimates (figures 58 and 59) of the rate at which that energy may be dissipated. Above all, those sections show how *wave speed* varies as a function of wavelength (figures 52, 56 and 57): the 'dispersive' property of water waves.

Reasons for paying such close attention to the apparently very special case of sinusoidal waves of small amplitude are given in sections 3.1 and 3.2. On the one hand, waves on a water surface are not uncommonly found with shapes *roughly* as in a sine wave of small amplitude; in particular, as remarked in section 3.2, they may have 'long crests' in relation to their wavelength so that they do approximately satisfy equations based on regarding crests as infinitely long.

A much stronger incentive to understanding sinusoidal waves of small amplitude on a water surface comes from the fact that an excellent way (and often the only way) of studying how the water surface may react to a small disturbance of more complicated shape is by *Fourier analysis*. This allows such a disturbance to be regarded as a linear combination of different sinusoidal disturbances; each of those, separately, should behave as calculated in sections 3.2–3.5. Furthermore, the fact that the equations governing small disturbances to the water surface are linear equations (Laplace's equation (5) with the boundary condition (13) satisfied on $z = 0$) means that such a *linear combination* of different sinusoidal solutions is still a solution.

There is an interesting link between the above two reasons for studying sinusoidal waves. Suppose, for example, that some localised storm of limited duration has generated ocean waves. Then we shall find that *different sinusoidal components*, into which the subsequent disturbance to the water surface may be analysed, are at some much later time *found in quite different places*. We can say that they are 'dispersed', according to that word's *ordinary meaning*, and that this is a consequence of water waves being 'dispersive' in the scientific sense (wave speed varying with wavelength). Waves of a particular length and direction of propagation are found, at that later time, around a particular place. This is one possible cause (others are given in sections 3.9 and 3.10) why, as remarked earlier, the waves observed at a particular place may be roughly sinusoidal.

Note that any such dispersion of the original energy over a large area reduces the energy *per unit area*; which in turn, by (28), reduces the wave amplitude. Small-amplitude theory *can* accordingly be used to describe the above-mentioned later stages of the disturbance's development (often described as the propagation of 'swell' away from the storm region). Disturbances during the storm itself may be large, causing the boundary condition to be far more complicated than its linearised form (13). Nevertheless, the wave energy does still get dispersed over a large area so that, from a certain time onwards, it is appropriate to use linear theory to analyse how the 'swell' is propagated.

A large stone thrown into a pond similarly causes a complicated initial disturbance, but soon afterwards we see a much more regular, concentric pattern of circular wave crests spreading outwards. The form of the water surface between two crests is approximately a sine wave. We find greater wavelengths around the outer edge of the pattern and smaller wavelengths nearer the centre. This is what we might expect from the fact (figure 52) that wave speed increases with wavelength.

A natural conjecture, indeed, might be that, at a time t after the initial disturbance, the waves of speed c will be found at a distance of about ct from the position of that disturbance Such a conjecture would, however, be seriously wrong. Surprisingly enough, it is an overestimate, for a not too shallow pond, by a factor of about 2. The small group of waves whose wave speed is c is found, in fact, at a different distance of about Ut from the initial disturbance, where U is a quantity called the *group velocity* which, for 'waves on deep water', is $\frac{1}{2}c$.

Chapter 3, as stated at its outset, is an introduction not only to the properties of water waves but also to the properties of group velocity. We shall find that this group velocity U is different from the wave speed c in any *dispersive* system (where c varies with wavelength). Sections 3.6, 3.7 and 3.8 give the general theory of group velocity for *isotropic* dispersive systems (where c does not vary also with direction). This is where the copious data on the dispersive (and other) properties of sinusoidal water waves given in the previous four sections find another major use: they are ideally suitable for illustrating this general theory.

Perhaps the most fundamental property of group velocity may be mentioned at once although a demonstration is postponed to section 3.8. The energy of sinusoidal waves is propagated, *not* at the wave speed c, but at the group velocity U.

This statement is in no way contradicted by the proof in section 1.3 that, for plane sound waves, the energy flux has magnitude c times the energy density, so that the energy of sound waves is propagated at the wave speed c. In fact, for all *nondispersive* waves, including sound waves, the group velocity U and the wave speed c coincide.

Anyone whose quantitative knowledge of waves is confined to nondispersive systems is likely to find the properties of group velocity quite surprising. After all, a striking feature of waves is their capability of carrying energy over long distances. With many waves, furthermore, the speed of movement of crests and troughs is extremely evident. It is natural to imagine, then, that this 'wave speed' c is also the velocity with which energy is propagated by the wave. For a majority of the wave motions occurring in nature, however, the wave speed at which crests and troughs are propagated is quite different from the group velocity at which energy is propagated.

Chapter 3 analyses isotropic systems where (as we shall see) energy is propagated at right angles to the crests. If the crests form a pattern of concentric circles, as when a large stone is thrown into a pond deep enough (section 3.2) for the resulting waves to be unaffected by non-uniformities of depth, the energy of the disturbance must be travelling outwards radially from the centre. Then, as stated earlier, the longest waves produced (typically, a few times bigger than the stone) appear at the outside of the expanding concentric pattern.

It is extremely instructive to watch those outside waves. Anyone observing, however carefully, the progress of one of the crests will suddenly lose sight of it! It seems an optical illusion at first; the possible result of mistaken identity between that crest and the next one coming along behind, to which the observer's gaze is now transferred; but then this next crest, too, disappears! Meanwhile, crests are coming along thick and fast behind. Indeed, at the inside edge of the concentric pattern, new crests are appearing 'from nowhere'; that is, from the now calmed central water.

The suddenness with which sizeable crests near the outside of the pattern disappear rules out gradual attenuation (section 3.5) as the mechanism. We shall see that the true explanation is that crests travel at a wave speed c twice as big (for 'waves on deep water') as the group velocity U with which the *energy* in waves of their length is moving forwards. Each crest of every wave is thus outstripping the associated energy, so that crests can survive *only* by evolving into crests of longer waves. That, however, is impossible for waves at the outside of the group (since the original disturbance produced no energy in waves of *greater* wavelength), and so crests there can only disappear.

At least four different methods of analysis help in giving understanding of this important subject of group velocity. Here, we postpone analysis by two methods already mentioned, the Fourier-analysis method and the energy-flux method, to section 3.7 and 3.8 respectively; postponing also a still more general approach to chapter 4, we briefly develop in this section a rather simple analysis that indicates, at least in a limited range of cases, the essential properties of group velocity and how its value is readily found.

The simple analysis applies only to those later stages of dispersion when, as described above, waves spreading out from an initial disturbance have become an extended group of crests with gradually varying wavelength (distance between successive crests). The question asked is how that spatial distribution of crests and wavelengths *varies with time*.

The method is concerned with isotropic wave propagation in general. Nevertheless, it is, in a certain sense, one-dimensional: it analyses spatial distribution only at right angles to the crests. Thus, for isotropic waves spreading two-dimensionally (say, water waves with concentric circular crests), it analyses the radial distribution of crests and wavelengths. It similarly analyses radial distribution for spherically propagating waves in an isotropic three-dimensional system. The method is applicable also to the common case of one-dimensional dispersive systems. Water waves themselves may form such a system, as when a train of water waves with parallel straight crests propagates at right angles to them (possibly, along a canal). In all of these cases, we use x as the coordinate measured at right angles to the crests.

The method assumes that, because the wavelength is varying only gradually (by a very small fraction of itself from one wave to the next), a local *phase* α can be defined. This is a quantity whose value at every crest is an even multiple of π; in fact, it is $2n\pi$, where the integer n *increases by* 1 for each successive crest that passes a particular point. Between crests, α varies smoothly; at troughs, for example, α is an odd multiple of π. The rate of change of α with time is 2π divided by the wave period, and this of course is the frequency ω in radians per second.

The phase α is a function not only of the time but also of the x-coordinate (distance at right angles to wave crests). At a given time, α shows a decrease with x, at a rate equal to the wavenumber k in radians per metre. (Since $k = 2\pi/\lambda$, this corresponds to the fact that α decreases with x by 2π between crests a distance λ apart; a *decrease* because α was defined so that crests arriving later have greater values of α.) Thus, $\alpha(x, t)$ is a function satisfying

$$\partial\alpha/\partial x = -k, \quad \partial\alpha/\partial t = \omega. \tag{88}$$

The use of this smoothly varying phase function $\alpha(x, t)$ implies a representation of (say) the water-surface displacement ζ in a form such as

$$\zeta = \zeta_1(x, t) \exp[i\alpha(x, t)], \tag{89}$$

where $\zeta_1(x, t)$ is a positive, slowly varying amplitude. Around a particular position x_0 and a particular instant t_0 the phase function is nearly linear; in fact,

$$\alpha(x, t) \doteqdot \alpha(x_0, t_0) - k_0(x - x_0) + \omega_0(t - t_0), \tag{90}$$

where $-k_0$ and ω_0 are the values of the derivatives (88) at (x_0, t_0). *Locally*,

9   L W F

then, equation (89) with ζ_1 slowly varying makes ζ *nearly a sine wave* (with wavenumber k_0 and frequency ω_0) as assumed earlier.

The equations (88) imply a relationship

$$\partial k/\partial t + \partial \omega/\partial x = 0 \tag{91}$$

between wavenumber and frequency. This can be thought of as an *equation of continuity* for phase. That is because k is a sort of phase density (phase per unit length) and ω is a sort of phase flow rate (phase passing a fixed point in unit time). The substitution $\omega = kc$, where c is wave speed, makes (91) particularly like a hydrodynamical equation of continuity (see section 2.1, where ρA corresponds to k and u to c); and, indeed, wave speed is sometimes called 'phase velocity'.

The dispersion relationship for an isotropic system specifies either c as a function of $k = 2\pi/\lambda$ or (as in (18) or (35)) $\omega = kc$ as a function of k. We use the latter form

$$\omega = \omega(k), \tag{92}$$

and assume that this dispersion relationship derived for exactly sinusoidal waves holds also to good approximation for the waves here considered (which locally are of nearly sinusoidal shape).

Substitution of (92) into (91) gives immediately

$$\partial k/\partial t + U\partial k/\partial x = 0, \tag{93}$$

where $$U = U(k) = d\omega/dk \tag{94}$$

will be called the group velocity. Equation (93) means that k is constant along paths in the (x, t) plane satisfying $dx/dt = U$; but if k is constant then U is constant, so these paths are straight lines

$$x - Ut = \text{constant.} \tag{95}$$

(Such reasoning is, of course, familiar from sections 2.8 and 2.9.)

Thus, in the case treated, the basic fact about group velocity has already been derived. As figure 60 indicates, waves of a given length $2\pi/k$ are found at points along such paths (95), and there is a different path for each wavelength. While a time t elapses, the position where waves of that length can be found moves on a distance Ut. Anyone who imagined that, because the crests travel at the wave speed $c = \omega/k$, those waves of length $2\pi/k$ should after time t be observed at the distance ct, would fail to find them! There he would find waves, if any, of quite different length. Only if he rigorously fixed his gaze on a point moving with the group velocity $U = d\omega/dk$ would he find that he was continually observing waves of the same length.

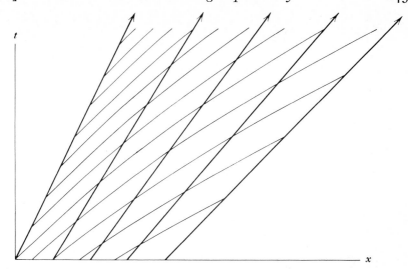

Figure 60. A wave group plotted on an (x, t) diagram. The thin curves indicate the paths of wave crests that have appeared 'from nowhere' at the back of the group and will disappear at the front. Along each thick straight line $x - Ut =$ constant the wavelength $\lambda = 2\pi/k$ remains constant.

In the special case of gravity waves on deep water, the equation (92) relating ω and k takes the simple form (18), leading to the conclusion

$$U = \mathrm{d}\omega/\mathrm{d}k = \mathrm{d}[(gk)^{\frac{1}{2}}]/\mathrm{d}k = \tfrac{1}{2}(g/k)^{\frac{1}{2}} = \tfrac{1}{2}(\omega/k) = \tfrac{1}{2}c. \tag{96}$$

Thus the group velocity in this case has the surprisingly simple form mentioned above: it is *half* the wave speed.

By contrast, any nondispersive waves have the wave speed c independent of k. Then $\omega = ck$ is a *constant* multiple of k and so

$$U = \mathrm{d}\omega/\mathrm{d}k = \mathrm{d}(ck)/\mathrm{d}k = c. \tag{97}$$

These nondispersive waves, for which there is no difference between group velocity and wave speed, include not only sound waves but also 'long' waves (with $c = (gh)^{\frac{1}{2}}$) on water of uniform depth h.

We can use the formulas of section 3.3 for general gravity waves on water of uniform depth to find the transition between the deep-water and long-wave results (96) and (97). Differentiating equation (35), we obtain

$$2\omega \frac{\mathrm{d}\omega}{\mathrm{d}k} = g\left(\tanh kh + \frac{kh}{\cosh^2 kh}\right), \tag{98}$$

so that

$$\frac{U}{c} = \frac{k}{\omega}\frac{\mathrm{d}\omega}{\mathrm{d}k} = \frac{1}{2}\left(1 + \frac{2kh}{\sinh 2kh}\right). \tag{99}$$

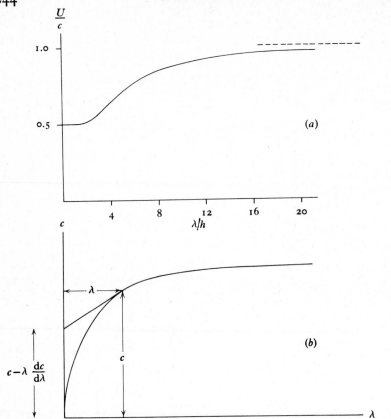

Figure 61. (*a*) Ratio of group velocity U to wave velocity c for gravity waves plotted against the wavelength–depth ratio λ/h. (*b*) A general geometrical construction to obtain U from the graph of c against λ, illustrated here for the case of gravity waves.

This ratio is plotted against λ/h, the ratio of wavelength to depth, in figure 61(*a*). Note that U/c rises only gradually from its deep-water value 0.5, and that even at the wavelength $14h$ used as the criterion for 'long' waves in section 3.3 the group velocity is still 6 % below the wave speed.

Another approach to these results is by rewriting the general expression (94) for U in terms of the relation between $c = \omega/k$ and $\lambda = 2\pi/k$ as

$$U = \mathrm{d}(kc)/\mathrm{d}k = c + k\,\mathrm{d}c/\mathrm{d}k = c - \lambda\,\mathrm{d}c/\mathrm{d}\lambda. \qquad (100)$$

This means that, given a curve of c against λ, we can find the group velocity U for given λ as the ordinate where the tangent at the point corresponding to waves of length λ intersects the vertical axis. This construction is shown in figure 61 (*b*), which throws further light on the gradual rise in U/c as λ increases.

One more implication of equation (100) is that any waves for which c *decreases* as λ increases have $U > c$. Thus, for ripples 'on deep water' (figure 56), the group velocity exceeds the wave speed whenever $\lambda < \lambda_m$. Indeed, for 'capillary waves' (ripples with $\lambda \ll \lambda_m$) equation (57) shows that

$$U = d[(\rho^{-1}Tk^3)^{\frac{1}{2}}]/dk = \tfrac{3}{2}(\omega/k) = \tfrac{3}{2}c. \tag{101}$$

This means that the energy in capillary waves of given length outruns the speed c of their crests by a factor $\tfrac{3}{2}$. Actually, the rather high attenuation rate for capillary waves (section 3.5) impedes easy observation of this relationship for waves generated by an initial local disturbance; however, a clear experimental demonstration that $U > c$ for capillary waves is described in section 3.9.

The conclusion that tangents to figure 56 intersect the vertical axis at an ordinate equal to the group velocity implies that the *minimum group velocity* corresponds to the inflexion point, where $\lambda = 2.54\lambda_m$, $c = 1.21c_m$ and $U = 0.767c_m$. By (56), then, waves on deep water have minimum group velocity $0.18\,\mathrm{m\,s^{-1}}$ (attained when $\lambda = 44\,\mathrm{mm}$ and $c = 0.28\,\mathrm{m\,s^{-1}}$). All the wave energy generated when a stone is thrown into a pond moves away faster than that, so that a calm region soon appears in the middle, surrounded by waves of lengths around 4 to 5 cm. Their crests seem (since $c > U$) to be emerging 'from nowhere'; that is, out of that central calmed region.

By contrast, in the shallow-water conditions of a ripple tank (see the curve marked $h = 5\,\mathrm{mm}$ in figure 57), the plot of c against λ is nearly horizontal for wavelengths exceeding 2 cm. Accordingly, ripples of such wavelengths exhibit negligible dispersion, which of course is what makes them suitable for visually representing sound waves.

This section has given just an introduction to group velocity, deriving rather simply, though not at all rigorously, the conclusion that the position where waves of given wavenumber k are found moves forward at velocity $U(k)$. Readers coming to group velocity for the first time can be assured that their grasp of the phenomenon will become much firmer if they go on to study the quite different types of analysis given in the next two sections.

3.7 The Fourier analysis of dispersive systems

Not only in water waves of small amplitude, but also in many other dispersive systems, the sinusoidal waves of each wavenumber have a definite wave speed (though not the same for each) and this suggests, as remarked at the beginning of section 3.6, the use of Fourier analysis to predict the development of disturbances of arbitrary shape. Such disturbances can be analysed, in fact, into a *linear combination* of sinusoidal waves; and we shall find that its *asymptotic evaluation* for large time both establishes rigorously the properties of group velocity stated in section 3.6 and goes further; determining, for example, the asymptotic behaviour of both the amplitude ζ_1 and the phase α in an expression like (89).

In order to develop the essential ideas without becoming entangled in the complexities of multi-dimensional Fourier analysis, we confine ourselves in this section to strictly one-dimensional propagation. Thus, we consider variations only in the x-direction. Also, we assume (ignoring attenuation until the end of the section) that a sinusoidal wave

$$a \exp\left[i(\omega t - kx)\right] \qquad (102)$$

(with constant amplitude a) is an exact solution of the system's linear equations of motion whenever the dispersion relationship (92) between ω and k holds.

For water waves, this implies that we analyse only disturbances independent of y, producing wave trains with straight crests all parallel to the y-axis. Waves like that might, for example, be generated in a deep canal (with sides parallel to the x-axis) by the sinking of a wide barge. For other examples of one-dimensional dispersive systems, see section 4.13. Although wave propagation in *more* than one dimension is of greater practical interest, a rigorous analysis of the one-dimensional case in this section may help to give quickly an improved and confident understanding of group velocity. It will also prepare readers for the fuller investigations of propagation in 2 or 3 dimensions outlined in chapter 4; those confirm, in the case of isotropic propagation as in a concentric pattern of water waves, an asymptotic behaviour closely as in the one-dimensional case; except that the amplitude includes an extra factor $x^{-\frac{1}{2}}$, corresponding to the wave energy moving outwards in ever widening circles of circumference $2\pi x$.

A very preliminary idea of how a linear combination of sinusoidal waves can have properties suggesting the concept of group velocity is obtained by

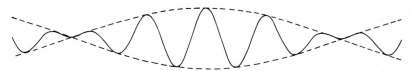

Figure 62. The linear combination (103) of two sinusoidal waves with the same amplitude and nearly the same wavenumber, plotted (full line) as a function of x. The broken line represents the wave envelope, enclosing 'packets' of waves that travel forward with the velocity (104).

combining just two waves with the same amplitude and *nearly* the same wavenumber. The addition formula for cosines shows that

$$a \cos(\omega_1 t - k_1 x) + a \cos(\omega_2 t - k_2 x)$$

$$= \{2a \cos[\tfrac{1}{2}(\omega_2 - \omega_1)t - \tfrac{1}{2}(k_2 - k_1)x]\} \cos[\tfrac{1}{2}(\omega_2 + \omega_1)t - \tfrac{1}{2}(k_2 + k_1)x].$$

$$(103)$$

Here, the factor in curly brackets is a slowly varying amplitude (with *small* wavenumber $\tfrac{1}{2}(k_2 - k_1)$) for the oscillations with much larger wavenumber $\tfrac{1}{2}(k_2 + k_1)$ represented by the last cosine. The combination therefore takes the form (figure 62) of a series of 'wave packets' travelling with the velocity

$$(\omega_2 - \omega_1)/(k_2 - k_1),$$

$$(104)$$

and incapable of exchanging energy with one another through the nodal points where that amplitude is zero. This suggests that, for a general disturbance represented by a Fourier integral, in which sinusoidal components of neighbouring wavenumbers do have approximately the same amplitude, the energy in those wavenumbers may travel with this velocity (104) whose limit as $k_2 \to k_1$ is the group velocity (94).

Although the above argument is far too special, it contains one idea applicable to much more general cases. A point such that

$$x/t = (\omega_2 - \omega_1)/(k_2 - k_1),$$

$$(105)$$

where the 'amplitude' (term in (103) in curly brackets) takes the full value $2a$, travels along with the velocity (104) for one very good reason: this allows the *phases* $\omega_1 t - k_1 x$ and $\omega_2 t - k_2 x$ of the two cosines on the left of (103) to remain *equal*, so that the two cosines always reach their maxima together; then, those maxima reinforce one another. Here we find once more (as at the end of section 1.11) the tentative idea, to be made precise in this section, that different sinusoidal waves may combine constructively, instead of interfering destructively, where their *phases are stationary*.

Now we advance from combining just two waves to combining an infinite number by considering the Fourier integral

$$\zeta = \int_0^\infty F(k) \exp\{i[\omega(k)\,t - kx]\}\,dk. \tag{106}$$

This combination of sinusoidal waves (102) takes the initial form*

$$f(x) = \int_0^\infty F(k) \exp(-ikx)\,dk \tag{107}$$

at time $t = 0$. In the case when this initial disturbance $f(x)$ is confined to a limited region, we investigate the asymptotic behaviour of (106) for large t.

Strictly speaking, we are not here considering the most general linear combination of one-dimensional waves (102): a dispersion relationship commonly has *more* than one solution for ω; often, as with either (18) or (35), *two solutions* equal in magnitude but opposite in sign. Then, to a general linear combination (106) of waves travelling in the positive x-direction, a combination

$$\int_0^\infty G(k) \exp\{i[-\omega(k)\,t - kx]\}\,dk \tag{108}$$

of waves travelling in the negative x-direction can be added; and, indeed, must be added if proper initial conditions (specifying *both* ζ and $\partial\zeta/\partial t$; that is, for water waves, the position and velocity of the surface) are to be satisfied. Since, however, the problems of asymptotically estimating the two sets of waves (106) and (108) for large t are mathematically equivalent, we henceforth concentrate entirely on the waves (106) travelling in the positive x-direction.

In that integral (106) we write the phase as $t\psi(k)$, where

$$\psi(k) = \omega(k) - (kx/t). \tag{109}$$

Then the integral becomes

$$\zeta = \int_0^\infty F(k) \exp[it\psi(k)]\,dk, \tag{110}$$

and we seek its behaviour as t becomes large *for a fixed value of x/t*. This corresponds to the idea (section 3.6) of 'the observer's gaze' moving at a fixed velocity.

* Note that, in accordance with the general practice of this book, the physical quantity $f(x)$ is given by the real part of the expression on the right-hand side of (107), where not only the exponential but also the amplitude $F(k)$ may be complex (see (112) below). Actually, this is the same as saying that $2f(x)$ is equal to the integral (107) taken from $-\infty$ to ∞, where $F(-k)$ is equal (again by (112)) to the complex conjugate of $F(k)$.

We shall prove that, as t becomes large, the integral (110) takes an asymptotic form determined entirely in terms of *where the phase* $t\psi(k)$ *is stationary*; that is, by the value (or possibly values) of k where $\psi'(k) = 0$. From (109), this is where

$$\omega'(k) = x/t; \tag{111}$$

corresponding, as expected, to where the observer's gaze moves at the group velocity (94). Physically, the idea of stationary phase is as hinted above: large numbers of different sinusoidal waves, when linearly combined into an integral like (110), may have rather a wide spread of phases, leading to a large measure of mutual cancellation by destructive interference, *except* where stationary phase allows mutual reinforcement of neighbouring wave components.

From the mathematical point of view, perhaps the clearest proof of this comes from Cauchy's theorem. The proof assumes that $\omega(k)$, and hence also $\psi(k)$, is an analytic function of k. When the initial disturbance $f(x)$ is confined to a limited region, then equation (107) implies that its Fourier transform,

$$F(k) = \pi^{-1} \int_{-\infty}^{\infty} f(x) \exp{(ikx)} \, dx, \tag{112}$$

being given by an integral over just a finite interval, is also an analytic function. Hence Cauchy's theorem can be applied in (110) to deform the path of integration from the real k-axis, on which $\psi(k)$ is real, to a path on which $\psi(k)$ has a positive imaginary part; *positive* so as to reduce the modulus of the integrand in (110) for large t.

It is, in fact, easy to make such a deformation of the path at all points except stationary points of $\psi(k)$. Suppose, for example, that there are no stationary points because $\psi'(k) > 0$ everywhere (this is commonly the case at, say, $x = 0$). Then we *raise* the path: that is, to each real k, we add on a small positive imaginary part which (because $\psi' > 0$) makes the imaginary part of $\psi(k)$ positive.

Note that, on a path of integration where the imaginary part of $\psi(k)$ is at least δ, the modulus of $\exp{[it\psi(k)]}$ is at most $\exp{(-t\delta)}$. Then the integral (110) also is exponentially small for large t; in fact, it is $O[\exp{(-t\delta)}]$, in the sense that its modulus satisfies

$$\left| \int_{0}^{\infty} F(k) \exp{[it\psi(k)]} \, dk \right| \leqslant [\exp{(-t\delta)}] \int_{0}^{\infty} |F(k)| \, dk \tag{113}$$

by the theorem that the modulus of any *integral* is less than or equal to the integral of the modulus of its integrand. Thus, in any such case when the

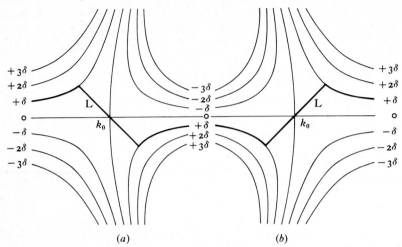

$$(a) \qquad\qquad\qquad (b)$$

Figure 63. Typical behaviour, near a stationary point k_0 of the phase function $\psi(k)$, of curves in the complex k-plane on which $\psi(k)$ has constant imaginary part (either -3δ, -2δ, $-\delta$, o, $+\delta$, $+2\delta$ or $+3\delta$).

    Case (a): $\psi''(k_0) < $ o (maximum).

    Case (b): $\psi''(k_0) > $ o (minimum).

We can deform the path of integration in (110) to one on which the imaginary part of $\psi(k)$ is $+\delta$ *except* near such a stationary point, where a link L must be used to cross from one such path to another (thick lines). Case of two stationary points: when $\psi(k)$ has a maximum followed by a minimum, cases (a) and (b) are combined. Then the path of integration in (110) may be deformed into one like that obtained by *joining up* the thick lines in both figures.

phase $\psi(k)$ has no stationary point, destructive interference is so effective that the disturbance becomes exponentially small.

    The argument is similar if $\psi'(k) < $ o everywhere (as may happen for *large* x/t), except that then the path has to be *lowered* in order to give the imaginary part of ψ a positive value δ. In fact, to achieve this, k needs to be given, approximately, the imaginary part

$$\delta/\psi'(k), \qquad\qquad\qquad (114)$$

which is negative if $\psi'(k) < $ o.

    The situation is fundamentally different, however, if $\psi(k)$ possesses a stationary point k_0, where $\psi'(k_0) = $ o. On one side of k_0, where $\psi'(k) < $ o, the path must be lowered; and on the other side, where $\psi'(k) > $ o, it must be raised. Therefore, continuity demands that the path shall go through k_0 itself, but this conflicts with the need for an imaginary part growing like (114) as $\psi'(k) \to$ o.

    Figure 63 shows, in fact, the typical hyperbolic behaviour of curves on which the imaginary part of ψ is equal to δ near such a stationary point k_0, both (*a*) if $\psi''(k_0) < 0$ and (*b*) if $\psi''(k_0) > 0$. Their shapes correspond to the fact that the Taylor expansion of $\psi(k) - \psi(k_0)$ is

$$\psi(k) - \psi(k_0) = \tfrac{1}{2}(k - k_0)^2 \psi''(k_0) + O(k - k_0)^3. \tag{115}$$

If the integral (110) is to be estimated effectively, the path of integration should be deformed as indicated in figure 63. Thus, at all positions away from a stationary point it is deformed into a path with the imaginary part of ψ at least equal to δ, giving very small contributions $O[\exp(-t\delta)]$ to (110) as a result of destructive interference; on the other hand, at any stationary point, an additional contribution (which is *not* exponentially small) is made from the link L between two such paths.

If the curves in figure 63 are regarded as the 'contour lines' on which the modulus of $\exp[it\psi(k)]$ has the reduced value $\exp(-t\delta)$, then the link L illustrated in each case may be regarded as a 'path of steepest descent' from k_0 to those contours. For this reason, the method of deforming the path is often called the *method of steepest descent*.

For example, in case (*b*) where $\psi''(k_0) > 0$, the link L takes the form

$$k - k_0 = s \exp(\tfrac{1}{4}\pi i) \tag{116}$$

with s real, and then (115) implies that

$$\exp[it\psi(k)] = \exp[it\psi(k_0) - \tfrac{1}{2}ts^2\psi''(k_0) + O(ts^3)], \tag{117}$$

of which the modulus *descends steeply* for both positive and negative s. The contribution to (110) from the integration across the link L can be written

$$\int_L \{F(k_0) + O(s)\} \exp[it\psi(k_0) - \tfrac{1}{2}ts^2\psi''(k_0)] \, (1 + O(ts^3)) \, ds \exp(\tfrac{1}{4}\pi i), \tag{118}$$

where $F(k)$ is replaced by the term in curly brackets and (116) and (117) have been used. Here, the range of integration extends to where the integrand is $O[\exp(-t\delta)]$, which we are already assuming is small, and so no error of greater order than that is made if we evaluate the integral of

$$\exp[-\tfrac{1}{2}ts^2\psi''(k_0)]$$

over the full range where it is significant; that is, as the Gaussian integral

$$\int_{-\infty}^{\infty} \exp[-\tfrac{1}{2}ts^2\psi''(k_0)] \, ds = [2\pi/t\psi''(k_0)]^{\frac{1}{2}}. \tag{119}$$

The corrections of orders $O(s)$ and $O(ts^3)$ by contrast give contributions $O(t^{-1})$ to (118). Finally, then, with errors $O[\exp(-t\delta)] + O(t^{-1})$, the integral across the link L is

$$F(k_0)\,[2\pi/t\psi''(k_0)]^{\frac{1}{2}}\exp\{\mathrm{i}[t\psi(k_0) + \tfrac{1}{4}\pi]\}. \tag{120}$$

There is such an asymptotic contribution to (110) from any stationary point of ψ where $\psi''(k_0) > 0$.

At a stationary point where $\psi''(k_0) < 0$, figure 63(a) shows that the link L needs to take the form

$$k - k_0 = s\exp(-\tfrac{1}{4}\pi\mathrm{i}). \tag{121}$$

Equations (117)–(120) then hold with $\exp(\tfrac{1}{4}\pi\mathrm{i})$ replaced by $\exp(-\tfrac{1}{4}\pi\mathrm{i})$ and $\psi''(k_0)$ by $-\psi''(k_0)$ (which in this case is the same as $|\psi''(k_0)|$). Using $\operatorname{sgn}\psi''(k_0)$ to mean $+1$ where $\psi''(k_0) > 0$ (so that k_0 is a minimum of ψ) and -1 where $\psi''(k_0) < 0$ (so that k_0 is a maximum), we can say that the general asymptotic contribution from a stationary point k_0 is

$$F(k_0)\,[2\pi/t|\psi''(k_0)|]^{\frac{1}{2}}\exp\{\mathrm{i}[t\psi(k_0) + \tfrac{1}{4}\pi\operatorname{sgn}\psi''(k_0)]\}. \tag{122}$$

The integral (110) is asymptotically equal to the value of (122) at the stationary point k_0 (where $\psi'(k_0) = 0$) if there is just one, or else to the sum of the values of (122) at all such stationary points, provided that $\psi''(k_0)$ is nonzero at each. For modifications where $\psi''(k_0)$ is zero (which means that the angles on the path of steepest descent have to be chosen differently) see section 4.11.

After the substitution (109), we conclude that the asymptotic form of (106) for large t is

$$F(k_0)\,[2\pi/t|\omega''(k_0)|]^{\frac{1}{2}}\exp\{\mathrm{i}[\omega(k_0)\,t - k_0x + \tfrac{1}{4}\pi\operatorname{sgn}\omega''(k_0)]\}, \tag{123}$$

where k_0 is the wavenumber for which

$$\omega'(k_0) = x/t. \tag{124}$$

Thus, as expected, waves of wavenumber k_0 are found at positions moving forward at the group velocity $\omega'(k_0)$. Note also that if equation (124) has two solutions (as for ripples with wavelengths on either side of that for minimum group velocity) then a contribution (123) from each is present: waves of different wavenumbers, but the same group velocity, are found at the same place.

In the notation (89), equation (123) implies that the amplitude ζ_1 of the waves generated becomes

$$\zeta_1(x, t) = |F(k_0)|\,[2\pi/t|\omega''(k_0)|]^{\frac{1}{2}}, \tag{125}$$

while their phase α becomes

$$\alpha(x, t) = \omega(k_0) t - k_0 x + \arg F(k_0) + \tfrac{1}{4}\pi \operatorname{sgn} \omega''(k_0) \tag{126}$$

in each case, with k_0 defined by (124). Note that this α does satisfy asymptotically the equations (88), since in a small change of x and t we have

$$d\alpha = [\omega'(k_0) \, dk_0] t + \omega(k_0) \, dt - (dk_0) x - k_0 \, dx + O(t^{-1}), \tag{127}$$

where the first and third terms on the right cancel by equation (124); that equation implies, also, that

$$\omega''(k_0) \, dk_0 = t^{-1}[dx - \omega'(k_0) \, dt], \tag{128}$$

yielding the error term in (127) from changes in $\arg F(k_0)$. (These *last* changes are, however, absent if the initial disturbance $f(x)$ is symmetrical about $x = 0$, which by (112) makes $F(k)$ real.) For an interpretation of the amplitude $\zeta_1(x, t)$, using energy considerations, see section 3.8.

The arguments were set out above for the case of waves without attenuation. It is easy, however, to produce a modified theory on the assumption that attenuation is present but small (as for most water waves). The conclusion takes the same form (123) but with an additional factor representing the decay during time t of waves of wavenumber k_0.

In fact, if $\exp[-t\sigma(k)]$ is the attenuation factor by which the amplitude of waves (102) of wavenumber k is reduced in time t, then this factor must be included in the integral (106), even though the initial form (107) of ζ remains unchanged. The analysis of the asymptotic behaviour of (110) then goes forward with the integrand's phase $t\psi(k)$ unchanged but its amplitude $F(k)$ replaced by $F(k) \exp[-t\sigma(k)]$, and the final conclusion is equation (123) with this change made in $F(k)$. The wave train's phase α, then, is unchanged from (126), but its amplitude (125) is changed to

$$\zeta_1(x, t) = |F(k_0)| \{\exp[-t\sigma(k_0)]\} [2\pi/t|\omega''(k_0)|]^{\frac{1}{2}}. \tag{129}$$

All the results of this section can be proved in a completely different way, using real-variable methods. The idea is to change the variable of integration in (110) to ψ so that the integral becomes the Fourier transform of $F(k) \, dk/d\psi$ with respect to ψ. The general theory of the asymptotic behaviour of Fourier transforms is then used to express that transform asymptotically as a sum of terms corresponding to the singularities of $F(k) \, dk/d\psi$, which are the points where $d\psi/dk = 0$. On balance, however, the proof by Cauchy's theorem was picked for inclusion in this section as appealing to ideas more widely accessible than that slightly more recondite general theory.

3.8 Energy propagation velocity

The results in the last two sections suggest that the group velocity

$$U(k) = d\omega/dk$$

is the velocity of propagation of the *energy* in waves of wavenumber k. Such an interpretation depends on the idea that, in a system satisfying linear equations with constant coefficients, there can be no transfer of energy from one part of the spectrum to another. Therefore, the path (95) along which waves of wave-number k are found must be the path along which the energy in waves of that wavenumber is propagated.

Equation (129) gives detailed support to this hypothesis. In any propagating wave satisfying linear equations, the total wave energy E_w must depend on the square of the disturbances, and therefore take a form such as

$$E_w = \int_{-\infty}^{\infty} C\zeta^2 \, dx, \tag{130}$$

where the multiplying factor C should be independent of x for a homogeneous system. (For gravity waves passing down a canal of breadth b, for example, $C = \rho g b$ since the wave energy per unit area is twice the potential energy (24).) Equation (107) for the initial waveform, therefore, gives

$$E_w = C\int_{-\infty}^{\infty} [f(x)]^2 \, dx = \tfrac{1}{2}\pi C \int_{-\infty}^{\infty} |F(k)|^2 \, dk = \pi C \int_{0}^{\infty} |F(k)|^2 \, dk \tag{131}$$

by Parseval's theorem; the right-hand side of (131) is the integral of an even function of k since (112) makes $F(-k)$ the complex conjugate of $F(k)$.

Equation (131) implies that the energy in wavenumbers between k_0 and $k_0 + dk$, after it has been attenuated by a factor $\exp[-2t\sigma(k_0)]$, is

$$\pi C |F(k_0)|^2 \exp[-2t\sigma(k_0)] \, dk. \tag{132}$$

If, asymptotically, this energy is found in the interval between $x = t\omega'(k_0)$ and $x = t\omega'(k_0 + dk)$, of length

$$|t\omega'(k_0) - t\omega'(k_0 + dk)| = t|\omega''(k_0)| dk, \tag{133}$$

then the energy per unit length is

$$\pi C |F(k_0)|^2 \{\exp[-2t\sigma(k_0)]\}/t|\omega''(k_0)|. \tag{134}$$

But, by (130), the energy per unit length is C times the mean square of ζ, or $\frac{1}{2}C\zeta_1^2$ in sinusoidal waves of amplitude ζ_1. Comparing this with (134) gives

$$\zeta_1 = |F(k_0)| \{\exp[-2t\sigma(k_0)]\} [2\pi/t|\omega''(k_0)|]^{\frac{1}{2}}: \tag{135}$$

a precise agreement with (129) which constitutes rather strong circumstantial evidence for the energy hypothesis!

Furthermore, the simple theory of section 3.6 can be extended to a nonhomogeneous system, and the results are still in accord with the interpretation of U as energy propagation velocity. Consider, for example, the case of waves travelling into shallower water mentioned already in section 3.3. Suppose that the depth h is reduced so gradually on a scale of wavelengths that the dispersion relationship (35) calculated for water of uniform depth h is a good approximation everywhere with the local value of h inserted in it. Suppose furthermore that $h = h(x)$, so that the depth varies only in the direction perpendicular to wave crests. Then equation (35) takes the form of a *nonhomogeneous dispersion* equation

$$\omega = \omega(k, x). \tag{136}$$

For any system satisfying an equation like (136), we define the group velocity as that associated with the wave properties *for fixed x*:

$$U(k, x) = \partial\omega(k, x)/\partial k. \tag{137}$$

Evidently, we cannot then deduce equation (93) for k from equation (91). We can, however, multiply equation (91) by U, using the fact that

$$U\partial k/\partial t = \partial\omega/\partial t \tag{138}$$

(since x is kept constant in both these time-derivatives *and* also in the partial derivative (137)), to give

$$\partial\omega/\partial t + U\partial\omega/\partial x = 0. \tag{139}$$

Equation (139) means that ω is constant along paths in the (x, t) plane satisfying

$$dx/dt = U; \tag{140}$$

although, in the nonhomogeneous case, these paths are not in general straight lines (figure 64) since we cannot infer from equations (136) and (137) that either U or, indeed, k is constant along a path.

The variation of U can be described conveniently if we rewrite (136) as an equation $k = k(\omega, x)$ for wavenumber in terms of frequency. Then

$$U^{-1} = \partial k(\omega, x)/\partial\omega, \tag{141}$$

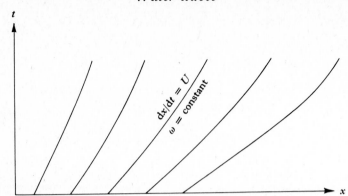

Figure 64. Nonhomogeneous dispersion depicted on an (x, t) diagram. The paths (140) are curves along which the *frequency* ω remains constant.

so that the paths (140) on which ω is constant can be calculated as

$$t = t_0(\omega) + \int_0^x U^{-1}\,\mathrm{d}x = t_0(\omega) + \int_0^x [\partial k(\omega, x)/\partial \omega]\,\mathrm{d}x, \qquad (142)$$

where $t_0(\omega)$ is the time at which waves of frequency ω passed $x = 0$.

The variation of k represents 'refraction' by the nonhomogeneous medium. For an illustration of this see section 3.3, where equation (40) is used to show that deep-water waves of 8-second period become reduced in length from 100 m to 25 m as the water depth is gradually reduced to 1 m.

Even in a spatially nonhomogeneous system, however, the general theory of normal modes of small oscillation suggests that there can be no transfer of energy between modes of different frequency. Accordingly, the conclusion that ω takes a constant value at positions moving forward with the group velocity U is consistent with the idea that U is the energy propagation velocity for sinusoidal waves.

We now verify this hypothesis by means of a direct calculation for sinusoidal gravity waves of *constant* wavenumber k in water of *uniform* depth h. This calculation's relevance to dispersion is that in a dispersed group of waves the wavenumber varies so gradually that locally the energy propagation velocity should be close to that for uniform k; similarly, theories like that just given assume that the depth h varies so gradually that the local wave properties are close to those for uniform h.

We write down the energy flux across a vertical plane $x = $ constant in the direction of motion of the crests, per unit length of crest, as

$$\int_{-h}^0 p_e(\partial \phi / \partial x)\,\mathrm{d}z, \qquad (143)$$

where p_e is the excess pressure (8) and $\partial\phi/\partial x$ the velocity component in the positive x-direction. The integral (143), representing rate of working by water with $x < 0$ on water with $x > 0$, is of the order of the squares of disturbances, as expected; whereas any additional integral from $z = 0$ to $z = \zeta$ (the height of the free surface above its undisturbed position $z = 0$) can be neglected as of the order of the cubes of disturbances.

In a wave travelling with wave speed c, equations (8) and (14) show that

$$p_e = -\rho\, \partial\phi/\partial t = +\rho c\, \partial\phi/\partial x, \tag{144}$$

so that the energy flux (143) per unit length of crest can be written as

$$2c \int_{-h}^{0} [\tfrac{1}{2}\rho(\partial\phi/\partial x)^2]\, dz; \tag{145}$$

that is, as $2c$ times the kinetic energy per unit horizontal area of just the *horizontal components* of water movement. The energy propagation velocity (which we write as U in anticipation of its being equal to the group velocity) is

$$U = \frac{\text{average energy flux per unit length of crest}}{\text{average wave energy per unit horizontal area}}, \tag{146}$$

where the averages are taken over a wave period. In (146), the wave energy is *twice* the kinetic energy of both horizontal *and* vertical movements (twice because the average potential and kinetic energies are equal). Therefore, by (145),

$$\frac{U}{c} = \frac{\text{average kinetic energy of horizontal components}}{\text{average kinetic energy of water movements}}. \tag{147}$$

Equation (147) makes readily intelligible both (i) the value $U/c = \tfrac{1}{2}$ for deep water, where horizontal and vertical motions have equal average kinetic energy as figure 50 shows; and (ii) the value $U/c = 1$ for long waves, where almost all the kinetic energy is in horizontal motions (section 2.2). Furthermore, the right-hand side is easily calculated for arbitrary depth, using (41) and (42), as

$$\int_{-h}^{0} \cosh^2 [k(z+h)]\, dz \bigg/ \int_{-h}^{0} \{\cosh^2 [k(z+h)] + \sinh^2 [k(z+h)]\}\, dz$$

$$= \int_{-h}^{0} \tfrac{1}{2}\{\cosh [2k(z+h)] + 1\}\, dz \bigg/ \int_{-h}^{0} \cosh [2k(z+h)]\, dz$$

$$= \tfrac{1}{2}\{1 + (2kh)/(\sinh 2kh)\}, \tag{148}$$

giving exactly the value of U/c obtained by a quite different method in equation (99).

The above calculation of energy flux, besides confirming the hypothesis that group velocity and energy propagation velocity are identical, permits calculations of amplitude variation in waves moving into gradually shallower water. We know that waves of 8-second period have a fourfold reduction in wave speed c from $12.5\,\mathrm{m\,s^{-1}}$ to $3.1\,\mathrm{m\,s^{-1}}$ as the depth is reduced from large values to $1\,\mathrm{m}$ (figure 53). Nevertheless, the reduction in group velocity is only twofold, from $U = \tfrac{1}{2}c = 6.2\,\mathrm{m\,s^{-1}}$ in deep water to $U = c = 3.1\,\mathrm{m\,s^{-1}}$ in long-wave conditions (with $\lambda = 25\,\mathrm{m}$ and $h = 1\,\mathrm{m}$). By (146), energy flux (per unit length of crest) is U times energy per unit area. Hence, for a *given* energy flux (per unit length of crest) towards the shore, the wave energy W per unit area is *increased* twofold as the depth is reduced to $1\,\mathrm{m}$. Therefore, by (28), the amplitude a (maximum elevation of the water surface) is also increased, although only by a factor $\sqrt{2}$. On the other hand, the maximum steepness of water surface is ka, which is increased by the much larger factor $4\sqrt{2}$. The eye is sensitive primarily to this water-surface steepness, which explains why waves appear to become a lot bigger on moving into shallow water.

One major conundrum remains. Why should energy propagation velocity, which is defined and calculated for perfectly sinusoidal waves of *fixed* wavenumber k and fixed frequency ω, be identical with group velocity, all of whose properties were derived in sections 3.6 and 3.7 from its definition as $\mathrm{d}\omega/\mathrm{d}k$: the ratio of *changes* in frequency and wavenumber in going to a neighbouring wave solution of the equations of motion? Our verification that (148) equals (99) has not resolved this conundrum since these expressions were arrived at quite differently: one by integrating $\cosh^2[k(z+h)]$ and $\sinh^2[k(z+h)]$, and the other by differentiating $k\tanh kh$.

Evidently, it is desirable to find a *general* proof that in a perfectly sinusoidal wave the energy propagation velocity is $\mathrm{d}\omega/\mathrm{d}k$. In giving this, we concentrate (as in section 3.7) on one-dimensional propagation in homogeneous systems without attenuation. Then a sinusoidal wave

$$\zeta = a\exp\left[\mathrm{i}(\omega t - kx)\right] \qquad (149)$$

is an exact solution of the linear equations of motion whenever ω and k are related by the dispersion relationship, which in this proof is used in the form

$$k = k(\omega). \qquad (150)$$

The difficult question is: in such a wave, how can the energy flux across $x = 0$, say, be determined in terms of the *changes* in k and ω in going to a neighbouring wave solution?

A good answer is as follows: 'by postulating the introduction of small forces generating energy dissipation which may bring about small, *pure imaginary* changes in k or ω.' One quite general way of achieving this can be easily imagined. The movement of every particle of matter in the system is supposed resisted by an additional force equal and opposite to β times its momentum. Here β is a constant, very much smaller than ω, that will be allowed to tend to zero at the end of the calculation.

For movements proportional to $\exp(i\omega t)$ with fixed real frequency ω, a gradual attenuation (reduction in amplitude as the wave travels in the positive x-direction) will be shown to result from this additional dissipative force, which alters the equation of motion of every particle of matter in the system as follows. The conservative forces acting on the particle are balanced not just by the mass-acceleration term

$$-M\omega^2 \mathbf{r} \tag{151}$$

(where M is the particle's mass and $\mathbf{r}$ its vector displacement) but by

$$-M\omega^2 \mathbf{r} + \beta M i \omega \mathbf{r}: \tag{152}$$

the mass-acceleration term minus a damping resistance equal and opposite to β times the momentum. The change from (151) to (152) is *exactly as if* ω^2 were changed everywhere (in a system otherwise unchanged) into $\omega^2 - i\beta\omega$, which when $\beta \ll \omega$ is the same as if ω were changed into $\omega - \frac{1}{2}i\beta$. For fixed real ω, then, the system is now such that k is changed into

$$k(\omega) - \tfrac{1}{2}i\beta k'(\omega); \tag{153}$$

and we may note that here the derivative $\mathrm{d}k/\mathrm{d}\omega$ makes a rather welcome appearance!

Now equation (149) shows the foreseen spatial attenuation, becoming

$$\zeta = a \exp\{i[\omega t - k(\omega)x] - \tfrac{1}{2}\beta k'(\omega)x\} \tag{154}$$

(by which, as always, the real part of the right-hand side is signified). This disturbance has only a finite energy for $x > 0$, which can be calculated from (130) as the average of

$$C\int_0^\infty \zeta^2 \mathrm{d}x, \tag{155}$$

which in turn (since the average of $\cos^2(\omega t - kx)$ is $\frac{1}{2}$) is

$$\tfrac{1}{2}C\int_0^\infty a^2 \exp\{-\beta k'(\omega)x\}\,\mathrm{d}x = \tfrac{1}{2}Ca^2/\beta k'(\omega). \tag{156}$$

The energy flow across $x = 0$ is now easy to calculate. Since the energy for $x > 0$ is finite and constant, the energy flow across $x = 0$ must exactly balance the rate of energy dissipation for $x > 0$. For each particle this is $\beta M \dot{r}^2$ since the resistive force is equal and opposite to $\beta M \dot{r}$. Therefore, the total rate of energy dissipation for $x > 0$ is 2β times the kinetic energy, or β times the total wave energy (kinetic and potential). This is β times (156), and in the limit as $\beta \to 0$ (representing the exact sinusoidal wave) we still have a finite value

$$\tfrac{1}{2} C a^2 / k'(\omega) \tag{157}$$

for this energy flux to compare with the energy per unit length $\tfrac{1}{2} C a^2$. The energy propagation velocity, which is the ratio of these two quantities, has thus been shown to be equal to

$$1/k'(\omega) = \mathrm{d}\omega/\mathrm{d}k, \tag{158}$$

which is the group velocity, in such a *general* dispersive system.

In sections 3.6–3.8 an attempt has been made to do justice to the richness of the concept of group velocity by analysing its properties from several different standpoints. A final unity is, perhaps, given to the diverse methods by the proof with which these analyses have just been concluded.

3.9 Wave patterns made by obstacles in a steady stream

The waves studied in most parts of this book are *unsteady* phenomena: the associated fluid movements vary with time; sometimes, nearly sinusoidally; sometimes, much more irregularly. We showed how sound may be generated by various unsteady movements of objects immersed in the fluid; similarly, water waves may be generated either by similar motions of immersed objects, or by various unsteady disturbances to the water surface (storms, entry of solid bodies, etc.).

By contrast, we describe in the present section the apparently paradoxical situation of waves that form part of a *completely steady stream flow*. At all points of the stream (including where the waves are) the flow is steady: the fluid velocity does not vary with time. Although the elevation of the surface may locally show a regular, nearly sinusoidal variation with position, it shows no change at all with time: the crests of the waves remain always in the same places as the stream sweeps by. The *stationary wave pattern* is generated by a completely stationary obstacle to the stream flow. This obstacle may be fixed in the stream or lie on the bottom; it may, indeed, be just an irregular feature in the bed.

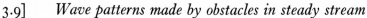

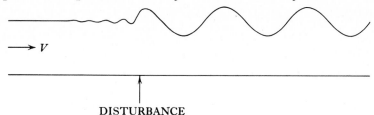

DISTURBANCE

Figure 65. Schematic indication of stationary waves generated by a steady disturb-ance to a steady stream of velocity V. The scale for vertical displacements of the undisturbed water surface has been exaggerated so that the ripples upstream of the disturbance are clearly to be seen as well as the larger gravity waves down-stream of it. The disturbance may take the form either of a step in the bed or of a cylindrical obstacle; the latter may be situated on the bed or in the midst of the water or on the surface.

Robert Frost described, in his poem 'West Running Brook', how

> The black stream, catching on a sunken rock,
> Flung backward on itself in one white wave,
> And the white water rode the black forever,
> Not gaining but not losing.

The steadiness of the flow caught his imagination:

> That wave's been standing off this jut of shore
> Ever since rivers, I was going to say,
> Were made in heaven.

He recognised how the upstream propagation of the crest could be exactly cancelling the downstream flow:

> Speaking of contraries, see how the brook
> In that white wave runs counter to itself.

He poetically explained the appeal of the phenomenon:

> It is this backward motion towards the source,
> Against the stream, that most we see ourselves in,
> The tribute of the current to the source.

Frost thought of the brook as time which sweeps individuals down to a sea of oblivion; from that standpoint, perhaps, his poetry is a sort of stationary wave pattern.

We analyse first a very simple stationary wave pattern: one with all wave crests perpendicular to the stream. These crests propagating at wave speed c in the opposite direction to a stream of speed V can be stationary if and only if

$$c = V. \tag{159}$$

Either a long 'cylindrical' obstacle, or a 'step' in the bed, spanning the stream at right angles to it, may generate this type of pattern (figure 65).

The dispersive property of water waves is crucial to the existence of the pattern: c varies with wavelength, so that, among all wavelengths, equation (159) fixes on one (or at most two) which can be stationary.

Admittedly, equation (56) shows that no such waves can be present in a *very slow* stream with $V < 0.23\,\mathrm{m\,s^{-1}}$. Indeed, stationary crests of waves propagating, not directly upstream but at some nonzero angle θ to the upstream direction are also ruled out in this case. That is because they must satisfy

$$c = V \cos \theta, \tag{160}$$

so that the component $V \cos\theta$ of stream velocity at right angles to the crest can cancel the crest's motion at the wave speed c. Equation (160) shows that $c \leqslant V$ for all stationary waves, so that no stationary wave pattern is possible on a stream with speed V less than the minimum wave speed c_m.

Returning to the case $\theta = 0$ (crests perpendicular to the stream), consider now a stream with V rather greater than c_m. Figure 56 then suggests that there are two different wavelengths for which condition (159) is satisfied. This appears to imply that we may observe, not a regular sinusoidal wave but the rather irregular pattern derived by superimposing two sine waves of incommensurate lengths.

Nevertheless, we find that dispersion acts (as before) to ensure that waves observed locally are practically sinusoidal. The property of dispersive waves that is critical here is that the group velocity U with which energy is propagated is different from the wave speed c. Waves with $\lambda > \lambda_m$ have $U < c$, so that the energy travels through the water at a speed U less than the speed $c = V$ of the stream. The energy in those waves is therefore always swept *downstream*. Accordingly, they are found downstream of the obstacle which is the source of the waves. By contrast, waves with $\lambda < \lambda_m$ have U greater than $c = V$ so that their wave energy travels *upstream* from the obstacle. Accordingly, the very short ripples and the longer gravity waves are found *in different positions*: on the upstream and downstream sides of the obstacle, respectively (figure 65).

A closely related way of phrasing the above argument uses a frame of reference *moving with the stream*. In that frame of reference, the obstacle is travelling at velocity V (in the 'upstream' direction) through water which is at rest except in so far as the obstacle's motion disturbs it. The longer waves generated, with wave speed $c = V$ *greater* than their energy propagation velocity U, are then left behind by the obstacle: every joule of wave energy after it is generated lags further and further behind it, and at each instant the wave energy lying closest behind it is that generated most recently. The energy in these longer waves (essentially gravity waves) is normally attenuated rather slowly (section 3.5) and therefore may be present even at considerable distances behind the obstacle. By contrast, the energy in the ripples, for which $c = V$ is less than U, runs ahead of the obstacle; but the rate of attenuation is much greater for them, so that they cannot get very far ahead of the obstacle before they are substantially attenuated (figure 65).

The power P_w which the moving obstacle must expend to generate these waves can be easily written down. We use W for wave energy per unit area, which by (28) is $\frac{1}{2}\rho g a^2$ for gravity waves of amplitude a, and has the same value for ripples if g is increased by the substitution (50). The energy W_d per unit area in the waves *downstream* of the obstacle is moving away from it at the velocity $V - U_d$ (where U_d is the group velocity for these longer waves with wave speed $c = V$). The energy W_u per unit area in the waves *upstream* of the obstacle is moving away from it at the velocity $U_u - V$. Therefore, the power generating both sets of waves is

$$P_w = b(V - U_d)W_d + b(U_u - V)W_u. \tag{161}$$

Here, b is the breadth of the stream, so that the rates of generation of new wave *area* downstream and upstream of the obstacle are $b(V - U_d)$ and $b(U_u - V)$ respectively.

This power P_w, of course, is not the total power required to propel the obstacle through the water; an additional 'frictional' power P_f is associated with generating the boundary layer and wake. The 'drag', or force resisting the obstacle's motion, is often written

$$D = D_w + D_f, \quad \text{where} \quad VD_w = P_w \quad \text{and} \quad VD_f = P_f, \tag{162}$$

as the sum of a 'wavemaking' drag and a frictional drag. The additional *thrusting* force on the obstacle, required to overcome the nonfrictional drag D_w, exerts the power $P_w = VD_w$ necessary to generate the waves.

In the *original* frame of reference with the obstacle at rest, the above expressions (161) and (162) remain potentially useful: they represent the force with which the stream acts on the obstacle (assumed, as above, to be spanning the stream). At the same time, the above calculation for an obstacle moving through still water may be directly applicable to questions of the resistance to motion along a canal of a broad barge that nearly fills its width.

In many practical cases, only the downstream gravity waves are significant. This is because, when V is substantially greater than $c_m = 0.23 \text{ m s}^{-1}$, the capillary waves satisfying (159) have extremely low wavelengths (figure 56).

Obstacles of substantial size cannot significantly excite such short waves (which, in fact, are most in evidence ahead of small obstacles in relatively slow streams). By contrast, we shall see that a large obstacle in a stream, or a broad barge moving along a canal, may excite behind itself the gravity waves satisfying (159) with rather large amplitude. This requires, as we indicate by calculation in one particular case at the end of this section, that their longitudinal dimensions are well matched to the length of those waves.

There are some interesting consequences of the fact that for water of depth h (figure 52) the wave speed cannot exceed $(gh)^{\frac{1}{2}}$. This means that no solution of (159), representing waves with crests perpendicular to the direction of motion, is possible when the stream speed $V > (gh)^{\frac{1}{2}}$. Obstacles of general shape can still make waves with oblique crests because equation (160) can still have solutions. On the other hand, an obstacle close in shape to a cylinder spanning the stream, which tends to make substantial wave with crests at right angles to the stream when $V < (gh)^{\frac{1}{2}}$, can make no such waves when $V > (gh)^{\frac{1}{2}}$ and typically generates oblique waves only weakly. Similarly, a broad barge almost spanning a canal along which it moves experiences a dramatic drop in resistance when its speed exceeds $(gh)^{\frac{1}{2}}$; when the main element in its wavemaking drag D_{w} disappears and only a much smaller element due to generation of oblique waves from the sides of the barge remains.

Scott Russell in 1844 wrote: 'As far as I am able to learn, this fact was discovered accidentally on the Glasgow and Ardrossan Canal of small dimensions. A spirited horse in the boat of William Houston, Esq., one of the proprietors of the works, took fright and ran off, dragging the boat with it, and it was then observed, to Mr Houston's astonishment, that the foaming stern surge which used to devastate the banks had ceased, and the vessel was carried on through water comparatively smooth, with a resistance very greatly diminished. Mr Houston had the tact to perceive the mercantile value of this fact to the Canal Company with which he was concerned.' He devoted himself to introducing on that canal vessels moving at velocities as high as 9 miles an hour, bringing 'a large increase of revenue to the Canal Proprietors'.

While the mechanisms of energy dissipation in waves of *small* amplitude are as listed in section 3.5, waves generated with an amplitude exceeding a certain value lose energy additionally by foaming at the crests, as in the marine phenomenon of 'white horses', and as Scott Russell mentions in this quotation. Often the wave nearest the obstacle is the only one that foams (as in Frost's 'one white wave'), losing thereby enough energy to reduce the amplitude of waves further downstream below the value at which foaming occurs.

Waves generated by cylindrical obstacles, either lying athwart a steady stream or, equivalently, spanning a canal full of still water up which they are travelling, have (as we have seen) many properties that are readily inferred from the dispersion relationship. The calculation of their amplitude, however, can be much harder. We conclude the account of those waves by giving such a calculation in a not too difficult case. We try to describe the calculation so that light may be thrown also on the general problem of stationary wave patterns in one-dimensional dispersive systems, and so that the nature of the relationship between wave amplitude and the obstacle's longitudinal dimensions is made clear.

We analyse the wave generated when the shape of the bed takes the form

$$z = -h + f_1(x), \quad \text{with} \quad f_1'(x) = f(x), \tag{163}$$

where we retain a linear problem by assuming that both $h^{-1}f_1(x)$ and the bed *slope* $f(x)$ are small. Evidently, this shape of bed is 'cylindrical' (independent of y) and makes only small disturbances to the stream. The *waves* generated should be associated (section 3.1) with the irrotational part of those disturbances; that is, the part external to a bottom boundary layer. For its velocity potential ϕ, a suitable boundary condition is

$$[\partial\phi/\partial z]_{z=-h} = Vf(x), \tag{164}$$

where V is the stream velocity outside the boundary layer and $f(x)$ is the small slope of the bottom; note that, because $h^{-1}f_1$ is also small, the boundary condition on ϕ can be applied accurately enough at the fixed level $z = -h$ instead of the variable level (163).

We indicate the method of analysis for a general one-dimensional dispersive system without attenuation, in which waves

$$\zeta = a \exp[i(\omega t + kx)] \tag{165}$$

travelling in the *negative* x-direction can be made stationary by a stream of velocity V in the *positive* x-direction. We suppose that for these waves (165), with arbitrary ω and k, a certain boundary value η is equal to

$$\eta = aB(\omega, k) \exp[i(\omega t + kx)], \tag{166}$$

and that the dispersive system is defined by the boundary condition $\eta = 0$ so that the dispersion relationship is

$$B(\omega, k) = 0. \tag{167}$$

Thus, for waves on water of uniform depth h, we could use the boundary value

$$\eta = [\partial\phi/\partial z]_{z=-h};\qquad(168)$$

and calculate easily from (10) and (12), using the methods of section 3.3, that

$$B(\omega,k) = i\omega^{-1}(\omega^2\cosh kh - gk\sinh kh)\qquad(169)$$

for gravity waves (with the substitution (50) made if surface tension is to be taken into account). As foreseen, the dispersion relationship (35) can be written in the form (167).

Now we consider the generation of stationary waves in a stream of velocity V by a nonzero value of η independent of the time:

$$\eta = Vf(x) = V\int_{-\infty}^{\infty} F(k)\exp(ikx)\,dk,\qquad(170)$$

as in (164). The waves (165) become stationary on a stream of velocity V if

$$\omega = kV,\qquad(171)$$

since superimposing a uniform velocity V on the system is equivalent to changing the coordinate x into $x-Vt$. Therefore, a stationary sinusoidal wave $\zeta = a\exp(ikx)$ implies a boundary value $\eta = aB(kV,k)\exp(ikx)$. Conversely, therefore, a boundary value $\eta = VF(k)\exp(ikx)$ generates a wave $\zeta = VF(k)[\exp(ikx)]/B(kV,k)$; and the more general boundary value (170) generates a wave

$$\zeta = \int_{-\infty}^{\infty} \frac{VF(k)\exp(ikx)}{B(kV,k)}\,dk.\qquad(172)$$

As in section 3.7, we find that the best way of evaluating this integral is by Cauchy's theorem. For $x > 0$, the path of integration should be *raised* a distance κ by giving k a positive imaginary part κ to make the integral $O[\exp(-\kappa x)]$. For $x < 0$, the path should be *lowered* by giving k an imaginary part $(-\kappa)$, again making the integral small; this time, $O[\exp(\kappa x)]$. With these errors, the integral is represented by the *residues* of the integrand at *poles* passed over in deforming the path: these, being values of k where $B(kV,k) = 0$, represent stationary waves.

In carrying out that apparently simple programme there is just one small, but very important and famous difficulty: the poles in question are *on* the original path of integration with respect to k (the real axis). This gives an apparent ambiguity to the answer, depending on whether the original integral (172) is taken to the left or to the right of particular poles (or even as a combination of both in some arbitrary proportions).

This mathematical ambiguity corresponds to a genuine physical ambiguity in the problem as analysed above. To the waves *actually* generated by the stationary disturbance (170) to the stream, there could be added on an *arbitrary* train of stationary waves satisfying $\eta = 0$ for all x (positive and negative). These are solutions of $B(kV, k) = 0$, and the energy in those waves might be generated either at $x = +\infty$ or at $x = -\infty$, depending on whether the group velocity was greater or less than V.

Evidently, the mathematical problem specified by (170), with solution (172), is not a complete representation of the physical problem; namely, to find the waves truly generated by the disturbance itself, subject to what is often called 'the radiation condition': that no wave energy is being generated 'at infinity'.

There are several efficient mathematical procedures equivalent to this radiation condition, some of which use group-velocity ideas directly. Others take into account the *attenuation* of wave energy which must always be present, and which does shift the poles unambiguously off the real axis. A process that will work even if attenuation is not taken into account is, however, desirable. Out of all such processes, the following turns out to be applicable in a usefully wide generality of cases (chapter 4).

The idea is that the waves truly generated by a stationary 'bump on the bottom' can be identified most readily if the bump is imagined to have *grown up slowly* to its present height, in a stream initially undisturbed. A slow exponential growth in the boundary value η gives a good general representation of this idea:

$$\eta = [\exp(\epsilon t)]\, Vf(x) = [\exp(\epsilon t)]\, V \int_{-\infty}^{\infty} F(k) \exp(ikx)\, dk. \qquad (173)$$

By looking exclusively for the associated waves, growing in proportion to $\exp(\epsilon t)$ but otherwise independent of time, we exclude the danger of the solution being contaminated by other wave energy generated 'at infinity'. (Here, ϵ is small and will ultimately be allowed to tend to zero.) The condition (171) is replaced by

$$\omega = kV - i\epsilon \qquad (174)$$

and the solution (172) by

$$\zeta = [\exp(\epsilon t)] \int_{-\infty}^{\infty} \frac{VF(k) \exp(ikx)}{B(kV - i\epsilon, k)}\, dk. \qquad (175)$$

This integral along the real k-axis can be evaluated unambiguously, because the poles have been slightly shifted off the path of integration. Corresponding to each real k_0 satisfying

$$B(k_0 V, k_0) = 0, \qquad (176)$$

the zero of $B(kV - i\epsilon, k)$ is approximately where

$$[(k - k_0) V - i\epsilon] [\partial B/\partial \omega]_{k=k_0} + (k - k_0) [\partial B/\partial k]_{k=k_0} = 0; \quad (177)$$

that is, at

$$k = k_0 + i\epsilon/(V - U), \quad (178)$$

where

$$U = [-(\partial B/\partial k)/(\partial B/\partial \omega)]_{k=k_0} \quad (179)$$

is the group velocity ($d\omega/dk$ in changes satisfying (167)).

For $x > 0$, then, raising the path of integration causes it to pass only over poles (178) with $U < V$ (group velocity less than stream velocity), giving a contribution of $(+2\pi i)$ times the residue from each. In the limit as $\epsilon \to 0$, this is

$$\zeta = \sum_{U<V} \frac{2\pi i V F(k_0) \exp(ik_0 x)}{[V\partial B/\partial \omega + \partial B/\partial k]_{k=k_0}} + O[\exp(-\kappa x)] \quad \text{for} \quad x > 0. \quad (180)$$

(The sum is over solutions of (176) with k_0 positive or negative and with $U < V$.) Similarly, lowering the path of integration for $x < 0$ causes it to pass over poles with $U > V$, giving a contribution of $(-2\pi i)$ times the residue from each. In the limit, this is

$$\zeta = \sum_{U>V} \frac{(-2\pi i) V F(k_0) \exp(ik_0 x)}{[V\partial B/\partial \omega + \partial B/\partial k]_{k=k_0}} + O[\exp(\kappa x)] \quad \text{for} \quad x < 0. \quad (181)$$

These equations confirm that waves with group velocity exceeding wave speed (for example, short ripples) are found upstream of the disturbance, while waves like gravity waves with $U < c$ are found downstream.

The only new information in equations (180) and (181) concerns the amplitude of the waves. The most significant factor in that amplitude is $F(k_0)$. This factor implies that waves generated with the particular wavenumber k_0 that allows them to remain stationary have an amplitude proportional to the Fourier component *with wavenumber k_0* of the wave-generating disturbance. This is why disturbances on a larger longitudinal scale generate only the gravity waves significantly.

The gravity waves generated by the bottom slope $f(x)$, by (180) and (169), are

$$\zeta = \frac{4\pi[F(k_0) \exp(ik_0 x) + F(-k_0) \exp(-ik_0 x)] \sinh(k_0 h)}{\sinh(2k_0 h) - (2k_0 h)},$$

where

$$\frac{V^2}{gh} = \frac{\tanh(k_0 h)}{k_0 h}. \quad (182)$$

Figure 66 shows that these tend to be significantly amplified for values of

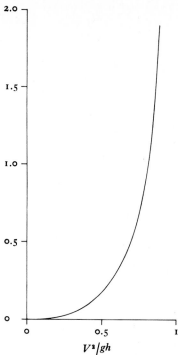

Figure 66. The factor $\sinh (k_0 h)/[\sinh (2k_0 h) - 2k_0 h]$ in the water-surface disturbance (182) due to a bottom irregularity in a steady stream of speed V and depth h, plotted against the Froude number V^2/gh.

V^2/gh just under 1 (for which $k_0 h$ is relatively small), even though when V^2/gh rises above 1 such wave crests perpendicular to the stream cannot (as William Houston, Esq. discovered) be observed.

3.10 Ship waves

We have analysed in section 3.9 both the wave pattern made by a stationary obstacle in a steady stream, and the (essentially identical) wave pattern made when the same obstacle is moving through still water, in a restricted set of cases: when the wave crests are effectively perpendicular to the direction of the stream (or of the obstacle's motion) because the obstacle practically spans the stream (or canal). In this section that restriction is lifted, and we begin to analyse much more general wave patterns, involving oblique waves which satisfy (160) and propagate at a wide variety of angles θ to the direction of the stream (or of the obstacle's motion). Especially, we investigate wave patterns that extend over large distances, both longitudinally and laterally, compared with the longitudinal and lateral dimensions of the obstacle.

Although the methods of analysis give information directly applicable to wave patterns made by a stationary obstacle in a stream (the case emphasised in section 3.9), this section's title 'Ship waves' implies that its main concern is with the (essentially identical) wave pattern generated by a ship *moving* at a constant velocity V over some wide expanse of *still* water. This subject is of great technological importance because P_w, the power required to generate the wave pattern carried along by the ship, can be a big fraction of the total power output of the ship's engines.

We analyse in this section various properties of ship waves which are simple consequences of the dispersion relationship for water waves. We also indicate some general features of P_w, including methods for determining it by model experiments. However, we postpone until after certain techniques have been developed in chapter 4 any discussion of the problems of calculating either P_w or the ship-wave amplitudes on which it depends.

Equation (160) states that waves of any wave speed $c \leqslant V$ can be stationary on a stream of speed V provided that they propagate at an angle

$$\theta = \cos^{-1}(c/V) \tag{183}$$

to the upstream direction. Then the velocity of the stream has a component perpendicular to a crest equal to $V \cos \theta$, cancelling out the crest's motion at the wave speed c.

The same condition (183) applies to any waves generated by a ship moving at constant velocity V through still water. The most obvious way of seeing this is by arguing that, in a frame of reference in which the ship is at rest, there is a steady flow past it of a stream of water with undisturbed velocity V, and any wave crests in this steady flow must be stationary. The condition (183) can also be derived, however, just as the equivalent condition for acoustic emissions from supersonic aircraft (section 2.14) was derived, from considerations of stationary phase: the phase of signals emitted by the ship at successive instants is stationary at a point P if the component $V \cos \theta$ of the ship's velocity in the direction of P is equal to the wave speed.

In any *nondispersive* case (with fixed c), including the acoustic case, the condition (183), with $V \geqslant c$, determines a *unique* direction of emission. In theory, there is a nondispersive case for ship waves: the 'long-wave' limit when all waves emitted are very long compared with the water depth h.

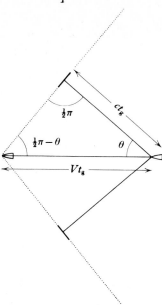

Figure 67. Waves generated in a ripple tank with constant wave speed c by a small model ship moving at velocity V. Waves are emitted at a fixed angle θ given by equation (183). Thick line: wave generated t_g seconds ago when the model was a distance Vt_g further back. Dotted line: locus of all such waves. (Note that the triangle with angles marked *must* be right-angled since (183) makes $\cos \theta$ the ratio of base to hypotenuse.)

In practice, however, the wavelength needs (see (38)) to be at least $14h$, and it proves impossible to operate on water of depth h a ship so long that it generates only waves of length $14h$ or more!

On the other hand, a ripple tank, which (section 3.4) should be of 5 mm depth, is quite suitable for demonstrating the unique direction of emission (183) of the nondispersive waves generated by a thin ship-like model about 50 mm long, pulled at a speed of about $0.30\,\mathrm{m\,s^{-1}}$ through the tank (in which the ripple speed is about $0.22\,\mathrm{m\,s^{-1}}$). This generates an approximately straight 'supersonic-boom-like' signal trailing from the model and moving out at the angle (183) on both sides of its path through the tank. Note, in fact, that at any one instant a portion of that wave generated t_g seconds earlier (where g stands for 'generated') has travelled a distance ct_g in the direction (183) while the model moved on a distance Vt_g. Simple trigonometry, therefore (figure 67), locates that portion of wave on a straight line trailing from the model at an angle $\frac{1}{2}\pi - \theta$ (known in aeronautics as the Mach angle).

For ship waves, the *other* extreme case (waves on deep water) is by far the most interesting, because normally the length of a ship moving in water of depth h is such that the ship generates only waves with $\lambda < 3.5h$. These waves (see (37)) satisfy the simple dispersion relationship

$$c = (g\lambda/2\pi)^{\frac{1}{2}}.$$

For such dispersive waves, equation (183) tells us that different wavelengths λ (and therefore different wave speeds c) are found for waves travelling in different directions θ. Can we conclude, however, just as in figure 67, that at any one instant a portion of wave generated t_g seconds earlier with wave speed c has travelled a distance ct_g in the direction (183) while the ship travelled a distance Vt_g from A to B (figure 68(a))? If so, then the wave is at (say) C_1, or C_2, or C_3, where in each case the angle ACB is a right angle (as is necessary if both $AB\cos\theta = Vt_g\cos\theta$ and AC are to have the same value, ct_g). By a well-known property of the diameter of a circle, the locus of points C such that ACB is a right angle is the circle with diameter AB (figure 68(a)).

The question posed in the last paragraph has undoubtedly been answered with a resounding 'No!' by every reader of the previous four sections. Such a reader is aware that, during t_g seconds, the energy generated in waves with wave speed c does *not* travel a distance ct_g; it travels a distance Ut_g, where U is the group velocity. For waves on deep water, $U = \frac{1}{2}c$. Therefore, all the wave energy generated t_g seconds earlier has travelled exactly half as far as indicated in figure 68(a).

The real locus of all that wave energy is obtained, therefore (figure 68(b)), if we contract about the point A, by a factor $\frac{1}{2}$, the locus previously obtained (circle with diameter AB). It becomes the circle with diameter AD, where D is halfway from A to B; and the waves are at positions (say) E_1, or E_2, or E_3, halfway towards the positions C_1, or C_2, or C_3 which they *would* have reached if the energy had travelled at the speed of the crests. Waves at E_1, propagating at a relatively small angle θ to the direction of the ship's motion, have relatively high wave speeds $c = V\cos\theta$ and therefore are the longer waves in the pattern, while the waves at E_3 have *smaller* c, and *much* smaller $\lambda = 2\pi c^2/g$.

At any one instant the complete ship-wave pattern comprises the waves generated at all previous instants. Those generated when the ship was at each particular point A, A', A'', ... of its path lie on the circles with diameters AD, A'D', A''D'', ..., where the points D, D', D'', ... are in each case exactly halfway to B, the present position of the ship.

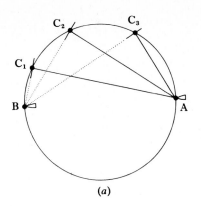

(a)

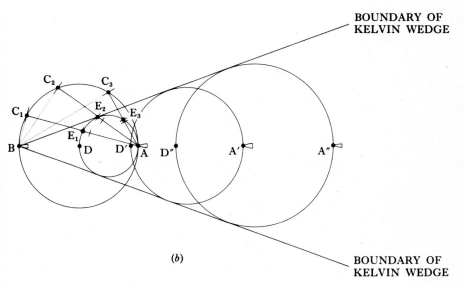

BOUNDARY OF
KELVIN WEDGE

BOUNDARY OF
KELVIN WEDGE

(b)

Figure 68. Waves generated in deep water by a ship B. Case (a): positions C_1, C_2, C_3 of any waves generated t_g seconds ago (when the ship was at A) if their energy had travelled a distance ct_g. Case (b): the real positions E_1, E_2, E_3 of the same waves, taking into account that their energy has only travelled a distance $\frac{1}{2}ct_g$. At each, the dependence of wavelength on direction of emission, as inferred from equation (183), is shown. The circle with diameter AD is the locus of all such waves. Other such circles, with diameters A′D′ and A″D″, are where waves generated when the ship was at A′ and A″ are now to be found. All such circles lie within the Kelvin wedge of semi-angle (184).

The tangent from B to the circle with diameter AD makes an angle

$$\sin^{-1}(\tfrac{1}{3}) = 19.5° \tag{184}$$

to the ship's path. This is because BD = AD, so that the circle's centre is at a distance from B of three times its radius $\tfrac{1}{2}$AD. The whole pattern of wave energy, comprising all the circles, lies therefore within a wedge (*the Kelvin ship-wave* wedge) of semi-angle 19.5°. All the circles carrying the wave energy generated at different times are tangent to this wedge.

The waves on the boundary of the wedge itself (as at E_2) are propagating at the angle

$$\theta = \tfrac{1}{2}[\tfrac{1}{2}\pi - \sin^{-1}(\tfrac{1}{3})] = 35° \tag{185}$$

to the ship's path with a wave speed $c = 0.816V$ and a wavelength equal to two-thirds of the pattern's maximum wavelength $2\pi V^2/g$. These farthest-out waves (with their crests at an angle $\tfrac{1}{2}\pi - \theta = 55°$ to the ship's path) are often the most clearly visible waves in the pattern, since within the interior of the wedge a more muddled appearance may be given by the super-imposition of longer waves (as at E_1) and much shorter waves (as at E_3).

Exceptions to this last statement may arise when the ship's length l is *large compared with* V^2/g. As shown in a simpler case in section 3.9, a ship generates preferentially waves of those wavelengths which predominate in a Fourier analysis of its disturbance to the water; these tend to be of the same order of magnitude as the ship's length l. If this is *large* compared with V^2/g, then the pattern tends to be dominated by waves around the maximum possible wavelength $2\pi V^2/g$, moving forwards behind the ship at *small* angles to the ship's path.

Another exception arises when l is small compared with V^2/g, as with speedboats. Then the pattern is dominated by waves much shorter than the maximum wavelength $2\pi V^2/g$. These are waves propagating at large angles to the ship's path; that is, waves like those at E_3 with their *crests* at small angles to the path.

The above remarks show clearly that the nature of the ship-wave pattern depends critically on the ratio

$$V^2/gl, \tag{186}$$

often called the *Froude number*. For low Froude numbers, longer waves with crests nearly perpendicular to the ship's path may predominate; for high Froude numbers, much shorter waves with crests at a small angle to the ship's path. On the other hand, for a wide range of intermediate Froude numbers, the entire wedge-shaped pattern may be evident, and the most clearly identifiable crests may be those on the boundary, making an angle 55° to the ship's path.

The conclusion that the nature of a ship-wave pattern depends on the Froude number (186) is of great technological importance. It means that in developing a new design for a ship's hull a good estimate of P_w, the power needed to generate the ship-wave pattern, can be made by tests on small-scale models of the hull shape. A geometrically similar model hull, with the length l scaled down by a large factor, gives the same Froude number (186) (so that it should give a geometrically similar wave pattern) if V^2 is scaled down by the same large factor; as when a one-hundredth-scale model is moved at one-tenth of full-scale speed.

The same conclusion is derived by the formal method of dimensional analysis. The power P_w, expended by geometrically similar ship models of different lengths l moving at different velocities V to generate a pattern of *gravity* waves on deep water of density ρ, should depend only on the 4 variables l, V, g and ρ. Accordingly, the ratio

$$P_w/\rho V^3 l^2 \qquad (187)$$

is a *nondimensional* quantity dependent on only those 4 variables. Now, the theory of dimensional analysis tells us that, because $4-3 = 1$, every nondimensional quantity dependent on only those 4 variables must be a function of any 1 such quantity. In particular, the ratio (187) must be a function of the Froude number (186).

For purposes of model tests, the ratio (187) is often rewritten as a wave-drag coefficient
$$D_w/\rho V^2 l^2, \qquad (188)$$

where $D_w = V^{-1}P_w$ (as in (162)) is the wavemaking component of the total resistance D to the ship's motion. To find D_w, it is necessary to measure D and subtract from it the frictional drag D_f; that is, the resistance which the model would experience if it made no waves. One way of obtaining the frictional drag for the immersed portion of the hull is by adding on to it an inverted but otherwise identical shape which is its mirror-image in the water surface, and moving the double model in a fully immersed condition in a deep tank at velocity V; then no waves are made, so that the measured resistance should be $2D_f$.

Figure 69 shows a typical measured dependence of the wave-drag coefficient (188) on the Froude number (186). For cargo ships, the economic advantages of increased speed (more journeys per year) begin to be cancelled by the large additional fuel costs from enhanced wavemaking resistance at a Froude number a little below the steeply rising part of the curve. This has put a premium on the aim of designing hull forms for which that steep rise is deferred to relatively *larger* values of the Froude number. For some of

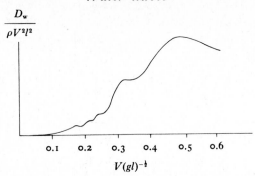

Figure 69. Typical measured dependence of the wave-drag coefficient (188), for a ship of length l moving at speed V, on $V(gl)^{-\frac{1}{2}}$, the square root of the Froude number (186). (We may note that the notation for the Froude number is variable, and $V(gl)^{-\frac{1}{2}}$ is itself often called the Froude number.) No scale for the ordinate is given because it varies far more with details of the shape of the ship (such as ratio of width to length) than does the indicated typical dependence on Froude number.

the theoretical considerations relevant to the design of such hulls, see chapter 4.

We conclude this account of what can be deduced simply from the dispersion relationship by calculating the *crest shapes* in the Kelvin ship-wave pattern. With the origin at the present position of the ship (supposed travelling in the negative x-direction), the waves generated when it was at

$$x = Vt_g = X, \quad \text{say,} \tag{189}$$

are now at points

$$x = X(1 - \tfrac{1}{2}\cos^2\theta), \quad y = \tfrac{1}{2}X\cos\theta\sin\theta, \tag{190}$$

where θ as before is the direction of wave propagation. This is because in figure 68(a) the distance AC is $X\cos\theta$, so that in figure 68(b) the distance AE is $\tfrac{1}{2}X\cos\theta$.

If we seek to trace the crest shapes in the ship-wave pattern, we must recognise that the crest at the point (190) makes an angle $\tfrac{1}{2}\pi - \theta$ with the x-axis. Therefore, its slope is

$$dy/dx = \cot\theta. \tag{191}$$

Looking continuously along that crest, we should observe portions of it with varying positions of generation $(X, 0)$ and directions of propagation θ; however, at each point, the crest slope must satisfy (191). Substituting for x and y in (191) from (190), we obtain after some reduction

$$dX/d\theta = -X\tan\theta. \tag{192}$$

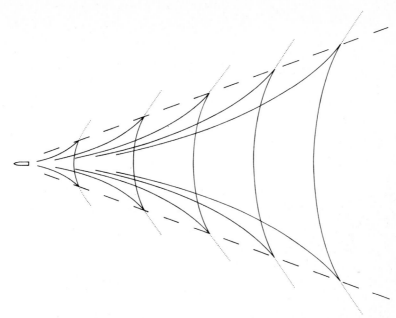

Figure 70. Plain lines: Kelvin ship-wave pattern. Broken lines: boundary of Kelvin wedge. Dotted lines: extension of waves beyond the Kelvin wedge indicated by the theory of sections 4.11 and 4.12.

This simple differential equation relating the changes in X and θ along the crest has the solution

$$X = X_1 \cos\theta, \tag{193}$$

where X_1 is constant. By (190), then, the crest shape is given parametrically by the equations

$$x = X_1 \cos\theta(1 - \tfrac{1}{2}\cos^2\theta), \quad y = \tfrac{1}{2}X_1 \cos^2\theta \sin\theta. \tag{194}$$

Figure 70 gives an indication of the crest shapes in the Kelvin ship-wave pattern by showing the curves (194) for several different constant values of X_1. All the crest shapes have cusps on the wedge boundary. The dotted-line extensions beyond that boundary are included because, as we shall see in chapter 4, the wave amplitude does not fall directly to zero beyond such a 'caustic' boundary, but tails away exponentially. In the interior, the longer waves propagating at small angles θ and the much shorter waves propagating at large angles θ are often superimposed for intermediate Froude numbers (figure 71). The longer waves predominate at low Froude numbers,

278

Figure 71. An observed ship-wave pattern. [*Courtesy of Aerofilms Ltd.*]

however; conversely, the much shorter waves predominate in the high-Froude-number conditions typical of speedboats. For methods of analysing how the Kelvin ship-wave pattern is modified by shallow-water effects (which allow the wedge to become wider), see chapter 4.

EXERCISES ON CHAPTER 3

1. A common characteristic of linear theories of wave motions is that waves of equal amplitude travelling in opposite directions can be combined linearly into a *standing wave*. (Such standing waves are not, of course, to be confused with the *stationary* waves of section 3.9.) Find the velocity potential for standing waves on deep water where the vertical displacement of the free surface takes the form

$$\zeta = a\mathrm{e}^{\mathrm{i}\omega t}\cos kx.$$

Show that the streamlines of the motion each take the form

$$\mathrm{e}^{kz}\sin kx = \text{constant},$$

and sketch these curves. With what amplitude does the particle whose average position is (x, y, z) oscillate on the streamline through that point?

2. Deep water fills a rectangular tank of length l and breadth $b \leqslant l$. Write down what dependence of the velocity potential ϕ on a product of trigonometrical functions of x and y will allow the boundary conditions at the tank's sides $x = 0$, $x = l, y = 0$ and $y = b$ to be satisfied. Show that the tank's resonant frequencies ω take the form

$$(g\pi)^{\frac{1}{2}} (n^2 l^{-2} + m^2 b^{-2})^{\frac{1}{4}},$$

where n and m are integers.

What is the least water depth for which *all* these resonant frequencies given by deep-water theory should be correct to within 3 %?

3. Other *quadratic* quantities besides energies may be calculated correctly from a linear theory. The reader is invited to prove that the travelling sinusoidal wave on deep water with surface elevation

$$\zeta = a\cos(\omega t - kx)$$

possesses an average momentum

$$\tfrac{1}{2}\rho\omega a^2$$

per unit horizontal area in the direction of wave propagation.

Such a calculation may at first sight appear of doubtful validity because a quantity like momentum, obtained by integrating the *first* power of a velocity component, could have any terms in a^2 modified by second-order improvements to a linear theory. Show, however, by the following method that this is impossible.

Prove first that periodic waves in deep water must have the velocity potential exactly periodic. Show, in fact, that if ϕ is a solution of Laplace's equation in the (x, z) plane with the velocity component $\partial\phi/\partial x$ periodic (taking the same value at $(x + \lambda, z)$ as at (x, z)), then ϕ could differ from a periodic function by at most a term of the form $xf(z)$, which must, however, vanish if it is to satisfy Laplace's equation and have zero gradient at $z = -\infty$. Hence prove that the total momentum at any fixed horizontal level $z = $ constant is zero. This means that the momentum, like the excess potential energy (24), arises entirely from the fact that the fluid's upper boundary is not a fixed horizontal level. Thus, the momentum depends on the *product* of small disturbances to a state with no motion and with the free surface horizontal, and can be correctly calculated from a linear theory as a second-order quantity.

The same method will yield still more information. Show that, on linear theory, fluid particles which in the undisturbed state were at the level $z = z_0 < 0$ lie at time t in the surface

$$z = z_0 + ae^{kz_0} \cos(\omega t - kx).$$

Deduce that those fluid particles that lay *below* $z = z_0$ in the undisturbed state possess an average momentum

$$\tfrac{1}{2}\rho\omega a^2 e^{2kz_0}$$

per unit horizontal area. Then obtain the average momentum per unit volume for fluid particles of undisturbed level z_0 by differentiation, giving their *mean horizontal velocity* as

$$\omega a^2 k e^{2kz_0}.$$

[This mean rate of movement, known as the 'Stokes drift', represents a second-order correction to the paths of fluid particles which are portrayed according to linear theory in figure 50. In each period $2\pi/\omega$, a fluid particle combines its motion round a circle of radius

$$ae^{kz_0}$$

with forward movement through a distance

$$2\pi ka^2 e^{2kz_0}$$

varying as the square of that radius. (Alternatively, though perhaps less convincingly, one may obtain this result by calculating to second order the paths of particles in the velocity field (21).)

The net forward movement of particles near the surface in deep-water waves is readily observed in tanks where waves generated at one end are dissipated upon a beach at the other. However, as the water depth becomes less, the above method for calculating the momentum distribution becomes less and less appropriate. Indeed, the proof that ϕ is exactly periodic fails; and there is nothing to prevent a second-order *uniform flow* from being added to the basic first-order fluid motion.

In, for example, a wave tank of the type just mentioned, continued 'Stokes drift' of surface water causes a build-up of fluid near the beach which brings about a pressure gradient generating a bulk flow backwards. This both reduces the surface drift and balances it with an opposing flow near the bottom. The situation is further complicated by the fact that the boundary layer associated with that unidirectional component of flow near the bottom can become far thicker than the 'Stokes boundary layer' associated (section 3.5) with the periodic component...All this means that the calculation of average momentum for periodic waves is a straightforward matter *only* in the deep-water case.]

4. Show that, in water of uniform depth h, the wave speed can be stationary as a function of wavenumber if and only if the wavenumber k has the following relationship to the corresponding deep-water value k_m given by equation (53):

$$\frac{k}{k_m} = \left[\frac{\sinh(2kh) - 2kh}{\sinh(2kh) + 2kh}\right]^{\frac{1}{2}}.$$

Sketch the function on the right-hand side, noting its behaviour for large and small values of kh. Use the information so derived to construct a proof that the wave speed has just one stationary value (a minimum) for any depth greater than $3^{\frac{1}{4}}k_{\mathrm{m}}^{-1}$ ($\doteqdot 5$ mm) but no stationary values for depths less than that.

5. For travelling sinusoidal waves in water of uniform depth h, the x-component of velocity in the Stokes boundary may be written

$$u = u_{\mathrm{ex}}\{1 - \exp\left[-(z+h)\,(i\omega/\nu)^{\frac{1}{2}}\right]\},$$

where the external-flow velocity u_{ex} is itself proportional to $\exp\left[i(\omega t - kx)\right]$. Use the equation of continuity to find the z-component of velocity, w, in and immediately outside the boundary layer.

Calculate w also in a case without any shear, where we simply take $u = u_{\mathrm{ex}}$, but apply a condition of zero normal velocity ($w = 0$) at $z = -h+\delta_1$. Verify that both calculations yield the same velocity distribution outside the boundary layer provided that δ_1 has the value (68).

6. A large number of identical *simple pendulums* of length l hang in equilibrium with the bobs (of mass M) spaced equally along a straight horizontal line L. A string is stretched to tension T along the line L and all the bobs are attached to it. The string's mass is negligible compared with that of the bobs. Show that *transverse* displacements of the bobs, in which the displacement of the nth bob perpendicular to L is y_n, satisfy the equations

$$M\ddot{y}_n = -Mgl^{-1}y_n + T(y_{n+1} - 2y_n + y_{n-1}),$$

if the spacing between bobs is taken as the unit of distance. Consider a travelling wave with y_n proportional to $\exp\left[i(\omega t - kn)\right]$, which can be thought of as having *phase velocity* $c = \omega/k$. Find this phase velocity c.

Show also that the group velocity takes the value

$$[gl^{-1} + 2M^{-1}T(1 - \cos k)]^{-\frac{1}{2}}M^{-1}T\sin k,$$

and prove that this is necessarily *less* than the phase velocity. Describe what this implies will happen to the row of pendulums when one of the bobs is given a small transverse blow.

[The experiment was invented by Reynolds, who used it to give audiences a clearly visible demonstration of a wave group moving forwards more slowly than individual crests ... The limiting case $T = 0$ is interesting since the group velocity vanishes. Reynolds would demonstrate that, when a wave was travelling to the right and suddenly the string was relaxed, the pendulums would continue to oscillate (now, at a frequency near their natural frequency $(g/l)^{\frac{1}{2}}$) so that there would still be an *appearance* of crests moving to the right; nevertheless, motion would cease to be transferred to any pendulums beyond those which had already been reached.]

7. Show that when the method of steepest descent for estimating the integral (110) for large t is carried to the next approximation, the result is that the asymptotic form (122) must be multiplied by a factor

$$1 + \frac{i}{2t}\left[\frac{5\psi''^2}{12\psi''^3} - \frac{\psi''''}{4\psi''^2} - \frac{F'\psi'''}{F\psi''^2} + \frac{F''}{F\psi''}\right],$$

with all functions calculated at the stationary point k_0 (where the values of F and ψ'' are to be assumed nonzero). Observe that this adds, to the amplitude in (122) of order $t^{-\frac{1}{2}}$, a correction of order $t^{-\frac{3}{2}}$. Why is this correction of smaller order than the *possible* maximum error $O(t^{-1})$ allowed for in the derivation of the estimate (122)?

[The reader is recommended to carry out the calculation in the case $\psi''(k_0) > 0$. The case $\psi''(k_0) < 0$ leads to the same result, as is most easily seen by changing the sign of $\psi(k)$ and taking the complex conjugate.]

8. In the linear theory of water waves, solve the following initial-value problem: given that the water surface's vertical displacement ζ and vertical velocity $w = \partial\zeta/\partial t$ take initial values depending on x alone, find the resulting waves. Show in fact that, if

$$\zeta = \int_0^\infty Z(k) \exp(-ikx)\,\mathrm{d}k, \quad w = \int_0^\infty W(k) \exp(-ikx)\,\mathrm{d}k$$

at time $t = 0$, then the consequent wave (106) travelling in the positive x-direction has

$$F(k) = \tfrac{1}{2}\{Z(k) - i[W(k)/\omega(k)]\}.$$

The water surface is released from rest in the position

$$\zeta = 2\epsilon x(x^2 + x_0^2)^{-1}.$$

Here, x_0 is large enough compared with λ_m for surface tension to be neglected (section 3.4), but small enough compared with the water depth (section 3.3) for deep-water results to be used. Having calculated $F(k)$ as $i\epsilon \exp(-kx_0)$, show that, when both x and t are large and positive,

$$\zeta \sim \epsilon t(\pi g)^{\frac{1}{2}} x^{-\frac{3}{2}} \exp\left[-\tfrac{1}{4}gt^2 x_0\, x^{-2} + i(\tfrac{1}{4}gt^2 x^{-1} + \tfrac{1}{4}\pi)\right].$$

Deduce that the position of maximum wave amplitude moves forward at velocity $(\tfrac{1}{3}gx_0)^{\frac{1}{2}}$. Also, verify that wave crests at that position move at twice that velocity.

Finally, check that the total wave energy for $x > 0$ accounts for half of the initial potential energy of the displaced water surface. Where has the other half gone?

9. Show that, on a linear theory of water waves, the part of the energy flux in the positive x-direction per unit length of crest contributed by the direct action of the surface tension T is

$$(-T\partial\zeta/\partial x)\,\partial\zeta/\partial t.$$

For travelling waves in water of depth h under the influence of both gravity and surface tension, check that this addition to the energy flux in the bulk of the fluid given by (145) raises the average energy flux to a value UW, where U and W are the group velocity and the average wave energy per unit horizontal area.

Write down the *proportion* of this average energy flux which direct energy transmission by surface tension contributes, and show that in the deep-water case this proportion can be written

$$2\lambda_m^2/(3\lambda_m^2 + \lambda^2).$$

[Thus, it is exactly $\tfrac{1}{2}$ when λ is equal to the wavelength λ_m for minimum wave speed, and rises to $\tfrac{2}{3}$ in the capillary-wave limit (λ/λ_m small).]

10. In a steady stream of speed V and depth h, stationary waves can be generated where a steady jet of air blows onto the surface. Consider the case when a plane air jet produces at the water surface a pressure distribution varying only with x (the coordinate in the direction of the stream). Use the theory of section 3.9 with

the boundary value η taken as the difference between the air pressure at the surface and the undisturbed atmospheric pressure p_a. Show that the quantity $B(kV, k)$ occurring in the analysis from equation (172) onwards is then equal to

$$B(kV, k) = (\tanh kh)^{-1} \rho V^2 k - \rho g - Tk^2.$$

If V exceeds the minimum wave speed $(4gT/\rho)^{\frac{1}{4}}$ but is substantially less than the long-wave speed $(gh)^{\frac{1}{2}}$, show that the equation $B(kV, k) = 0$ is satisfied by two positive values of k close to the values

$$k_1 = (2T)^{-1} [\rho V^2 - (\rho^2 V^4 - 4\rho gT)^{\frac{1}{2}}], \quad k_2 = (2T)^{-1} [\rho V^2 + (\rho^2 V^4 - 4\rho gT)^{\frac{1}{2}}]$$

obtained by replacing $\tanh kh$ by 1.

When the air jet exerts a total force G per unit breadth of stream by means of an air pressure distribution $\pi^{-1} G x_0 (x^2 + x_0^2)^{-1}$, show that waves close to

$$\zeta = -2G(\rho^2 V^4 - 4\rho gT)^{-\frac{1}{2}} e^{-k_1 x_0} \sin(k_1 x)$$

are generated downstream of the air jet, while waves with ζ close to

$$\zeta = -2G(\rho^2 V^4 - 4\rho gT)^{-\frac{1}{2}} e^{-k_2 x_0} \sin(k_2 x)$$

are generated upstream. For a stream of speed 0.30 m s^{-1}, show that the lengths of the waves generated are about 0.05 m for the downstream waves and about 0.006 m for the upstream waves, whose amplitude is smaller by a factor $\exp(-970x_0)$, where x_0 is in metres.

[Thus, the upstream waves may become hard to observe for x_0 greater than about 5 mm.]

11. We show in sections 3.4 and 3.6 respectively that water waves with length less than 4 mm are capillary waves and that the ratio of their group velocity to their wave speed is $\frac{3}{2}$. The reader is invited to carry out an analysis similar to that of section 3.10 for the generation of such capillary waves by the steady motion through water, at velocity V, of a solid body (whose dimensions must not exceed 1 or 2 mm for the analysis to be a good approximation).

Show that waves are found all round the obstacle. In coordinates with the present position of the obstacle as origin and the negative x-axis along its direction of motion, find the present position of the waves that were emitted at an angle θ to that direction when the obstacle was at $(X, 0)$.

Show that, along any single crest, the quantities X and θ vary in such a way that $X \cos^3 \theta$ takes a constant value, say X_1. Deduce that the crest shape is given parametrically by the equations

$$x = X_1 \sec \theta (\tan^2 \theta - \tfrac{1}{2}), \quad y = \tfrac{3}{2} X_1 \sec \theta \tan \theta.$$

Sketch this curve, noting that it goes through the points $(-0.5X_1, 0)$ and $(0, \pm 1.3X_1)$ and asymptotes to $y = \pm 1.5 X_1^{\frac{1}{3}} x^{\frac{2}{3}}$.

[These crest shapes concave to the obstacle are easy to observe when (say) a knitting needle is moved through water well lit from above.]

4. INTERNAL WAVES

4.1 Introduction to internal gravity waves

The ideas of dispersion and of group velocity are developed in chapter 3 for *isotropic* cases, where the wave speed c though varying with wavelength shows no separate dependence on the direction of propagation. Then, wave energy is propagated at right angles to the crests at the group velocity U.

Chapter 4 has been written with three aims in mind. The first is to give a general analysis of dispersive systems which satisfy linear equations allowing the wave speed c to vary with direction of propagation as well as with wavelength; in these *anisotropic* systems the group velocity has to be considered as a vector U and is not necessarily perpendicular to the crests. A second aim is to illustrate the general theory, mainly by the example of internal waves in a stratified medium; a case important in oceanography and meteorology (section 4.3), and one so different from isotropic cases that the wave energy is propagated at a group velocity U *parallel* to the crests! A third aim is to give unity to the book's subject matter by using the general analysis to *substantiate and extend* (i) the ideas of geometrical acoustics given in chapters 1 and 2, as well as (ii) some approaches to calculating water waves given in chapter 3.

We begin chapter 4 by describing internal gravity waves, as one example of a highly anisotropic wave system, important to the understanding of our natural environment. We have seen in chapter 3 how gravity acts to restore the flatness of a surface which separates water from air, and how the balance between that restoring force and the inertia of the water governs the propagation of waves on such a surface. There are many other systems with light fluid overlying heavy fluid that are *stable* in the sense that disturbances (in which heavy fluid may penetrate light fluid, and vice versa) are opposed by the action of gravity. In any such stable systems, wave propagation is made possible through a balance between the gravitational restoring force and the total fluid inertia.

Before describing the main subject of this section (gravity waves in continuously stratified fluid) we mention a much simpler case of internal waves, easy to study by the methods of chapter 3. In many deep estuaries (such as Norwegian fjords) the fresh river water tends to move seawards *above* the heavier salt water; essentially, because tidal motions are too weak for the gravitational stability of the interface between fresh and salt water to be overcome by 'turbulent mixing' effects (at least while the sea surface is relatively calm). We consider the dynamics of such an interface, idealised by regarding it as a thin surface of separation between a heavier fluid of uniform density ρ_2 and a lighter fluid of uniform density $\rho_1 < \rho_2$. Thus, any effects of *diffusion* of salt into the overlying fresh water are neglected.

No complicated analysis is required: the ideas of generalised stiffness and generalised inertia explained in section 3.2 yield at once the dispersion relationship. If $z = 0$ is the undisturbed position of the interface, then its elevation to $z = \zeta$ increases the potential energy of the lower fluid by $\frac{1}{2}\rho_2 g\zeta^2$ per unit area as in section 3.2. However, it *decreases* that of the upper fluid by $\frac{1}{2}\rho_1 g\zeta^2$, as the same method of calculation shows. The generalised stiffness of the interface (coefficient of $\frac{1}{2}\zeta^2$ in the potential energy) is therefore

$$(\rho_2 - \rho_1)g \tag{1}$$

per unit horizontal area; it depends on the *difference* of density between the two fluids.

In addition, any movement of the interface generates motion in the lower fluid with kinetic energy $\frac{1}{2}\rho_2 k^{-1}(\partial\zeta/\partial t)^2$ per unit area, as in section 3.2. Simultaneously, it generates an identical (though inverted) motion in the upper fluid with kinetic energy $\frac{1}{2}\rho_1 k^{-1}(\partial\zeta/\partial t)^2$. These expressions are good approximations provided that both the bottom and the free surface are at distances of at least 0.28λ from the interface so that they do not interfere significantly with the motions. The generalised inertia of the interface (coefficient of $\frac{1}{2}(\partial\zeta/\partial t)^2$ in the kinetic energy) is therefore

$$(\rho_2 + \rho_1)k^{-1} \tag{2}$$

per unit horizontal area; this depends on the *sum* of the densities of the two fluids. To obtain the dispersion relationship we write ω^2, the square of the radian frequency, as the ratio of generalised stiffness (1) to generalised inertia (2); that is, as

$$\omega^2 = gk[(\rho_2 - \rho_1)/(\rho_2 + \rho_1)]. \tag{3}$$

Evidently, the wave motions for given wavelength $2\pi/k$ are exactly as in surface gravity waves but with the frequency ω (and therefore also the wave speed $c = \omega/k$) *reduced* by a factor equal to the square root of the ratio in square brackets. That reduction factor is around 0.11 to 0.12 for river water overlying sea-water. Therefore, although a ship may be moving much too slowly in relation to its length to excite any surface waves (section 3.10), it may be moving *fast* in relation to the speeds of waves (with wavelengths related to the ship's length) on such an estuarial interface. This means that the ship may experience a substantially enhanced drag on entering the estuary. The extra drag is strictly a wavemaking resistance, associated with the power required to generate the internal waves (even though the ship makes no *visible* waves on the free surface).

We move on now from waves on an interface between two homogeneous fluids to waves in a fluid with continuous stratification. Later (section 4.3), we check that the result (3) does agree with a limiting form of the results for continuously stratified fluid.

We analyse small disturbances to an equilibrium distribution of density $\rho_0(z)$ in an atmosphere or an ocean. Here, ρ_0 is a continuously decreasing function of the height z. The equilibrium distribution of pressure $p_0(z)$ must also decrease with height, according to the hydrostatic law

$$\mathrm{d}p_0(z)/\mathrm{d}z = -\rho_0(z)g. \tag{4}$$

This pressure distribution affects the gravitational restoring force. Suppose, for example, that a particle of fluid at height z, with density $\rho_0(z)$, makes a small movement so that its height increases to $z + \zeta$. This brings it to a region where the equilibrium density takes the smaller value

$$\rho_0(z) + \zeta \rho_0'(z), \quad \text{with} \quad \rho_0'(z) < 0, \tag{5}$$

and where the pressure is reduced to

$$p_0(z) - \rho_0(z)g\zeta. \tag{6}$$

In a reversible process, the particle must experience this pressure drop at constant entropy. Therefore, its own density is reduced to

$$\rho_0(z) - \rho_0(z)g\zeta[c_0(z)]^{-2}, \tag{7}$$

where $[c_0(z)]^2$, the square of the undisturbed sound speed at pressure $p_0(z)$ and density $\rho_0(z)$, is the ratio of pressure changes to density changes at constant entropy.

The particle's *excess* density over that of the surrounding fluid is therefore the difference of (7) from (5), namely

$$\{-\rho_0(z)g[c_0(z)]^{-2} - \rho_0'(z)\}\zeta. \tag{8}$$

Multiplying this by g gives the particle's excess of weight over buoyancy, per unit volume of fluid; in other words, the gravitational restoring force

$$\{-\rho_0(z)g[c_0(z)]^{-2} - \rho_0'(z)\}g\zeta. \tag{9}$$

This simple argument shows that gravitational stability is *not* assured by simply requiring $\rho_0'(z) < 0$ as in (5); the restoring force (9) is positive only if

$$\rho_0'(z) < -\rho_0(z)g[c_0(z)]^{-2}. \tag{10}$$

Thus, a stratified atmosphere or ocean is stable only if the relative rate of decrease of density with height $(-\rho_0'/\rho_0)$ exceeds gc_0^{-2}. For a fuller discussion of this important stability condition, see section 4.3.

When the condition (10) for stable stratification is satisfied, the restoring force per unit volume (9) may be written

$$\rho_0(z)\,[N(z)]^2\zeta, \tag{11}$$

where the equation

$$N(z) = \{[-g\rho_0'(z)/\rho_0(z)] - g^2[c_0(z)]^{-2}\}^{\frac{1}{2}} \tag{12}$$

defines a positive quantity $N(z)$ with the dimensions of frequency. Evidently, the potential energy and generalised stiffness associated with this gravitational restoring force (11) are

$$\tfrac{1}{2}\rho_0(z)\,[N(z)]^2\zeta^2 \quad \text{and} \quad \rho_0(z)\,[N(z)]^2, \tag{13}$$

each per unit volume.

It is easily seen that $N(z)$, which is often called the Väisälä–Brunt frequency, represents a sort of maximum frequency for oscillations under gravity. That is because, with the vertical displacement ζ as generalised coordinate, the generalised inertia has a *minimum* value $\rho_0(z)$ per unit volume. This minimum is attained when the oscillations are purely vertical, with kinetic energy

$$\tfrac{1}{2}\rho_0(z)\,(\partial\zeta/\partial t)^2 \tag{14}$$

per unit volume. Any horizontal components in the motions must simply add to the kinetic energy (14). That would increase the generalised inertia above $\rho_0(z)$, and reduce the frequency squared, ω^2 (ratio of generalised stiffness to generalised inertia), below $[N(z)]^2$. For example, in oscillations where all fluid motions are inclined at an angle θ to the vertical, the generalised inertia would be $\rho_0(z)\sec^2\theta$ and so the frequency should be

$$\omega = N(z)\cos\theta. \tag{15}$$

The essential correctness of these tentative conclusions will be demonstrated in two stages. The first stage uses equations of motion in an approximate form due to Boussinesq. This assumes that, in internal gravity waves, as in surface gravity waves (section 3.1), the equation of continuity can be simplified by ignoring compressibility. The second stage, given in section

4.2, uses the full equations for a compressible fluid, and therefore implies the possible presence of sound waves as well as internal waves; on the other hand, it proves that they propagate independently under a certain condition (indicated already at the end of section 1.2) which is commonly satisfied.

In this section we give only the analysis based on the Boussinesq approximation. The idea of this is that although the excess density (8) generates a significant gravitational restoring force (9) which must be taken into account in the fluid's momentum equation, the consequent oscillations take place at such low frequency that the *rate of change* of excess density negligibly affects the equation of continuity.

If p_e and ρ_e stand for the excess pressure $p - p_0$ and excess density $\rho - \rho_0$, the equation of momentum in linearised form is

$$\rho_0 \, \partial \mathbf{u}/\partial t + \nabla p_e = \rho_e \, \mathbf{g}, \tag{16}$$

where the vector gravitational acceleration $\mathbf{g} = (0, 0, -g)$ is directed downwards. In this form (16), the hydrostatic balance equation $\nabla p_0 = \rho_0 \, \mathbf{g}$ has been subtracted out. The equation of continuity becomes

$$\nabla \cdot (\rho_0 \, \mathbf{u}) = 0 \tag{17}$$

if the term $\partial \rho_e / \partial t$ is neglected (as discussed above). By taking the divergence of (16), and using (17), we deduce that

$$\nabla^2 p_e = -g \, \partial \rho_e / \partial z. \tag{18}$$

If now we write q as the upward component of mass flux (that is, the z-component of $\rho_0 \, \mathbf{u}$), then equation (16) implies that

$$\partial q / \partial t + \partial p_e / \partial z = -g \rho_e; \tag{19}$$

and, from equations (18) and (19), we deduce that

$$\nabla^2 (\partial q / \partial t) = g \, \partial^2 \rho_e / \partial z^2 - g \nabla^2 \rho_e = -g(\partial^2 \rho_e / \partial x^2 + \partial^2 \rho_e / \partial y^2). \tag{20}$$

Note that the right-hand side of (20) is positive at those points on a given level (fixed z) where ρ_e has a local maximum. The effect of such a local positive value of $\nabla^2 (\partial q / \partial t)$ is to generate at nearby positions *negative* values of $\partial q / \partial t$ itself: that is, downward accelerations, such as might be expected around a maximum of the excess density.

Also, the value (8) for the excess density ρ_e implies, with the definition (12) of $N(z)$, that

$$g \partial \rho_e / \partial t = [N(z)]^2 \rho_0(z) \, \partial \zeta / \partial t = [N(z)]^2 q. \tag{21}$$

By (20), therefore,

$$\nabla^2 (\partial^2 q / \partial t^2) = -[N(z)]^2 (\partial^2 q / \partial x^2 + \partial^2 q / \partial y^2). \tag{22}$$

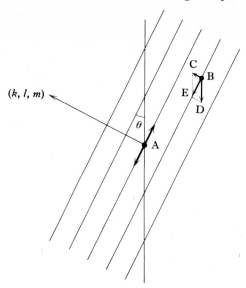

Figure 72. The equation of continuity implies that the motions of each fluid particle A in a sinusoidal internal wave must be perpendicular to the wavenumber vector (k, l, m). Therefore, they lie in surfaces of constant phase. The net force on the fluid is a combination of a vertical gravitational force BD and of a pressure-gradient force BC perpendicular to surfaces of constant phase. Accordingly its resultant BE can itself lie in those surfaces but must act up or down the direction of steepest ascent.

This is an interesting partial differential equation for the upward mass flux q. Its *anisotropic* character is evident from the appearance of not only the three-dimensional Laplace operator ∇^2 but also the two-dimensional Laplace operator $\partial^2/\partial x^2 + \partial^2/\partial y^2$.

While we postpone to later sections any study of the solutions of (22) for general $N(z)$, we note at once that when $N(z)$ is independent of z the equation possesses simple plane-wave solutions; namely,

$$q = q_1 \exp\left[i(\omega t - kx - ly - mz)\right], \tag{23}$$

where
$$\omega^2 = N^2(k^2 + l^2)/(k^2 + l^2 + m^2). \tag{24}$$

These are waves whose crests and other surfaces of constant phase are planes $kx + ly + mz = $ constant, perpendicular to the wavenumber vector (k, l, m). Equation (24) shows that, as predicted earlier, the frequency ω of such a wave is at most equal to N.

Actually, figure 72 shows that equation (24) is exactly consistent with the result $\omega = N \cos \theta$ given in (15). The equation of continuity (17) implies that the velocity vector **u** in a plane-wave solution (23) is *perpendicular* to the wavenumber vector (k, l, m). This means that the fluid motions are all parallel to the surfaces of constant phase. We can, however, say more: the oscillatory fluid motions are all up and down the direction of steepest ascent in such a constant-phase surface. This is because equation (16) makes $\rho_0 \, \partial \mathbf{u}/\partial t$ coplanar with the wavenumber vector (direction of ∇p_e) and the vertical (direction of $\rho_0 \, \mathbf{g}$); in other words, although gradients of excess pressure perpendicular to a constant-phase surface *are* able to combine with vertical gravity forces to give fluid accelerations parallel to that surface, these *must* be in the line of steepest ascent. Accordingly, plane waves involve unidirectional oscillations of fluid at an angle θ to the vertical, where

$$\sin \theta = |m| \, (k^2 + l^2 + m^2)^{-\frac{1}{2}}. \tag{25}$$

This makes equation (24) agree with the result (15) obtained for oscillations at such an angle θ to the vertical from considerations of generalised stiffness and inertia.

The above geometrical argument is easily verified by a detailed calculation of p_e and $\rho_0 \mathbf{u}$ from (21), (18) and (16). This gives

$$p_e = -\omega m (k^2 + l^2)^{-1} q_1 \exp \left[\mathrm{i}(\omega t - kx - ly - mz) \right], \tag{26}$$

and

$$\rho_0 \mathbf{u} = [-km(k^2 + l^2)^{-1}, \, -lm(k^2 + l^2)^{-1}, \, 1] \, q_1 \exp \left[\mathrm{i}(\omega t - kx - ly - mz) \right], \tag{27}$$

where **u** is perpendicular to the wavenumber vector (k, l, m), and coplanar with it and the z-direction.

The fact that p_e and **u** are in phase means that the waves produce a nonzero *wave energy flux*,

$$\mathbf{I} = p_e \mathbf{u} = (p - p_0) \, \mathbf{u}. \tag{28}$$

We know from section 1.3 that for sound waves such a wave energy flux (28) is called acoustic intensity, and, like the particle velocity **u**, is directed at right angles to crests (surfaces of constant phase). By contrast, in internal waves the particle velocity **u** is *parallel* to surfaces of constant phase, so that the wave energy flux (28) must equally be directed parallel to surfaces of constant phase. This surprising conclusion is confirmed in the next section, and many of its implications are pursued in some detail in the rest of the chapter.

4.2 Combined theory of sound and internal waves

The methods of inferring the dispersion relationship for internal waves given in section 4.1 (using equations of motion to derive (24), or energy arguments with the geometrical considerations of figure 72 to derive (15)) are conveniently simple. Nevertheless, these methods that take into account the excess density resulting from vertical displacement in *one* context (gravity effect in the momentum equation) but ignore its rate of change in *another* context (equation of continuity) are not, perhaps, fully convincing.

One route to finding conditions under which they may give a good approximation is by comparing the neglected term $\partial \rho_e / \partial t$, which by (21) is $g^{-1} N^2 q$, with just one of the terms retained in the equation of continuity (17); namely, $\partial q / \partial z$. This suggests the need for *wavenumbers to be large compared with* $g^{-1} N^2$ if the retained term is to be large compared with the neglected term. It may also be thought necessary to ask if we are justified in assuming excess densities ρ_e to be derived *only* from vertical displacement to a level of different hydrostatic pressure p_0. The fluid's compressibility would allow *additional* density changes $c_0^{-2} p_e$ resulting from pressure changes p_e at a fixed level. Equation (18) suggests, however, that this might make a negligible modification to ρ_e for *wavenumbers large compared with* $g c_0^{-2}$ (because the solution of (18) for p_e would then be of smaller order of magnitude than $c_0^2 \rho_e$).

These two tentative conditions for the accuracy of the Boussinesq approximation are closely related, since by (12) the quantities $g^{-1} N^2$ and $g c_0^{-2}$ add up to

$$g^{-1} N^2 + g c_0^{-2} = [-\rho_0'(z)/\rho_0(z)], \tag{29}$$

the relative rate of decrease of density with height. For *wavenumbers large compared with this sum* (29), *both* conditions are therefore satisfied.

The two effects just mentioned as neglected in the Boussinesq approximation are, of course, the main effects governing the propagation of *sound*; in which compressibility associates with each local pressure change a local density change, which in turn affects the divergence of the velocity field. Accordingly, it was *only* by neglecting those effects in section 4.1 that sound waves could be excluded from the solutions to our equations.

There are three incentives to studying in this section the full linearised equations for a stratified compressible fluid, in order to find solutions that may include *both* internal gravity waves *and* sound waves. One incentive is to check our tentative conditions on wavenumber for the correctness of the dispersion relationship (24) for internal waves. A second incentive is to find conditions for sound waves to be unaffected by gravity. The case of uniform entropy per unit mass was considered at the end of section 1.2: this is the case $p_0'(z) = [c_0(z)]^2 \rho_0'(z)$, when $N = 0$ by equations (4) and (12), so that there are no internal waves. Then the equations of sound were inferred to be accurate for wavenumbers $2\pi/\lambda$ large compared with gc_0^{-2}. In the present section we extend this conclusion, showing that for general N the wavenumber must in addition be large compared with $g^{-1}N^2$ if sound waves are to be affected negligibly by gravity. Thus, one and the same pair of conditions on wavenumber permits us to ignore *both* compressibility effects on internal waves *and* gravity effects on sound: under those conditions, sound and internal waves are completely 'decoupled'. This suggests a third incentive to the present study: to obtain a basis for analysing, especially in section 4.13, cases when (because those conditions are violated) there is a coupling between sound and internal waves.

We begin by writing the full linearised equation of continuity as

$$\partial \rho_e / \partial t + \nabla \cdot (\rho_0 \mathbf{u}) = 0. \tag{30}$$

When we use (30) instead of (17) after taking the divergence of the linearised momentum equation (16), we obtain

$$\nabla^2 p_e = \partial^2 \rho_e / \partial t^2 - g \partial \rho_e / \partial z. \tag{31}$$

We can recognise the first term on the right-hand side as the value of $\nabla^2 p_e$ for pure sound waves and the second term as its value (18) for pure internal waves.

We need also the full linearised equation stating that in a reversible process the rate of change of pressure following a particle is the square of the sound speed times the corresponding rate of change of density. The exact form (*not* linearised) of this equation is

$$\partial p / \partial t + \mathbf{u} \cdot \nabla p = c^2 (\partial \rho / \partial t + \mathbf{u} \cdot \nabla \rho), \tag{32}$$

where c is the local sound speed. If in this equation $\mathbf{u}$ and $p - p_0$ and $\rho - \rho_0$ and $c - c_0$ are small quantities, products of any two of which are neglected, and p_0, ρ_0 and c_0 depend on z alone, the equation takes on the linearised form

$$\partial p_e / \partial t + w \, dp_0 / dz = [c_0(z)]^2 [\partial \rho_e / \partial t + w \, d\rho_0 / dz], \tag{33}$$

where $w = q/\rho_0$ is the upward component of $\mathbf{u}$. With the hydrostatic condition (4) and the definition (12) of $N(z)$ this gives

$$\partial \rho_e / \partial t = [c_0(z)]^{-2} \partial p_e / \partial t + g^{-1} [N(z)]^2 q. \tag{34}$$

As before, we recognise the first term on the right-hand side as the value of $\partial \rho_e / \partial t$ for pure sound waves in homogeneous fluid, and the second term as its value given for pure internal waves by (21).

We can eliminate ρ_e from our equations by using (19), the z-component of the momentum equation. In (34) this gives

$$\partial^2 q / \partial t^2 + [N(z)]^2 q = - \partial^2 p_e / \partial z \partial t - g[c_0(z)]^{-2} \partial p_e / \partial t, \qquad (35)$$

while (31), with $\partial^2 \rho_e / \partial t$ substituted from (34) and $g \partial \rho_e / \partial z$ from (19), gives

$$\partial^2 p_e / \partial x^2 + \partial^2 p_e / \partial y^2 - [c_0(z)]^{-2} \partial^2 p_e / \partial t^2 = \partial^2 q / \partial z \partial t + g^{-1}[N(z)]^2 \partial q / \partial t. \quad (36)$$

Equations (35) and (36) govern the combined propagation of sound and internal waves in a stratified fluid.

Before looking for conditions for waves of the two types to be decoupled, we verify that our equations are consistent with conservation of wave energy. For reasons first explained in section 1.3, we use equation (28) to define $\mathbf{I} = p_e \mathbf{u}$ as the wave energy flux; that is, a vector whose component $\mathbf{I} \cdot \mathbf{n}$ in the direction of any unit vector $\mathbf{n}$ is the rate at which wave energy is being transported in the direction of $\mathbf{n}$ across a small plane element at right angles to $\mathbf{n}$, per unit area of that plane element, through the rate $p_e(\mathbf{u} \cdot \mathbf{n})$ of working by the excess pressures p_e. The associated wave energy per unit volume may be expected to be

$$W = \tfrac{1}{2}[\rho_0(z)] (\mathbf{u} \cdot \mathbf{u}) + \tfrac{1}{2}[c_0(z)]^{-2}[\rho_0(z)]^{-1} p_e^2 + \tfrac{1}{2}[\rho_0(z)] [N(z)]^2 \zeta^2; \quad (37)$$

that is, the sum of the kinetic energy, the acoustic potential energy, and the internal-wave potential energy (13). We now check that the rate of change of this wave energy density W can always be exactly accounted for as minus the divergence of the wave energy flux (28).

To this end, we need an expression

$$\zeta = \int w \, dt = [\rho_0(z)]^{-1} \int q \, dt \qquad (38)$$

for the vertical displacement ζ. Then we integrate equation (34) with respect to time to give

$$[N(z)]^2 \zeta = g[\rho_0(z)]^{-1} \{\rho_e - [c_0(z)]^{-2} p_e\}. \qquad (39)$$

Also, in (33) we replace the second square bracket by $-\rho_0 \nabla \cdot \mathbf{u}$ from the equation of continuity (30), to give

$$\partial p_e / \partial t = w g \rho_0(z) - [c_0(z)]^2 [\rho_0(z)] \nabla \cdot \mathbf{u}. \qquad (40)$$

The rate of change of wave energy (37) is

$$\partial W/\partial t = \mathbf{u} \cdot [\rho_0(z)\, \partial \mathbf{u}/\partial t] + [c_0(z)]^{-2}[\rho_0(z)]^{-1} p_e\, \partial p_e/\partial t$$
$$+ [\rho_0(z)]\, [N(z)]^2\, \zeta w, \qquad (41)$$

which can be evaluated through adding up (i) the scalar product of $\mathbf{u}$ with the momentum equation (16), (ii) the product of $[c_0(z)]^{-2}\,[\rho_0(z)]^{-1} p_e$ with (40), and (iii) the product of $[\rho_0(z)]\, w$ with (39). The result is

$$\partial W/\partial t = \{-\mathbf{u} \cdot \nabla p_e - wg\rho_e\} + \{[c_0(z)]^{-2} wgp_e - p_e \nabla \cdot \mathbf{u}\}$$
$$+ \{wg\rho_e - [c_0(z)]^{-2} wgp_e\}, \qquad (42)$$

where the three expressions in curly brackets correspond to (i), (ii) and (iii) respectively. On the right-hand side, four terms cancel, and since

$$\nabla \cdot \mathbf{I} = \nabla \cdot (p_e \mathbf{u}) = \mathbf{u} \cdot \nabla p_e + p_e \nabla \cdot \mathbf{u}, \qquad (43)$$

the remaining terms give the expected equation

$$\partial W/\partial t = -\nabla \cdot \mathbf{I}. \qquad (44)$$

Thus, we have verified that the linearised equations we are using satisfy an energy conservation equation with the wave energy flux defined as $\mathbf{I} = p_e \mathbf{u}$ (as in sound waves) and with the wave energy density W given by an expression (37) where the potential energies associated with sound and internal waves are simply added together.

These linearised equations, with all variables except p_e and q eliminated, become the pair of second-order equations (35) and (36). Now we investigate properties of solutions of those equations, looking in particular for conditions under which their solutions take the form of decoupled sound waves and internal gravity waves.

It is not possible to obtain exact solutions in which the functions $N(z)$ and $c_0(z)$ occurring in (35) and (36) take a quite general form. Much, however, can be found out about wavelike solutions to the equations from the *local dispersion relationship*. This is a relationship

$$\omega = \omega(k, l, m, z), \qquad (45)$$

varying with the height z, between the frequency and the three components of wavenumber, in solutions that locally vary as

$$p_e = p_1 \exp[i(\omega t - kx - ly - mz)], \quad q = q_1 \exp[i(\omega t - kx - ly - mz)]. \quad (46)$$

The local dispersion relationship is calculated by ignoring the rates of change of the amplitudes p_1 and q_1 with position compared with the rate of change of the sinusoidally fluctuating factor Then equation (35) becomes

$$(\omega^2 - N^2)q_1 = \omega(m + igc_0^{-2})p_1, \tag{47}$$

and equation (36) becomes

$$(c_0^{-2}\omega^2 - k^2 - l^2)p_1 = \omega(m + ig^{-1}N^2)q_1. \tag{48}$$

There are interesting cases when equations (35) and (36) have the *exact* solution (46), with p_1, q_1, ω, k, l and m constants satisfying (47) and (48). Those are cases with $c_0(z)$ and $N(z)$ constant. They include the case of an *atmosphere at uniform temperature* T_0, and constant ratio of specific heats γ. Then $c_0 = (\gamma R T_0)^{\frac{1}{2}}$ is constant (section 1.2) and the hydrostatic balance equation (4) with $p_0 = R T_0 \rho_0$ gives

$$\rho_0'(z)/\rho_0(z) = -g/R T_0 = -\gamma g c_0^{-2}, \tag{49}$$

so that (12) becomes $N = (\gamma - 1)^{\frac{1}{2}} g c_0^{-1}. \tag{50}$

For example, this expression for the isothermal Väisälä–Brunt frequency takes value ranging from 0.018 to 0.015 s^{-1} (corresponding to oscillation periods from 6 to 7 *minutes*) for air at temperatures from 20 °C to − 60 °C.

Furthermore, in many other cases when $c_0(z)$ and $N(z)$ do vary, but only gradually on a scale of wavelengths, we shall find in section 4.5 that, as in the one-dimensional case (section 3.8), we can trace the propagation of the wave energy if the local dispersion relationship (45) is known everywhere. In these cases too, then, the nature of the waves can be found out by the kind of study of that dispersion relationship which now follows.

Eliminating p_1 and q_1 from (47) and (48) gives the dispersion relationship as

$$c_0^{-2}\omega^4 - [(k^2 + l^2 + m^2) + im(gc_0^{-2} + g^{-1}N^2)]\omega^2 + (k^2 + l^2)N^2 = 0. \tag{51}$$

We study the solutions to (51), regarded as a quadratic equation for ω^2, under the conditions suggested at the beginning of this section. We assume, in other words, that

the wavenumber $(k^2 + l^2 + m^2)^{\frac{1}{2}}$ is large compared with $[-\rho_0'(z)/\rho_0(z)]$, (52)

which by (29) implies that it is large compared *both* with $g^{-1}N^2$ *and* with gc_0^{-2}; and, therefore, also large compared with their geometric mean $c_0^{-1}N$.

Under these conditions, the quadratic equation (51) for ω^2 has the property that the middle coefficient is approximately $k^2 + l^2 + m^2$ *and* is much larger than the geometric mean $c_0^{-1}N(k^2 + l^2)^{\frac{1}{2}}$ of the other two. Now, to good approximation, any quadratic equation with middle coefficient much larger than the geometric mean of the other two has roots as follows: the larger root is given by ignoring its last term, and the smaller by ignoring its first term. In this case, therefore, the larger root is approximately

$$\omega^2 = c_0^2(k^2 + l^2 + m^2) \tag{53}$$

and the smaller root is approximately

$$\omega^2 = N^2(k^2 + l^2)/(k^2 + l^2 + m^2). \tag{54}$$

Here, (53) is the ordinary relationship representing the *nondispersive* property of sound waves (with their wave speed c_0 independent of wavenumber), and equation (54) is the dispersion relationship (24) for internal waves. We have proved, therefore, that under the condition (52) (that is, when the undisturbed density $\rho_0(z)$ varies gradually on a scale of wavelengths) the two types of waves are completely decoupled: neither is influenced by the presence of the other.

It may also be useful, as we found in sections 3.3 and 3.8, to calculate what value the wavenumber must take for a given frequency ω. We shall show (section 4.5) that waves propagating through a stratified atmosphere retain (just like water waves moving towards a beach) a constant frequency ω; and, also, that, in an atmosphere where the undisturbed density ρ_0 and other properties vary only with the vertical coordinate z, it is only the z-component of wavenumber, m, that can vary with position; the other components k and l remain constant like ω. There is interest, therefore, in solving equation (51) for m to see how it may vary with $c_0(z)$ and $N(z)$ for fixed ω, k and l.

Using equation (29), we can write the solution of (51) for m as

$$m = \tfrac{1}{2}\mathrm{i}\frac{\rho_0'(z)}{\rho_0(z)} + \left\{(k^2 + l^2)\frac{[N(z)]^2 - \omega^2}{\omega^2} + \frac{\omega^2}{[c_0(z)]^2} - \frac{1}{4}\left[\frac{\rho_0'(z)}{\rho_0(z)}\right]^2\right\}^{\tfrac{1}{2}}. \tag{55}$$

The appearance of an imaginary part $\tfrac{1}{2}\mathrm{i}\rho_0'/\rho_0$ in m, even though under condition (52) it is negligibly small, may seem surprising, but has a simple explanation. In the case when (46) is an exact solution of the linearised equations (the case with $N(z)$ and $c_0(z)$ constant) this term $\tfrac{1}{2}\mathrm{i}\rho_0'(z)/\rho_0(z)$ is also a constant by (29), and accordingly contributes to $\exp(-\mathrm{i}mz)$ a factor *proportional to* $[\rho_0(z)]^{\tfrac{1}{2}}$; but energy conservation, as we show below, does in this case require the amplitudes of p_e and q to vary with z exactly in proportion to $[\rho_0(z)]^{\tfrac{1}{2}}$.

In the meantime, the real part of m gives the true vertical wavenumber as the square root of the expression in curly brackets in (55). In this expression, at frequencies ω *below* the Väisälä–Brunt frequency $N(z)$, the first term dominates, giving the simple internal-wave dispersion relationship (54). By comparison, the other terms are of the order of the *squares* of the small quantities $c_0^{-1}N$ and (ρ_0'/ρ_0). Actually, equation (29) shows that when $\omega < N$ the sum of those other terms is algebraically less than

$$c_0^{-2}N^2 - \tfrac{1}{4}(g^{-1}N^2 + gc_0^{-2})^2 = -\tfrac{1}{4}(gc_0^{-2} - g^{-1}N^2)^2, \tag{56}$$

so that it is *negative*. This in theory sets a lower limit on the horizontal wavenumber $(k^2 + l^2)^{\frac{1}{2}}$ for internal waves to be possible; however, wavenumbers lie far above that lower limit when condition (52) is satisfied.

Conversely, when ω is large compared with N, the expression in curly brackets in (55) is dominated by the terms $-(k^2 + l^2) + \omega^2[c_0(z)]^{-2}$ leading to the acoustic relationship (53). In theory, for values of ω only *moderately* greater than $N(z)$, acoustic propagation is rendered *anisotropic* by the additional term

$$(k^2 + l^2)[N(z)]^2\omega^{-2}; \tag{57}$$

however, this finding of anisotropy does not contradict condition (52) since at those frequencies the wavenumbers are merely of the same order as $[-\rho_0'(z)/\rho_0(z)]$. Whenever the wavenumber is in a large ratio to this, and so also to Nc_0^{-1}, the term $-(k^2 + l^2)$ in curly brackets in (55) exceeds (57) by the *square* of that ratio.

The above analysis of the term in curly brackets in (55) shows the decoupling of sound and gravity waves to be very precise; the error is always *quadratic* in the small ratio of $[-\rho_0'(z)/\rho_0(z)]$ to wavenumber. By contrast, the imaginary part of (55) may be worth taking more seriously, as being linearly proportional to that ratio. Careful analysis as follows, however, indicates that such a small imaginary part in a local dispersion relationship may (if used too simple-mindedly) give misleading results in general cases; although, of course, it gives a correct answer in the special case when (46) is an exact solution; and is able, if properly interpreted, to suggest useful results in a wider range of cases.

Waves of fixed real frequency ω must have their energy density W independent of t at any fixed position. By (44), therefore, the divergence $\nabla \cdot \mathbf{I}$ is zero. This means, in an atmosphere whose properties vary only in the z-direction, that there can be no variation with z in the *upward component*,

$$p_e w = [\rho_0(z)]^{-1}p_e q \tag{58}$$

of the wave energy flux $\mathbf{I}$. But by (48), for waves that are locally of the form (46), the part of p_e in phase with q is

$$\{[c_0(z)]^{-2}\omega^2 - k^2 - l^2\}^{-1}\omega m q, \tag{59}$$

which specifies an average value of the upward energy flux (58) as

$$\tfrac{1}{2}[\rho_0(z)]^{-1}\{[c_0(z)]^{-2}\omega^2-k^2-l^2\}\,\omega m q_1^2. \tag{60}$$

This is independent of z if and only if the amplitude q_1 varies as

$$[\rho_0(z)]^{\frac{1}{2}}|[c_0(z)]^{-2}\omega^2-k^2-l^2|^{\frac{1}{2}}m^{-\frac{1}{2}}. \tag{61}$$

In the case when (46) is an *exact* solution, with $N(z)$, $c_0(z)$ and m constant, this does imply proportionality to $[\rho_0(z)]^{\frac{1}{2}}$ as the imaginary part of (55) suggests.

In general cases with fluid properties varying slowly on a scale of wavelengths, we shall indeed find (section 4.5) that arguments deriving amplitude variation from energy flux as above (or as in section 2.6) give reliable answers; furthermore, although for internal waves the result is not a simple proportionality to $[\rho_0(z)]^{\frac{1}{2}}$, nevertheless that factor constitutes, as (55) implies, the leading modification needed to results calculated from the Boussinesq approximation. In other words, errors in the Boussinesq approximation are greatly reduced if the values for p_e and q which it gives are multiplied by factors proportional to $[\rho_0(z)]^{\frac{1}{2}}$.

4.3   Internal waves in the ocean and in the atmosphere

Against the background of the basic theory of internal waves (section 4.1) and of the conditions for them to be decoupled from sound waves (section 4.2), we sketch first the nature and significance of internal waves in the ocean and then their somewhat contrasted nature and significance in the atmosphere.

All oceanic internal waves are completely decoupled from sound waves. In fact, condition (52) is amply satisfied even at wavelengths comparable with the oceanic depth, since the total change of water density in the ocean is limited to about 4 %. This maximum change is mainly the result of the huge pressures, around $10^8\,\mathrm{N\,m^{-2}}$, found at the greatest oceanic depths, around 10 km.

The density of sea-water is effectively a function

$$\rho(p, T, \chi) \tag{62}$$

of pressure p, temperature T and *salinity* χ: the proportion by mass of dissolved salts. The composition of those salts does vary, but too little to produce density changes significant for ocean dynamics.

Temperature variations in the ocean, from the freezing-point of sea-water, 271 K, to values around 300 K, are responsible for density changes of about 0.5 %. Salinity changes, from about $\chi = 0.034$ to $\chi = 0.037$, are responsible for density changes of about 0.2 %.

Paradoxically, these density changes due to variations in temperature and salinity are much more important for ocean dynamics than the bigger changes of up to 4 % due to variations in pressure. In fact, the dynamics of the ocean is dominated by the distribution, not of the actual density (62), but of the quantity

$$\rho_a(T, \chi) = \rho(p_a, T, \chi); \tag{63}$$

defined as the density which water of given temperature T and salinity χ would have if it were brought to atmospheric pressure p_a without change of T and χ. From research ships, oceanographers lower instruments called temperature–salinity–depth recorders to determine the variation of T and χ with depth. The distribution of ρ_a with depth can then be read off from graphs like those in figure 73.

Here we explain the dominance of ρ_a for internal waves, whose dynamics, by (22), is governed by the distribution of the Väisälä–Brunt frequency (12). If $T_0(z)$ and $\chi_0(z)$ are the undisturbed distributions of temperature and salinity with the depth $(-z)$, and

$$\rho_{a0}(z) = \rho_a(T_0(z), \chi_0(z)) \tag{64}$$

is the corresponding density distribution corrected to atmospheric pressure, we show that to close approximation

$$N(z) = [-g\rho'_{a0}(z)/\rho_{a0}(z)]^{\frac{1}{2}}. \tag{65}$$

This means that the dynamics of internal waves depends only on the vertical distribution of ρ_a.

Physically, this is because, at oceanic temperatures, reversible changes in *water* are very nearly isothermal (section 1.2). Fluid rising to a higher level retains its salinity and, to close approximation, its temperature. Therefore, it keeps its initial value of ρ_a, so that the restoring force (11) depends primarily on any excess of ρ_a over its undisturbed value at the new pressure-level.

Formally, we can express this by writing

$$d(\ln \rho_0)/dz = p'_0(z)\, \partial(\ln \rho)/\partial p + T'_0(z)\, \partial(\ln \rho)/\partial T + \chi'_0(z)\, \partial(\ln \rho)/\partial \chi, \tag{66}$$

where $p'_0(z) = -g\rho_0(z)$ and where $\partial\rho/\partial p$ at constant T and χ is c_N^{-2}, the inverse square of the 'Newtonian' speed of sound (section 1.2). Equation (12) for $N(z)$ gives, therefore,

$$[N(z)]^2 = g^2(c_N^{-2} - c_0^{-2}) - g\{T'_0(z)\, \partial(\ln \rho)/\partial T + \chi'_0(z)\, \partial(\ln \rho)/\partial \chi\}. \tag{67}$$

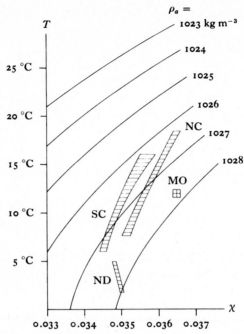

Figure 73. The density ρ_a of water of given temperature T and salinity χ reduced to atmospheric pressure. Hatched areas: regions of the temperature–salinity diagram typical of some particular water masses found in the North Atlantic: the N. Atlantic Deep Water ND, the N. Atlantic Central Water NC, the S. Atlantic Central Water SC and the Mediterranean outflow MO.

The approximation (65) is derived by (i) neglecting the first term on the right-hand side and (ii) replacing the curly bracket in (67) by its value at atmospheric pressure, $\rho'_{a0}(z)/\rho_{a0}(z)$.

We can express the error in (i), from expressions in section 1.2 for differences between the sound speed c_0 and its Newtonian value c_N, as

$$g^2\alpha^2 T/c_p \qquad (68)$$

in terms of the specific heat of sea-water (about 3950 J kg^{-1} K^{-1}) and its coefficient of expansion α. The greatest value of α in the ocean is about 3×10^{-4} K^{-1}, found in the warmest waters, and making (68) about 7×10^{-7} s^{-2}. When we compare this with a typical contribution to the value of N^2 in such warmer waters of around 3×10^{-5} s^{-2}, we see that the maximum resulting error in N is only just over 1%.

For cold water at great depths, the error in the substitution (ii) predominates: the coefficient of expansion $\alpha = -\partial(\ln\rho)/\partial T$, although never negative for sea-water, takes reduced values about $1 \times 10^{-4}\,\mathrm{K^{-1}}$ at such low temperatures, and is increased by about $2 \times 10^{-5}\,\mathrm{K^{-1}}$ under pressures of under $10^8\,\mathrm{N\,m^{-2}}$. Although as a result the *percentage* error in N rises to 10% at those depths, the absolute error in N^2 is less than before (at most $4 \times 10^{-7}\,\mathrm{s^{-2}}$, obtained when we multiply the error in $g\alpha$ by a typical deep-water temperature gradient of under 2 degrees per km). These estimates suggest that the dynamics of the ocean can be sufficiently accurately studied by means of the simplified form (65) for $N(z)$.

The restoring force (11) is positive only if $[N(z)]^2$ is positive. The distributions $T_0(z)$ and $\chi_0(z)$ of temperature and salinity can be stable, therefore, only if

$$\rho'_{a0}(z) = T'_0(z)\,\partial\rho_a/\partial T + \chi'_0(z)\,\partial\rho_a/\partial\chi \tag{69}$$

is negative. Stability is facilitated either by a positive temperature gradient $T'_0(z)$, or by a negative salinity gradient $\chi'_0(z)$.

Texts on oceanography explain that important clues regarding ocean circulation are given by worldwide observations of the distribution of temperature and salinity with depth. Such observations have identified numerous water masses represented by particular limited regions of a temperature–salinity diagram: in the North Atlantic, for example, the N. Atlantic Deep Water, the N. Atlantic Central Water, the S. Atlantic Central Water, the Mediterranean outflow (figure 73). The geographical distribution of each of these may give a key to circulation movements at various depths; where, also, the existence of relatively concentrated currents may be indicated as a 'geostrophic' effect of horizontal gradients in T and χ. Oceanographers study also how various effects (of cooling, heating, evaporation, fresh-water inflow, and ice formation) act to modify the distributions of T and χ.

An important aspect of all these studies is the knowledge of when a vertical distribution of T and χ will be stable, and of when (by contrast) its instability to vertical movements will promote mixing and a tendency towards greater homogeneity. With $T'_0(z) < 0$ and $\chi'_0(z) > 0$ instability is certain since both terms in (69) are positive. When just *one* of those conditions is satisfied the sum (69) may still be positive, leading to gross instability: any vertical displacements of fluid are accelerated by gravity forces. In the Gulf of Toulon, for example, cooling of the surface by the Mistral wind has been observed to generate fluid overturning and homogenisation in the whole depth of the Gulf.

Actually, a vertical distribution of temperature and salinity appears to be fully stable only when *both* $T'_0(z) \geqslant 0$ *and* $\chi'_0(z) \leqslant 0$. Recent researches have shown that when only one of those conditions is satisfied, even though the sum (69) is negative, the ocean stratification remains unstable to certain rather specialised kinds of vertical displacement. For example, when $T_0'(z) > 0$ and $\chi_0'(z) > 0$, characteristic types of downward displacement of the upper water (which is relatively salty and warm) are observed. These, known as 'salt fingers', have such a form that heat (with its diffusivity about 100 times greater than that of salt) diffuses out of them as they descend; then, their buoyancy due to being hotter than the surrounding fluid ceases to overcome the excess weight due to their being saltier.

It is common, for a variety of reasons like this, to find in observed distributions of T and χ an upward increase of temperature *and* decrease of salinity. Both effects make positive contributions to $[N(z)]^2$, so that internal waves can be present.

An important variant on this pattern is as follows. Changes of temperature and salinity initiated at the surface may often bring about instability within an uppermost layer of the ocean which accordingly becomes well mixed. In such a *well-mixed layer* (which surface waves often help to stir) the value of $N(z)$ may be practically zero (because T_0 and χ_0 are nearly uniform). Below the layer (whose thickness is of the order of 10^2 m) there may be a region of sharp transition, called the *thermocline*, where the temperature drops steeply with increasing depth, and there can also be some increase in salinity. In the thermocline, $N(z)$ takes substantial values around 10^{-2} s^{-1} or more, decreasing to more normal positive values at greater depths (figure 74).

Under these conditions, internal waves may be *trapped* within the thermocline region. We saw in section 4.1 that $N(z)$ can be viewed as a greatest possible frequency of local oscillations. Accordingly, if some oscillatory forcing effect had a frequency ω among the larger values in figure 74, we might expect it to excite oscillations only in the thin layer where $N(z) \geqslant \omega$. The oscillations would be trapped within that thin layer and could only propagate horizontally.

Analysis by equation (22) shows that these expectations are essentially correct, although the disturbance does penetrate a little into the region where $N(z) < \omega$. Equation (22) has solutions

$$q = Q(z) \exp[\mathrm{i}(\omega t - kx)], \tag{70}$$

representing waves propagating horizontally in the x-direction with wave speed ω/k, provided that

$$\omega^2 Q''(z) + k^2\{[N(z)]^2 - \omega^2\} Q(z) = 0. \tag{71}$$

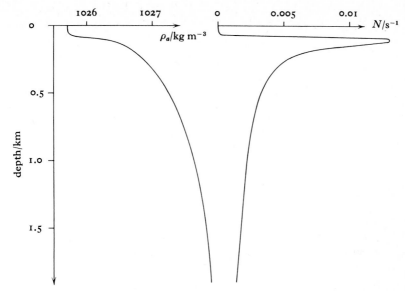

Figure 74. A typical average variation of ρ_a in the ocean as a function of depth, together with the associated variation in the Väisälä-Brunt frequency N.

(To simplify the analysis we exclude dependence on the y-coordinate; note, however, that waves with any direction of horizontal propagation could be written in the form (70) by suitable choice of the x-axis.)

Equation (71) is of the right form to represent trapped waves. Wherever $N(z) > \omega$, it has a form like the equation of simple harmonic motion; it must, in fact, have 'wavy' solutions *always concave to the z-axis* since Q'' necessarily has the opposite sign to Q. On the other hand, where $N(z) < \omega$, Q'' necessarily has the same sign as Q and solutions show an exponential-type behaviour (increasing or decreasing). Trapped waves are those where $Q(z)$ *decreases* exponentially on both sides of the interval of z where $N(z) > \omega$.

Texts on differential equations show that any equation of this form (71), where $N(z) > \omega$ in just one interval of values of z, possesses such trapped-wave solutions for values of k belonging to a certain increasing sequence $k_0, k_1, k_2, \ldots$. Furthermore, the graph of the solution $Q = Q_n(z)$ corresponding to $k = k_n$ makes just n crossings of the axis (at each of which, by (71), it has a point of inflexion).

We omit the details of that classical analysis, indicating only (in figure 75) the essential reasons for the existence of such a sequence. In the region $N(z) > \omega$ where the solution is curved concavely to the z-axis, the value of

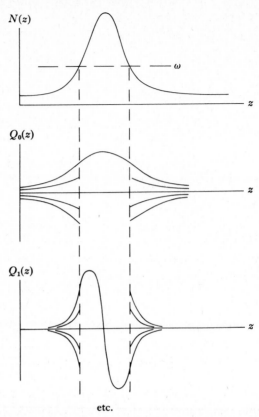

Figure 75. Illustrating the solutions of (71) vanishing for large z (positive or negative). The broken lines enclose the region $N(z) > \omega$ of 'wavy' solutions curved concavely to the axis. For particular values $k_0, k_1, \ldots$ of k it is possible to join one of the solutions vanishing for large negative z to one of those vanishing for large positive z by means of such a wavy solution with 0, 1, $\ldots$ axis crossings.

k determines how steeply its curvature increases with displacement from the axis, which in turn determines its degree of waviness. Now, any solution *vanishing exponentially* for large negative z (like the surface waves of chapter 3) must reach a point of inflexion where $N(z)$ becomes equal to ω. Note that all such solutions must be constant multiples of one another, and so have a well-determined value of $Q'(z)/Q(z)$ at that point of inflexion. The question is whether such a solution on reaching the other side of the interval where $N(z) > \omega$ will have achieved the right value of $Q'(z)/Q(z)$ to fit a solution decreasing exponentially as z increases beyond that interval. For various values of k, determining the solution's degree of waviness (increase of concave curvature with displacement from the axis), it is possible to achieve this (i) with no crossing of the axis, using a moderate value k_0; (ii) with one crossing, using a considerably larger value k_1; (iii) with two crossings and a much larger value k_2; and so on.

The greatest horizontal wave speed ω/k is found with the smallest wave number, $k = k_0$. The corresponding distribution of vertical mass flow, $Q_0(z)$, is positive for all z. This means that the whole thermocline region rises and falls in phase; then, it is said to make *sinuous* oscillations. We should note, incidentally, that the wavenumber k_0 varies with ω: these horizontally propagating waves, therefore, exhibit isotropic dispersion, governed by a group velocity $U = [k_0'(\omega)]^{-1}$ in accordance with the general theory set out in the second half of chapter 3.

A considerably lower wave speed ω/k is found for $k = k_1$. The corresponding distribution $Q_1(z)$ changes sign in the middle of the thermocline region. The oscillations are then called *varicose*: where the upper part of the thermocline is rising, the lower part is falling, and vice versa; so that the thermocline region is varying in thickness. The increasingly slower oscillations with $k = k_2$, k_3, ... are increasingly more complicated; for each, the group velocity is $[k_n'(\omega)]^{-1}$.

Readers familiar with quantum mechanics may recognise equation (71) as similar to Schrödinger's equation

$$(\hbar^2/2M)\psi''(z) + [E - V(z)]\psi(z) = 0 \tag{72}$$

governing the wave function $\psi(z)$ for a particle of mass M and energy E in a 'potential well' $V(z)$. For a given energy-level between the top and bottom of the well a similar infinite sequence of trapped-wave 'states' is possible for increasing values of the mass M. More commonly, physicists take M fixed and identify a sequence of energy-levels E such that trapped waves exist; although that may be just a finite sequence. It is worth noting that, for equation (71), this latter procedure is equally possible: it corresponds to finding a sequence of frequency-levels ω_0, ω_1, ... for fixed k; where $\omega_0(k)$, for example, is simply the inverse function to $k_0(\omega)$.

In the limit of an exceedingly steep thermocline, separating regions with little or no density gradient, internal waves in the 'sinuous' mode $k = k_0$ become identical with the waves on a density discontinuity mentioned at the beginning of this chapter. To see this, note that the required solutions to (71) must be proportional to $\exp(kz)$ below the layer and to $\exp(-kz)$ above it. Accordingly, Q'/Q must change from $+k$ to $-k$ through the layer, and for the $k = k_0$ mode this change must be a monotonic decrease from $+k$ to $-k$. If the layer thickness is small compared with $1/k$, then this distribution of Q'/Q can make only small proportional changes to Q. Therefore, equation (71) makes the change in Q'/Q closely equal to the integral through the layer of Q''/Q; namely,

$$-k^2 \int \{\omega^{-2}[N(z)]^2 - 1\}\,dz \doteqdot -k^2\omega^{-2} \int [N(z)]^2\,dz, \qquad (73)$$

where the integral of 1 is the layer thickness, neglected already in comparison with $1/k$. Equating (73) to the change $(-2k)$ due to the decrease from $+k$ to $-k$, and using (12), we obtain

$$\omega^2 = \tfrac{1}{2}gk \int [-\rho_0'(z)/\rho_0(z)]\,dz = \tfrac{1}{2}gk \ln(\rho_2/\rho_1) \qquad (74)$$

in terms of the densities ρ_1 and ρ_2 above and below the layer.

It is interesting that the expression (74), based on the approximations of internal-wave theory, is so close to the exact value (3): the error is easily verified to be of the order of the cube $(\rho_2 - \rho_1)^3$ of the density difference. This indicates that even extremely large local values of $\rho_0'(z)/\rho_0(z)$ need not disturb the accuracy of the Boussinesq approximation provided that the total relative change of density remains small.

The above discussion of internal waves in the ocean has been concerned with those waves of relatively higher frequency (periods of 30 minutes or less) which tend to be trapped around the thermocline. We must note, however, that waves of much lower frequencies are also present. These are freer to propagate throughout the depth of the ocean, and in general show greater similarities to the atmospheric internal waves which we describe next. For indications of their role in exchanging momentum between current movements at different levels, see section 4.6.

The atmosphere differs from the ocean in many ways relevant to internal waves. Above all, the density becomes reduced by large factors (almost without limit) as the altitude z increases. Nevertheless, since $[-\rho_0'(z)/\rho_0(z)]$ takes values in the range from 0.08 to 0.16 km^{-1}, the vast majority of internal waves (with wavelengths up to 1 or 2 km) satisfy comfortably the condition (52) under which they are completely decoupled from sound waves. In just one important case, discussed in section 4.13, there is very strong coupling between internal gravity waves and sound waves: this is where pressure disturbances propagate horizontally in such a way that above each point on the Earth the air movements at all heights z are in phase. Such waves have wavelengths so great that, far from satisfying condition (52), they are rather analogous to the 'long waves' of chapter 2.

In further contrast to the ocean, where $N(z)$ may have a distribution strongly 'peaked' near the thermocline (figure 74), many stably stratified regions of the atmosphere have a more gradually varying distribution of $N(z)$; as, for example, in the *stratosphere*, an extensive region with its base (the *tropopause*) just over 10 km (or, in the tropics, rather more) above sea-level. In such cases, internal waves with wavelengths of order 10^2 to 10^3 m can be analysed by the ray-tracing techniques of the next two sections, based on the assumption that $N(z)$ varies slowly on a scale of wavelengths.

In the atmosphere, again, winds modify and interact with internal waves still more than do currents in the ocean. Especially for this reason, we include a section 4.6 on ray tracing in a wind, with applications both to internal-wave and to sound propagation in non-uniform airflows.

Stationary internal waves can be generated by obstacles (mountains, etc.) in a steady wind. Like surface gravity waves in a steady stream (section 3.9) they are found downstream of the obstacle and are therefore known as 'lee waves'. Often, condensation in the crests makes them clearly visible. They are briefly analysed in section 4.12.

The mention of condensation may remind us of a last major difference between the atmosphere and the ocean in respect of internal waves. The water-vapour content of the atmosphere is subject to phase changes which release or absorb latent heat. These heat exchanges can modify the expression (8) for excess density and therefore also the stability condition (10).

Under 'dry' conditions (where some water vapour may be present but no condensation), the stability criterion (10) can readily be transformed into a condition on the 'lapse rate' (rate of decrease of *temperature* $T_0 = p_0/R\rho_0$ with height):

$$- T_0'(z) < c_p^{-1} g; \tag{75}$$

here, the specific heat of air at constant pressure is $c_p = 1000$ J kg^{-1} K^{-1}. Still more simply, (75) can be derived directly from the thermodynamic result for a perfect gas that, in reversible changes, $c_p \, dT = \rho^{-1} dp$; accordingly, air suffering the pressure drop $\rho g \zeta$ cools by an amount $c_p^{-1} g \zeta$ and falls back if and only if this exceeds the drop $[- T_0'(z)] \zeta$ in ambient temperature. This limiting lapse rate (75) for stability is close to 1 degree per 100 m. When that rate is exceeded, vertical displacements are vigorously promoted; they generate rapid mixing, which proceeds until it has reduced the lapse rate below the limit.

Under 'moist' conditions (as in clouds), where the partial pressure of water vapour in the air equals the *saturation vapour pressure* p_v and droplets of condensed water are also present, the critical lapse rate for stability is much smaller. Most simply, this is seen by recognising that the effective c_p for such a mixture is considerably increased. That is because an increase of temperature increases p_v, and so demands a heat input not only to warm the air but also to evaporate enough water to produce the necessary increase in partial pressure of water vapour. Texts on meteorology show how this approximately *doubles* c_p in sea-level conditions at 20 °C, and therefore *halves* the critical lapse rate for stability (the reduction is somewhat less at lower temperatures, while showing a slight increase with falling pressure).

Cumulus clouds are regions of vertical movement, vigorously promoted under saturated conditions when the lapse rate $-T_0'(z)$ exceeds its critical value for moist air. The visible clouds mark where the air is rising and becoming cooler leading to condensation; the gaps between them mark where air is falling and becoming warmer leading to evaporation. Often, a stable layer above a region of cumulus cloud may be a region of active internal-wave propagation, forced by the vigorous vertical motions below it.

4.4 Introduction to anisotropic dispersion

Two principal methods of analysis are convenient for describing internal gravity waves. Where the Väisälä–Brunt frequency $N(z)$ has a sharply peaked distribution, as in thermoclines, a trapped-wave analysis is often helpful (section 4.3). Propagation of trapped waves is horizontal, so that the dispersion is isotropic; for a given frequency, the greatest wave speed is found in sinuous oscillations, which are a natural generalisation of gravity waves on a discontinuous interface.

The other method of analysis, applicable where $N(z)$ is slowly varying on a scale of wavelengths, makes a truly three-dimensional extension of the ideas of group velocity developed in chapter 3. This second method, described in full in section 4.5, is rather widely useful; indeed, even with a peaked distribution of $N(z)$ it can give a surprisingly good account (section 4.11) of all the trapped-wave oscillations with wave speeds less than that for sinuous oscillations; in addition, it is still more widely applied to cases of gradually varying density gradients.

Readers to whom the nature of isotropic dispersion has become familiar may still find the properties of the vector group velocity for internal waves and other anisotropic systems quite surprising. For this reason (as in chapter 3) we precede the general treatments in subsequent sections by a particularly simple analysis of a typical case; in fact, of the *homogeneous* system, with

N uniform, for which sinusoidal plane-wave solutions (23) were obtained in section 4.1.

Reasons for paying special attention to sinusoidal waves may be noted just as in section 3.6. In particular, although we postpone to section 4.8 a three-dimensional Fourier analysis of wave generation by a local disturbance of complicated shape, we can anticipate its conclusions; namely, that different sinusoidal plane-wave components are at some much later time *found in quite different places*. Accordingly, the waves observed at a particular place may be roughly sinusoidal.

For uniform N, we have seen (section 4.1) that waves of a particular frequency $\omega \leqslant N$ have their surfaces of constant phase at a definite angle $\cos^{-1}(\omega/N)$ to the vertical. If we were to suppose that the waves which a localised source emits in any direction have their constant-phase surfaces at right angles to that direction, we should expect to find waves of frequency ω in directions making that angle $\cos^{-1}(\omega/N)$ to the *horizontal* . . . However, we know from chapter 3 that such crude arguments by analogy from non-dispersive systems are untrustworthy, and in this case we shall find that they could hardly be more wrong! A first indication that this is so has been given by the result (end of section 4.1) that wave energy flux is directed *parallel* to surfaces of constant phase. This suggests, correctly as we shall see, that the waves found in a particular direction from the source have their surfaces of constant phase parallel to that direction; accordingly, waves of frequency ω are found in directions making the angle $\cos^{-1}(\omega/N)$ to the *vertical*.

This section combines, for anisotropic dispersion, the roles played by sections 3.6 and 3.8 for isotropic dispersion. We begin by deriving the properties of the vector group velocity as simply as possible by a method appropriate to the later stages of dispersion, when waves of widely different wavenumbers have become widely separated; then, they are so much dispersed that the wavenumber vector varies between them only gradually on a scale of wavelengths. The method applies to any *homogeneous* anisotropic system; that is, the frequency may depend arbitrarily on the magnitude and direction of the local wavenumber vector but cannot be separately dependent on position. (It is this last assumption which, for internal waves governed by the dispersion relationship (24), demands that N must be uniform.) We end by verifying (as in section 3.8) that the group velocity so derived is the same as the energy propagation velocity for sinusoidal waves.

The simple method of analysis, like that of section 3.6, depends upon the definition of a local *phase* α. In fact, because the wavenumber vector (k, l, m) is assumed to vary only gradually (by just a small fraction of its magnitude $2\pi/\lambda$ in one wavelength λ) a physical quantity q in the waves can be represented in the approximately sinusoidal form

$$q = Q(x, y, z, t) \exp[i\alpha(x, y, z, t)]; \qquad (76)$$

here, $Q(x, y, z, t)$ is a positive, slowly varying amplitude, and $\alpha(x, y, z, t)$ is a phase. At a fixed point (x, y, z), equation (76) requires that this phase increases by 2π in one wave period, so that $\partial\alpha/\partial t$ can be regarded as the local frequency ω in radians per second. Similarly, at a given time, α shows a decrease with x at a rate equal to k, the x-component of the local wavenumber in radians per metre. With analogous results regarding y and z, this gives

$$\partial\alpha/\partial x = -k, \quad \partial\alpha/\partial y = -l, \quad \partial\alpha/\partial z = -m, \quad \partial\alpha/\partial t = \omega. \qquad (77)$$

Equations (77) for α ensure that locally the wave is nearly of the sinusoidal form (23). In fact, around a particular position (x_0, y_0, z_0) and a particular instant t_0 the phase function is nearly linear, with

$$\alpha(x, y, z, t) \doteq \alpha(x_0, y_0, z_0, t_0) - k_0(x - x_0) - l_0(y - y_0) - m_0(z - z_0) + \omega_0(t - t_0); \qquad (78)$$

here, $-k_0$, $-l_0$, $-m_0$ and ω_0 are the values of the derivatives (77) at (x_0, y_0, z_0, t_0). Thus, *locally*, equation (76) with $\alpha(x, y, z, t)$ slowly varying makes q nearly proportional to $\exp[i(\omega_0 t - k_0 x - l_0 y - m_0 z)]$ as in a sinusoidal plane wave.

Whenever the derivatives (77) of a phase function α satisfy a dispersion relationship

$$\omega = \omega(k, l, m), \qquad (79)$$

specifying the local frequency ω as a function of the magnitude and direction of the local wavenumber vector (k, l, m), we can deduce the value and the main properties of the vector group velocity 'in two lines' (equations (83) and (84) below). For internal waves, this dispersion relationship (79) takes a somewhat special form (24), which depends *only* on the direction and not on the magnitude of (k, l, m), but we find later many examples of more general dependence. In every case the phase α is defined so that one particular physical quantity q comes to its 'crests' (maxima) where α is a multiple of 2π; some *other* physical quantity may have its maxima at other phases but always at a fixed position in the cycle (for example, in internal waves the vertical displacement ζ or the excess density ρ_e would have its maxima where $\alpha - \frac{1}{2}\pi$ is a multiple of 2π).

The analysis to be given now (and generalised in the following section) is like that of section 1.10 in that it is easiest to follow if a suffix notation is used. Accordingly, whenever we describe anisotropic waves in general, we use coordinates (x_1, x_2, x_3) and a wavenumber vector (k_1, k_2, k_3) and the convention that a suffix occurring twice within any term of an equation is automatically summed from 1 to 3. Only when, as an illustration, we use internal waves (for which the vertical z-direction is special) do we go back to the coordinates (x, y, z).

In the suffix notation, the phase α can be written $\alpha(x_1, x_2, x_3, t)$ with

$$\partial\alpha/\partial x_i = -k_i, \quad \partial\alpha/\partial t = \omega. \tag{80}$$

The dispersion relationship $\quad \omega = \omega(k_1, k_2, k_3) \tag{81}$

becomes $\quad \partial\alpha/\partial t = \omega(-\partial\alpha/\partial x_1, -\partial\alpha/\partial x_2, -\partial\alpha/\partial x_3). \tag{82}$

Then the two-line proof of the properties of group velocity starts by differentiating this function ω of three variables with respect to x_i. This gives

$$\frac{\partial^2\alpha}{\partial x_i\,\partial t} = \left(\frac{\partial\omega}{\partial k_j}\right)\left(-\frac{\partial^2\alpha}{\partial x_i\,\partial x_j}\right), \tag{83}$$

where the summation convention makes the right-hand side signify the sum of three terms, for $j = 1, 2$ and 3. These represent, respectively, the rate of change of $\omega(k_1, k_2, k_3)$ resulting from the rates of change in k_1, k_2 and k_3 in response to a change in x_i (keeping the other coordinates and the time constant). By (80), equation (83) can be written

$$\frac{\partial k_i}{\partial t} + U_j \frac{\partial k_i}{\partial x_j} = 0, \tag{84}$$

where $\qquad\qquad U_j = \partial\omega/\partial k_j \tag{85}$

is defined to be the vector group velocity.

Equation (84) is the three-dimensional equivalent of the dispersion rule $\partial k/\partial t + U\partial k/\partial x = 0$ for one-dimensional systems. Its interpretation is similar: it means that the wavenumber vector k_i is constant in changes satisfying $\qquad\qquad dx_j/dt = U_j. \tag{86}$

As in section 3.6, we can think of an observer letting the point (x_1, x_2, x_3) at which he is gazing move according to equation (86); that is, move at the group velocity U_j; then equation (84) means that he will observe always waves of the same wavenumber k_i. Furthermore, equations (81) and (85) show that when k_i is constant then the group velocity U_j is also constant. Thus, the path (86) is necessarily a straight line

$$x_j - U_j t = \text{constant}, \tag{87}$$

traversed at a constant velocity U_j.

For homogeneous systems, the basic fact about the group velocity U_j has already been derived. Waves of a given wavenumber vector k_i (whose magnitude is $2\pi/\lambda$ and whose direction is normal to surfaces of constant phase) are found at points moving at constant velocity U_j along such straight-line paths (87), and there is a different path for each wavenumber.

These paths (87) in three-dimensional systems are called *rays*; the word used in chapters 1 and 2 already for similar straight-line paths along which sound energy moves in a homogeneous acoustic medium. In fact, rays in a nondispersive system are a simple special case of the general idea of rays now being developed. Thus, sound waves with constant speed c_0 and wavenumber vector k_i have frequency

$$\omega = c_0(k_1^2 + k_2^2 + k_3^2)^{\frac{1}{2}} \tag{88}$$

so that (85) implies $$U_j = c_0 k_j(k_1^2 + k_2^2 + k_3^2)^{-\frac{1}{2}}; \tag{89}$$

this group velocity vector has constant magnitude c_0 and is directed along the wavenumber vector (that is, perpendicular to surfaces of constant phase) in agreement with the facts on the speed and direction of acoustic energy propagation which we know from section 1.3.

For internal waves, we return to the coordinates (x, y, z), with (k, l, m) as wavenumber vector. Then the group velocity is

$$\mathbf{U} = (\partial\omega/\partial k,\ \partial\omega/\partial l,\ \partial\omega/\partial m): \tag{90}$$

the *gradient of the frequency in wavenumber space*. When the dispersion relationship is given by (24), the group velocity (93) can be calculated as

$$\mathbf{U} = \frac{Nm}{k^2 + l^2 + m^2}\left[\frac{(km,\ lm,\ -k^2 - l^2)}{(k^2 + l^2)^{\frac{1}{2}}(k^2 + l^2 + m^2)^{\frac{1}{2}}}\right]; \tag{91}$$

here, the factor in square brackets is readily seen to be a *unit vector*, perpendicular to the wavenumber vector (k, l, m).

Without calculation, we could have predicted that $\mathbf{U}$, the gradient of the frequency (24) in wavenumber space, would be perpendicular to the wavenumber vector. For that gradient's component in the direction of (k, l, m) itself must be the rate of change of ω when the magnitude but *not* the direction of (k, l, m) changes; and (24) shows this to be zero.

The fact that waves of a given wavenumber (k, l, m) continue to be found at positions moving with the constant velocity $\mathbf{U}$ implies that the *energy* in waves of that wavenumber should move with that velocity. We now check that in sinusoidal waves $\mathbf{U}$ is indeed the velocity of energy propagation, first for the special case of internal waves and then for general anisotropic systems.

For sinusoidal internal waves, the wave energy flux $\mathbf{I} = p_e \mathbf{u}$ is clearly seen by (27) to lie in the same *direction* as the group velocity (91); a direction perpendicular to the wavenumber vector and coplanar with it and the vertical. Furthermore, the wave energy flux averaged over a period is the mean product of $\mathbf{u}$ with the excess pressure (26); and this, since the square of a cosine has the average value $\frac{1}{2}$, is

$$\mathbf{I} = \tfrac{1}{2}\rho_0^{-1} q_1^2 \frac{Nm}{k^2+l^2} \left[\frac{(km, lm, -k^2-l^2)}{(k^2+l^2)^{\frac{1}{2}} (k^2+l^2+m^2)^{\frac{1}{2}}} \right], \tag{92}$$

where the value of ω has been substituted from (24). The average wave energy per unit volume, W, must be the average of $\rho_0(\mathbf{u} \cdot \mathbf{u})$ (twice the kinetic energy), which (27) gives as

$$W = \tfrac{1}{2}\rho_0^{-1} q_1^2 \frac{k^2+l^2+m^2}{k^2+l^2}. \tag{93}$$

(The concluding fraction in (93) is the factor $\sec^2\theta$ by which the kinetic energy of the vertical motions alone is enhanced when the oscillations are inclined at an angle θ to the vertical.) Equations (91), (92) and (93) exhibit clearly the fact that the energy propagation velocity $\mathbf{I}/W$ is the same as the group velocity $\mathbf{U}$.

The fact that the group velocity of internal waves is parallel to surfaces of constant phase means that internal waves generated by a localised source could never have the familiar appearance (figure 7) of concentric circular crests centred on the source. Instead, the crests and other surfaces of constant phase *stretch radially outward* from the source (figure 76) because wave energy travels along rays, which are parallel to those surfaces.

For a source of definite frequency $\omega \leqslant N$ (section 3.1), those surfaces are all at a definite angle $\theta = \cos^{-1}(\omega/N)$ to the vertical; therefore, all the wave energy generated in the source region travels at that angle to the vertical. Accordingly, it is confined to a double cone with semi-angle θ.

Laboratory experiments on internal waves are often carried out in a stratified salt solution with uniform vertical salinity gradient $[-\chi_0'(z)]$, giving a constant Väisälä–Brunt frequency N. The optical system known as schlieren is very sensitive to two-dimensional variations of density. Figure 76 is a schlieren picture of the two-dimensional propagation (in the plane of the paper) of waves generated by the vertical oscillation, at frequency ω, of a horizontal cylinder (with axis perpendicular to the paper). The

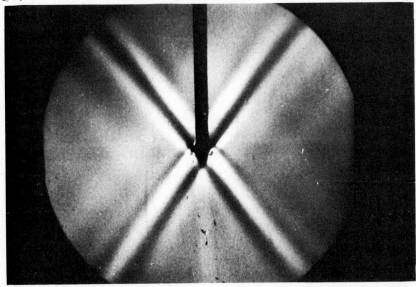

Figure 76. Schlieren picture of waves generated in stratified fluid of uniform Väisälä–Brunt frequency N by oscillation of a horizontal cylinder at frequency 0.70N. Note that surfaces of constant phase stretch out radially from the source. [*Photograph by D. H. Mowbray.*]

surfaces of constant phase are then planes at an angle $\cos^{-1}(\omega/N)$ to the vertical, sometimes described as forming a 'St Andrew's Cross'.

Although a still picture like figure 76 shows well the radial character of the surfaces of constant phase, it does not indicate their movement. A motion picture confirms that, just as in all travelling waves of wavelength λ, the crests and other surfaces of constant phase are moving perpendicular to themselves at the wave speed $c = \omega\lambda/2\pi$ which for these waves is

$$c = (N\lambda/2\pi)\cos\theta. \qquad (94)$$

Note that the *sense* of this perpendicular movement is easily predicted: equation (91) shows that the *vertical* component of the group velocity has always the *opposite* sign to the vertical component of wavenumber. Internal waves, then, obey the paradoxical rule that the energy propagation has an *upward* component when the motion of the crests has a *downward* component, and vice versa. In figure 76 for example, the motion of crests perpendicular to themselves is *downwards* in the upper half of the picture, where the energy is propagating *upwards*; and upwards in the lower half where the rays point downwards. At any one time, only one or two wavelengths can be seen because, when the crests have travelled quite a short

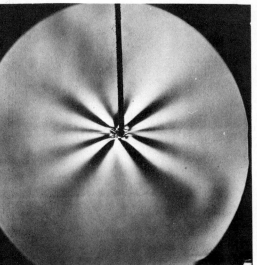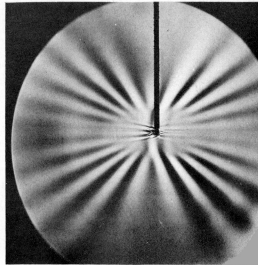

Figure 77. Schlieren picture of waves generated by a brief horizontal displacement (b) of a circular cylinder (a) after 10 seconds, (b) after 25 seconds. Note that the angle between crests decreases with time and is greatest, at any one time, for crests nearest to the vertical. [*Photograph by T. N. Stevenson.*]

distance, they have left the St Andrew's Cross (with dimensions comparable with the cylinder) within which all the energy is confined.

The magnitude of the group velocity (91) is

$$U = N|m|(k^2 + l^2 + m^2)^{-1} = (N\lambda/2\pi)\sin\theta. \tag{95}$$

The sin θ dependence on the angle θ between surfaces of constant phase and the vertical means that the group velocity tends to zero for those vertical oscillations of a fluid column which can take place at the Väisälä–Brunt frequency itself. By contrast, any very slow changes with $\omega \ll N$ involve nearly horizontal motions whose energy propagates nearly horizontally at a group velocity $(N\lambda/2\pi)$ proportional to the distance λ between adjacent (nearly horizontal) crests.

The proportionality of group velocity to λ for any given θ means that when a localised disturbance of limited duration generates waves of various wavelengths λ and frequencies ω, those found in a particular direction θ after time t (with, of course, frequency $\omega = N\cos\theta$) have become spread out with respect to wavelength; in fact, those of wavelength λ have travelled a distance $(N\lambda t/2\pi)\sin\theta$, proportional to λ itself. Neighbouring crests are a distance λ apart; therefore, they subtend an angle

$$(2\pi/Nt)\csc\theta \tag{96}$$

at the source. The fact that this angle is independent of λ is consistent with the crests being radial. Note also that the angle (96) between crests *decreases with time* from when the disturbance was generated; and that it is greatest, at any one time, for crests nearest to the vertical. Figure 77, showing the wave pattern at two different times after waves were initially excited by a brief displacement of a cylinder in stratified salt solution, exhibits all these features. For a more quantitative analysis of this pattern, see section 4.8.

We conclude this section by proving that, for sinusoidal waves,

$$q = a \exp\left[\mathrm{i}(\omega t - k_1 x - k_2 x_2 - k_3 x_3)\right], \tag{97}$$

in any homogeneous anisotropic system, the group velocity (85) is the velocity of energy propagation. It suffices to prove that their x_1-components are equal; obviously, for this very general system, the proofs that the x_2-components, or the x_3-components, are equal can be derived from that by permuting the suffixes.

We are required, then, to prove that the x_1-component of the energy propagation velocity is $\partial\omega/\partial k_1$; namely, the rate of change of ω with k_1 *keeping k_2 and k_3 constant*. Actually, these last words suggest a very simple proof indeed. If the system is constrained to fixed values of k_2 and k_3 (the components of wavenumber perpendicular to the x_1-direction) it becomes effectively a one-dimensional system with q proportional to $\exp\left[\mathrm{i}(\omega t - k_1 x_1)\right]$ and a one-dimensional dispersion relationship connecting ω and k_1. The proof at the end of section 3.8 then goes through with k_1 and x_1 for k and x and without any other change. (For example, since k_2 and k_3 remain fixed, it is now k_1 which has to change into $k_1(\omega) - \frac{1}{2}\mathrm{i}\beta k_1'(\omega)$.) Finally, the energy propagation velocity perpendicular to the plane $x_1 = 0$ is proved to be the rate of change $\mathrm{d}\omega/\mathrm{d}k_1$, still keeping k_2 and k_3 constant; which, by definition, is the partial derivative $\partial\omega/\partial k_1$.

Thus, in homogeneous anisotropic systems, the properties of the vector group velocity (including its identity with the energy propagation velocity) are very straightforward generalisations of the properties of the scalar group velocity characteristic of isotropic systems. By contrast, we show in the next section that wave dispersion in nonhomogeneous anisotropic systems shows features that are by no means mere generalisations of isotropic behaviour.

4.5 General theory of ray tracing

We come now to the book's centre of gravity: the section describing the general theory of ray tracing in nonhomogeneous anisotropic dispersive wave systems. The section is central because (i) it comprehends as special cases much of what has gone before: the theory of nonhomogeneous one-dimensional systems (sections 2.6 and 3.8), of ray tracing in geometrical acoustics (sections 1.11 and 2.14), and of homogeneous systems in isotropic cases (section 3.6) and anisotropic cases (section 4.4); and (ii) it derives rather simply a general theory which finds many applications to various wave systems in chapter 4 and the first part of the epilogue, as well as being extended further to include interactions with steady flows (sections 4.6 and 4.7), some additional information that Fourier analysis can give (sections 4.8–4.11), and (in the epilogue, part 2) the incorporation of nonlinear effects.

The present theory, like that of section 4.4, analyses waves that have become so much dispersed that the wavenumber vector varies between them only gradually on a scale of wavelengths. In a homogeneous system this permits the use (equation (76)) of a standard form for the waves,

$$q = Q(x_1, x_2, x_3, t) \exp\left[i\alpha(x_1, x_2, x_3, t)\right], \tag{98}$$

which to good approximation is locally sinusoidal; in (98), the phase α satisfies

$$\partial\alpha/\partial x_i = -k_i, \quad \partial\alpha/\partial t = \omega, \tag{99}$$

and the local frequency ω and wavenumber k_i are connected by a dispersion relationship (81). The same form (98) for the waves remains appropriate in a nonhomogeneous system, provided that the fluid properties defining the dispersion relationship (for example, N and c_0) also vary with position only gradually on a scale of wavelengths. This allows the waves, locally, to be approximately sinusoidal, with wavenumber k_i and frequency ω given by (99), but with ω specified for given k_i in a way that varies also with position x_i; in fact, the dispersion relationship takes the form

$$\omega = \omega(k_1, k_2, k_3, x_1, x_2, x_3). \tag{100}$$

Locally, again, the energy propagation velocity must have approximately the value

$$U_j = \partial\omega/\partial k_j, \tag{101}$$

which we have just shown it to take for exactly sinusoidal waves. The expression (101) has, however, a slightly different meaning from (85); by

(100), it is the rate of change of ω with respect to k_j in changes keeping constant not only the other components of wavenumber but also the *position* (x_1, x_2, x_3). At that fixed position, for waves which are approximately sinusoidal and satisfy the local dispersion relationship, (101) gives to good approximation the velocity of energy propagation.

In a nonhomogeneous system, we may expect (section 3.8) that wave energy propagated at this group velocity (101) will suffer 'refraction'; that is, changes of wavenumber due to the nonhomogeneity. By contrast, the frequency ω should remain unchanged; indeed, Fourier analysis shows that no energy transfer between different frequencies can occur in any system satisfying linear equations whose coefficients have no explicit dependence on time. These expectations are verified by the following simple analysis.

Using (99) to write (100) in the form

$$\partial \alpha / \partial t = \omega(-\partial \alpha / \partial x_1, -\partial \alpha / \partial x_2, -\partial \alpha / \partial x_3, x_1, x_2, x_3), \tag{102}$$

we first differentiate this with respect to x_i. The result is nearly the same as in equation (83), but includes one additional term: as well as the three terms in (83) resulting from the rates of change of k_1, k_2 and k_3 in response to a change in x_i, there is the additional term $\partial \omega / \partial x_i$ due to the direct dependence of ω on x_i for *constant* k_1, k_2 and k_3. Therefore,

$$\frac{\partial^2 \alpha}{\partial x_i \, \partial t} = \left(\frac{\partial \omega}{\partial k_j}\right)\left(-\frac{\partial^2 \alpha}{\partial x_i \, \partial x_j}\right) + \frac{\partial \omega}{\partial x_i}. \tag{103}$$

By (99) and (101), equation (103) can be written

$$\frac{\partial k_i}{\partial t} + U_j \frac{\partial k_i}{\partial x_j} = -\frac{\partial \omega}{\partial x_i}. \tag{104}$$

Here, the left-hand side represents the rate of change of k_i with time at a position moving with the group velocity U_j; that is, in changes satisfying

$$\mathrm{d}x_j / \mathrm{d}t = U_j. \tag{105}$$

In such changes, then, $\quad \mathrm{d}k_i / \mathrm{d}t = -\partial \omega / \partial x_i. \tag{106}$

Equation (106) specifies the *refraction of wave energy*; that is, the rate of change of wavenumber along paths (105) traversed at the energy propagation velocity U_j; in a word, along *rays*.

The equations governing these, in general, *curved* rays take a more symmetrical form if we write (105) as

$$\mathrm{d}x_i / \mathrm{d}t = +\partial \omega / \partial k_i. \tag{107}$$

The proof that, as expected, the *frequency* ω remains constant along a ray then follows by (i) writing down the rate of change of (100) with time for arbitrary rates of change of k_i and x_i as

$$\mathrm{d}\omega/\mathrm{d}t = (\partial\omega/\partial k_i)(\mathrm{d}k_i/\mathrm{d}t) + (\partial\omega/\partial x_i)(\mathrm{d}x_i/\mathrm{d}t); \tag{108}$$

and then (ii) using (106) and (107) to see that for the actual rates of change the right-hand side of (108) is zero.

We now insert a brief parenthesis on parallels with other fields of study. Readers familiar with classical dynamics may recognise in equations (106) and (107) the characteristic *form* (symmetrical except for the minus sign in (106)) of Hamilton's equations for a conservative dynamical system. Those are written in terms of a *Hamiltonian function*

$$H(p_1, p_2, ..., p_n, q_1, q_2, ..., q_n), \tag{109}$$

which is the system's total energy (kinetic plus potential) expressed as a function of generalised coordinates $q_1, q_2, ..., q_n$ and associated generalised momenta $p_1, p_2, ..., p_n$; Hamilton's equations then take the form

$$\mathrm{d}p_i/\mathrm{d}t = -\partial H/\partial q_i, \quad \mathrm{d}q_i/\mathrm{d}t = +\partial H/\partial p_i, \tag{110}$$

for $i = 1, ..., n$. Equations (106) and (107) are precisely of this form with $n = 3$ and with

$$k_i, \quad x_i \quad \text{and} \quad \omega \quad \text{replacing} \quad p_i, \quad q_i \quad \text{and} \quad H. \tag{111}$$

Furthermore, the verification via equation (108) that ω remains constant along a ray is an exact parallel to the standard verification that for any solution of the equations of motion (110) the total energy H remains constant.

One somewhat obvious feature of the parallelism (111) is that the three Cartesian coordinates x_i for the 'packet' of wave energy travelling along its ray correspond to the generalised coordinates q_i in the dynamical system; these x_i are, indeed, the natural choice of generalised coordinates to describe the wave packet's position. Two much less obvious features are that

$$\text{the wavenumber } k_i \text{ corresponds to the momentum } p_i \tag{112}$$

and that   the frequency ω corresponds to the energy H. $\qquad$ (113)

However, readers familiar with quantum mechanics will recognise the statements (112) and (113) as part of the 'correspondence principle'. Quantum mechanics is especially concerned to emphasise that *particles behave like waves*; specifically, waves whose radian wavenumber k_i and radian frequency ω when multiplied by a certain constant $\hbar$ give the particle's momentum and energy. Reciprocally, this book stresses that *waves behave like particles* (moving along rays); indeed, a wave packet changes position and wavenumber according to equations (106) and (107) which can be regarded as those for a 'particle' whose energy–momentum relationship at every position exactly parallels the frequency–wavenumber relationship (dispersion relationship) for the waves.

That concludes the brief parenthesis, to which in the rest of this book we make little further reference. Nevertheless, it may be worth remembering that the system of six first-order differential equations (106) and (107) for following a wave packet's motion along rays, and the refraction of wave energy, have many of the practically useful features of the equations of motion of a particle; especially, the ready *computability* of solutions given the initial values of x_i and k_i.

An essential difference of the present theory from the corresponding theory for *one*-dimensional nonhomogeneous systems lies in the fact that the equation

$$\omega = \text{constant along a ray,} \qquad\qquad (114)$$

deduced above from (108), and equally true in one-dimensional systems (figure 64) has a different significance in the two cases. In three-dimensional systems, a single equation (114) cannot possibly replace the *three* equations (106) describing the refraction of wave energy. Thus, it is not possible to trace a ray merely by using knowledge of its *direction* (equations (107)) together with the rule (114). This works for one-dimensional nonhomogeneous systems (section 3.8) where it constitutes 2 equations for 2 unknowns k and x. It is inadequate, however, for three-dimensional systems because it constitutes only 4 equations for 6 unknowns k_i and x_i ($i = 1, 2, 3$). In general, then, rays are defined by a sixth-order system of equations (106) and (107), and they can at most be reduced to a fifth-order system by using (114).

It may also be noted that equations (106) and (107) are already in a form suitable for the application of standard computer routines for ordinary differential equations to obtain their solutions quickly under a wide variety of initial conditions. In such computations, the rule (114) may have its greatest value as a check on accuracy.

Ray tracing is of value not only because it leads to information on the spatial distribution of the wavenumber vector k_i but also because wave *energy* moves along rays; therefore, ray tracing can be used to determine the distribution of wave amplitude. The concept of 'ray tubes' introduced already in section 1.11 and much used in section 2.14 is often an excellent means for doing this.

In dispersive systems, the ray-tube method can be used whenever waves of fixed frequency ω are being investigated. This may arise *either* because all the waves in the system are being generated by oscillations at a fixed frequency ω, *or* because a preliminary Fourier analysis with respect to frequency has been made.

For waves of fixed frequency ω, the wave pattern is stationary with respect to time. Then the equation

$$\partial W/\partial t + \nabla \cdot \mathbf{I} = 0, \tag{115}$$

relating for a general dispersive system the wave energy density and the wave energy flux

$$\mathbf{I} = W\mathbf{U}, \tag{116}$$

is reduced to an equation

$$\nabla \cdot (W\mathbf{U}) = 0 \tag{117}$$

stating that $\mathbf{I}$ is a solenoidal vector field. In terms of the cross-sectional area A of a thin ray tube (tubular surface made up of rays, to which of course the group velocity vector $\mathbf{U}$ is everywhere tangential), equation (117) can be written as follows:

$$WUA = \text{constant along a ray tube.} \tag{118}$$

Physically, this expresses conservation of the energy flowing along the tube, in terms of variations in the wave-energy density W, the magnitude U of the energy propagation velocity, and the ray-tube area A.

For nondispersive systems, the method is simplified in that the ray tubes are the same for all frequencies. For example, in sound waves satisfying (88) with c_0 a function of position x_i, the equations (106) and (107) for the rays become

$$\mathrm{d}k_i/\mathrm{d}t = -(k_1^2 + k_2^2 + k_3^2)^{\frac{1}{2}}\,\partial c_0/\partial x_i, \quad \mathrm{d}x_i/\mathrm{d}t = c_0 k_i (k_1^2 + k_2^2 + k_3^2)^{-\frac{1}{2}}; \tag{119}$$

and, given any ray pattern for one fixed value of ω, the same ray pattern must satisfy the equations (88) and (119) with ω, k_1, k_2 and k_3 all multiplied by a common constant factor.

For dispersive systems that have not been Fourier-analysed with respect to time, the full equation (115) must be used. This can be written, using (116), as

$$\mathrm{d}W/\mathrm{d}t = -W\nabla \cdot \mathbf{U} = -W\,\partial U_j/\partial x_j, \tag{120}$$

where the left-hand side is $\partial W/\partial t + \mathbf{U} \cdot \nabla W$, the rate of change of W along a ray. Equation (120), however, is not quite straightforward to solve; for example, it cannot be directly integrated along rays together with the other ordinary differential equations (106) and (107) because knowledge of *neighbouring* solutions is required to determine the partial derivatives on the right. The inconvenience of this approach has led to the ray-tube method (preceded where necessary by Fourier analysis) being generally preferred; except for homogeneous systems, where a good alternative method, to be described in section 4.8, is available.

At the end of this section, we illustrate the general theory by the example of internal waves in stratified fluid, satisfying equation (24) with the Väisälä–Brunt frequency $N(z)$ varying gradually with z on a scale of wavelengths. We precede that, however, by showing what a major simplification of the general theory occurs *whenever* only one coordinate (say, $x_3 = z$) occurs explicitly in the dispersion relationship (100). Then the equations (106) for k_1 and k_2 become $dk_1/dt = 0$ and $dk_2/dt = 0$, showing that k_1 and k_2 as well as ω remain constant along rays. These *three* conditions are actually equivalent to the *three* equations (106); not only is the number of conditions right but the precise equivalence follows readily from (108). Thus, three-dimensional systems whose nonhomogeneity involves just one coordinate retain much of the simplicity of one-dimensional systems.

In a stratified atmosphere or ocean we use (x, y, z) coordinates with the z-axis vertical. Then, waves of any type whose properties are independent of x and y can have their rays traced by means of the equations

$$\omega = \text{constant}, \quad k = \text{constant}, \quad l = \text{constant}. \tag{121}$$

For a dispersion relationship

$$\omega = \omega(k, l, m, z) \tag{122}$$

these determine m as a function of z. Then the rule that the ray direction is that of the group velocity (90) can be written

$$\frac{dx}{dz} = \frac{\partial \omega/\partial k}{\partial \omega/\partial m}, \quad \frac{dy}{dz} = \frac{\partial \omega/\partial l}{\partial \omega/\partial m}, \tag{123}$$

with right-hand sides known as functions of z; these may be integrated to obtain the variation of x and y along the ray.

In such cases, the ray-tube method for oscillations of fixed frequency ω takes a particularly simple form because *sections of a ray tube by each horizontal plane have the same area*. This follows from the fact that solutions of (123) with arbitrary initial values x_0, y_0 at height z_0 take the form

$$x = x_0 + \int_{z_0}^{z} \left(\frac{\partial \omega/\partial k}{\partial \omega/\partial m} \right) dz, \quad y = y_0 + \int_{z_0}^{z} \left(\frac{\partial \omega/\partial l}{\partial \omega/\partial m} \right) dz \tag{124}$$

at height z; where the integrals represent always the same horizontal translations $(x - x_0, y - y_0)$ for different rays. We conclude, then, by applying the divergence theorem within the part of a ray tube bounded by two horizontal planes to the divergence-free vector field $W\mathbf{U}$, that its z-component is independent of z:

$$W\,\partial\omega/\partial m = \text{constant along a ray.} \tag{125}$$

In physical terms, for oscillations of fixed frequency ω, each ray carries a definite value, not only of the horizontal components of wavenumber k and l but also of the upward component of wave energy flux.

For sound waves in a stratified atmosphere, the equations (123) with

$$\omega = (k^2 + l^2 + m^2)^{\frac{1}{2}}\, c_0(z) \tag{126}$$

give the rather elementary information

$$\mathrm{d}x/\mathrm{d}z = k/m, \quad \mathrm{d}y/\mathrm{d}z = l/m, \tag{127}$$

meaning that rays are parallel to the wavenumber vector. They can be traced by *Snell's law*, that the sine of the angle θ between a ray and the vertical takes a value

$$\sin\theta = (k^2 + l^2)^{\frac{1}{2}}(k^2 + l^2 + m^2)^{-\frac{1}{2}} = (k^2 + l^2)^{\frac{1}{2}}\,\omega^{-1}c_0(z), \tag{128}$$

which (by (121)) varies with altitude in proportion to the sound speed $c_0(z)$. Sound amplitudes are inferred from the fact that the upward component of wave energy flux, $\quad W c_0 \cos\theta, \tag{129}$

is constant along a ray.

For internal waves, the rays being parallel to the group velocity (91) satisfy equations

$$\mathrm{d}x/\mathrm{d}z = -km(k^2 + l^2)^{-1}, \quad \mathrm{d}y/\mathrm{d}z = -lm(k^2 + l^2)^{-1}, \tag{130}$$

meaning that each ray is perpendicular to the wavenumber vector and coplanar with it and the vertical. The analogue to Snell's law is that the angle θ between a ray and the vertical satisfies

$$\sec\theta = [1 + (\mathrm{d}x/\mathrm{d}z)^2 + (\mathrm{d}y/\mathrm{d}z)^2]^{\frac{1}{2}} = (k^2 + l^2 + m^2)^{\frac{1}{2}}(k^2 + l^2)^{-\frac{1}{2}} = \omega^{-1}N(z); \tag{131}$$

essentially a restatement of equation (15) which shows that $\sec\theta$ varies with altitude in proportion to the Väisälä–Brunt frequency $N(z)$. The constant upward component (125) of wave energy flux becomes

$$-WNm(k^2 + l^2)^{\frac{1}{2}}(k^2 + l^2 + m^2)^{-\frac{3}{2}}, \tag{132}$$

which by (93) can be written as

$$-\tfrac{1}{2}\rho_0^{-1}q_1^2\,Nm(k^2 + l^2)^{-\frac{1}{2}}(k^2 + l^2 + m^2)^{-\frac{1}{2}} \tag{133}$$

in terms of the wave amplitude q_1; or, using (24), as

$$-\tfrac{1}{2}\rho_0^{-1} q_1^2 m\omega(k^2 + l^2)^{-1}. \tag{134}$$

Evidently, this expression is an excellent approximation, under condition (52) for sound waves and internal waves to be decoupled, to the value (60) of the upward component of energy flux in the fully coupled case.

In the ocean, the relative variations in ρ_0 are negligible and so the constancy of (134) means that $q_1^2 m$ is constant along a ray. With the result $\partial\alpha/\partial z = -m$ for the phase α, this implies an approximate solution

$$q = Cm^{-\frac{1}{2}} \exp\left[\mathrm{i}\left(\omega t - kx - ly - \int_{z_0}^{z} m\,\mathrm{d}z\right)\right], \tag{135}$$

where C is constant, and by (24) we have

$$m = (k^2 + l^2)^{\frac{1}{2}} \{\omega^{-2}[N(z)]^2 - 1\}^{\frac{1}{2}}. \tag{136}$$

We may check this result, in the case $l = 0$, against solutions of the form (70). The function $Q(z)$, by (71) and (136), should satisfy

$$Q''(z) = -m^2 Q(z), \tag{137}$$

but (135) implies that

$$Q(z) = Cm^{-\frac{1}{2}} \exp\left(-\mathrm{i}\int_{z_0}^{z} m\,\mathrm{d}z\right). \tag{138}$$

Rather as in section 2.6, we assess the error in (138) by substituting it into the left-hand side of (137), giving

$$[-m^2 - \tfrac{1}{2}(m''/m) + \tfrac{3}{2}(m'/m)^2]\,Q(z). \tag{139}$$

This satisfies (137) with an error of order the *square* of the ratio of the relative rate of change of m (that is, m'/m or $(m''/m)^{\frac{1}{2}}$) to m itself; note that the $m^{-\frac{1}{2}}$ factor in (138) was necessary for the error to be of the order of the *square* of this ratio. When m changes by a small fraction of itself within a distance $1/m$, the approximation should be excellent.

Note, however, that the approximation (138) cannot be used for trapped waves; the error becomes unacceptable wherever $N(z)$ falls to the value ω so that m tends to zero (giving the rays vertical cusps). We postpone to section 4.11 the analysis of a refinement of ray tracing which allows it to be used also in such cases.

For internal waves in the atmosphere, where ρ_0 may vary substantially, we see that the implications of the Boussinesq approximation (137) that amplitude varies approximately like $m^{-\frac{1}{2}}$ may be unreliable; indeed, they need to be modified (because (134) must remain constant) to a variation as $\rho_0^{\frac{1}{3}} m^{-\frac{1}{2}}$. This confirms the suggestion at the end of section 4.2 that the omission of a factor $\rho_0^{\frac{1}{3}}$ from q and from p_e constitutes the leading error in the Boussinesq approximation.

4.6 Ray tracing in a wind

Waves in fluids differ from other kinds of waves in one important respect: the fluid medium through which the waves propagate may be involved in flow on a large scale. Such a flow is commonly generated by some cause independent of the waves; in this section, for example, we analyse effects of atmospheric winds on the propagation of sound and of internal waves. We already note, however, some reciprocal influence of the wave propagation on the flow. In the next section, we study streaming motions entirely generated by wave propagation.

We have earlier (section 3.9) described water waves on a stream, which however is just a uniform flow on which the waves are superimposed. In a certain frame of reference, the water is at rest except for those waves, made by what in that frame of reference is a moving object.

In this section we are concerned with wave propagation through non-uniform flows, which exhibit a fluid's power of indefinite deformation. Although at each point a frame of reference can be chosen so that locally there is no mean flow and the waves are propagating through fluid at rest, the required frame of reference is different at different points. This, as we shall see, modifies the energetics of the wave propagation.

At the same time, we confine ourselves to flows whose nonuniformity is only gradual on a scale of wavelengths. First, we show how the general theory of ray tracing (section 4.5) can be extended so as to apply to wave propagation of a rather general type through a fluid in motion, provided that its properties (including the flow field as well as properties that would influence the waves even in fluid at rest) vary gradually on a scale of wavelengths. As our main illustrations we use the propagation of sound and internal waves in a nonuniform wind. We also mention results for water-wave propagation in a nonuniform current.

In all cases we find that the rules regarding energy (section 4.5) are modified because there is exchange of energy between the waves and the mean flow. We give a simple theory of this energy exchange, applicable to sound, water waves and internal gravity waves, and suitable also for studying wave-generated streaming (section 4.7); later (epilogue, part 2), we see that one of its main conclusions (conservation of wave *action*) is a principle of very great generality which can be used as the key to analysing energy exchange in a much wider range of cases.

We use
$$\mathbf{V} = (V_1, V_2, V_3) \tag{140}$$

to signify the mean-velocity field through which the waves propagate. Often, we consider the waves in a local frame of reference moving at a particular value (140) of the mean flow velocity. In that frame of reference there is locally no wind, and the waves have the frequency

$$\omega_r = \omega_r(k_1, k_2, k_3, x_1, x_2, x_3) \tag{141}$$

appropriate to fluid at rest. The suffix r in (141) can be regarded as standing for *rest*, or else for *relative*: ω_r is the frequency of the motions relative to the local mean velocity. As in section 4.5, we assume that the properties of the fluid vary so gradually that only their local values affect the local dispersion relationship (141). Similarly, we assume that changes in the mean flow within a wavelength are so slight as to affect the dispersion relationship negligibly.

A simple relationship exists between the absolute frequency ω (frequency of oscillations occurring at a fixed point in space) and the relative frequency ω_r (frequency of oscillations occurring at a point moving with velocity V_j). In terms of a phase function $\alpha(x_1, x_2, x_3, t)$, the absolute frequency ω and wavenumber vector k_i are given by its partial derivatives (80). The relative frequency ω_r is the rate
$$\partial\alpha/\partial t + V_j \,\partial\alpha/\partial x_j \tag{142}$$

at which the phase changes for a point moving with velocity V_j. Therefore,

$$\omega_r = \omega - V_j k_j. \tag{143}$$

Equation (143) relating the frequencies of the same waves in two different frames of reference is of crucial importance for ray tracing in a wind, and has other applications. For example, waves generated in a fluid at rest by a *moving source* which oscillates at frequency ω have a frequency ω_r relative to the fluid which satisfies (143) provided that V_j is *minus* the source velocity; in other words, V_j (as above) is the velocity of the fluid relative to the frame of reference in which ω is measured. In the context of moving sources, equation (143) has long been called the Doppler relationship, and it is natural to call it by this name also for waves propagating through non-uniformly moving fluids.

For sound waves whose wavenumber vector k_i (perpendicular to wave crests) makes an angle Θ with the wind velocity V_j, the usual acoustic relationship

$$\omega_{\mathrm{r}} = c_0(k_1^2 + k_2^2 + k_3^2)^{\frac{1}{2}} \tag{144}$$

satisfied by ω_{r} implies that

$$V_j k_j = V(k_1^2 + k_2^2 + k_3^2)^{\frac{1}{2}} \cos\Theta = \omega_{\mathrm{r}}(V/c_0)\cos\Theta, \tag{145}$$

where V is the wind speed. Equation (143) therefore gives

$$\omega_{\mathrm{r}} = \frac{\omega}{1 + (V/c_0)\cos\Theta}. \tag{146}$$

This form of the Doppler relationship means that for acute angles Θ (downwind propagation) waves of given frequency ω have relative frequencies ω_{r} that become progressively smaller (corresponding to *increasing* wavelengths) as the wind speed increases. The same equation (146) gives the familiar diminished frequency ω_{r} of sound heard by an observer in still air from a sound source of frequency ω moving *away* from him at speed V (in a direction making an angle Θ to the line joining source and observer). Conversely, the Doppler effect makes $\omega_{\mathrm{r}} > \omega$ either (i) when a moving source approaches an observer in still air or (ii) when waves propagate upwind.

Ray tracing in a wind starts from equations (141) and (143) in the combined form

$$\omega = k_j V_j(x_1, x_2, x_3) + \omega_{\mathrm{r}}(k_1, k_2, k_3, x_1, x_2, x_3). \tag{147}$$

Then this relationship between the derivatives (80) of a phase function α leads, as in the general theory starting from equation (100), to equations (106) and (107) for rates of change along rays. The velocity (107) of a wave packet along its ray becomes

$$dx_i/dt = V_i + \partial\omega_{\mathrm{r}}/\partial k_i = V_i + U_i; \tag{148}$$

simply the *vector sum* of the wind velocity and the group velocity in fluid at rest. As we might expect, the wave packet travels at the local group velocity relative to the local wind.

Equation (147) shows also that, in a wind, the law (106) governing refraction takes the form

$$dk_i/dt = -k_j\,\partial V_j/\partial x_i - \partial\omega_{\mathrm{r}}/\partial x_i. \tag{149}$$

Here, the first term represents refraction due to wind gradients, and the second term that due to changing fluid properties (such as c_0 and N).

As in the general theory the result (108) shows that the frequency ω remains constant along rays. However, the *relative* frequency ω_r may vary; in fact,

$$d\omega_r/dt = (\partial\omega_r/\partial k_i)\, dk_i/dt + (\partial\omega_r/\partial x_i)\, dx_i/dt$$

$$= - U_i k_j\, \partial V_j/\partial x_i + V_i\, \partial\omega_r/\partial x_i. \tag{150}$$

There is a surprising connection between changes in relative frequency and in wave energy. From section 4.5 we know that in fluid at rest the wave energy is carried unchanged along rays. This is not so for waves in non-uniformly moving fluid, essentially because the waves by themselves do not constitute a conservative system: they can exchange energy with the mean flow.

A quick way to see this is to use at every point a local frame of reference moving with the local mean velocity V_j. We write the wave energy density W_r to remind us that it is the value of W for the wave motions *relative* to the local wind. Thus, W_r is related to amplitude and wavenumber as in fluid at rest, since it is the value of W in that local frame of reference in which the undisturbed fluid is at rest.

A general principle of mechanics (used already in section 1.9) states that a particle of mass M moving in an *accelerating* frame of reference behaves as if, in addition to all the ordinary forces acting on it, an extra force called the *inertial force* were present. This inertial force equals $(-M)$ times the acceleration of the frame of reference. (The force is essentially a mass-acceleration term which, having been omitted in the expression of Newton's second law of motion when only accelerations *relative* to the frame were included, needs to appear on the 'force' side of the equation with opposite sign.)

When a fluid is analysed with respect to a frame of reference different at each point, whose velocity at (x_1, x_2, x_3) is $V_j(x_1, x_2, x_3)$, then a particle to which the wave motions give a velocity u_i is subject to all the forces appropriate to such wave motions in a fluid at rest plus one additional force. This is the inertial force

$$(-\rho)\, u_i\, \partial V_j/\partial x_i \tag{151}$$

per unit volume, where ρ is the particle's mass per unit volume and

$$u_i\, \partial V_j/\partial x_i \tag{152}$$

is the acceleration of the frame of reference associated with the particle's velocity u_i. Accordingly, the local rate of change $\partial W_r/\partial t$ of wave energy per unit volume is equal to the *sum* of the usual value,

$$-\nabla \cdot \mathbf{I} = -\partial(W_r\, U_i)/\partial x_i \tag{153}$$

and the rate of working $\qquad -\overline{\rho u_i u_j}\, \partial V_j/\partial x_i \tag{154}$

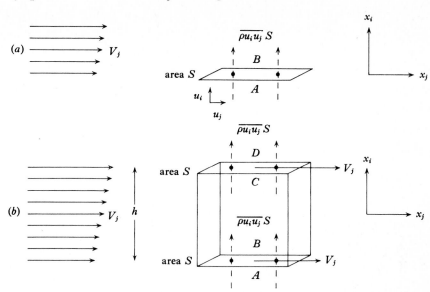

Figure 78. Illustrating wave motions with velocities u_i relative to a nonuniform mean flow velocity V_j. (a) These act to transport the x_i-component of momentum, ρu_j per unit volume, in the x_i-direction across area S at a mean rate $\overline{\rho u_i u_j}\, S$. Therefore $\overline{\rho u_i u_j}$ acts as a mean stress (force per unit area). (b) The net force on a block associated with this component of stress is the difference $-h[\partial(\overline{\rho u_i u_j})/\partial x_i]\, S$ between the values of that force on two opposing faces of the block of area S a distance h apart. The corresponding mean rate of working on the block is

$$-V_j[\partial(\overline{\rho u_i u_j})/\partial x_i]$$

per unit volume. Mean *exchange* of energy across the boundaries accounts for a part of this; namely, $-\partial[V_j(\overline{\rho u_i u_j})/\partial x_i]$ per unit volume due to the difference between the rates of working $V_j(\overline{\rho u_i u_j}\, S)$ by the block on the two opposing faces a distance h apart. The remainder, given by (155), represents the net mean gain of energy per unit volume by the mean flow.

by the inertial force per unit volume (151) acting on a fluid particle with velocity u_j. The bar in (154) signifies an average over a wave period.

From section 1.10 we recognise $\rho u_i u_j$ as a *momentum flux tensor*, representing, for example, rate of transport of the x_j-component of momentum (ρu_j) in the x_i-direction. The *mean momentum flux* occurring in (154) can redistribute the momentum of the mean flow: figure 78 shows how this quantity acts on the mean flow like an externally imposed *stress* (force per unit area) which does work at a rate $+\overline{\rho u_i u_j}\, \partial V_j/\partial x_i$ (155)

per unit volume. We conclude that local wave energy is changed at a rate per unit volume given by two processes, both energy-conserving: (i) the term (153) represents transport of the wave energy W_r at the group velocity U_i relative to the local mean flow; (ii) the term (154) represents an exchange with the mean flow, which experiences an equal and opposite rate of change of energy (155).

We call $\overline{\rho u_i u_j}$ the *Reynolds stress*. Reynolds first showed how turbulent variations in fluid velocity about a mean flow act with an effective stress on that mean flow; waves propagating through a mean flow act similarly and the expression for the effective stress is exactly the same.

In a linear theory, cubes of disturbances can be neglected in an energy or its rate of change, so the Reynolds stress occurring in (154) can be written

$$\rho_0 \overline{u_i u_j}. \tag{156}$$

In general, as shown in section 1.10, a momentum flux tensor may include also an isotropic term $(p - p_0)\,\delta_{ij}$; accordingly, waves that generate a second-order mean excess pressure (as can be true for sound waves, because the graph of pressure against density is concave upwards) may generate an additional stress-like interaction with a mean flow. In this chapter, however, we confine ourselves to mean flows of low Mach number with zero divergence $\partial V_i / \partial x_i$. Then no work can be done against the velocity gradient $\partial V_j / \partial x_i$ by stresses that are isotropic (carrying the δ_{ij} factor); therefore, the interaction with such a mean flow is completely represented by the Reynolds stress (156).

We furthermore confine ourselves in this chapter to systems where the external force acting (namely, gravity) can either be neglected as in sound waves (section 4.2) or has no component in the direction of the mean flow; thus, the stratification assumed in the theory of internal waves is compatible only with horizontal mean winds, while any steady currents that may interact with water waves must be horizontal. Therefore, for every component of Reynolds stress (156) which could affect the rate of working (154) because V_j is nonzero, the external force acting in the j-direction vanishes so that the local rate of change of $\rho_0 u_j$ in the frame of reference moving with the local wind is

$$\partial(\rho_0 u_j)/\partial t = -\partial p_e/\partial x_j. \tag{157}$$

Accordingly,
$$\rho_0 u_j = \omega_r^{-1} k_j p_e \tag{158}$$

and the Reynolds stress (156) takes the value

$$\omega_r^{-1} k_j \overline{p_e u_i} = \omega_r^{-1} k_j I_i. \tag{159}$$

It is interesting that this expression for mean momentum flux recalls the correspondences (112) and (113): to obtain mean momentum flux we take out a frequency factor from the mean energy flux I_i and substitute a wave-number factor.

In all these cases, then, the equation for rate of change of wave energy density in the local frame of reference (sum of (153) and (154)) is

$$\partial W_{\mathrm{r}}/\partial t = -\partial(W_{\mathrm{r}} U_i)/\partial x_i - \omega_{\mathrm{r}}^{-1} k_j I_i \partial V_j/\partial x_i. \tag{160}$$

This equation can also be written in terms of rates of change along a ray:

$$\mathrm{d}W_{\mathrm{r}}/\mathrm{d}t = -W_{\mathrm{r}} \partial U_i/\partial x_i + W_{\mathrm{r}} \omega_{\mathrm{r}}^{-1} \mathrm{d}\omega_{\mathrm{r}}/\mathrm{d}t, \tag{161}$$

where the second term is obtained from (150) in the frame of reference moving with the local mean flow (so that $V_i = 0$ and $I_i = W_{\mathrm{r}} U_i$).

Equation (161) gives the remarkably simple law

$$\mathrm{d}(W_{\mathrm{r}} \omega_{\mathrm{r}}^{-1})/\mathrm{d}t = -(W_{\mathrm{r}} \omega_{\mathrm{r}}^{-1}) \partial U_i/\partial x_i, \tag{162}$$

stating that the *wave action*, defined so that its value per unit volume is $W_{\mathrm{r}} \omega_{\mathrm{r}}^{-1}$ (wave energy divided by relative frequency), is carried unchanged along rays. Although various restrictive assumptions have been made in the above derivation, we indicate in the epilogue that the principle is one of very great generality; in particular, it plays for waves propagating through a mean flow the same role that the principle of conservation of wave energy plays for waves in fluid at rest.

Conservation of wave action implies that wave energy *increases* (at the expense of the mean flow) wherever the rays move into regions of greater ω_{r}; on the other hand, wave energy is lost (fed into the mean flow) wherever ω_{r} decreases. Equation (143) shows that changes in the mean-flow component parallel to k_i (that is, perpendicular to crests and in the direction of their motion) produce opposite changes in ω_{r} and hence in the wave energy. For example, water waves moving from still water into a region of opposing current gain in wave energy.

The principle (162) is used most easily for waves of fixed frequency so that the oscillations at any fixed point in space have a steady periodic character (compare section 4.5). Then, in the original frame of reference where the mean flow velocity is V_i, the principle takes the form

$$\partial[W_{\mathrm{r}} \omega_{\mathrm{r}}^{-1}(U_i + V_i)]/\partial x_i = 0. \tag{163}$$

Equation (163) states that the vector in square brackets, called the *wave action flux*, is solenoidal; in other words, its magnitude varies along ray tubes in inverse proportion to their cross-sectional area. Waves of fixed frequency ω have a variable flow of energy along a ray tube and a variable relative frequency ω_r but their *ratio* (the flow of wave action along the ray tube) is constant.

We conclude this section (like section 4.5) by showing the simplified form taken by the theory in a stratified atmosphere where the dispersion relationship takes a form

$$\omega_r = \omega_r(k, l, m, z) \qquad (164)$$

depending only on the vertical coordinate z. We set out the theory for the case of a unidirectional but sheared wind

$$\mathbf{V} = (V(z), 0, 0). \qquad (165)$$

The reader will find it easy to extend the theory to the case when the mean flow has a y-component as well as an x-component; however, as explained above, a z-component would be incompatible with a constant undisturbed stratification.

The expression (147) for ω becomes

$$\omega = kV(z) + \omega_r(k, l, m, z). \qquad (166)$$

As in the general theory for waves satisfying a relation (122) between frequency and wavenumber depending on only one coordinate z, the refraction equations are precisely equivalent to the rules (121) stating that ω, k and l take constant values along a ray. Then equation (166) determines m as a function of z, and equations (148) give the ray direction as

$$\frac{dx}{dz} = \frac{V(z) + \partial\omega_r/\partial k}{\partial\omega_r/\partial m}, \quad \frac{dy}{dz} = \frac{\partial\omega_r/\partial l}{\partial\omega_r/\partial m}, \qquad (167)$$

with right-hand sides known as functions of z; these may be integrated to obtain the variation of x and y along a ray.

As in section 4.5 we infer that sections of a ray tube by each horizontal plane have the same area, and deduce from equation (163) that the upward component of wave action flux,

$$W_r\,\omega_r^{-1}\,\partial\omega_r/\partial m, \qquad (168)$$

is constant along a ray. Wave amplitudes are readily inferred by this principle when the rays have been found.

For sound waves we write the components of the wavenumber vector as

$$k = k_H \cos\psi, \quad l = k_H \sin\psi, \quad m = k_H \cot\theta, \qquad (169)$$

so that k_H is its constant horizontal resultant, ψ is its constant azimuthal angle to the wind direction, and θ is its *variable* angle to the vertical. Then

$$\omega_r = c_0(z)(k^2 + l^2 + m^2)^{\frac{1}{2}} = c_0(z)\,k_H \operatorname{cosec}\theta, \qquad (170)$$

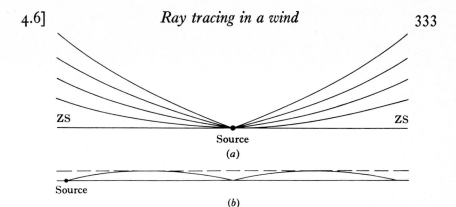

Figure 79. (*a*) Upward curvature of rays of sound emitted by a source at ground level into an atmosphere whose temperature decreases with height. The region ZS, lying below the ray emitted tangentially to the ground, is often called 'zone of silence'. (*b*) Downward curvature of rays of sound emitted by a source at ground level into an atmosphere whose temperature increases with height. Rays emitted at small enough angles can curve back to the ground.

so that equation (166) gives

$$\sin \theta = \frac{c_0(z)}{\omega k_{\mathrm{H}}^{-1} - V(z) \cos \psi} \qquad (171)$$

as the analogue of Snell's law (128). Refraction by temperature variations and by wind variations are represented, respectively, by the numerator and denominator of (171), Ray tracing, when θ has been determined as a function of altitude z, proceeds from equations (167) in the form

$$\mathrm{d}x/\mathrm{d}z = V(z)[c_0(z)]^{-1} \sec \theta + \cos \psi \tan \theta, \quad \mathrm{d}y/\mathrm{d}z = \sin \psi \tan \theta. \quad (172)$$

Sound amplitudes are inferred from the fact that the upward component of wave action flux,

$$(W_r \omega_r^{-1}) c_0 \cos \theta = k_{\mathrm{H}}^{-1} W_r \sin \theta \cos \theta, \qquad (173)$$

remains constant along a ray.

In the absence of significant wind, temperature variations can have an appreciable effect on sound propagation. A positive lapse rate satisfying (75) involves a fall in $c_0(z)$ with height by up to $6\,\mathrm{m\,s^{-1}}$ per km. This causes (171) to decrease with height, so that rays are curved upwards. Although the curvature is only slight (the *radius* of curvature can be written $\mathrm{d}z/\mathrm{d}(\sin \theta)$ and comes to at least 60 km) it may significantly interfere with horizontal propagation over flat ground: it lifts the energy-carrying rays by as much as 8 m in 1 km horizontal distance (or by 33 m in 2 km). Although the region without any rays (figure 79) is not a zone of total silence, we shall find (section 4.11) that amplitudes decrease steeply with depth inside it.

In an 'inversion', as when overnight cooling of the ground has produced an *increase* of temperature (and so of (171)) with height, rays (again in the absence of wind) are curved downward. Then sound propagation which is nearly horizontal may exhibit 'trapped-wave' behaviour. This is suggested by the fact that rays approaching the height where $c_0(z)$ is equal to $\omega k_{\mathrm{H}}^{-1}$ (so that $\theta = \frac{1}{2}\pi$) become horizontal, and then can only curve back to the ground (figure 79); after reflection from the ground they can repeat the process. As remarked at the end of section 4.5, the detailed analysis of such trapped waves requires a refinement of ray tracing (section 4.11), but there is an indication here that horizontal propagation can be enhanced under such 'inversion' conditions.

The effects of wind are qualitatively similar but can be quantitatively greater. The wind velocity $V(z)$ may increase with height considerably more steeply than $c_0(z)$ either increases or decreases. Then upwind propagation (with cos ψ negative) involves a substantial decrease of (171) with height: the rays curve upwards and sound amplitudes at ground level are correspondingly reduced. By contrast, downwind propagation (with cos ψ positive) is enhanced at ground level, again through trapped-wave behaviour, as suggested by the fact that rays approaching the critical level, where

$$V(z)\cos\psi = \omega k_{\mathrm{H}}^{-1} - c_0(z), \qquad (174)$$

become horizontal and then curve back towards the ground. Common observation suggests, indeed, that a wind causes sound amplitudes at ground level to be substantially increased on the downwind side of a source and decreased on the upwind side. We now see that these are effects of the wind's *nonuniformity* (increase with altitude).

For internal waves we write

$$k = k_{\mathrm{H}}\cos\psi, \quad l = k_{\mathrm{H}}\sin\psi, \quad m = -k_{\mathrm{H}}\tan\theta \qquad (175)$$

so that k_{H} and ψ have the same meanings as in (169) and θ, though defined differently, is still the angle which rays make with the upward vertical in the absence of wind. Then (166) becomes

$$\omega = k_{\mathrm{H}} V(z)\cos\psi + \omega_{\mathrm{r}} = k_{\mathrm{H}} V(z)\cos\psi + N(z)\cos\theta, \qquad (176)$$

giving
$$\sec\theta = \frac{N(z)}{\omega - k_{\mathrm{H}} V(z)\cos\psi} \qquad (177)$$

as the analogue to the still-air rule (131). Ray tracing, when θ has been determined from (177), proceeds from equations

$$\mathrm{d}x/\mathrm{d}z = k_{\mathrm{H}} V(z)\,[N(z)]^{-1}\sec^2\theta\,\mathrm{cosec}\,\theta + \cos\psi\tan\theta, \quad \mathrm{d}y/\mathrm{d}z = \sin\psi\tan\theta,$$
$$(178)$$

derived from (167) with the group velocity (91). Amplitude variations are inferred from the fact that the upward component of wave action flux

$$\tfrac{1}{2}\rho_0^{-1} q_1^2 k_{\mathrm{H}}^{-1} \tan\theta \tag{179}$$

(being expression (133) divided by $\omega_{\mathrm{r}} = N\cos\theta$) remains constant along a ray.

Refraction effects due to variations in $N(z)$ can be large (section 4.5) and so can those due to variations in $V(z)$. For internal waves, however, a *qualitative* difference between the two kinds of refraction effect exists.

Admittedly, just as in sections 4.3 and 4.5, trapped waves can result wherever θ tends to 0, so that the relative frequency ω_{r} reaches its maximum possible value $N(z)$. The required condition

$$\omega = N(z) + k_{\mathrm{H}} V(z)\cos\psi \tag{180}$$

can be satisfied in regions where *either* $N(z)$ *or* $V(z)\cos\psi$ is reduced (the latter is the wind component in the direction of the horizontal component of the propagation relative to the air). Rays reaching such a critical level can only turn back on themselves (due to a change of sign of θ). When $V(z) = 0$ (that is, in still air), both derivatives in (178) tend to zero at the critical level and the rays form vertical cusps (figure 80(*a*)). In a wind, however, this is not so; then, the first term in (178) dominates dx/dz near the critical level (where the wind dominates because the group velocity falls to zero). This is an *integrable* singularity in dx/dz (because the cosec θ in (178) behaves like the *inverse square root* of a small departure of $\sec\theta$ from 1) and the rays are tangents to the critical level (figure 80(*b*)) as in the acoustic case (figure 79).

With internal waves, however, a special type of critical level can additionally appear. This is where θ tends to $\tfrac{1}{2}\pi$ so that ω_{r} tends to 0 and oscillations relative to the wind cease (which is impossible in still air since $\omega_{\mathrm{r}} = \omega$ is constant). This occurs where

$$V(z)\cos\psi = \omega k_{\mathrm{H}}^{-1}; \tag{181}$$

that is, where the component of the wind in the direction of the horizontal component of the wave propagation relative to it *rises* to the value $\omega k_{\mathrm{H}}^{-1}$.

Rays are asymptotic (figure 81) to such a critical level, where dx/dz has a nonintegrable singularity since the $\sec^2\theta$ in (178) behaves like the *square* of (177) as $\theta \to \tfrac{1}{2}\pi$; thus, rays take an infinite time to get there. In the meantime, the wave energy goes to zero, since conservation of wave *action* (wave energy divided by ω_{r}) implies that where the relative frequency ω_{r} goes to zero *the wave energy is all lost to the mean flow*.

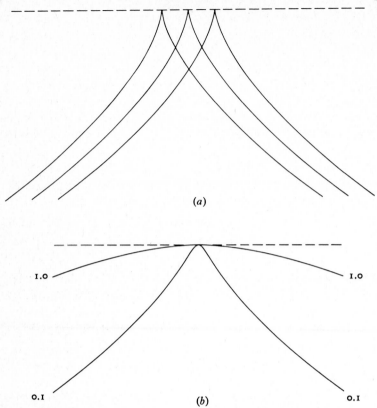

Figure 80. Behaviour of internal waves approaching a critical level (broken line) where equation (180) is satisfied as a result of $N(z)$ having fallen linearly with height z.

Case (a): without wind, the rays have vertical cusps at the critical level (here, this is where $N(z)$ becomes equal to the wave frequency ω).

Case (b): with wind, the rays are tangent to the critical level. They are drawn for the case of a uniform wind V in the plane of the paper (so that $\psi = 0$). They satisfy $dx/dz = \tan\theta + [kV/(\omega - kV)]\sec\theta\,\mathrm{cosec}\,\theta$ and are drawn for values 1.0 and 0.1 of the coefficient in square brackets; note that case (a) represents the limiting behaviour when this coefficient is zero.

Indeed, by (159) the Reynolds stress transferring horizontal mean momentum upwards is equal to the horizontal component of wavenumber times the upward component of wave action flux, both of which are constant along rays. Therefore, the mean flow around the critical level is accelerated through the waves providing a constant stress below that level which is absent above it.

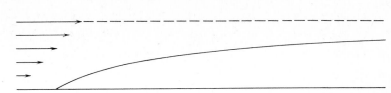

Figure 81. Behaviour of internal waves approaching that special type of critical level (broken line) where the relative frequency ω_r tends to zero. The wave energy is all lost to the mean flow before the ray asymptotes to such a critical level. The case shown has the wind $V(z)$ increasing linearly with height, with downwind propagation ($\psi = 0$) and uniform Väisälä–Brunt frequency N.

The above discussion gives a general indication of how internal waves may play a major role in vertical exchange of wind momentum. The critical level (181) where the energy and momentum transported by the waves is absorbed varies with ω, k_H and ψ. Thus, it is different for different Fourier components, and many of them may contribute significantly to momentum transport. Effects generally similar are thought to occur also in the ocean.

4.7 Steady streaming generated by wave attenuation

We have just analysed how a steady nonuniform wind may modify the propagation of sound waves and internal waves. We showed that such waves can exchange energy and momentum with the mean flow. Although (strictly speaking) this may interfere with the mean flow's assumed steadiness, the energy exchanges are normally much too slow for the resulting slight departures from steadiness to affect the propagation laws. Nevertheless, even in relation to the large energy of the mean winds those exchanges may (in cases like that treated at the end of section 4.6) produce significant redistribution in the longer term.

The Reynolds stress (156) represents the action of the waves on any divergenceless mean flow. In this section we consider how far the propagation of sound waves and internal waves through fluid initially at rest may, by the action of that Reynolds stress, generate a steady streaming motion. First, we prove that no such effect can occur unless the waves are being *attenuated*. Then, we analyse streaming generated when sound waves are attenuated by processes either in the body of the fluid (section 1.13) or at a boundary (section 2.7). Finally we give the rather elementary theory of viscous attenuation of internal waves and show how that process may, or may not, lead to streaming.

The mean force with which waves act on a fluid element (figure 78) results from *differences* between values of the Reynolds stress on opposite sides of it. Thus, the mean force acting in the x_j-direction, per unit volume of fluid, is

$$F_j = -\partial(\rho_0 \overline{u_i u_j})/\partial x_i: \tag{182}$$

the *gradient* of Reynolds stress.

For unattenuated internal waves, this force is zero, since by (17) it can be written

$$-\rho_0 \overline{u_i \, \partial u_j/\partial x_i} \tag{183}$$

and the fluid velocity u_i is parallel to surfaces of constant phase while the gradient $\partial/\partial x_i$ of any quantity is perpendicular to them. This conclusion that (182) is zero is consistent with the fact that, in the system described at the end of section 4.6, the Reynolds stress took a uniform value in the region occupied by the waves.

For unattenuated sound waves, the force F_j is not necessarily zero but nevertheless can generate no streaming. This is because it is the gradient of a scalar; therefore, it will not generate divergenceless flow because it can be completely cancelled by the gradient of a mean pressure.

The result may be proved as follows. Unattenuated sound waves are irrotational so that the derivative in (183) is the same as $\partial u_i/\partial x_j$, making that expression a gradient of

$$-\tfrac{1}{2}\rho_0 \overline{u_i^2}. \tag{184}$$

Admittedly, in sound waves (182) includes not only the term (183) but also a term

$$-\overline{u_j \, \partial(\rho_0 u_i)/\partial x_i} = \overline{u_j \, \partial \rho_e/\partial t}, \tag{185}$$

but this has the same value as

$$-\overline{\rho_e \, \partial u_j/\partial t} = \overline{\rho_e \rho_0^{-1} \, \partial p_e/\partial x_j} \tag{186}$$

because the mean value of the time-derivative of any fluctuating quantity such as $\rho_e u_j$ must be zero; and (186) is equal to a gradient of

$$\tfrac{1}{2}\rho_0^{-1} c_0^{-2} \overline{p_e^2}. \tag{187}$$

Thus, F_j is the gradient of a scalar quantity equal to the sum of (184) and (187); which is the potential energy density minus the kinetic energy density. Their difference is of course zero under conditions close to those in a plane wave, including 'far-field' conditions (chapter 1): then the force F_j actually vanishes. Even in a near field, however, the force F_j produces no streaming because it can be exactly balanced by the gradient of a mean pressure equal to the difference between the kinetic and potential energy densities.

In the laboratory, sources of sound are often observed to generate a steady airflow. This 'sonic wind' became familiar when powerful sources (often at ultrasonic frequencies) based on the piezoelectric properties of quartz began to be commonly used; and so was sometimes known as the 'quartz wind'. The analysis just given suggests strongly that sonic wind must depend on the *attenuation* of an acoustic beam; a view supported by the fact that the flow is most marked at those very high frequencies where attenuation in the body of the fluid is substantial. Indeed, the fraction of acoustic energy so attenuated in each wave period is shown in section 1.13 to be

$$2\pi\omega\delta c_0^{-2}, \tag{188}$$

where the diffusivity of sound δ includes constant contributions from viscosity and heat conduction as well as a frequency-dependent contribution from thermodynamic 'lags'. Thus, the fraction of energy lost per unit distance of propagation, which is (188) divided by the wavelength $2\pi c_0 \omega^{-1}$, takes a value

$$\beta = \delta\omega^2 c_0^{-3} \tag{189}$$

that becomes especially significant at high frequencies.

These are frequencies at which a plane source of sound such as the face of an oscillating quartz crystal is able to generate a narrow beam (section 1.12). We consider first the sonic wind associated with the attenuation of such a beam.

The total force on the fluid exerted by a beam of power P (in watts) is Pc_0^{-1}, distributed so that the force per unit length at a distance X from the source is

$$Pc_0^{-1}\beta e^{-\beta X}. \tag{190}$$

Two alternative ways of seeing this are as follows.

The Reynolds stress (156) in a beam of sound is evidently dominated by the component parallel to the beam (the $i = j = 1$ term $\rho_0\overline{u_1^2}$ if the beam is along the x_1-axis). Accordingly, the gradient (182) is dominated by rate of change with respect to x_1 due to attenuation. Indeed, we have just shown that any effects of the beam not being exactly parallel (section 1.12) would be without effect in the absence of attenuation. Accordingly, the force per unit volume (182) is closely parallel to the beam and takes a value associated with attenuation at the rate β per unit distance; namely,

$$\beta\rho_0 u_1^2 = \beta W = \beta I c_0^{-1}, \tag{191}$$

where W is the wave energy density and I the intensity (wave energy flux). Therefore, the force per unit length takes the form (190) because the integral of I over the cross-sectional area of the beam is equal to the power

$$Pe^{-\beta X} \tag{192}$$

remaining in the beam at a distance X from the source after attenuation at a rate β per unit distance.

Even more simply (perhaps) we can use the result (159) proved for sound waves in the last section to see that the momentum flux in the beam is equal to the energy flux I multiplied by $\omega^{-1}k_1$ which is c_0^{-1}. The total flow of momentum at a distance X from the source is therefore equal to the flow of energy (192) multiplied by c_0^{-1}. The force (190) per unit length generating the steady flow is then seen as the decrease per unit length in this acoustic rate of flow of momentum.

The distribution of steady force due to sound waves (F_j per unit volume in general, or for the beam (190) per unit length) generates a flow field which itself becomes steady as soon as it has increased to the level where viscous resistance to it prevents further fluid acceleration. Then the equations determining the divergenceless mean velocity field $\bar{u}_j$ take the form

$$\rho_0 \bar{u}_i \, \partial \bar{u}_j / \partial x_i = F_j - \partial \bar{p} / \partial x_j + \mu \nabla^2 \bar{u}_j. \tag{193}$$

The right-hand side of (193) includes forces per unit volume due to the waves, to the mean pressure distribution and to viscous stresses (see section 3.5) and the left-hand side is the mass-acceleration term under steady-flow conditions.

Acoustic streaming motions are commonly calculated by neglecting the left-hand side of (193), so that $\bar{u}_j$ is taken to satisfy

$$F_j - \partial \bar{p} / \partial x_j + \mu \nabla^2 \bar{u}_j = 0, \quad \partial \bar{u}_j / \partial x_j = 0. \tag{194}$$

This simplification may appear logical because mean flows (being produced by forces proportional to acoustic power) are proportional to the squares of disturbances; the left-hand side of (193) is therefore proportional to the fourth power of disturbances and should be neglected in determining any second-order quantity like a mean flow field. We give first the results of thus linearising the equation (193) for $\bar{u}_j$ so that it reduces to the equations (194) governing low-Reynolds-number hydrodynamics; later, we show how very weak the disturbances must, however, be for this to be a good approximation.

Texts on low-Reynolds-number hydrodynamics prove that solutions of the equations (194) in unbounded fluid can be written as

$$\bar{u}_j = \int \frac{F_j r^2 + (F_i r_i) r_j}{8\pi\mu r^3} \, d\tau, \tag{195}$$

where the vector r_i (with magnitude r) stands for the displacement from the volume element $d\tau$, where the force $(F_j \, d\tau)$ acts, to the point where the velocity $\bar{u}_j$ is required. The integrand in (195) is called the *stokeslet* velocity field generated by the force $(F_j \, d\tau)$.

Under conditions of extremely high attenuation rate β, the distribution of force per unit length (190) in a beam of sound can be represented well enough as the stokeslet velocity field generated by a single concentrated force $F = Pc_0^{-1}$ acting at the centroid $X = \beta^{-1}$ of the distribution. This velocity field is easily plotted (figure 82(a)) by the rule that the volume flow through any 'hoop' of radius s centred on (and perpendicular to) the line of action of that concentrated force is

$$(4\mu)^{-1} F s^2 r^{-1}. \tag{196}$$

Evidently, the volume flow (196) across every cross-section of one of the streamtubes in figure 82(a) must take the same constant value. Equally spaced values are used for these volume flows through the different streamtubes of figure 82(a).

A more accurate representation is given in figure 82(b), showing the velocity field of the exponential distribution of force (190) along the axis of the beam. At distances more than about $3\beta^{-1}$ from the centroid C of the force distribution it coincides closely with the velocity field of a single stokeslet at C (figure 82(a)).

Sonic winds are often substantially modified by solid boundaries, and the methods of low-Reynolds-number hydrodynamics can be used for analysing this. When an acoustic beam is generated by an oscillating diaphragm flush with a plane wall (compare section 1.12) that wall, by providing additional viscous resistance to the flow, may reduce both its magnitude and its extent. In low-Reynolds-number hydrodynamics, the velocity field generated by a concentrated force F is modified as follows by the presence of a plane wall normal to that force: if I is the mirror-image of the stokeslet's position in the plane, the volume flow (196) is reduced to

$$(4\mu)^{-1} F s^2 [r^{-1} - r_I^{-1} - \tfrac{1}{2}(r_I^2 - r^2) r_I^{-3}], \tag{197}$$

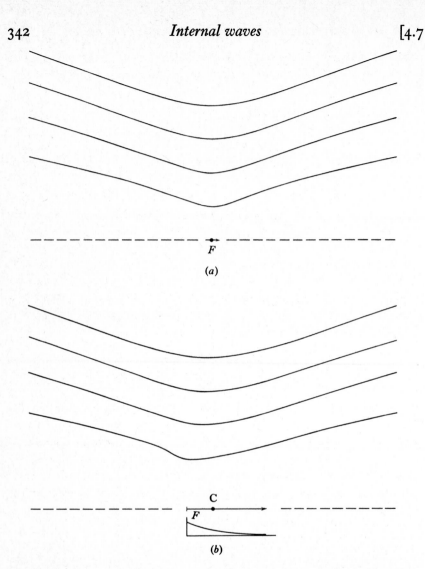

Figure 82. Slow steady streaming produced by the mean force F with which an acoustic beam acts on the fluid. (*a*) Streamtubes for a 'stokeslet' (the solution for slow flow generated by a force F concentrated at a point). (*b*) Streamtubes for an exponential distribution of force (190) characteristic of an attenuated acoustic beam. In each case the broken line represents the axis of symmetry. The volume flow rate between each pair of adjacent streamtubes drawn here is the same, and equal to that within the innermost streamtube.

where r_I is distance from the image I. (Evidently, the simple subtraction of r_I^{-1} would make the solid wall a streamsurface, but the further subtraction of the dipole-like term $\frac{1}{2}(r_I^2 - r^2)r_I^{-3}$ is needed to give the unique solution satisfying the no-slip condition.)

We can use (197) to calculate the steady flow generated by a distribution (190) of force per unit length at a distance X from the wall in which the sound source is placed (figure 83). We see that the presence of that single boundary suffices to confine the flow to a vortex ring driven by the sonic beam and resisted primarily by wall friction. The total flow is proportional to $P(c_0\mu\beta)^{-1}$ and an average wind velocity is proportional to $P\beta(c_0\mu)^{-1}$.

The difficulties in going beyond low-Reynolds-number hydrodynamics to a full use of the nonlinear equation (193) are in general enormous, but fortunately an exact similarity solution of (193) is known for the case of a single concentrated force F. Its nature depends on the value of the non-dimensional parameter $\rho_0 F\mu^{-2}$ (which is the *square* of an effective Reynolds number). The solution reduces to a stokeslet of strength F for values of this parameter less than about 10.

In this solution the volume flow through a 'hoop' of radius s perpendicular to the line of action of the force takes the form

$$4\pi\mu\rho_0^{-1}s^2(\sigma r - x)^{-1},\tag{198}$$

where x is the distance of the hoop in the direction of the force, measured from its point of application (thus, $r^2 = x^2 + s^2$). The constant σ satisfies

$$\rho_0 F\mu^{-2} = 16\pi[\tfrac{4}{3}\sigma(\sigma^2 - 1)^{-1} + \sigma - \sigma^2\tanh^{-1}(\sigma^{-1})],\tag{199}$$

where the term in square brackets behaves like σ^{-1} for large σ. The asymmetry due to the x term in (198) becomes insignificant for σ greater than about 5, corresponding to $\rho_0 F\mu^{-2} < 10$.

This suggests that a stokeslet representation of an acoustic beam's capability of generating steady flow is acceptable only for acoustic power outputs less than

$$10\mu^2\rho_0^{-1}c_0 \doteqdot 10^{-6}\,\text{W for air.}\tag{200}$$

For acoustic power outputs exceeding 10^{-6} W, the jet-like asymmetry in the sonic wind, resulting from fluid *inertia*, becomes substantial.

Figure 84 shows three configurations of streamtubes for different values of $\rho_0 F\mu^{-2}$. Departures from the symmetry characteristic of the low-Reynolds-number case (figure 82(a)) are moderate in figure 84(a) with $\rho_0 F\mu^{-2} = 35$ (corresponding to a source of 3×10^{-6} W). A jet-like flow,

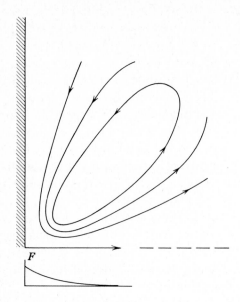

Figure 83. Slow steady streaming produced by an acoustic beam which emerges from a plane wall generating the distribution of force (190). The full lines are meridian sections of streamtubes dividing the total flow $0.080P(c_0\,\mu\beta)^{-1}$ into four equal parts.

spreading however at a large angle, forms in figure 84(b) with $\rho_0\,F\mu^{-2} = 314$ (acoustic power 3×10^{-5} W) and a narrow jet in figure 84(c) with

$$\rho_0\,F\mu^{-2} = 3282 \text{ (acoustic power } 3 \times 10^{-4}\text{ W).}$$

In this last case inertia is carrying the jet momentum far away from the source.

Jets at still higher Reynolds numbers than that are turbulent. Observations have shown that the mean flow then retains the general appearance of figure 84(c), because the turbulence in a round jet delivering momentum at the rate F redistributes it at a rate roughly equivalent to the action of an 'effective viscosity' μ_e with $\rho_0\,F\mu_e^{-2}$ of the order of 3×10^3.

At the higher acoustic powers, then, the sonic wind takes the form of a narrow jet which at powers of 10^{-3} W or more is actually turbulent. In any case of a narrow jet there should be practically no boundary interference of the type illustrated in figure 83. Energy dissipation takes place principally within the jet and the necessary entrainment of fluid into it suffers little resistance from any boundaries much wider than the jet.

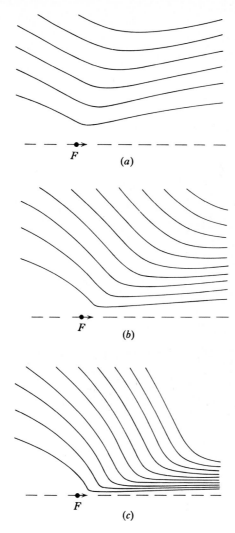

Figure 84. Not so slow streaming generated in fluid of density ρ_0 and viscosity μ by a concentrated force F, for three values of $\rho_0 F \mu^{-2}$: (a) 35; (b) 314; (c) 3282. In each case the broken line represents the axis of symmetry. The volume flow rate between any two adjacent streamtubes is equal to that within the innermost streamtube.

Sound propagated in narrow tubes, such that most of the acoustic energy dissipation takes place in thin boundary layers on the walls of the tube (section 2.7), generates a completely different type of streaming. A distribution of mean force (190) along the tube is still present, although the proportional energy attenuation per unit length β is given not by (189) but by the expression

$$\beta = (2\nu\omega)^{\frac{1}{2}} (c_0 a)^{-1} \tag{201}$$

for a tube of radius a. However, this force, having a uniform distribution (191) across the tube, generates no streaming: it is automatically cancelled by an opposing mean pressure gradient.

Accordingly, only the departure *within the boundary layer* of the mean force (182) from its value outside it generates streaming. We analyse this with the special coordinates used in section 2.7 to describe the boundary layer; thus, z is distance from the wall and x is distance along the tube in the direction of propagation.

The gradient $\partial/\partial z$ is dominant within the boundary layer, where therefore it is not surprising that

$$-\rho_0 \, \partial(\overline{uw})/\partial z \tag{202}$$

turns out to be a good approximation to the axial component of the force per unit volume (182). Acting in that narrow region, it sets up a steady streaming resisted by viscous shear stress, so that

$$\mu \partial \bar{u}/\partial z = (\rho_0 \overline{uw}) - (\rho_0 \overline{uw})_{\text{ex}}. \tag{203}$$

Here, the suffix ex is used as in section 2.7 for a value external to the boundary layer, so that (203) equates the viscous shear stress resisting mean flow at a point in the boundary layer to the net forcing by Reynolds stress acting between that point and the edge of the boundary layer. Accordingly, the sheared mean flow in the boundary layer takes the form

$$\bar{u} = \nu^{-1} \int_0^z [\overline{uw} - (\overline{uw})_{\text{ex}}] \, dz. \tag{204}$$

We calculate this when the axial velocity u in the sound waves has the form

$$u = u_{\text{ex}}\{1 - \exp[-z(i\omega/\nu)^{\frac{1}{2}}]\} \tag{205}$$

derived in section 2.7. The velocity w perpendicular to the tube wall is then

$$w = -i(\omega/c_0)(i\omega/\nu)^{-\frac{1}{2}} u_{\text{ex}}\{1 - \exp[-z(i\omega/\nu)^{\frac{1}{2}}]\} \tag{206}$$

in the boundary layer, so that the divergence $\partial u/\partial x + \partial w/\partial z$ is independent of z (as is necessary to make $\partial\rho/\partial t$ constant across the boundary layer).

The mean product of the real parts of (205) and (206) is

$$\overline{uw} = -\tfrac{1}{2}(\omega/c_0)(\nu/2\omega)^{\frac{1}{2}}|u_{ex}\{1 - \exp[-z(i\omega/\nu)^{\frac{1}{2}}]\}|^2, \qquad (207)$$

from which the integral (204) may be written as

$$\bar{u} = |u_{ex}|^2(2c_0\sqrt{2})^{-1}\int_0^z \{1 - |1 - \exp[-z(i\omega/\nu)^{\frac{1}{2}}]|^2\}(\omega/\nu)^{\frac{1}{2}}\,dz. \quad (208)$$

The value of $\bar{u}$ at the edge of the boundary layer (where $z(\omega/\nu)^{\frac{1}{2}}$ is large) is readily found as

$$(\bar{u})_{ex} = |u_{ex}|^2(4c_0)^{-1}. \qquad (209)$$

It is interesting that the kinematic viscosity ν disappears when the value (209) of (204) at the edge of the boundary layer is evaluated.

To sum up, forces within the boundary layer generate a wind (209) just outside the layer. If no net flow through the tube is possible (as when the sound waves are being generated at a closed end) the mean pressure gradient must exactly suffice to generate a central return flow (with Poiseuille distribution) such that there is no mean transport of fluid across any cross-section. Then the mean flow at distance s from the axis of the tube has the parabolic form

$$\bar{u} = |u_{ex}|^2(4c_0)^{-1}[-1 + 2(s/a)^2]. \qquad (210)$$

For a beam of intensity I, the value of $|u_{ex}|^2$ is $2I/\rho_0 c_0$. Accordingly, maximum 'sonic wind' velocities in a narrow tube are

$$I(2\rho_0 c_0^2)^{-1}: \qquad (211)$$

a value reaching $4\,\mathrm{mm\,s^{-1}}$ at atmospheric pressure when $I = 10^3\,\mathrm{W\,m^{-2}}$ (a sound intensity of $150\,\mathrm{dB}$ within the tube).

The above calculation is for a tube so long that frictional attenuation makes negligible any wave reflected from the far end. A different extreme case is that of a tube acting as a resonator excited into some standing wave pattern. Then the axial velocity u takes the form (205) where, apart from the $\exp(i\omega t)$ factor, u_{ex} is some *real* function of x (instead of showing the $\exp(-i\omega x/c_0)$ dependence characteristic of a travelling wave). The corresponding w is

$$(i\omega/\nu)^{-\frac{1}{2}}(du_{ex}/dx)\{1 - \exp[-z(i\omega/\nu)^{\frac{1}{2}}]\} \qquad (212)$$

for the same reason as in (206). In this case the calculation of the effect of

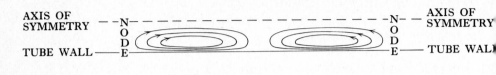

Figure 85. Streaming generated in a pipe by a sinusoidal standing wave (for which $u_{\mathrm{ex}} = u_1 \sin(2\pi x/\lambda)$ apart from the exp $(i\omega t)$ factor). The relationship between the radial scale (tube radius) and axial scale (wavelength) can be arbitrary. The total streaming flow is divided into four equal parts by the streamtubes shown. The streaming tends to make fine dust accumulate at the nodes $x = 0$ and $x = \frac{1}{2}\lambda$. This may allow a measurement of the distance $\frac{1}{2}\lambda$ for waves of given frequency, leading to a determination of the velocity of sound.

the mean-force term (202) proceeds as above, giving a contribution

$$-u_{\mathrm{ex}}(\mathrm{d}u_{\mathrm{ex}}/\mathrm{d}x)(4\omega)^{-1} \tag{213}$$

to the mean velocity $(\bar{u})_{\mathrm{ex}}$ on the edge of the boundary layer. However, this is a motion where a significant effect

$$-\rho_0\,\partial(\overline{u^2})/\partial x \tag{214}$$

of longitudinal Reynolds stress arises from the stronger *axial variation* of the mean square axial velocity u. Calculation of that effect by closely similar methods gives a contribution to $(\bar{u})_{\mathrm{ex}}$ just twice as much as (213). Finally, then, adding the two together,

$$(\bar{u})_{\mathrm{ex}} = -\tfrac{3}{4}\omega^{-1}u_{\mathrm{ex}}(\mathrm{d}u_{\mathrm{ex}}/\mathrm{d}x). \tag{215}$$

Equation (215) is a famous conclusion of Rayleigh which shows that in standing waves the streaming motion at the edge of the boundary layer is always *towards the nodes** (positions where the velocity amplitude u_{ex} is zero). This explains why dust particles tend to accumulate at the nodes: a phenomenon used in the nineteenth century to determine accurately the distance between nodes and hence the velocity of sound (figure 85).

Standing waves outside a boundary layer are observed also when a travelling sound wave is incident upon a fixed compact obstacle. We know (section 1.10) that, if in the sound wave the velocity

$$\mathbf{u} = \mathbf{u}_1 \exp(i\omega t) \tag{216}$$

would be found at the position of the obstacle if that were absent, then with it present the nearby velocity field is practically the divergenceless flow of

* A motion balanced as before, of course, by a central return flow.

an oscillating stream (216) around the obstacle. This velocity field possesses an oscillating boundary layer of the general form (205), where z is distance from the obstacle's surface. Here again, apart from the $\exp(i\omega t)$ factor, the external flow velocity u_{ex} is a real function of position on the surface.

This is a boundary layer with an extra term $-z(du_{ex}/dx)$ in w, needed additionally to the previous term (212) so that the local velocity field has divergence zero (instead of being just *constant* across the boundary layer). Actually, however, this extra term is found to make no difference to the calculated mean velocity $\bar{u}$ on the edge of the boundary layer (although it alters the distribution *within* the boundary layer, producing an interesting reversed flow very near the surface). Accordingly, Rayleigh's law (215) still applies. It means that the streaming motion is always *towards the stagnation points*.

For example, a fixed circular cylinder of radius a responding to plane waves perpendicular to its axis has an external-flow distribution

$$u_{ex} = 2u_1 \sin(x/a) = 2u_1 \sin\theta \qquad (217)$$

if $x = a\theta$ is a boundary-layer coordinate measured along the surface from the front stagnation point. Then the mean flow at the edge of the boundary layer is given by (215) as

$$(\bar{u})_{ex} = -\tfrac{3}{2}(\omega a)^{-1}u_1^2 \sin 2\theta; \qquad (218)$$

this flow, as usual, is directed towards the stagnation points ($\theta = 0$ and π).

In low-Reynolds-number hydrodynamics the associated general streaming motion around the cylinder takes the form illustrated in figure 86, with azimuthal velocity falling off as the inverse cube of distance from the axis. By contrast, if the Reynolds number based on the streaming velocity (218) were substantial, inertial effects would tend to confine that streaming motion within a rather thinner region...

The energy of internal waves is attenuated (figure 87) by viscous action at a rate

$$\nu(k^2 + l^2 + m^2) \qquad (219)$$

per unit time, or in terms of the energy propagation velocity (91) at a rate

$$\beta = \nu(k^2 + l^2 + m^2)^2 (Nm)^{-1} \qquad (220)$$

per unit distance along rays. This leads to a mean force per unit volume (182) which is directed along rays and has magnitude βW in terms of the

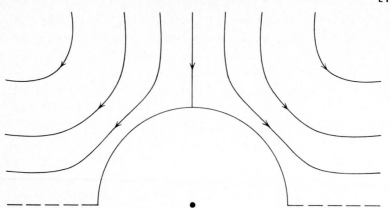

Figure 86. Slow streaming around a fixed circular cylinder due to a sound wave propagating in the direction indicated by a broken line (about which the complete streaming pattern is symmetrical). The total streaming flow in each quadrant is divided into four equal parts by the streamlines shown.

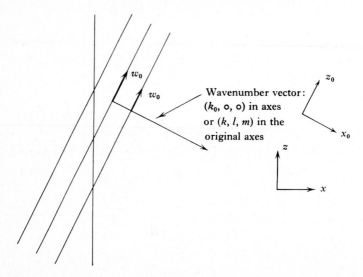

Figure 87. Illustrating viscous attenuation of internal waves. The fluid motion, by figure 72, is in the direction of steepest ascent in surfaces of constant phase. In axes (x_0, y_0, z_0) for which this is the z_0-direction and the wavenumber vector is in the x_0-direction, the fluid velocity is $(0, 0, w_0)$ and the shear is $(-\partial w_0/\partial x_0)$. Therefore, the rate of energy dissipation per unit volume is the mean of $\mu(\partial w_0/\partial x_0)^2$, where μ is the viscosity. The total wave energy is the mean of $\rho_0 w_0^2$ (twice the kinetic energy) per unit volume. The ratio of energy dissipation rate to total wave energy is therefore νk_0^2, where $\nu = \mu/\rho_0$ is the kinematic viscosity. In the original axes, this takes the value (219).

wave energy density W. The vertical component of this mean force is readily balanced by that gravitational restoring force which opposes a small mean upward displacement of fluid. The horizontal component is then available as a powerful source of horizontal streaming.

Under laboratory conditions, however, it is easy for horizontal pressure gradients to be set up in a finite container that will cancel such a horizontal force. In this case, additional vertical displacements generate gravitational restoring forces to cancel also that pressure distribution's vertical gradient, and streaming ceases!

4.8   Stationary phase in three dimensions

The principal aim of chapter 4 is to analyse the general properties of waves in anisotropic dispersive systems subject to linear equations. Our introduction to anisotropic dispersion (section 4.4) deals with systems that are homogeneous but otherwise of a rather general nature. For such systems, it concentrates on waves that have become rather substantially dispersed, so that those at each position are roughly sinusoidal with a more or less well-defined local wavenumber. Then a phase function can be defined, with properties leading to rules for ray tracing and energy propagation. That method is generalised in sections 4.5, 4.6 and 4.7; first, to nonhomogeneous systems, and then to waves that interact with a mean flow.

Now it is time to investigate more fully the nature of anisotropic dispersion. We aim to be able to analyse a complicated disturbance which, initially, may have no similarity at all to a sine wave, although as time goes on we may expect its components with different wavenumbers to become substantially separated from one another (dispersed). Such an aim, for strictly one-dimensional propagation, was achieved in section 3.7 through a Fourier analysis of the disturbance as a linear combination of sinusoidal components, followed by an asymptotic evaluation for large t by the method of stationary phase.

The present section makes a similar use of three-dimensional Fourier analysis and of a theory of stationary phase in three dimensions to determine the asymptotic development of waves from a complicated initial disturbance in an anisotropic system subject to linear equations. Just as in section 3.7, however, the fact that Fourier analysis is to be used restricts us to *homogeneous* systems (typically, those satisfying equations with *constant* coefficients) so that each individual Fourier component (a sine wave of constant amplitude) can separately be a solution of the equations of motion.

Subject to that limitation we shall find that the method describes satisfactorily the effects of those early stages of dispersion to which the method
of section 4.4 could not be applied; and, thereafter, gives results in agreement with ray-tracing methods. These facts suggest, perhaps, a hybrid
approach to the analysis of the dispersion of an initially complicated
localised disturbance in a nonhomogeneous system: the method of the
present section can disentangle the effects of the early stages of dispersion
(neglecting any nonhomogeneity present within the limited region occupied
by waves during those stages), and then the theory of rays in nonhomogeneous systems (section 4.5) can trace the waves' subsequent development.

As an illustration, we use the general theory to analyse how a complicated
initial disturbance in a stably stratified fluid is dispersed in the form of
internal gravity waves. Also, we derive analogous results for two-dimensional
propagation and use them to demonstrate certain properties (foreshadowed
in chapter 3) of the dispersion of a storm-like disturbance to the ocean
surface.

Mathematically, we are concerned with *initial-value problems* (for a
complementary type of problem where waves are generated by a source
oscillating at a steady frequency, see section 4.9). We assume that a sinusoidal oscillation

$$q = a \exp\left[i(\omega t - k_j x_j)\right] \tag{221}$$

with constant amplitude a is a solution of the equations of motion whenever
the homogeneous dispersion relationship

$$\omega = \omega(k_1, k_2, k_3) \tag{222}$$

is satisfied. Then the three-dimensional Fourier integral

$$q = \tfrac{1}{2} \int_{-\infty}^{\infty} \int_{-\infty}^{\infty} \int_{-\infty}^{\infty} Q_0(k_1, k_2, k_3) \exp\left\{i[\omega(k_1, k_2, k_3)\, t - k_j x_j]\right\} dk_1\, dk_2\, dk_3 \tag{223}$$

represents a rather general solution of those equations, taking the initial
form

$$q_0(x_1, x_2, x_3) = \tfrac{1}{2} \int_{-\infty}^{\infty} \int_{-\infty}^{\infty} \int_{-\infty}^{\infty} Q_0(k_1, k_2, k_3) \exp\left(-ik_j x_j\right) dk_1\, dk_2\, dk_3 \tag{224}$$

at time $t = 0$. In the case when this initial disturbance is confined to a
limited region, we investigate the asymptotic behaviour of (223) for large t.

Here we are including the factors $\tfrac{1}{2}$ because in the right-hand sides of
(223) and (224) each wavenumber (k_1, k_2, k_3) is *duplicated*, effectively

reappearing as $(-k_1, -k_2, -k_3)$. In any practical application it is advisable therefore to work with only half the wavenumber space (just as in section 3.7 we worked with the half $0 < k < \infty$ of the one-dimensional wavenumber space): any half will do which excludes $(-k_1, -k_2, -k_3)$ whenever it includes (k_1, k_1, k_3). Then, because the inverse Fourier transform

$$Q_0(k_1, k_2, k_3) = \tfrac{1}{4}\pi^{-3} \int_{-\infty}^{\infty} \int_{-\infty}^{\infty} \int_{-\infty}^{\infty} q_0(x_1, x_2, x_3) \exp(ik_j x_j) \, dx_1 \, dx_2 \, dx_3$$

(225)

has $Q_0(-k_1, -k_2, -k_3)$ as its complex conjugate, q_0 is equal to the *real part* of the integral in (224) taken only over that half-space. For example, with internal waves it may be convenient to use the half-space $-\infty < k_3 < 0$; this gives

$$q_0 = \int_{-\infty}^{\infty} \int_{-\infty}^{\infty} \int_{-\infty}^{0} Q_0(k_1, k_2, k_3) \exp(-ik_j x_j) \, dk_1 \, dk_2 \, dk_3 \qquad (226)$$

in accordance with the convention that any physical quantity equated to a complex expression is given by its real part. Exactly the same remarks are true of (223), where we may always take

$$\omega(-k_1, -k_2, -k_3) = -\omega(k_1, k_2, k_3) \qquad (227)$$

since necessarily the complex conjugate of (221) is also a solution.

We also need to remark, as in section 3.7, that for *given* (k_1, k_2, k_3) a dispersion relationship does not necessarily have just one solution (222) for ω. For example, with internal waves satisfying (24) it has two solutions equal in magnitude but opposite in sign; indeed, for a wavenumber in the half-space $-\infty < k_3 < 0$ the solutions with ω positive and negative correspond to wave energy propagating upwards and downwards respectively. For this system, we prove at the end of the section that general initial conditions can always be satisfied by a linear combination (223) of waves with positive ω plus a similar combination of waves with negative ω; together with (in general) the addition of a *steady* motion which, having velocities purely horizontal and divergenceless, cannot excite any restoring forces. Apart from any such steady component, the asymptotic evaluation of each of the contributions (223) corresponding to two different solutions (222) of the dispersion relationship poses identical mathematical problems, and we can concentrate on just one of them (for example, the internal-wave solution with positive ω) in developing the theory.

The method is a direct generalisation of that in section 3.7: we write the phase as $t\psi$, where

$$\psi(k_1, k_2, k_3) = \omega(k_1, k_2, k_3) - (k_j x_j/t); \qquad (228)$$

then (223) becomes

$$q = \tfrac{1}{2} \int_{-\infty}^{\infty} \int_{-\infty}^{\infty} \int_{-\infty}^{\infty} Q_0(k_1, k_2, k_3) \exp\left[it\psi(k_1, k_2, k_3)\right] dk_1\, dk_2\, dk_3, \quad (229)$$

which we seek to evaluate as t becomes large for a fixed value of x_j/t; that is, at a point moving away from the origin at a fixed vector velocity. We assume that ω, and hence ψ, is an analytic function of its variables. We note as before that Q_0, being the inverse Fourier transform (225) of a function which vanishes outside the region of the initial disturbance, is also an analytic function.

We estimate (229) by making a small imaginary shift in the ensemble of integration so that wherever possible the imaginary part of ψ becomes at least $+\delta$. Then the modulus of the exponential in (229) becomes at most $\exp(-t\delta)$, making the integral exponentially small for large t.

For this purpose, a deformation in which

$$k_j \quad \text{becomes} \quad k_j + g_j(k_1, k_2, k_3)\, i\delta \quad (230)$$

can give ψ, approximately, the imaginary part

$$(\partial\psi/\partial k_j)\, g_j\, \delta. \quad (231)$$

In any range of integration where the magnitude of the vector $(\partial\psi/\partial k_j)$ has a positive lower bound (so that *stationary* values of the phase ψ are shunned), the magnitude of g_j, needed to make (231) at least $+\delta$, is bounded above.

This means, first, that where ψ has no stationary values the whole integral (229) vanishes exponentially for large t. Furthermore, where ψ does have one or more stationary values, the integral can be estimated in terms of a contribution from a small ensemble of integration around each, outside which the magnitude of the vector $(\partial\psi/\partial k_j)$ has a positive lower bound.

We assumed above that the integral is unchanged in value by the deformation (230); this follows from general results for analytic functions of several complex variables, but we may perhaps indicate a simple *ad hoc* proof. For small δ, the deformation changes the volume element $dk_1\, dk_2\, dk_3$ by a factor

$$1 + (\partial g_j/\partial k_j)\, i\delta + O(\delta^2) \quad (232)$$

as is familiar to students of fluid dynamics from the physical significance of the divergence $\partial g_j/\partial k_j$, or may be obtained as the expansion to order δ of the Jacobian

$$\partial(k_1 + g_1\, i\delta, k_2 + g_2\, i\delta, k_3 + g_3\, i\delta)/\partial(k_1, k_2, k_3) \quad (233)$$

of the transformation (230). Accordingly, with $O(\delta^2)$ neglected, the integral

$$\int_{-\infty}^{\infty}\int_{-\infty}^{\infty}\int_{-\infty}^{\infty} f(k_1, k_2, k_3)\, dk_1\, dk_2\, dk_3 \qquad (234)$$

of any analytic function $f(k_1, k_2, k_3)$ changes as a result of the deformation (230) to

$$\int_{-\infty}^{\infty}\int_{-\infty}^{\infty}\int_{-\infty}^{\infty} [f + (\partial f/\partial k_j)g_j\, i\delta]\,[1 + (\partial g_j/\partial k_j)\, i\delta]\, dk_1\, dk_2\, dk_3. \qquad (235)$$

The rate of change with respect to δ is therefore

$$i\int_{-\infty}^{\infty}\int_{-\infty}^{\infty}\int_{-\infty}^{\infty} [\partial(fg_j)/\partial k_j]\, dk_1\, dk_2\, dk_3, \qquad (236)$$

which vanishes provided (as we assume) that at infinity fg_j is small of sufficient order (actually, $o(k_1^2 + k_2^2 + k_3^2)^{-1}$). The fact that at each stage of any deformation the rate of change of the integral with respect to δ is zero means that no change at all in the value of the integral can occur.

Having established by such a deformation that the only contribution to (229) is from a small ensemble of integration E around any stationary value of ψ, we calculate that contribution as in section 3.7. Near a stationary value

$$k_j = k_j^{(0)}, \quad \text{where} \quad \partial\psi/\partial k_j = 0, \qquad (237)$$

the expansion of ψ is as

$$\psi = \psi^{(0)} + \tfrac{1}{2}(k_i - k_i^{(0)})(k_j - k_j^{(0)})(\partial^2\omega/\partial k_i\, \partial k_j)^{(0)} + O(|k_j - k_j^{(0)}|^3), \qquad (238)$$

where the index (0) on ψ or its second derivative (which by (228) is the *same* as the second derivative of ω) indicates the value at $(k_1^{(0)}, k_2^{(0)}, k_3^{(0)})$. Then, as in section 3.7, the integral in (229), taken over the local ensemble of integration E, can be approximated as

$$\iiint_E Q_0^{(0)} \exp\left[it\psi^{(0)} + \tfrac{1}{2}it(k_i - k_i^{(0)})(k_j - k_j^{(0)})(\partial^2\omega/\partial k_i\, \partial k_j)^{(0)}\right] dk_1\, dk_2\, dk_3, \qquad (239)$$

with error factors $1 + O(|k_j - k_j^{(0)}|)$ from replacing Q_0 by its value at the position $k_j^{(0)}$ of stationary phase, and $1 + O(t|k_j - k_j^{(0)}|^3)$ from the error in (238).

The exponential of the quadratic function of k_j occurring in (239) does not lend itself to easy integration as it stands. However, a simple *rotation of axes* makes the integration very easy indeed. We rotate the axes of k_1, k_2, k_3 so that they become the principal axes for the tensor

$$(\partial^2\omega/\partial k_i\, \partial k_j)^{(0)}; \qquad (240)$$

that is, the axes in which it is a diagonal matrix (all its values for $i \neq j$ are zero). Then the integral (239) splits up into a *product*

$$Q_0^{(0)} \exp(it\psi^{(0)})\prod_{j=1}^{3}\left\{\int \exp\left[\tfrac{1}{2}it(k_j - k_j^{(0)})^2(\partial^2\omega/\partial k_j^2)^{(0)}\right] dk_j\right\} \qquad (241)$$

of three one-dimensional integrals. Each of these should be evaluated over a deformed range of integration which brings the modulus of the integrand in (241) down to less than $\exp(-t\delta)$ so that the whole boundary of the local ensemble of integration E can be continuous with the remaining deformed ensemble of integration in which that is true.

We saw in section 3.7 that each one-dimensional integration of the type occurring in (241) is carried out most expeditiously by the deformation

$$k_j - k_j^{(0)} = s_j \exp\left[\tfrac{1}{4}\pi i \operatorname{sgn}(\partial^2\omega/\partial k_j^2)^{(0)}\right], \tag{242}$$

which turns it into a Gaussian integral with error $O[\exp(-t\delta)]$. This gives the value

$$Q_0^{(0)} \exp(it\psi^{(0)}) \prod_{j=1}^{3} \left\{[2\pi/t|\partial^2\omega/\partial k_j^2|^{(0)}]^{\frac{1}{2}} \exp\left[\tfrac{1}{4}\pi i \operatorname{sgn}(\partial^2\omega/\partial k_j^2)^{(0)}\right]\right\} \tag{243}$$

for (241) with an error factor $1 + O(t^{-\frac{1}{2}}) + O[\exp(-t\delta)]$ in each term of the product. (We postpone a discussion of modifications when any of the principal second derivatives $(\partial^2\omega/\partial k_j^2)^{(0)}$ at the stationary point is zero to section 4.11.)

Although the result (243) has been derived in special axes only (the principal axes of the tensor (240)) it is easily written in a form invariant under rotation of axes and *therefore* equally valid in the original system of wavenumber axes. We write the product of the three phase factors

$$\exp\left[\tfrac{1}{4}\pi i \operatorname{sgn}(\partial^2\omega/\partial k_j^2)^{(0)}\right] \quad \text{as} \quad \exp(i\Theta),$$

where Θ characterises the nature of the stationary point $k_j^{(0)}$ as follows:

$\Theta = \tfrac{3}{4}\pi$ at a minimum, $-\tfrac{3}{4}\pi$ at a maximum,
and $\tfrac{1}{4}\pi$ or $-\tfrac{1}{4}\pi$ at a saddle-point according as the function (228)
increases along two or only one of the three principal directions. (244)

Furthermore, we write the product of the three second derivatives in (243) as the value $J^{(0)}$ at $k_j = k_j^{(0)}$ of the Jacobian determinant

$$J = \partial(U_1, U_2, U_3)/\partial(k_1, k_2, k_3) \tag{245}$$

of the group velocity vector $U_j = \partial\omega/\partial k_j$ with respect to the vector k_j. This is the determinant of the matrix (240), which in principal axes has zero off-diagonal elements, making $J^{(0)}$ just the product of its three diagonal elements $(\partial^2\omega/\partial k_j^2)^{(0)}$. At the same time, $J^{(0)}$ is invariant under rotation of axes because it represents the expansion of a volume element in a change of variables from k_j to U_j.

Using Θ and $J^{(0)}$ in (243), and substituting also for $\psi^{[0]}$ from (228), we derive the asymptotic form of the integral in (229) for large t as

$$q = Q_0^{(0)}(2\pi/t)^{\frac{3}{2}}|J^{(0)}|^{-\frac{1}{2}}\exp\{i[\omega(k_1^{(0)}, k_2^{(0)}, k_3^{(0)})\,t - k_j^{(0)}\,x_j + \Theta]\} \qquad (246)$$

if there is just one stationary point $k_j^{(0)}$; and, otherwise, as the sum of contributions (246) from each stationary point.

Actually, (229) makes q equal to $\frac{1}{2}$ of the complete infinite integral which has been thus estimated. Since, however, we do not wish to regard $-k_j^{(0)}$ as a truly different stationary point from $k_j^{(0)}$, we suppress the factor $\frac{1}{2}$ and take q as equal to the sum of terms (246) (signifying in each case the real part) for only those wavenumbers with ψ stationary that lie within a prescribed half-space.

By (235), these stationary points are where

$$x_j/t = \partial\omega/\partial k_j = U_j; \qquad (247)$$

that is, they move at the group velocity. This gives a simple explanation of why the Jacobian (245) appears in the asymptotic solution (246).

In, for example, a system where the wave energy density for a sinusoidal wave (221) is obtained by multiplying the mean square ($\frac{1}{2}|a|^2$) of q by some factor $W_0(k_1, k_2, k_3)$, the initial wave energy for the disturbance (224) associated with an element $dk_1\,dk_2\,dk_3$ of the wavenumber half-space is

$$W_0(k_1, k_2, k_3)\,4\pi^3|Q_0|^2\,dk_1\,dk_2\,dk_3. \qquad (248)$$

This is because Parseval's theorem makes $2\pi^3|Q_0|^2\,dk_1\,dk_2\,dk_3$ the contribution to the mean square of (224) from each element of the *whole* wavenumber space. At time t, when waves of wavenumber k_j are found at the position (247), those within the element $dk_1\,dk_2\,dk_3$ fill a region of volume

$$|\partial(U_1 t, U_2 t, U_3 t)/\partial(k_1, k_2, k_3)|^{(0)}\,dk_1\,dk_2\,dk_3 = t^3|J^{(0)}|\,dk_1\,dk_2\,dk_3. \qquad (249)$$

The ratio of (248) to (249) specifies the energy *density* as

$$W_0(k_1, k_2, k_3)\,4\pi^3|Q_0|^2\,t^{-3}|J^{(0)}|^{-1}. \qquad (250)$$

An excellent check on the work is given by the fact that (250) is $W_0(k_1, k_2, k_3)$ times the mean square of the asymptotic solution (246); where, indeed, only the information on the phase shift Θ was inaccessible by such energy arguments. However, the more rigorous approach of this section does confirm that results derived by assuming energy to be propagated at the group velocity are not vitiated by the need for a complicated disturbance to become 'disentangled' in the early stages of dispersion.

Internal waves give an excellent illustration of our results. The rule (244) makes Θ always equal to $+\frac{1}{4}\pi$ for internal waves. This follows from the fact that in wavenumber space, by (24), the surfaces $\omega = $ constant are cones with vertical axis, and with the value of ω increasing as a function of cone angle. Everywhere, therefore, the *azimuthal* direction is a principal direction in which ω is *increasing*. In the meridional plane, we can calculate the two principal directions but it is unnecessary: the function (228) (ω minus a linear approximation) must *increase in one of them and decrease in the other* because there is a direction (the cone's generator) along which it remains constant. To sum up, the phase increases in just two out of the three principal directions so that $\Theta = \frac{1}{4}\pi$.

For internal waves we use a wavenumber vector (k, l, m). A straight-forward, although lengthy, calculation of J, defined in (245), from the group velocity components (91), gives

$$J = - N^3 m^4 (k^2 + l^2)^{-\frac{1}{2}} (k^2 + l^2 + m^2)^{-\frac{9}{2}}. \tag{251}$$

The calculation is shorter if we represent wavenumbers in the half-space by spherical polar coordinates so that

$$k = K \cos\theta \cos\psi, \quad l = K \cos\theta \sin\psi, \quad m = -K \sin\theta \tag{252}$$

(just as in (175) but with k_{H} replaced by $K \cos\theta$). This simplifies the group velocity vector (91), so that at time t the position $\mathbf{U}t$ of the waves with wavenumber (k, l, m) has coordinates

$$x = NtK^{-1}\sin^2\theta \cos\psi, \quad y = NtK^{-1}\sin^2\theta \sin\psi, \quad z = NtK^{-1}\sin\theta \cos\theta. \tag{253}$$

Then

$$t^3 J = \partial(x, y, z)/\partial(k, l, m) = [\partial(x, y, z)/\partial(K, \theta, \psi)][\partial(k, l, m)/\partial(K, \theta, \psi)]^{-1}$$

$$= [-(Nt)^3 K^{-4}\sin^4\theta][K^2\cos\theta]^{-1} = -(Nt)^3 K^{-6}\sin^4\theta \sec\theta, \tag{254}$$

in agreement with (251).

Equation (246) now implies that

$$q = [Q^{(0)}_0(K\cos\theta\cos\psi, K\cos\theta\sin\psi, -K\sin\theta)](2\pi/Nt)^{\frac{3}{2}}(K^3\operatorname{cosec}^2\theta\cos^{\frac{1}{2}}\theta)$$

$$\times \exp\{i[Nt\cos\theta - (x\cos\psi + y\sin\psi)K\cos\theta + zK\sin\theta + \tfrac{1}{4}\pi]\}. \tag{255}$$

For waves of given wavenumber this emphasises an amplitude decrease like $t^{-\frac{3}{2}}$, as equations (253) for the position of those waves show their energy filling a volume expanding like t^3. Their crests move, of course, transversely to the radius vector (253).

It is interesting to express the local amplitude of q (say, q_1, given by the first line of (255)) in terms of position coordinates, themselves written in polar form

$$x = r \sin \theta \cos \psi, \quad y = r \sin \theta \sin \psi, \quad z = r \cos \theta. \tag{256}$$

Here, by (253),

$$r = NtK^{-1} \sin \theta. \tag{257}$$

Then the wave amplitude is

$$q_1 = Q^{(0)}_0(Ntr^{-1} \sin \theta \cos \theta \cos \psi, Ntr^{-1} \sin \theta \cos \theta \sin \psi, -Ntr^{-1} \sin^2 \theta)$$
$$\times (2\pi Nt)^{\frac{3}{2}} r^{-3} \sin \theta \cos^{\frac{1}{2}} \theta. \tag{258}$$

This formula indicates a slight preference for propagation at angles θ to the vertical intermediate between 0 and $\frac{1}{2}\pi$. At any one time t, it shows how the amplitude has an inverse-cube rate of decay for large r, where the first line becomes $Q^{(0)}_0(0, 0, 0)$, proportional by (225) to the *integral* of the initial disturbance. At a fixed position, on the other hand, the signal *grows* with time until some wavenumber cutoff $K_{\max}$ (beyond which $Q^{(0)}_0$ is negligible) is reached at time $rN^{-1}K_{\max} \operatorname{cosec} \theta$.

Here we may remark that even if no strong cutoff is present in the initial conditions it will be provided nevertheless by an additional multiplying factor $\exp(-\frac{1}{2}\nu K^2 t)$ in (258) representing energy attenuation at the rate (219). (Such an attenuation factor can, as in section 3.7, be directly incorporated into the integral (223) and hence into its asymptotic form (246).)

We have described above the upward-travelling internal waves (specified by positive ω). However, the solution of equations (16)–(22) for general initial conditions involves also downward-travelling internal waves (specified by negative ω). For example, given the initial values of *both* the vertical displacement *and* the vertical velocity (or, equivalently, of ρ_e and q, which are related by (21)), we can determine the subsequent development of q as the sum of a term (223) with positive ω and another such term with different amplitude function and with negative ω; and (21) implies a similar form for ρ_e.

Such solutions for ρ_e and q determine also p_e through (18). At the same time they determine u and v, provided that the linear equations of motion may be assumed to have been satisfied while the initial disturbance was being set up. This is because equations (16) imply that

$$\partial(\partial v/\partial x - \partial u/\partial y)/\partial t = 0; \tag{259}$$

thus, the vertical component of vorticity cannot change and, being zero in the undisturbed state, must remain zero. Then, for each z, the two-dimensional vector field (u, v) is specified uniquely by its zero curl and by its divergence

$$\partial u/\partial x + \partial v/\partial y = -\rho_0^{-1}\,\partial q/\partial z \qquad (260)$$

known from (17).

On the other hand, if more general initial conditions are allowed, they can generate, by (259), an arbitrary value of $\partial v/\partial x - \partial u/\partial y$ independent of the time. Then, u and v can differ from the irrotational solution of (260) by an arbitrary horizontal motion which is *both* divergenceless *and* steady. There is no restoring force associated with that motion which can continue undisturbed while the wave energy is propagated away as analysed above.

Surface waves, and certain other waves in fluids like those on interfaces (section 4.1), are propagated two-dimensionally. For such cases, the analysis of this section goes through with the products in (241) and (243) involving two terms instead of three. This makes

$$\Theta = \tfrac{1}{2}\pi \text{ at a minimum, } -\tfrac{1}{2}\pi \text{ at a maximum,}$$
$$\text{o at a saddle-point;} \quad \text{and} \quad J = \partial(U_1, U_2)/\partial(k_1, k_2); \qquad (261)$$

while (246) is replaced by

$$q = Q_0^{(0)}(2\pi/t)|J^{(0)}|^{-\frac{1}{2}}\exp\{i[\omega(k_1^{(0)}, k_2^{(0)})\,t - k_j^{(0)}\,x_j + \Theta]\}, \qquad (262)$$

with $k_j^{(0)}$ defined by (247).

It is interesting to note the simple form of these results for isotropic waves satisfying

$$\omega = \omega(k), \quad \text{with} \quad k = (k_1^2 + k_2^2)^{\frac{1}{2}} \quad \text{and} \quad \omega'(k) = U(k). \qquad (263)$$

Then the condition (247) for $k_j = k_j^{(0)}$ becomes

$$x_j = U(k)\,t(k_j/k), \quad x = (x_1^2 + x_2^2)^{\frac{1}{2}} = U(k)\,t, \qquad (264)$$

confirming that the wavenumber is in the direction of propagation for isotropic waves, whose energy travels a distance $U(k)\,t$ in time t. Also, (261) becomes

$$\Theta = \tfrac{1}{4}\pi \operatorname{sgn}[U(k)] + \tfrac{1}{4}\pi \operatorname{sgn}[U'(k)]; \quad J = k^{-1}U(k)\,U'(k); \qquad (265)$$

in particular, $\Theta = 0$ for surface gravity waves. Finally, (262) can be written

$$q = x^{-\frac{1}{2}}(Q_0^{(0)}(2\pi k^{(0)})^{\frac{1}{2}})|2\pi/t\omega''(k^{(0)})|^{\frac{1}{2}}\exp\{i[\omega(k^{(0)})\,t - k^{(0)}x + \Theta]\}, \qquad (266)$$

confirming the statement at the beginning of section 3.7 that wave amplitudes in isotropic two-dimensional propagation vary with position and time exactly as in one-dimensional propagation but with an extra $x^{-\frac{1}{2}}$ factor. Here, the expression in large round brackets, determining the energy *per radian* in waves whose wavenumber has magnitude $k^{(0)}$, replaces the $F(k_0)$ of section 3.7.

4.9   General theory of oscillating sources of waves

We have just used three-dimensional Fourier analysis to indicate the asymptotic response of a homogeneous linear system in *free vibrations* resulting from a complicated initial disturbance in a limited region. The complementary analysis of *forced vibrations* (waves generated by an oscillating source of fixed frequency ω_0) is given in sections 4.9 and 4.10. The work is again restricted to homogeneous systems because Fourier analysis is being used. The asymptotic estimation is with respect, not to time, but to *distance from the source*. We shall see that at large distances the propagation is as described by ray-tracing methods; suggesting as before that non-homogeneous systems might be treated by a hybrid method: the present analysis would be used to generate a solution within a region of moderate dimensions around the source where nonhomogeneity is of little significance, while the waves' development beyond that region would be traced by the theory of rays in nonhomogeneous systems (section 4.5).

The following general approach gives useful results on the *directional distribution* of the wave energy generated by sources for many types of wave system: not only those exhibiting anisotropic dispersion (like internal waves, discussed in detail in section 4.10) but also *isotropic* systems like water waves, and even nondispersive systems like sound waves. In particular, the results of chapter 1 on compact sources of sound are recovered (and generalised to other systems) after which new results are derived on radiation from noncompact sound sources.

The method is extended later (section 4.12) to waves generated by *travelling* forcing effects oscillating at a fixed frequency ω_0. Their analysis necessitates a method catering for anisotropy, since (whatever the wave system) an effective anisotropy is introduced by the motion of the source. The case $\omega_0 = 0$ (a nonoscillating moving source) includes the problems of sections 3.9 and 3.10, and in particular allows an estimate of the distribution of wave energy associated with the steady motion of a ship. Conversely, our present method leans heavily on an idea first introduced at the end of section 3.9.

We consider forcing by an oscillating source of frequency ω_0 whose spatial distribution is given as

$$f(x_1, x_2, x_3) \exp(i\omega_0 t) \qquad (267)$$

in a sense to be made precise. We use a Fourier analysis of f over the *whole* wavenumber space,

$$f(x_1, x_2, x_3) = \int_{-\infty}^{\infty} \int_{-\infty}^{\infty} \int_{-\infty}^{\infty} F(k_1, k_2, k_3) \exp(-ik_j x_j) \, dk_1 \, dk_2 \, dk_3, \qquad (268)$$

because when the frequency ω_0 is fixed (and no other frequency, even $-\omega_0$, is allowed) the wavenumbers (k_1, k_2, k_3) and $(-k_1, -k_2, -k_3)$ are truly different. For various methods of forcing we find below that a characteristic quantity q specifying the waves is given as a related Fourier integral

$$q = [\exp(i\omega_0 t)] \int_{-\infty}^{\infty} \int_{-\infty}^{\infty} \int_{-\infty}^{\infty} \frac{F(k_1, k_2, k_3) \exp(-ik_j x_j)}{B(\omega_0, k_1, k_2, k_3)} \, dk_1 \, dk_2 \, dk_3, \qquad (269)$$

where the function B is such that the equation

$$B(\omega, k_1, k_2, k_3) = 0 \qquad (270)$$

is a form of the dispersion relationship.

For water waves, the propagation is in two dimensions, so that x_3 and k_3 (and the integration with respect to k_3) should be suppressed above. Actually, in a somewhat general system of two-dimensionally propagating waves, we can envisage that for oscillations of the form

$$q = a \exp[i(\omega t - k_1 x_1 - k_2 x_2)], \qquad (271)$$

with arbitrary ω, k_1 and k_2, a certain boundary value η (on some bounding plane $x_3 = $ constant) might be calculated as

$$\eta = aB(\omega, k_1, k_2) \exp[i(\omega t - k_1 x_1 - k_2 x_2)]. \qquad (272)$$

Then, if the *undisturbed* dispersive system is defined by the boundary condition $\eta = 0$, the dispersion relationship can be written

$$B(\omega, k_1, k_2) = 0. \qquad (273)$$

On the other hand, forcing by an oscillatory boundary disturbance

$$\eta = f(x_1, x_2) \exp(i\omega_0 t) = \int_{-\infty}^{\infty} \int_{-\infty}^{\infty} F(k_1, k_2)$$
$$\times \exp[i(\omega_0 t - k_1 x_1 - k_2 x_2)] \, dk_1 \, dk_2 \qquad (274)$$

would (as in section 3.9) generate the response

$$q = [\exp(i\omega_0 t)] \int_{-\infty}^{\infty} \int_{-\infty}^{\infty} \frac{F(k_1, k_2) \exp[-i(k_1 x_1 + k_2 x_2)]}{B(\omega_0, k_1, k_2)} \, dk_1 \, dk_2 \qquad (275)$$

since (271) and (272) imply that the value of η for each frequency and wavenumber must be divided by B to give q.

In three-dimensional wave systems, forcing effects are more commonly described by a differential equation. This leads, however, to the same Fourier-integral form of solution (269) if we can write the equation as

$$B\left(-i\frac{\partial}{\partial t}, i\frac{\partial}{\partial x_1}, i\frac{\partial}{\partial x_2}, i\frac{\partial}{\partial x_3}\right)q = f(x_1, x_2, x_3)\exp(i\omega_0 t), \qquad (276)$$

where $B(\omega, k_1, k_2, k_3)$ is a *polynomial* in its four variables (or, more generally, a ratio of polynomials). Then free vibrations satisfy the differential equation (276) with zero on the right-hand side, so that any plane-wave solution

$$q = a\exp[i(\omega t - k_j x_j)] \qquad (277)$$

is subject to the dispersion relationship (270). In the general case, however, when the source term defined by (267) and (268) appears as a forcing function on the right-hand side of (276), a solution may be written in the form (269).

Exactly the same 'important and famous difficulty' as was described at the end of section 3.9 arises when we seek to evaluate (269) or (275). The integrations with respect to the real variables k_j encounter *poles* wherever

$$B(\omega_0, \mathbf{k}) = 0. \qquad (278)$$

We describe this equation (the dispersion relationship for $\omega = \omega_0$) as defining the *wavenumber surface* S for frequency ω_0 (a surface in wavenumber space); or, for two-dimensional wave systems, as defining a *wavenumber curve* S in the (k_1, k_2) plane. A serious ambiguity in the integral (269) arises from such poles within the ensemble of integration: different values of the integral can be obtained according to whether the k_1-integration, for example, is taken in the complex k_1-plane along a path to the left or right of a particular pole (or even as a combination of both in arbitrary proportions).

This mathematical ambiguity corresponds to a genuine physical ambiguity. To the 'forced' waves *actually* radiated from the source there might be added on an arbitrary linear combination of 'free' waves (solutions of the equations with $f = 0$). Their energy might be moving radially inwards *towards* the source region, having been generated 'at infinity'. Therefore, the mathematical problem specified by (274) or (276) fails to represent completely the physical problem of finding the waves generated by the source itself, subject to the '*radiation condition*' that no wave energy is being generated at infinity.

We should remark (as in section 3.9) that various alternative mathematical procedures are equivalent to this radiation condition: some use group-velocity ideas directly; others the fact that *attenuation* makes the wave-number complex for free waves of fixed real frequency ω_0. A process that will work even if attenuation is not taken into account is, however, desirable; and the idea (section 3.9) of the source of waves having *grown up slowly* to its present level turns out to be easily and generally applicable.

This means that the mathematical problem is solved with

$$\omega_0 \quad \text{replaced by} \quad \omega_0 - \mathrm{i}\epsilon. \tag{279}$$

Then, in particular,

$$\exp(\mathrm{i}\omega_0 t) \quad \text{is replaced by} \quad \exp(\epsilon t + \mathrm{i}\omega_0 t), \tag{280}$$

giving the source (267) a slow exponential growth from zero at $t = -\infty$ up to its present level. By looking exclusively for the associated growing waves, varying with time like $\exp(\epsilon t + \mathrm{i}\omega_0 t)$, we exclude the danger of the solution being contaminated by other wave energy generated 'at infinity'. (Here, ϵ is a small positive number which is later allowed to tend to zero.)

As in section 4.8, a *rotation of axes* proves convenient for deriving the results quickly and clearly. To find the directional distribution of wave energy we must obtain the asymptotic behaviour of (269) along an arbitrary straight line L, stretching out from the source region. To this end, whatever the direction of L, we make a preliminary rotation of axes so that L *becomes the positive x_1-axis*. Then we seek the asymptotic behaviour as $x_1 \to +\infty$ of the expression (269) for $x_2 = x_3 = 0$; that is, of

$$q = [\exp(\mathrm{i}\omega_0 t)] \int_{-\infty}^{\infty} \int_{-\infty}^{\infty} \mathrm{d}k_2 \, \mathrm{d}k_3 \int_{-\infty}^{\infty} \frac{F(k_1, k_2, k_3) \exp(-\mathrm{i}k_1 x_1)}{B(\omega_0, k_1, k_2, k_3)} \, \mathrm{d}k_1. \tag{281}$$

We estimate the inner integral in (281) by Cauchy's theorem, as in section 3.9. In the complex k_1-plane, the path of integration is *lowered* a distance κ_1; that is, k_1 is given a negative imaginary part $(-\kappa_1)$ so as to make the integral $O[\exp(-\kappa_1 x_1)]$. With an error of this order, the integral is equal to $(-2\pi\mathrm{i})$ times the sum of the residues of the integrand at any poles (where $B = 0$) passed over in deforming the path (figure 88).

The substitution (279) moves these poles off the real k_1-axis. Further-more, it moves them *below* the axis if and only if the dispersion relationship (270) defines ω as a function of $\mathbf{k}$ *such that* $\partial\omega/\partial k_1 > 0$. Then and only then do the poles contribute a residue term to the inner integral in (281) when the path is lowered. Physically, the condition limits us to waves where $\partial\omega/\partial k_1$, the component along L of the energy propagation velocity, is *outwards*

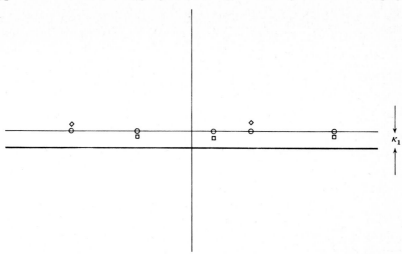

Figure 88. Displaced path of integration (thick line) in the complex k_1-plane for the inner integral in (281). ○ Positions of poles when $\epsilon = 0$; □ displaced position, when $\epsilon > 0$, of poles such that $\partial \omega / \partial k_1 > 0$; ◇ displaced position, when $\epsilon > 0$, of poles such that $\partial \omega / \partial k_1 < 0$. Only the poles □ contribute a residue term when the path is lowered.

from the source; thus, it excludes the possibility of energy propagating inwards.

We can now let $\epsilon \to 0$: the introduction of the $\exp(\epsilon t)$ factor in (280) has fulfilled its function of limiting the waves generated to solutions of the dispersion relationship (278) satisfying the condition for outward propagation of energy. We use S_+ to mean *that part of* S (the wavenumber surface or curve for $\omega = \omega_0$) on which

$$\partial \omega / \partial k_1 > 0. \tag{282}$$

Then the inner integral in (281) consists, with error $O[\exp(-\kappa_1 x_1)]$, of a sum of contributions $(-2\pi i)$ times the residue

$$F(k_1, k_2, k_3) \, [\partial B(\omega_0, k_1, k_2, k_3) / \partial k_1]^{-1} \exp(ik_1 x_1) \tag{283}$$

of the integrand from any pole satisfying (282); that is, from any point on S_+ with the given values of k_2 and k_3. The integral with respect to k_2 and k_3 can therefore be regarded as an integral over S_+:

$$q = -2\pi i [\exp(i\omega_0 t)] \iint_{S_+} F(k_1, k_2, k_3) \, [\partial B(\omega_0, k_1, k_2, k_3) / \partial k_1]^{-1}$$
$$\times \exp(-ik_1 x_1) \, dk_2 \, dk_3. \tag{284}$$

Equation (284) is a fundamental expression for the waves generated in a particular direction stretching outwards from the oscillating source, expressed in axes rotated so that that direction is the positive x_1-axis. In the rest of this section we approximate it further by the method of stationary phase in two important cases: (i) two-dimensional propagation, when the k_3 variable does not occur in (284) and S is a curve; (ii) three-dimensional propagation with S a surface curved in two directions (k_2 and k_3). In section 4.10, special treatments are given for cases when S either *lacks* curvature, being (iii) a plane; or has curvature in one direction only, being (iv) a generalised cylinder or (v) a generalised cone (as with internal waves). In the meantime, we defer to section 4.11 any particular problems of estimation along directions L that are *exceptional* in some sense for the surface S.

In case (i), *two-dimensional propagation*, with S a curve in the (k_1, k_2) plane, the integral (284) is readily estimated for large x_1 by the one-dimensional method of stationary phase described in section 3.7. The phase $(-k_1 x_1)$ is stationary on S_+ at any point

$$(k_1^{(0)}, k_2^{(0)}) \quad \text{where} \quad dk_1/dk_2 = 0. \tag{285}$$

Near such a stationary point $(k_1^{(0)}, k_2^{(0)})$, S_+ is defined by the equation

$$k_1 = k_1^{(0)} + \tfrac{1}{2}(d^2 k_1/dk_2^2)^{(0)} (k_2 - k_2^{(0)})^2 + O(|k_2 - k_2^{(0)}|^3). \tag{286}$$

Then the method of section 3.7 shows that from any such point there is an asymptotic contribution

$$-2\pi i[\exp(i\omega_0 t)] F(\mathbf{k}^{(0)}) [(\partial B(\omega_0, \mathbf{k})/\partial k_1)^{(0)}]^{-1} \exp(-i k_1^{(0)} x_1)$$
$$\times [2\pi/|d^2 k_1/dk_2^2|^{(0)} x_1]^{\frac{1}{2}} \exp[-\tfrac{1}{4}\pi i \operatorname{sgn}(d^2 k_1/dk_2^2)^{(0)}] \tag{287}$$

to (284) for large x_1 (cases where the curvature $|d^2 k_1/dk_2^2|^{(0)}$ is locally zero are postponed to section 4.11).

For convenience, this result should now be expressed in a form invariant under rotation of axes; then it can be applied to wave generation problems directly without any preliminary change of axes. In relation to an arbitrary direction L, the condition (285) for stationary phase specifies that $(k_1^{(0)}, k_2^{(0)})$ is a point of the wavenumber curve S where a normal $\mathbf{n}$ is in the direction of L. At any point of a curve there are two normal directions but, by the definition (282) of S_+, we see that L must be in the direction of

the normal $\mathbf{n}$ to S along which
ω is *increasing* above its value ω_0 taken on S. (288)

With this definition, we can replace $\partial B/\partial k_1$ by $\partial B/\partial n$. We also replace $|d^2 k_1/dk_2^2|^{(0)}$ by the *curvature* $\kappa^{(0)}$ of the wavenumber curve at $(k_1^{(0)}, k_2^{(0)})$; and define the phase term in (287), incorporating the $(-i)$ factor as well as the final exponential, as $\exp(i\Theta)$, where

$$\Theta = -\tfrac{1}{4}\pi \text{ or } -\tfrac{3}{4}\pi \text{ according as } S \text{ is curved convexly or concavely to } L. \quad (289)$$

With this notation,

$$q = F(k_1^{(0)}, k_2^{(0)}) \, [(\partial B/\partial n)^{(0)}]^{-1} (8\pi^3/\kappa^{(0)}x)^{\frac{1}{2}} \exp\{i[\omega_0 t - k_j^{(0)} x_j + \Theta]\} \quad (290)$$

is the contribution to q from any point $(k_1^{(0)}, k_2^{(0)})$ on the wavenumber curve satisfying (288).

Note that the waves found along the line L are identified by (288) as *those for which L is a ray*, since their group velocity $\partial \omega / \partial k_j$ is in the direction of L. Also, expression (290) is consistent with the idea of a uniform energy flux between adjacent rays since (figure 89) the distance between them varies with distance x from the source in proportion to $\kappa^{(0)}x$.

Water waves are a special example of two-dimensional propagation in which the wavenumber curve is a circle (with $\kappa^{(0)}$ taking the constant value $g\omega_0^{-2}$). We postpone using them for illustration, however, until the discussion of ship waves in section 4.12 which uses (290) in a less special form because motion of the source imposes an effective anisotropy.

In case (ii) *three-dimensional propagation*, with S a surface curved in two directions, the integral (284) is estimated by a simple extension of the above method. The phase $(-k_1 x_1)$ is stationary on S_+ at any point $(k_1^{(0)}, k_2^{(0)}, k_3^{(0)})$ where

$$\partial k_1/\partial k_2 = 0 \quad \text{and} \quad \partial k_1/\partial k_3 = 0. \quad (291)$$

Near such a stationary point the expansion of k_1 would in general include *not only* a term in $(k_2 - k_2^{(0)})^2$ as in (286), together with a corresponding term in $(k_3 - k_3^{(0)})^2$ *but also* a term in the cross-product $(k_2 - k_2^{(0)})(k_3 - k_3^{(0)})$. However, just as in section 4.8 we can first rotate the k_2- and k_3-axes (about the k_1-axis) so that the coefficient $(\partial^2 k_1/\partial k_2 \, \partial k_3)^{(0)}$ of that cross-product becomes zero. In such 'principal axes of curvature' for S, we have

$$k = k_1^{(0)} + \tfrac{1}{2}(\partial^2 k_1/\partial k_2^2)^{(0)} (k_2 - k_2^{(0)})^2 + \tfrac{1}{2}(\partial^2 k_1/\partial k_3^2)(k_3 - k_3^{(0)})^2 + O(|k_j - k_j^{(0)}|^3). \quad (292)$$

The asymptotic form of the integral (284) then breaks up into a product of integrals with respect to k_2 and to k_3. Its value consists of expression (287) *with the second line repeated* after a substitution of $(\partial^2 k_1/\partial k_3^2)^{(0)}$ for $(\partial^2 k_1/\partial k_2^2)^{(0)}$.

In a form invariant under rotation of axes, the condition (291) states that $k_j^{(0)}$ is a point of S where a normal **n** is in the direction L; and, by the definition

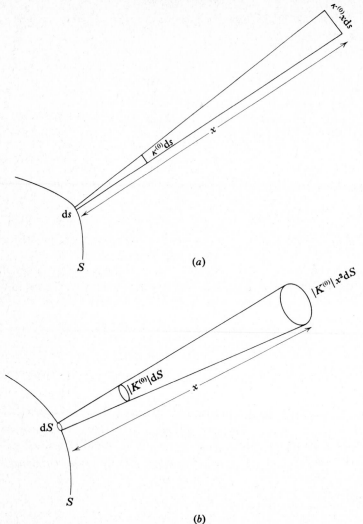

Figure 89. Spreading of the energy that travels between neighbouring directions normal to S.

(a) Two-dimensional case: two normal directions include an angle $\kappa^{(0)}ds$, where ds is a length-element of the wavenumber curve S and $\kappa^{(0)}$ is its curvature; the separation between them is $\kappa^{(0)}x\,ds$ at a large distance x.

(b) Three-dimensional case: a cone of normals to the wavenumber surface S includes a solid angle $|K^{(0)}|\,dS$, where dS is an area-element of S and $K^{(0)}$ is its Gaussian curvature; the cross-sectional area of the cone is therefore $|K^{(0)}|x^2\,dS$ at a large distance x.

(282) of S_+ this normal direction must, as in (288), be one in which ω is *increasing* above its value ω_0 taken on the wavenumber surface. A new feature, however, is that the product

$$(\partial^2 k_1/\partial k_2^2)^{(0)} (\partial^2 k_1/\partial k_3^2)^{(0)} = K^{(0)} \tag{293}$$

of the two principal curvatures of the wavenumber surface appears (its modulus being raised to the $(-\frac{1}{2})$th power) when the second line of (287) is repeated as specified above. This product $K^{(0)}$ is the famous measure of the local curvedness of the surface S known as its *Gaussian curvature*; we postpone to section 4.11 consideration of places on the surface where $K^{(0)} = 0$. The phase term in (287), incorporating the $(-i)$ factor as well as the sgn term and the repeated sgn term, becomes $\exp(i\Theta)$ with

$\Theta = 0$ or $-\frac{1}{2}\pi$ or $-\pi$ according as both or one or none of
the principal curvatures of S are curved convexly to L. (294)

With this notation,

$$q = F(k_1^{(0)}, k_2^{(0)}, k_3^{(0)}) \, [(\partial B/\partial n)^{(0)}]^{-1} \, (4\pi^2/|K^{(0)}|^{\frac{1}{2}} x) \exp\{i[\omega_0 t - k_j^{(0)} x_j + \Theta]\} \tag{295}$$

is the contribution to q from any point $(k_1^{(0)}, k_2^{(0)}, k_3^{(0)})$ on the wavenumber surface satisfying (288).

This very general result for three-dimensional propagation from a source oscillating at frequency ω_0 identifies the waves found in a direction L as those for which L is a ray, since by (288) the group velocity vector $\partial\omega/\partial k_j$ is in the direction of L. Then, it specifies their amplitude as a product of the source function's Fourier transform $F(\mathbf{k}^{(0)})$, a term $[(\partial B/\partial n)^{(0)}]^{-1}$ depending on the dispersion relationship (270), and a term $4\pi^2/|K^{(0)}|^{\frac{1}{2}} x$ which ensures conservation of energy flux since the cross-sectional area of a ray tube made up of normals to an elementary area dS of the wavenumber surface is $|K^{(0)}|x^2 dS$ (figure 89).

To exhibit this in more detail we suppose as in section 4.8 that the energy density for waves of wavenumber $\mathbf{k}$ can be written as

$$W = W_0(\mathbf{k}) \overline{q^2}. \tag{296}$$

Then (295) gives the directional distribution of wave energy as

$$W = W_0(\mathbf{k}^{(0)}) |F(\mathbf{k}^{(0)})|^2 [(\partial B/\partial n)^{(0)}]^{-2} (8\pi^4/|K^{(0)}|x^2). \tag{297}$$

The energy flux is $\mathbf{I} = W\mathbf{U}$, where $\mathbf{U}(\mathbf{k}^{(0)})$ is the group velocity, with magnitude $U = \partial\omega/\partial n$, which takes the value

$$U = -(\partial B/\partial n)/(\partial B/\partial\omega) \tag{298}$$

for a dispersion relationship (270). The power output of the source can be obtained from (297) by multiplying the magnitude of the energy flux by the cross-sectional area $|K^{(0)}|x^2\mathrm{d}S$ of the ray tube associated with an element $\mathrm{d}S$ of the wavenumber surface, and integrating over S to give

$$P = 8\pi^4 \int\!\!\int_S UW_0(\mathbf{k})|F(\mathbf{k})|^2(\partial B/\partial n)^{-2}\,\mathrm{d}S; \tag{299}$$

here, the index (o) used to pick out a point on S with normal $\mathbf{n}$ in a particular direction L may obviously be dropped. By (298), an equivalent form of the power output P that avoids calculating a normal derivative is

$$P = 8\pi^4 \int\!\!\int W_0(\mathbf{k})|F(\mathbf{k})|^2\, U^{-1}(\partial B/\partial\omega)^{-2}\,\mathrm{d}S. \tag{300}$$

Sound waves give a simple illustration of these results. The equation

$$\partial^2\rho/\partial t^2 - c_0^2\,\partial^2\rho/\partial x_j^2 = f(x_1, x_2, x_3)\exp(\mathrm{i}\omega_0 t) \tag{301}$$

describes sound generation by an oscillating source distribution (see section 1.10, where the right-hand side of (301) is written as the source strength per unit volume, $\partial Q/\partial t$). In this case,

$$B = -\omega^2 + c_0^2 k_j^2, \quad U = c_0, \quad W = c_0^2\rho_0^{-1} \tag{302}$$

and the surface S is a sphere (centre the origin) of radius ω_0/c_0. Then equation (300) for the power output becomes

$$P = 8\pi^5(\rho_0 c_0)^{-1}\langle|F(\mathbf{k})|^2\rangle_S \tag{303}$$

in terms of the average

$$\langle|F(\mathbf{k})|^2\rangle_S = (4\pi\omega_0^2/c_0^2)^{-1}\int\!\!\int_S |F(\mathbf{k})|^2\,\mathrm{d}S \tag{304}$$

of $|F|^2$ over the spherical surface S with radius ω_0/c_0.

Acoustically compact source distributions are those on a length scale small compared with c_0/ω_0. Therefore, their Fourier transform $F(\mathbf{k})$ changes over a much *larger* wavenumber scale than ω_0/c_0. Normally this makes the average in (304) close to the value $|F(\mathbf{o})|^2$ at the sphere's centre:

$$\langle|F(\mathbf{k})|^2\rangle_S \doteq |F(\mathbf{o})|^2 = (2\pi)^{-6}|f_{\mathrm{tot}}|^2, \tag{305}$$

where

$$f_{\mathrm{tot}} = \int_{-\infty}^{\infty}\int_{-\infty}^{\infty}\int_{-\infty}^{\infty} f(x_1, x_2, x_3)\,\mathrm{d}x_1\,\mathrm{d}x_2\,\mathrm{d}x_3 \tag{306}$$

is the total source strength. Then (303) becomes

$$P = (\tfrac{1}{2}|f_{\mathrm{tot}}|^2)/(4\pi\rho_0 c_0), \tag{307}$$

exactly as would be inferred from the method of section 1.6 (the numerator being the mean square total source strength).

On the other hand, for a distribution with zero total source strength, $F(\mathbf{k})$ is dominated when $\mathbf{k}$ is small by a term

$$F(\mathbf{k}) \doteqdot i k_j G_j (2\pi)^{-3}, \quad \text{where} \quad G_j = \int_{-\infty}^{\infty} \int_{-\infty}^{\infty} \int_{-\infty}^{\infty} x_j f(x_1, x_2, x_3) \, dx_1 \, dx_2 \, dx_3 \tag{308}$$

is the equivalent dipole strength (moment of the source distribution). This means that

$$\langle |F(\mathbf{k})|^2 \rangle_S \doteqdot \tfrac{1}{3} (\omega_0/c_0)^2 |G_j|^2 (2\pi)^{-6}, \tag{309}$$

so that (303) becomes

$$P = (\tfrac{1}{2}|G_j|^2) \, \omega_0^2/(12\pi\rho_0 c_0^3), \tag{310}$$

exactly as would be inferred from the method of section 1.7 for compact source regions with dipole far fields.

For quite general wave systems, a source is similarly called compact if $F(\mathbf{k})$ varies on a wavenumber scale much larger than the dimensions of the wavenumber surface S. Then equations (305) and (306) hold on S, provided that the total source strength f_{tot} is nonzero, and the power output (300) has a value

$$P = (8\pi^2)^{-1} (|f_{\text{tot}}|^2) \iint_S W_0(\mathbf{k}) \, U^{-1} (\partial B/\partial \omega)^{-2} \, dS \tag{311}$$

proportional to the mean square of that total source strength. The integral in (311) is a coefficient which need be evaluated only once for each wave system: it is independent of the source distribution and varies only with the frequency ω_0. The corresponding formula when f_{tot} is zero but the equivalent dipole strength (308) is nonzero is

$$P = (8\pi^2)^{-1} \left| G_j \iint_S k_j W_0(\mathbf{k}) \, U^{-1} (\partial B/\partial \omega)^{-2} \, dS \right|^2 ; \tag{312}$$

this time the integral is a *vector* which can be calculated once and for all as a function of ω_0 for the wave system in question.

For sound generated by noncompact source distributions the average in (303) may differ considerably from such simple estimates. Also, the directional distribution of acoustic intensity $I = W c_0$ may be complicated; thus, its value in the direction L specified by a unit vector $\mathbf{n}$ is given by (297), where

$$\mathbf{k}^{(0)} = \omega_0 c_0^{-1} \mathbf{n} \tag{313}$$

is the point on S with normal $\mathbf{n}$ in the direction ω increasing. From (302), this is

$$I = 2\pi^4 (\rho_0 c_0)^{-1} x^{-2} |F(\omega_0 c_0^{-1} \mathbf{n})|^2. \tag{314}$$

For the purpose of shaping a desired directional distribution (314), a planar source distribution as described in section 1.12 generally suffices, since its variation can match the essentially two-dimensional variation with respect to a unit vector $\mathbf{n}$. Actually, any *effectively planar* source distribution, meaning one whose dimensions in *just one* direction (say the x_1-direction) are compact, satisfies on S the approximate equation

$$F(\mathbf{k}) \doteq F(0, k_2, k_3) = (2\pi)^{-1} F_1(k_2, k_3), \tag{315}$$

where $F_1(k_2, k_3)$ is the two-dimensional Fourier transform of the planar source-strength distribution

$$f_1(x_2, x_3) = \int_{-\infty}^{\infty} f(x_1, x_2, x_3)\, dx_1. \tag{316}$$

Then (314) takes the form

$$I = \tfrac{1}{2}\pi^2 (\rho_0 c_0)^{-1} x^{-2} |F_1(\omega_0 x_2 / x c_0, \omega_0 x_3 / x c_0)|^2, \tag{317}$$

which is seen to express in the present suffix notation the conclusions of section 1.12 on planar source distributions if we take into account that the F_1 there used is $2\pi^2 (\rho_0 i\omega_0)^{-1}$ times that defined above.

Many more illustrations of the use of the general wave-generation result (295) for noncompact source distributions are indicated in the epilogue. In every case the crucial facts are that the normal to the wavenumber surface $\omega = \omega_0$ in the direction ω increasing shows the direction in which waves of a particular wavenumber will be found, and that their amplitude is proportional to the Fourier component of the source distribution with that wavenumber. In some cases there is also interest in the shape of surfaces of constant phase

$$\mathbf{k}^{(0)} \cdot \mathbf{x} = \text{constant}. \tag{318}$$

Since $\mathbf{k}^{(0)}$ is a point on S with normal $\mathbf{n}$ in the direction of $\mathbf{x}$, equation (318) means that

$$\mathbf{x} = x\mathbf{n} \quad \text{with} \quad x = \frac{\text{constant}}{\mathbf{k}^{(0)} \cdot \mathbf{n}}. \tag{319}$$

In the language of geometry, this specifies any surface (319) of constant phase (figure 90) as a *reciprocal polar* of the wavenumber surface with respect to the origin; that is, the locus of the poles of its tangent planes.

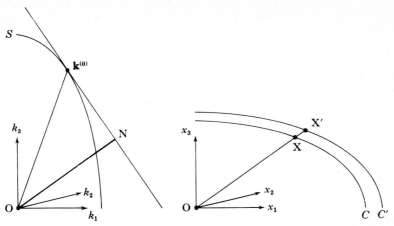

Figure 90. A surface of constant phase is the locus of a point given by equation (319). Here, $\mathbf{k}^{(0)} \cdot \mathbf{n}$ is ON, the length of the perpendicular dropped on to the tangent plane at $\mathbf{k}^{(0)}$. This makes $\mathbf{x}$ equal to (say) $\overrightarrow{\text{OX}}$, with OX parallel to ON and with the product of their lengths taking a constant value C (thus, X is the 'pole' of the tangent plane). A neighbouring surface of constant phase is the locus of X', with OX' once more parallel to ON and with the product of *their* lengths taking a slightly different constant value C'.

4.10   Internal waves generated by an oscillating source

The above general theory of oscillating sources of waves in two or three dimensions leads to results (290) in terms of the curvature of the wavenumber curve or (295) in terms of the Gaussian curvature of the wavenumber surface. Those equations are meaningless, however, for some wave systems, including the internal waves that give this chapter its title. The wavenumber surface for internal waves of fixed frequency ω_0 is a cone; at any point, therefore, one of the principal axes of curvature is along a generator, with associated curvature zero. Accordingly, the Gaussian curvature (product of the two principal curvatures) is everywhere zero.

This section shows the substantial modifications to the theory of oscillating sources of waves that are necessary in any wave system where the Gaussian curvature of the wavenumber surface S is identically zero; the results being illustrated by the case of internal waves. We defer, however, to section 4.11 analysis of the less radical (because purely local) modifications that are necessary if the Gaussian curvature vanishes on an isolated locus within the wavenumber surface.

We give first the relatively simple results that hold when S is the least curved of surfaces: a plane. These results are illustrated in many later sections. In the meantime, the partly similar results for two-dimensional propagation, in cases when S is a straight *line*, are illustrated here by internal waves *constrained to propagate two-dimensionally* as a result of generation by a source which is uniform in one horizontal direction (say, the y-direction). This keeps $l = 0$ in the dispersion relationship (24) so that the wavenumber curve (relating k and m for $\omega = \omega_0$) becomes a *pair* of straight lines. For this case, results are given also with viscous attenuation incorporated, and compared with experiment.

After that, wavenumber surfaces S curved in just one direction (the *developable surfaces*, that can be opened out into a plane without stretching) are considered. In particular, detailed analysis is given for conical surfaces, and illustrated by internal waves.

In every case we can start from the fundamental equation (284) for the waves appearing in a particular direction L stretching outwards from an oscillating source. That equation describes the waves in axes rotated so that L becomes the positive x_1-axis; then, S_+ is the part of S defined by (282).

A special feature of the wavenumber surfaces discussed in this section is the absence of any isolated point on S_+ where the phase $(-k_1 x_1)$ is stationary. Consequently, the wave amplitude distribution is no longer proportional (as in section 4.9) to a value of the spatial Fourier transform of the source distribution, $F(\mathbf{k})$. We shall find instead that it depends on a *partial* Fourier transform, where the source distribution is operated on in different orthogonal directions by *either* Fourier transformation *or* some alternative operation.

Cases when S_+ is a plane invite the use of special axes (x, y, z) with, say, the x-axis perpendicular to the plane. Then the corresponding wavenumber (k, l, m) satisfies $k = $ constant on the plane.

Consider, for example, the case when

$$B = B(\omega, k) \tag{320}$$

independently of l and m. This is a rather simple case when no partial derivatives with respect to y and z enter the differential equation (276) for the forced oscillations. The wavenumber surface S consists of a set of planes $k = k_0$ where

$$B(\omega_0, k_0) = 0. \tag{321}$$

For radiation in directions L with x *increasing*, the condition (282) defines S_+ as a subset of those planes such that equation (321) defines a function $\omega_0(k_0)$ with

$$d\omega_0/dk_0 > 0. \tag{322}$$

With this notation, the integral (284) over any plane $k = k_0$ which is part (or the whole) of S_+ can be written

$$- 2\pi\mathrm{i}[\exp{(\mathrm{i}\omega_0 t)}] \int_{-\infty}^{\infty} \int_{-\infty}^{\infty} F(k_0, l, m) \left[\partial B(\omega_0, k)/\partial k \right]_{k=k_0}^{-1}$$

$$\times \exp{[-\mathrm{i}(k_0 x + ly + mz)]}\, \mathrm{d}l\, \mathrm{d}m. \qquad (323)$$

Here, we have reverted from the special axes in which the direction L is the x_1-axis to the fixed axes (x, y, z) with the x-axis perpendicular to the plane $k = k_0$. If the angle between L and this x-axis is Φ then $(\mathrm{d}k_2\, \mathrm{d}k_3)$ on S_+ becomes $(\mathrm{d}l\, \mathrm{d}m) \cos\Phi$ but also $(\partial B/\partial k_1)$ becomes $(\partial B/\partial k) \cos\Phi$ and the two factors $\cos\Phi$ cancel out. Alternatively, (323) could have been derived directly by applying the method of section 4.9 in the (x, y, z) axes, with a radiation condition which, where x is positive, forbids energy propagation in a direction with x decreasing.

Evidently, the integration with respect to l and m in (323) 'unscrambles' the Fourier transformation; that is, it restores the original functional dependence on y and z in the source distribution, leaving only a partial Fourier transform with respect to x. This is because equation (268) can be written

$$f(x, y, z) = \int_{-\infty}^{\infty} F_{(x)}(k, y, z) \exp{(-\mathrm{i}kx)}\, \mathrm{d}k, \qquad (324)$$

where the partial Fourier transform $F_{(x)}$ is equal to

$$F_{(x)}(k, y, z) = \int_{-\infty}^{\infty} \int_{-\infty}^{\infty} F(k, l, m) \exp{[-\mathrm{i}(ly + mz)]}\, \mathrm{d}l\, \mathrm{d}m. \qquad (325)$$

The radiation (323) in directions x increasing is given, then, as

$$q = - 2\pi\mathrm{i}F_{(x)}(k, y, z) [\partial B(\omega_0, k)/\partial k]_{k=k_0}^{-1} \exp{[\mathrm{i}(\omega_0 t - k_0 x)]}, \qquad (326)$$

if there is just one wavenumber k_0 satisfying (321) and (322), or otherwise as the sum of contributions (326) from every such wavenumber.

Equation (326) represents a strictly parallel beam, because the partial Fourier transform $F_{(x)}(k, y, z)$ given by (324) is zero for any y and z such that the source distribution $f(x, y, z)$ vanishes for all x. The beam is simply a *projection* of the source region in the x-direction. Since its cross-sectional area is independent of x, the wave amplitude similarly remains constant.

Note the contrast between such a strictly parallel beam, resulting from a planar wavenumber surface, and the acoustic beam generated by a planar source distribution (sections 1.12 and 4.9); the latter is initially parallel but spreads out always into a conical form (however small the cone angle may be) while the distribution changes from the original functional dependence on y and z to a Fourier transform thereof. No such Fourier transformation with respect to y and z, or conversion to a conical beam, can occur when the wavenumber surface itself is planar.

Indeed, another derivation of (326) for a partial differential equation (276) including no y- or z-derivatives would be based on solving it as a one-dimensional problem for each constant pair of values of y and z. This might be the quickest derivation, but here we have preferred to show how the result can be obtained as an extreme special case of our general treatment which introduces well the discussion of intermediate special cases like that of forced internal waves.

A source distribution generates internal waves according to the equation

$$\nabla^2(\partial^2 q/\partial t^2) + [N(z)]^2(\partial^2 q/\partial x^2 + \partial^2 q/\partial y^2) = \partial^3 Q/\partial t^2 \partial z, \qquad (327)$$

where $\partial Q/\partial t$ (rate of change of mass outflow rate per unit volume) represents source strength in the sense of chapter 1. Equation (327) is derived by adding a mass outflow rate Q to the right-hand side of (17), and so changing the right-hand sides of (18), (20) and (22) by $-\partial Q/\partial t$, by $+\partial^2 Q/\partial t \partial z$ and by $+\partial^3 Q/\partial t^2 \partial z$. Here we consider (327) in the case $N = $ constant so that Fourier analysis can be used.

A source distribution

$$\partial Q/\partial t = f(x,z)\exp(i\omega_0 t) \qquad (328)$$

independent of y generates two-dimensional wave propagation, somewhat similar in character to that just discussed because the wavenumber 'curve' lacks curvature, being a pair of straight lines. Equation (327), with $N = $ constant, then becomes

$$(N^2 - \omega_0^2)\,\partial^2 q/\partial x^2 - \omega_0^2\,\partial^2 q/\partial z^2 = i\omega_0(\partial f/\partial z)\exp(i\omega_0 t). \qquad (329)$$

This is a special case of (276) with

$$B(\omega, k, m) = [\omega^2 m^2 - (N^2 - \omega^2)k^2]/(\omega m), \qquad (330)$$

where the dispersion relationship $B = 0$ takes the form

$$\omega = [N^2 k^2/(k^2 + m^2)]^{\frac{1}{2}}. \qquad (331)$$

The wavenumber curve $\omega = \omega_0$ is a pair of straight lines in the (k, m) plane, at the angle $\theta = \cos^{-1}(\omega_0/N)$ to the k-axis. These are shown in

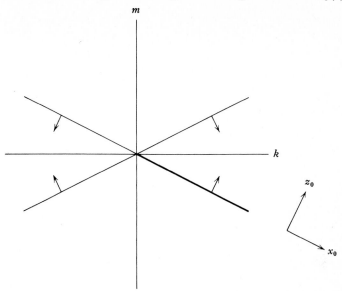

Figure 91. The wavenumber curve S for two-dimensional propagation of internal waves (in the (x, z) plane). The arrows point in the directions in which waves corresponding to the four different half-lines in S are found. For those corresponding to the thick half-line, special axes (as in figure 87) are used, with the z_0-axis in the direction of energy propagation.

figure 91, which also indicates by arrows the directions of those normals along which ω is increasing. These are the directions of the rays: four different directions making the St Andrew's Cross pattern (figure 76) associated with internal waves of fixed frequency. Each direction is associated with a *half-line* in the wavenumber pattern; for example, as we know from section 4.4, waves propagating in a direction with x and z increasing must have k positive and m negative.

The main difference from the previous case when a *whole* plane in the wavenumber space contributes to the radiation in a particular direction is that only *half* a straight line so contributes. This result of group-velocity analysis (with the important consequences described below) can easily be missed in a treatment of (329) by conventional factorisation of the differential operator on the left-hand side without proper application of the radiation condition.

We seek here an expression for those waves, with k positive and m negative, whose energy propagates upwards and to the right. Yet another change of axes helps them to be evaluated conveniently. We use (x_0, z_0)

with x_0 in the direction of the wavenumber vector (the diagonally down-ward direction of travel of the crests) and z_0 at right angles (in the diagonally upward direction of the rays). The half-line $k_0 > 0$, $m_0 = 0$ generates the waves, and the two-dimensional form of (284) becomes

$$q = -2\pi i[\exp(i\omega_0 t)] \int_0^\infty F(k_0, 0)[\partial B/\partial m_0]_{m_0=0}^{-1} \exp(-ik_0 x_0) \, dk_0. \quad (332)$$

Here, it is easily verified from (330) that the value of $\partial B/\partial m_0$ on the half-line is a constant, $2N$; in fact, since $B = 0$ on the half-line, the normal derivative $\partial B/\partial m_0$ in the m_0-direction, which is at an angle $\frac{1}{2}\pi - \theta$ to the k-direction, must be (cosec θ) times $\partial B/\partial k$, giving

$$\partial B/\partial m_0 = (\text{cosec}\,\theta)(N^2 - \omega_0^2)(-2k/\omega_0 m) = (2N\tan\theta)(-k/m) = 2N$$

$$(333)$$

because $m = -k\tan\theta$. Therefore

$$q = -\pi i N^{-1}[\exp(i\omega_0 t)] \int_0^\infty F(k_0, 0) \exp(-ik_0 x_0) \, dk_0. \quad (334)$$

Just as with the plane wavenumber surface, the rays for $\omega = \omega_0$ are in a definite direction (the one, at an angle θ to the vertical, here taken as the z_0-direction); accordingly, the beam (334) is again a parallel one and there is no reduction in amplitude as z_0 increases. Another similarity to the plane wavenumber surface is that the half-line picks out a partial Fourier trans-form with respect to the direction of propagation (here, the z_0-direction). It selects, in fact, the component $m_0 = 0$, which involves merely an *integra-tion* with respect to z_0. Indeed, if the integral in (334) *were* taken from $-\infty$ to ∞, it would give an amplitude factor

$$\int_{-\infty}^\infty F(k_0, 0) \exp(-ik_0 x_0) \, dk_0 = (2\pi)^{-1} \int_{-\infty}^\infty f(x_0, z_0) \, dz_0 = f_S(x_0), \quad (335)$$

which is $(2\pi)^{-1}$ times the source strength integrated in the direction normal to S (that is, in the z_0-direction).

For a half-line, however, the fact that the integral in (334) is only from 0 to ∞ brings about an important modification which gives the solution its wave-like character. A component of (334) proportional to

$$\exp[i(\omega_0 t - k_0 x_0)]$$

for $0 < k_0 < \infty$ has wave crests moving in the expected direction x_0 increasing (diagonally *downwards*, at right angles to the rays). The combined wave

(334) takes a form dependent on a classical *splitting*, by which any function $f_S(x_0)$ given by a Fourier integral as in (335) is expressed as the sum

$$f_S^+(x_0) + f_S^-(x_0) = \int_0^\infty F(k_0, 0) \exp(-ik_0 x_0) \, dk_0$$
$$+ \int_{-\infty}^0 F(k_0, 0) \exp(-ik_0 x_0) \, dk_0. \quad (336)$$

Physically, this corresponds to the splitting of a *standing wave*

$$f_S(x_0) \exp(i\omega_0 t)$$

into the sum of two *travelling waves* moving in the positive and negative x_0-directions respectively. The stratified medium takes the integrated source distribution (335), *splits it* as in (336) and sends waves of amplitude $f_S^+(x_0)$ diagonally upwards and to the right and those of amplitude $f_S^-(x_0)$ diagonally downwards and to the left.

For any *rational* function $f_S(x_0)$, the split (336) is achieved most easily by expressing it in partial fractions, and including those with poles in the upper half of the complex x_0-plane in f_S^+ and the rest in f_S^-. For example,

when $$f_S(x_0) = 2Ch(x_0^2 + h^2)^{-1}, \quad f_S^+(x_0) = C(h + ix_0)^{-1}; \quad (337)$$

as the reader may verify by this partial-fraction rule or by direct calculation using $F(k_0, 0) = C \exp(-hk_0)$. In this case, the waves (334) can be written as

$$q = -\pi i N^{-1} [\exp(i\omega_0 t)] C(h + ix_0)^{-1} \quad (338)$$

or, in modulus and argument form, as

$$q = \pi C N^{-1} (h^2 + x_0^2)^{-\frac{1}{2}} \exp\{i[\omega_0 t - \tan^{-1}(x_0/h) - \tfrac{1}{2}\pi]\}, \quad (339)$$

which emphasises the wavelike dependence on x_0 although indicating the absence of any one complete wavelength. Figure 92 shows the nature of successive waveforms in this case. In the absence of energy dissipation, this pattern of wave motion at right angles to rays would be propagated along the rays indefinitely and without attenuation.

It is, however, straightforward to incorporate in the above analysis the effects of viscous attenuation at the rate (220) per unit distance z_0 along rays. This rate can be written

$$\beta = 2\beta_0 k_0^3, \quad \text{where} \quad \beta_0 = \tfrac{1}{2}\nu N^{-1} \operatorname{cosec} \theta \quad (340)$$

and k_0 as above is the *magnitude* of the wavenumber vector. Such a rate of energy attenuation introduces an exponential factor $\exp(-\beta_0 k_0^3 z_0)$ into the

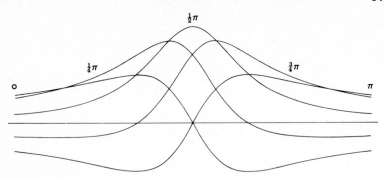

Figure 92. Successive waveforms (real part of (338) or (339) plotted for successive values o, $\frac{1}{4}\pi$, $\frac{1}{2}\pi$, $\frac{3}{4}\pi$ and π of $\omega_0 t$, as a function of x_0/h), in unattenuated two-dimensional internal waves generated by an oscillating source specified by (337). The crest movement is, of course, perpendicular to the z_0-direction along which the wave energy travels.

amplitude of waves of wavenumber k_0. This must be incorporated in the integral (332) and hence in (334), giving

$$q = -\pi i N^{-1}[\exp{(i\omega_0 t)}] \int_0^\infty F(k_0, o) \exp{(-ik_0 x_0 - \beta_0 k_0^3 z_0)}\, dk_0. \quad (341)$$

Here we may note that, for unattenuated waves, the general concept (section 4.9) of compact sources, whose Fourier transforms vary on a wavenumber scale much larger than the dimensions of S, is meaningless when S extends to infinity as in all the wave systems studied in this section. Meaning is restored to it, however, by viscous attenuation; which, for internal waves at distances greater than z_0 from the source, eliminates all wavenumbers significantly greater than $(\beta_0 z_0)^{-\frac{1}{3}}$; then, sources whose Fourier transform varies on a scale much larger than that are effectively compact.

If, for example, a cylindrical wire oscillating horizontally in the fluid has its cross-section compact in this sense, then it may be represented by a horizontal dipole of strength $G\exp{(i\omega_0 t)}$ equal to the force per unit length which it exerts on the fluid. This gives

$$F(k, m) \doteqdot ikG \quad (342)$$

as in (308) so that the value of $F(k_0, o)$ in the rotated wavenumber axes (k_0, m_0) is $i(k_0 \cos \theta) G$ and

$$q = (\pi N^{-1} \cos \theta)\, G[\exp{(i\omega_0 t)}] \int_0^\infty k_0 \exp{(-ik_0 x_0 - \beta_0 k_0^3 z_0)}\, dk_0. \quad (343)$$

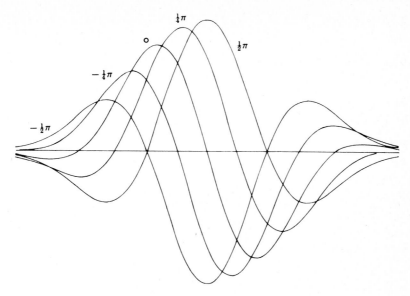

Figure 93. Successive waveforms (real part of (343)) computed by Thomas and Stevenson for successive values $-\frac{1}{2}\pi$, $-\frac{1}{4}\pi$, o, $\frac{1}{4}\pi$ and $\frac{1}{2}\pi$ of $\omega_0\,t$ in two-dimensional internal waves being attenuated by viscosity after generation by a compact oscillating source.

Figure 93 shows the nature of successive waveforms in this case, computed from (343) by Thomas and Stevenson, while figure 94 indicates a highly satisfactory comparison between their theoretical results and their meticulous experimental observations.

Between those wave systems with S doubly curved, analysed in section 4.9 by the method of stationary phase, and the above systems where stationary phase is useless because S is flat, lie intermediate cases with a singly curved form of S. Then there is one principal direction of *nonzero* curvature in which the method of stationary phase can be used, but the above methods requiring evaluation of an infinite integral must be used in the other direction. Such a surface with Gaussian curvature K everywhere zero is necessarily developable (deformable into a plane without stretching); and the normals to it form only a *singly* infinite set of directions: a 'fan' of rays along which the wave energy is constrained to travel. We indicate the general method of treatment by carrying it out for the case of internal waves, with S a cone, and with the normals to that cone forming the complementary conical 'fan' of rays.

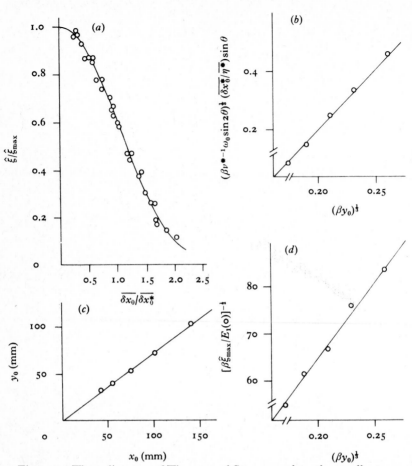

Figure 94. These diagrams of Thomas and Stevenson show the excellent agreement between their measurements (the points) and their calculations (the lines). Without explaining their notation in detail we note that the matters checked are: (*a*) the envelope of the waveform; (*b*) the variation of wave width as the ($\frac{1}{3}$)th power of distance from the source; (*c*) the inclination of the central ray to the vertical; (*d*) the variation of maximum wave amplitude as the ($-\frac{2}{3}$)th power of distance from the source.

A three-dimensional oscillating source

$$\partial Q / \partial t = f(x, y, z) \exp(\mathrm{i}\omega_0 t) \tag{344}$$

generates internal waves in stratified fluid satisfying (327) with N uniform according to the equation

$$(N^2 - \omega_0^2)(\partial^2 q / \partial x^2 + \partial^2 q / \partial y^2) - \omega_0^2 \, \partial^2 q / \partial z^2 = \mathrm{i}\omega_0(\partial f / \partial z) \exp(\mathrm{i}\omega_0 t). \tag{345}$$

Any linear partial differential equation like (345) where the independent variable occurs only in spatial derivatives of a single order (here 2) has a wavenumber surface which is 'conical' in a general sense: that is, it is made up of straight lines through the origin and therefore has zero Gaussian curvature; essentially, because such an equation makes B a *homogeneous* function of k, l and m. With (345), for example, it is

$$B(\omega, k, l, m) = [\omega^2 m^2 - (N^2 - \omega^2)(k^2 + l^2)]/(\omega m), \qquad (346)$$

which gives the dispersion relationship $B = 0$ the form (24), so that the wavenumber surface $\omega = \omega_0$ is a cone with semi-angle $\theta = \cos^{-1}(\omega_0/N)$.

For estimating the waves radiated in a particular direction L we use special axes very similar to those of figure 91. First, we take the vertical plane through L as $y_0 = 0$; then, the other axes are as before: x_0 is along a generator of the cone so that it points in the direction of the wavenumber vector, and z_0 (at right angles, of course) is along a ray. These axes are suitable because the method of stationary phase picks out the generator in the vertical plane through L as the sole contributor to radiation along L; at the same time, the whole half-line represented by that generator contributes as before to the radiation.

Equation (284) becomes in this case

$$q = -2\pi i[\exp(i\omega_0 t)] \iint_{S_+} F(k_0, l_0, m_0)[\partial B/\partial m_0]^{-1}$$
$$\times \exp[-i(k_0 x_0 + m_0 z_0)]\, dk_0\, dl_0; \qquad (347)$$

the difference from (332) is that the cone's curvature prevents m_0 from being zero except where $l_0 = 0$. The curvature of the cone in a plane $k_0 = $ constant at the point $l_0 = 0$ is easily calculated as

$$\kappa_0 = -(\partial^2 m_0/\partial l_0^2) = k_0^{-1}\tan\theta; \qquad (348)$$

indeed, the curvature of the cone's circular cross-section by a plane $m = $ constant is $k^{-1} = (k_0 \cos\theta)^{-1}$ and this must be $(\operatorname{cosec}\theta)$ times the curvature κ_0 in a plane normal to the generator. The method of stationary phase in one dimension (section 3.7) applied to the integration with respect to l_0 in (347) now gives

$$q = -2\pi i\{\exp[i(\omega_0 t + \tfrac{1}{4}\pi)]\}\int_0^\infty F(k_0, 0, 0)(2N)^{-1}(2\pi/\kappa_0 z_0)^{-\frac{1}{2}}$$
$$\times \exp(-ik_0 x_0)\, dk_0 \qquad (349)$$

since as before equation (333) gives $\partial B/\partial m_0$ on $l_0 = m_0 = 0$. The asymptotic form of the waves, then, is

$$q = N^{-1}(2\pi^3/z_0 \tan\theta)^{\frac{1}{2}} \{\exp[i(\omega_0 t - \tfrac{1}{4}\pi)]\} \int_0^\infty k_0^{\frac{1}{2}} F(k_0, 0, 0) \exp(-ik_0 x_0) \, dk_0. \tag{350}$$

In (350), the reduction of amplitude like $z_0^{-\frac{1}{2}}$ arises from the fact that energy spreads out from the source in a conical 'fan' of rays; the separation between two such rays increases in proportion to the distance along them. Nevertheless, although the energy spreads out tangentially to the fan, the width of the beam normal to the fan does not increase.

An interesting feature appears in the amplitude distribution normal to the fan of rays, given by the integral in (350). We know already that limiting the range of integration to $(0, \infty)$ performs a conversion from the standing-wave character of an oscillating source into something like a travelling wave (figures 92 and 93). The new factor $k_0^{\frac{1}{2}}$ brings in, additionally, the operation known as *taking the derivative of* $(\frac{1}{2})$*th order*. We met this last in relation to the asymptotic behaviour of sound waves propagated two-dimensionally (section 1.4); and we should not, perhaps, be surprised at its reappearance in the asymptotic form of solutions of a partial differential equation (345) which, with z replaced by a multiple of the time, would *become* the two-dimensional wave equation. The solution of (345) satisfying the radiation condition, however, involves not only a derivative of $(\frac{1}{2})$th order but also the splitting of a standing wave into waves travelling in opposite directions; and this might well be missed in any direct application of the analogy just mentioned.

In terms of the source distribution integrated in planes normal to the generator,

$$\int_{-\infty}^\infty F(k_0, 0, 0) \exp(-ik_0 x_0) \, dk_0$$
$$= (2\pi)^{-2} \int_{-\infty}^\infty \int_{-\infty}^\infty f(x_0, y_0, z_0) \, dy_0 \, dz_0 = f_S(x_0), \tag{351}$$

the integral in (350) is proportional to the derivative of $(\frac{1}{2})$th order of $f_S^+(x_0)$. For example, if $f_S(x_0)$ takes the form (337) then we find that

$$\int_0^\infty k_0^{\frac{1}{2}} F(k_0, 0, 0) \exp(-ik_0 x_0) \, dk_0 = \tfrac{1}{2} C\pi^{\frac{1}{2}}(h + ix_0)^{-\frac{3}{2}} \tag{352}$$

so that

$$q = \pi^2 C N^{-1}(2z_0 \tan\theta)^{-\frac{1}{2}} (h^2 + x_0^2)^{-\frac{3}{4}} \exp\{i[\omega_0 t - \tfrac{3}{2}\tan^{-1}(x_0/h) - \tfrac{1}{4}\pi]\}. \tag{353}$$

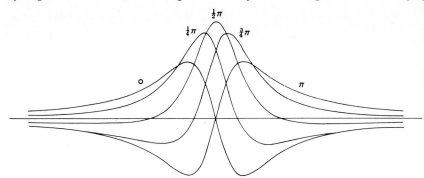

Figure 95. Successive waveforms (real part of (353) plotted, for successive values o, $\frac{1}{4}\pi$, $\frac{1}{2}\pi$, $\frac{3}{4}\pi$ and π of $\omega_0 t + \frac{1}{4}\pi$, as a function of x_0/h) in unattenuated three-dimensional internal waves generated by an oscillating source specified by (337). The energy travels away from the source along the surface of a cone in this case, with the crest motions normal to that conical surface.

Figure 95 shows the nature of successive waveforms in this case, for a source of total strength $(2\pi)^3 C$.

We have seen that the asymptotic form of internal waves from an oscillating source is a relatively simple one even in three dimensions. Note that, with dipole forcing, the derivatives of the waveforms in figure 95 would appear; those differentiated waveforms are just a little 'wavier'. As before, the additional factor $\exp\left(-\beta_0 k_0^3 z_0\right)$ in the integral would be needed to allow for viscous attenuation.

4.11 Caustics

Throughout the development of group-velocity ideas and ray theory from section 3.6 onwards we have deferred until the present section any consideration of the local behaviour of waves near *caustics*. Here, we introduce this word for the first time: a caustic is a boundary between a region with a complicated wave pattern, due to interference between *two* groups of waves, and a neighbouring region including *no* waves.* Caustics are prominent local features of many diverse patterns of waves in fluids, all of which can be treated together because the key to their understanding is a single mathematical concept: the Airy integral. Straightforward ray theory breaks down near a caustic, but the Airy integral allows us to produce a 'healed' version in which the local difficulty is overcome.

* Perhaps a more universally applicable definition would allow groups of waves from different sources to be also present; even then, however, the number of groups of waves in the latter region would be *two fewer* than in the former.

In homogeneous systems analysed by the method of stationary phase, the local difficulty arises (sections 3.7, 4.8, 4.9) wherever a *second* derivative of the phase (or, in more than one dimension, a principal second derivative) vanishes along with the gradient itself. Near such a point, the Airy integral plays the same role as does the Gaussian integral at a general point. *Rays run together* near such points, because the group velocity is stationary. The locus of those points is the caustic, which separates a region without rays from a region twice covered by rays. However, the *assumptions* of ray theory break down in the neighbourhood of a caustic, because local gradients of wavenumber become large.

In nonhomogeneous systems such as sound waves in a stratified atmosphere or in a wind (section 4.6), it is similarly possible for rays to run together, forming an 'envelope' beyond which ray theory suggests a zone of silence. It is also common for internal waves to show a caustic, with two systems of rays below it and none above; this may be either an envelope (figure 80(*b*)) or a cusp-locus (figure 80(*a*)) of rays. In all of these cases, a 'healed' version of ray theory can be derived from the properties of the Airy integral: this time, its differential properties.

The boundary of any trapped-wave region is a caustic. We shall find that ray theory, in its 'healed' version, is a convenient and simple method of approximate analysis of trapped-wave systems.

First of all, we take up once more the method of stationary phase for tracing how a limited disturbance develops in a one-dimensional wave system (section 3.7). We show how it must be modified around a wavenumber $k = k_c$ where the group velocity $U = d\omega/dk$ is stationary; that is, where

$$U'(k_c) = \omega''(k_c) = 0, \tag{354}$$

as happens with water waves at their minimum group velocity (figure 56). The locus

$$x/t = U(k_c) \tag{355}$$

carrying waves of wavenumber k_c is a caustic, because values of U on one side of such a stationary value $U(k_c)$ are found for *two* wavenumbers k, and values on the other side for *none*.

As in section 3.7 we study the asymptotic behaviour of

$$\zeta = \int_0^\infty F(k) \exp\left[it\psi(k)\right] dk, \tag{356}$$

with

$$\psi(k) = \omega(k) - (kx/t) \tag{357}$$

by concentrating on where the phase $t\psi(k)$ has zero first derivative; that is, where

$$x/t = \omega'(k) = U(k). \tag{358}$$

The detailed method of section 3.7 cannot be used for $k = k_c$ where, by (354) and (357), the second derivative of ψ is also zero. That method must even give a bad approximation *near* $k = k_c$, where $\psi''(k)$ is small so that its contribution to the Taylor expansion of ψ need not dominate that of $\psi'''(k)$.

The *idea* of the method of section 3.7 can, however, be used. Away from the stationary point, we deform the path to one on which the imaginary part of ψ is at least $+\delta$. Near the stationary point, however (where this is impossible), a short link L is made on which a Taylor expansion is used to estimate the integral's asymptotic value.

Under the condition (355), equations (357) and (354) give

$$\psi'(k_c) = 0 \quad \text{and} \quad \psi''(k_c) = 0. \tag{359}$$

Contours on which the imaginary part of ψ is $+\delta$ then take the forms shown in figure 96(a) when

$$\psi'''(k_c) = \omega'''(k_c) > 0, \tag{360}$$

and are their mirror images when $\omega'''(k_c)$ is negative. (We omit consideration of the unusual case when $\omega'''(k_c)$ is zero.) The link L can best be taken as a 'path of steepest descent' of $\exp[it\psi(k)]$ from $k = k_c$ to those contours. This path is a pair of straight segments each at 30° to the real axis, because (359) gives

$$\psi(k) = \psi(k_c) + \tfrac{1}{6}(k-k_c)^3 \psi'''(k_c) + O(k-k_c)^4, \tag{361}$$

and $(k-k_c)^3$ is pure imaginary when $\arg(k-k_c)$ is either $\tfrac{1}{6}\pi$ or $\tfrac{5}{6}\pi$.

We use the same link L throughout the *neighbourhood* of the caustic; that is, when (355) is almost, but not exactly, satisfied. Then $\psi(k)$ includes a linear term

$$(k-k_c)\psi'(k_c) = (k-k_c)[\omega'(k_c) - (x/t)], \tag{362}$$

which is a significant addition very close to $k = k_c$. For small values of the coefficient in square brackets, however, the cubic term in (361) must dominate at the extremities of L. Therefore, the link L remains effective in raising the imaginary part of ψ to $+\delta$. With error $O[\exp(-t\delta)]$, then, the integral (356) can be taken as an integral over the link L, giving

$$\zeta = \int_L [F(k_c) + O(|k-k_c|)] \exp\{it[\psi(k_c) + (k-k_c)\psi'(k_c)$$
$$+ \tfrac{1}{6}(k-k_c)^3 \psi'''(k_c) + O(|k-k_c|^4)]\}\, dk. \tag{363}$$

To evaluate (363) we need, not the *Gaussian integral* of the exponential of a quadratic, but another standard integral: the *Airy integral* of the exponential of a sum of a linear term and a cubic term. The integral (363) is reduced to standard form by the substitution

$$k - k_c = [\tfrac{1}{2} t \psi'''(k_c)]^{-\frac{1}{3}} s, \tag{364}$$

which simplifies the cubic term to $\tfrac{1}{3} i s^3$, and by then putting

$$X = [\tfrac{1}{2} t \psi'''(k_c)]^{-\frac{1}{3}} t \psi'(k_c), \tag{365}$$

which simplifies the linear term to $i s X$, giving

$$\zeta = F(k_c) \{ \exp [i t \psi(k_c)] \} [\tfrac{1}{2} t \psi'''(k_c)]^{-\frac{1}{3}} \int \exp [i(s X + \tfrac{1}{3} s^3)] \, ds \tag{366}$$

with an error factor $1 + O(t^{-\frac{1}{3}})$ since (364) makes

$$O(|k - k_c|) = O(t |k - k_c|^4) = O(t^{-\frac{1}{3}}). \tag{367}$$

The integral in (366) is taken along a bent path in the s-plane corresponding to L in the k-plane. At the extremities of this, the modulus of the integrand falls to $O[\exp(-t\delta)]$. The integral therefore differs by an error of only this order from the same integral taken all the way to infinity. We define

$$\mathrm{Ai}(X) = (2\pi)^{-1} \int_{\infty \exp(\frac{5}{6}\pi i)}^{\infty \exp(\frac{1}{6}\pi i)} \exp [i(s X + \tfrac{1}{3} s^3)] \, ds, \tag{368}$$

and infer that the asymptotic form of ζ is

$$\zeta \sim 2\pi F(k_c) \{ \exp [i t \psi(k_c)] \} [\tfrac{1}{2} t \psi'''(k_c)]^{-\frac{1}{3}} \mathrm{Ai}(X). \tag{369}$$

The above derivation was related to figure 96, showing the form of L for $\psi'''(k_c) > 0$, but the conclusion is correct also when $\psi'''(k_c) < 0$, changing the L of figure 96 (a) into its mirror image in the real axis. The cube root in (364) should then be taken negative, so that in the s-plane the path reverts to the form stretching from $\infty \exp(\frac{5}{6}\pi i)$ to $\infty \exp(\frac{1}{6}\pi i)$ used in (368).

The Airy integral (368) is readily computed, with the results shown in figure 97. The nature of this graph can be explained by the theory of stationary phase. When X has a substantial positive value the phase $s X + \tfrac{1}{3} s^3$ has no stationary point near the real axis so that the integral becomes exponentially small, corresponding as we shall see to the side of the caustic without waves. When X has a substantial negative value there are two well-separated stationary points for real s, given by

$$s = +|X|^{\frac{1}{2}} \quad \text{and} \quad s = -|X|^{\frac{1}{2}}; \tag{370}$$

at these, the phase's second derivative $2s$ is positive and negative, respectively. Then it is natural to seek to estimate the Airy integral as a sum

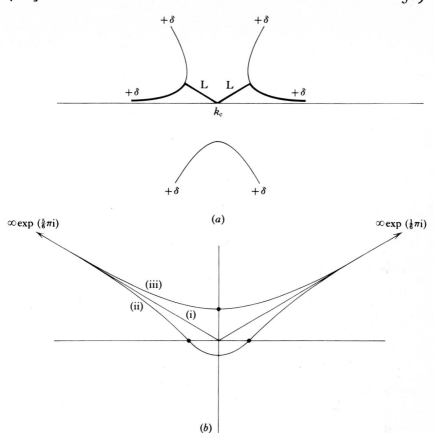

Figure 96. (*a*) Typical behaviour, near any stationary point k_c of the phase function $\psi(k)$, where $\psi''(k_c) = 0$ and $\psi'''(k_c) > 0$, of curves in the complex plane on which $\psi(k)$ has constant imaginary part $+\delta$. We deform the path of integration in (356) to one on which the imaginary part of $\psi(k)$ is $+\delta$ except near k_c, where a link L is used to cross from one such path to another (thick line).
(*b*) The complex s-plane defined by the substitution (364): (i) path of integration corresponding to L, used to define the Airy integral Ai (X); (ii) deformed path of integration used to obtain the asymptotic form (371) of Ai(X) for X large and negative; (iii) deformed path of integration used to obtain the asymptotic form (373) of Ai(X) for X large and positive.

of two Gaussian integrals by deforming the path (rather as in figure 63) into a path passing through these at $+45°$ and at $-45°$ to the real axis (figure 96). This estimation for $X < 0$ gives

$$\text{Ai}(X) \sim (2\pi)^{-1}\{\exp[-\tfrac{2}{3}i|X|^{\frac{3}{2}}+\tfrac{1}{4}\pi i]\}[2\pi/2|X|^{\frac{1}{2}}]^{\frac{1}{2}}$$
$$+(2\pi)^{-1}\{\exp[+\tfrac{2}{3}i|X|^{\frac{3}{2}}-\tfrac{1}{4}\pi i]\}[2\pi/2|X|^{\frac{1}{2}}]^{\frac{1}{2}}, \qquad (371)$$

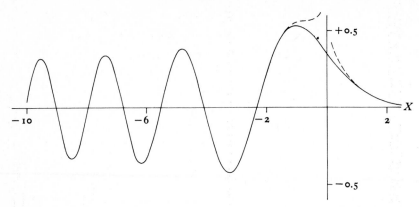

Figure 97. Full line: the Airy integral Ai(X). Broken lines: its asymptotic forms (373) for $X > 0$ and (374) for $X < 0$.

including two different wave-like contributions. These modulate the $\exp[it\psi(k_c)]$ in (369), corresponding as we shall see to two groups of waves on the side of the caustic with $X < 0$.

Actually, for substantial positive X the path can similarly be deformed (figure 96) so as to pass through an imaginary position of stationary phase $s = +iX^{\frac{1}{2}}$, where the integrand takes the value $\exp(-\frac{2}{3}X^{\frac{3}{2}})$ and the phase has second derivative $2iX^{\frac{1}{2}}$. The path of steepest descent through this point is parallel to the real axis and allows an estimate of Ai(X) as the Gaussian integral

$$(2\pi)^{-1}[\exp(-\tfrac{2}{3}X^{\frac{3}{2}})]\int \exp[-X^{\frac{1}{2}}(s-iX^{\frac{1}{2}})^2]\,ds. \tag{372}$$

This form (372) for substantial positive X, readily evaluated as

$$\mathrm{Ai}(X) \sim \tfrac{1}{2}\pi^{-\frac{1}{2}}X^{-\frac{1}{4}}\exp(-\tfrac{2}{3}X^{\frac{3}{2}}) \quad \text{when} \quad X > 0, \tag{373}$$

and the form given by (371) for substantial negative X, namely

$$\mathrm{Ai}(X) \sim \pi^{-\frac{1}{2}}|X|^{-\frac{1}{4}}\cos(\tfrac{2}{3}|X|^{\frac{3}{2}}-\tfrac{1}{4}\pi) \quad \text{when} \quad X < 0, \tag{374}$$

are also plotted in figure 97. These broken-line curves show how the Airy integral makes a quick and smooth transition from its wavy asymptotic form (374) for negative X to its exponentially vanishing asymptotic form (373) for positive X.

When $\psi(k)$ has the form (357), equations (365) and (369) give

$$\zeta \sim \frac{2\pi F(k_c)}{[\tfrac{1}{2}t\omega'''(k_c)]^{\frac{1}{3}}} \mathrm{Ai}\left\{\frac{t\omega'(k_c)-x}{[\tfrac{1}{2}t\omega'''(k_c)]^{\frac{1}{3}}}\right\} \exp\{i[\omega(k_c)t - k_c x]\}. \tag{375}$$

Equation (375) shows how waves are significant on the side of the caustic
where $x - t\omega'(k_c)$ has the same sign as $\omega'''(k_c)$;   (376)
there, two groups of waves are present because the Airy function for
negative X involves the sum (371) of two exponentials. Those two groups
of waves of nearby wavenumber 'beat' together: figure 97 describing the
slow variation of their amplitude shows, indeed, how it first vanishes at
$X = -2.34$.

On the other side of the caustic we see (from figure 97 for *positive* X)
that the wave amplitude falls away exponentially, to reach only 3 % of its
maximum already within a distance

$$|x - t\omega'(k_c)| = 2.5 |\tfrac{1}{2} t\omega'''(k_c)|^{\frac{1}{3}}$$ (377)

of the caustic. This decay distance grows gradually with time but becomes
an increasingly insignificant fraction of the total spread of the wave group
which grows in proportion to t itself.

The general theory (section 3.7) of stationary phase in one dimension
gives an amplitude decaying like $t^{-\frac{1}{2}}$ as the energy is so spread out; however,
that theory's inapplicability near the caustic is evident from the factor
multiplying $t^{-\frac{1}{2}}$, namely, $|\omega''(k)|^{-\frac{1}{2}}$ which becomes infinite there. The true
amplitude at the caustic itself is by no means infinite since Ai(o) has the
finite value 0.355. That amplitude does, however, decay more slowly than
the $t^{-\frac{1}{2}}$ rate found elsewhere; namely by (369), in proportion to $t^{-\frac{1}{3}}$.

Actually, equation (365) shows how, for fixed x/t satisfying the condition
(376) for waves to be present, ζ asymptotically reverts to the $t^{-\frac{1}{2}}$ form
because the $|X|^{-\frac{1}{4}}$ in (374) decays with time in proportion to $t^{-\frac{1}{6}}$. At the
caustic boundary of the wave region, then, a slight enhancement of ampli-
tude (from a value of order $t^{-\frac{1}{2}}$ to one of order $t^{-\frac{1}{3}}$) occurs just *before* its
exponential decay to zero.

A caustic is a sort of 'wound' for straightforward ray theory, which
predicts an amplitude rising to infinity at the caustic and then falling
discontinuously to zero. The Airy-integral solution (375) 'heals' the wound,
allowing a perfectly finite and continuous transition between the two
different regimes.

Even within the limits of homogeneous one-dimensional wave systems,
the caustic need not itself be a ray. We showed in figure 60 how, if the
initial disturbance is *not* confined to a limited region, the rays in the (x, t)
plane can be nonconcurrent straight lines. These can have a curved envelope
where they run together (figure 98); and this is a caustic with two groups
of rays on one side and none on the other.

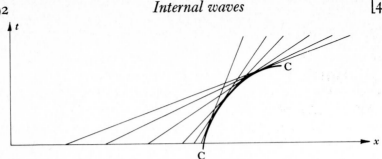

Figure 98. A caustic C as the envelope of straight-line rays $x - Ut =$ constant in the (x, t) diagram for a homogeneous one-dimensional wave system.

We can analyse this case by taking the disturbance in the form (356) with

$$t\psi(k) = t\omega(k) + g(k) - kx, \tag{378}$$

so that the stationary-phase condition is

$$x = g'(k) + t\omega'(k). \tag{379}$$

Equation (379) describes a system of straight rays, with the waves of wavenumber k situated initially at $x = g'(k)$. The caustic is defined by the equations (359), giving

$$x = g'(k_c) + t\omega'(k_c) \quad \text{with} \quad g''(k_c) + t\omega''(k_c) = 0. \tag{380}$$

The equations (380) are simply the parametric equations for the *envelope* of the straight lines (379); the parameter k_c (the wavenumber on the caustic) is defined as a function of time by the second equation. For such a curved caustic, equation (369) with ψ as in (378) gives

$$\zeta \approx \frac{2\pi F(k_c)}{[\frac{1}{2}g'''(k_c) + \frac{1}{2}t\omega'''(k_c)]^{\frac{1}{3}}} \, \text{Ai} \left\{ \frac{g'(k_c) + t\omega'(k_c) - x}{[\frac{1}{2}g'''(k_c) + \frac{1}{2}t\omega'''(k_c)]^{\frac{1}{3}}} \right\}$$
$$\times \exp\{i[g(k_c) + t\omega(k_c) - k_c x]\} \tag{381}$$

with properties similar to those discussed above.

Another straightforward extension of the one-dimensional theory is to propagation in two or three dimensions as analysed in section 4.8. It will suffice to give the details for an initially limited disturbance propagating in two dimensions.

The problem is then to estimate asymptotically an integral

$$q = \frac{1}{2} \int_{-\infty}^{\infty} \int_{-\infty}^{\infty} Q_0(k_1, k_2) \exp[it\psi(k_1, k_2)] \, dk_1 \, dk_2. \tag{382}$$

The general theory of stationary phase gives the estimate in the form (262), which however becomes infinite where the Jacobian

$$J = \partial(U_1, U_2)/\partial(k_1, k_2) \tag{383}$$

vanishes. The infinity is inevitable in a theory where energy travels along rays; in fact, these must have run together where J vanishes, because the volume occupied at time t by energy in the wavenumber element $dk_1\,dk_2$ is then

$$|\partial(U_1 t, U_2 t)/\partial(k_1, k_2)|\,dk_1\,dk_2 = t^2 |J|\,dk_1\,dk_2. \tag{384}$$

However, the assumptions of ray theory (slow variation of $\mathbf{k}$) have also broken down. Once more, the Airy integral 'heals the wound'.

As in section 3.8, we estimate (382) by using local axes in which the tensor

$$\partial^2\psi/\partial k_i\,\partial k_j = \partial^2\omega/\partial k_i\,\partial k_j \tag{385}$$

has only diagonal elements. Now, at any point (k_1^c, k_2^c) of the caustic, since the *determinant* J of that tensor vanishes, one of those diagonal elements must itself be zero; we suppose that *only* one of them is zero, and we number the axes so that it is $\partial^2\omega/\partial k_1^2$. Near (k_1^c, k_2^c) we estimate (382) as the product

$$q = \int \exp\{it[(k_1 - k_1^c)(\partial\psi/\partial k_1)^c + \tfrac{1}{6}(k_1 - k_1^c)^3(\partial^3\psi/\partial k_1^3)^c]\}\,dk_1$$
$$\times \tfrac{1}{2}Q_0(k_1^c, k_2^c)\{\exp[it\psi(k_1^c, k_2^c)]\}\int \exp[\tfrac{1}{2}it(k_2 - k_2^c)^2(\partial^2\psi/\partial k_2^2)^c]\,dk_2, \tag{386}$$

where the first line estimates the integral with respect to k_1 in an Airy-integral form, and the second line comprises all the other factors including a Gaussian integral with respect to k_2. The result is a product of an Airy-integral term and an ordinary stationary-phase expression:

$$q = \frac{2\pi Q_0(k_1^c, k_2^c)}{[\tfrac{1}{2}t(\partial^3\omega/\partial k_1^3)^c]^{\frac{1}{3}}}\,\mathrm{Ai}\left\{\frac{t(\partial\omega/\partial k_1)^c - x_1}{[\tfrac{1}{2}t(\partial^3\omega/\partial k_1^3)^c]^{\frac{1}{3}}}\right\}\exp[i(\omega^c t - \mathbf{k}^c\cdot\mathbf{x})]$$
$$\times \left[\frac{2\pi}{t|(\partial^2\omega/\partial k_2^2)^c|}\right]^{\frac{1}{2}}\exp\left[\tfrac{1}{4}\pi i\,\mathrm{sgn}\left(\frac{\partial^2\omega}{\partial k_2^2}\right)^c\right]. \tag{387}$$

(Here, the real part is to be taken as in section 4.8 because the $\tfrac{1}{2}$ in (386) has been suppressed and the complex-conjugate term with $(-k_1^c, -k_2^c)$ for (k_1^c, k_2^c) omitted. In three dimensions, the second line would be repeated with k_3 for k_2.)

We may apply (387) to the waves generated when a stone is thrown into a pond. A circular calm region in the centre grows at the minimum group velocity $U_c = U(k_c)$ corresponding to waves of length 44 mm (see the end of section 3.6). Near the circumference of that circle, equation (387) applies. The principal directions are radial, with

$$\partial\omega/\partial k_1 = U_c, \quad \partial^2\omega/\partial k_1^2 = 0, \quad \partial^3\omega/\partial k_1^3 = U_c^2/k_c\,\omega_c, \tag{388}$$

Figure 99. Waves present at a large time t after a stone is thrown into a pond. The point C is at a distance $U_c t$ from the centre of the disturbance, where U_c is the minimum group velocity 0.18 m s^{-1}. The wave amplitude decays exponentially behind C but fluctuates ahead of C. Wave crests travel forward at a speed 58% faster than that of C itself.

and azimuthal, with

$$\partial\omega/\partial k_2 = 0, \quad \partial^2\omega/\partial k_2^2 = U_c/k_c, \tag{389}$$

giving

$$q = \frac{2\pi Q_0(k_c, 0)}{(\tfrac{1}{2} t U_c^2/k_c\,\omega_c)^{\frac{1}{3}}} \, \text{Ai}\left[\frac{U_c t - x_1}{(\tfrac{1}{2} t U_c^2/k_c\,\omega_c)^{\frac{1}{3}}}\right]\left(\frac{2\pi k_c}{U_c t}\right)^{\frac{1}{2}} \exp\left[i(\omega_c t - k_c x_1 + \tfrac{1}{4}\pi)\right]. \tag{390}$$

Figure 99 shows the form of this expression after a large time t; then it makes a transition from wave amplitudes of order t^{-1}, modulated as a result of 'beats' between gravity waves and capillary waves, through a peak amplitude of order $t^{-\frac{5}{6}}$, to a calm inner region with waves of exponentially vanishing amplitude. The circular caustic grows at velocity U_c while the wave crests travel at the faster speed $\omega_c/k_c = 1.58 U_c$. Accordingly, in the representation relative to the caustic (figure 99), the waves are travelling forwards (emerging 'from nowhere') at speed $0.58 U_c$.

The results of section 4.9 on waves from an oscillating source can be similarly generalised. Again we set them out for two-dimensional propagation; a case with important applications to ship waves (section 4.12).

The full expression (275) for two-dimensional waves from an oscillating source is approximated with an exponentially small error as the two-dimensional form of equation (284); there, the x_1-axis is taken in the direction in which the value of the solution is being estimated. If the curvature $\kappa^{(0)}$ at the point of S_+ with normal $\mathbf{n}$ in that direction is nonzero, then the asymptotic form of (284) is as in (290).

However, a point of inflexion C on the wavenumber curve S (that is, a point of zero curvature) generates a *caustic*. This stretches out from the source in the direction of the *normal at the point of inflexion* C. Because rays in the directions of normals from points on the curve at either side of C lie on the *same* side of the caustic, there are two groups of waves on that side and none on the other.

We estimate the waves at points either on *or near* the caustic by taking the x_1-axis exactly along the caustic (that is, along the normal at C). Then the two-dimensional form of (284) becomes

$$q = -2\pi i[\exp(i\omega_0 t)] \int_{S_+} F(k_1, k_2) [\partial B/\partial k_1]^{-1} \exp[-i(k_1 x_1 + k_2 x_2)] \, dk_2,$$

(391)

where x_2, measuring displacement from the caustic, is nonzero, although much smaller than x_1.

On the wavenumber curve S, an expansion of k_1 as a function of k_2 near the point of inflexion $k_2 = k_2^c$ is not as in (286) because $(d^2 k_1/dk_2^2)^c$ vanishes. Indeed, it is

$$k_1 = k_1^c + \tfrac{1}{6}(k_2 - k_2^c)^3 (d^3 k_1/dk_2^3)^c + O(k_2 - k_2^c)^4.$$

(392)

The exponential inside the integral in (391) can therefore be approximated as

$$\exp\{-i[k_1^c x_1 + k_2^c x_2 + (k_2 - k_2^c) x_2 + \tfrac{1}{6}(k_2 - k_2^c)^3 (d^3 k_1/dk_2^3)^c x_1]\}, \quad (393)$$

which is the exponential of a term independent of $k_2 - k_2^c$ plus a linear term plus a cubic term. This allows (391) to be estimated in terms of the Airy integral as

$$q = -\frac{4\pi^2 i F(k_1^c, k_2^c) \exp[i(\omega_0 t - k_1^c x_1 - k_2^c x_2)]}{[\partial B/\partial k_1]^c [\tfrac{1}{2} x_1 (d^3 k_1/dk_2^3)^c]^{\frac{1}{3}}} \text{Ai}\left\{\frac{x_2}{[\tfrac{1}{2} x_1 (d^3 k_1/dk_2^3)^c]^{\frac{1}{3}}}\right\}.$$

(394)

The 'wound', separating a ray-theory solution (290) varying as $x_1^{-\frac{1}{2}}$, but with a coefficient $|\kappa^{(0)}|^{-\frac{1}{2}}$ becoming infinite at C, from a region beyond the caustic where no rays penetrate, is 'healed' by this solution; it allows the wave amplitude to rise, in fact, to a value of order $x_1^{-\frac{1}{3}}$ before exponentially decaying to zero beyond the caustic. Ship waves are used in section 4.12 to illustrate this behaviour.

We conclude this section with some account of caustics in non-homogeneous wave systems, although we confine ourselves to the stratified case (122) where only one coordinate z appears explicitly in the dispersion relationship. Then, along a ray, the frequency ω and the horizontal components of wavenumber k and l remain constant. Accordingly, the dispersion relationship defines the vertical component of wavenumber m as a function

$$m = m(z).$$

(395)

For example, with sound waves

$$[m(z)]^2 = \omega^2[c_0(z)]^{-2} - k^2 - l^2, \tag{396}$$

while with internal waves

$$[m(z)]^2 = (k^2 + l^2)\{\omega^{-2}[N(z)]^2 - 1\}. \tag{397}$$

In both cases $[m(z)]^2$ may vanish (with nonzero first derivative) at a *caustic*, where either $c_0(z)$ takes the value $\omega(k^2 + l^2)^{-\frac{1}{2}}$ or $N(z)$ takes the value ω. On one side of the caustic there are two groups of waves, corresponding to opposite signs for $m(z)$ (figures 79 and 80), and on the other (where there is no real solution for $m(z)$) no waves.

In each case, for waves of given ω, k and l, an equation of the form

$$q = Q(z)\exp[i(\omega t - kx - ly)], \quad \text{where} \quad Q''(z) + [m(z)]^2 Q(z) = 0 \tag{398}$$

must be satisfied by a suitably chosen physical quantity q. This was proved in (137) for the upward mass flux q in internal waves; and follows for sound waves from the equations of chapter 2 if the quantity q is

$$p_e/\rho_0^{\frac{1}{2}}. \tag{399}$$

Ray theory (see equations (138)) suggests solutions

$$Q(z) \doteq K[m(z)]^{-\frac{1}{2}}\exp\left[\pm i\int_{z_0}^z m(z)\,dz\right], \tag{400}$$

where the rays carrying energy upwards, for example, have the negative sign for sound waves but the positive sign for internal waves. Those solutions, however, have an unacceptable discontinuity from a value tending to infinity as the caustic $m = 0$ is approached to a zero value immediately beyond it.

Near a caustic $z = z_1$ forming an upper boundary to the wave region, where

$$[m(z)]^2 = \beta_1(z_1 - z) + O(z_1 - z)^2 \tag{401}$$

with β_1 positive, it is appropriate to use a solution of the equation

$$Q''(z) + \beta_1(z_1 - z)Q(z) = 0 \tag{402}$$

to 'heal' this unacceptable discontinuity. The Airy integral (368) again fulfils this function.

In fact, (368) shows that

$$\mathrm{Ai}''(X) - X\,\mathrm{Ai}(X) = (2\pi)^{-1}\int_{\infty\exp(\frac{5}{6}\pi i)}^{\infty\exp(\frac{1}{6}\pi i)}(-s^2 - X)\exp[i(sX + \tfrac{1}{3}s^3)]\,ds$$

$$= (2\pi)^{-1}i\int_{\infty\exp(\frac{5}{6}\pi i)}^{\infty\exp(\frac{1}{6}\pi i)}d\{\exp[i(sX + \tfrac{1}{3}s^3)]\} = 0 \tag{403}$$

because the exponential in curly brackets vanishes at both limits. Therefore,

$$Q(z) = Q_1 \operatorname{Ai}[\beta_1^{\frac{1}{3}}(z - z_1)] \tag{404}$$

is a solution of (402) which involves no waves where $\beta_1^{\frac{1}{3}}(z - z_1)$ is large and positive. On the other hand, where $\beta_1^{\frac{1}{3}}(z - z_1)$ is large and negative, equation (374) shows that the solution $Q(z)$ behaves like

$$Q_1 \pi^{-\frac{1}{2}} \beta_1^{-\frac{1}{12}} (z_1 - z)^{-\frac{1}{4}} \cos\left[\tfrac{2}{3}\beta_1^{\frac{1}{2}}(z_1 - z)^{\frac{3}{2}} - \tfrac{1}{4}\pi\right]. \tag{405}$$

This may be compared with the local form of ray-theory solution given by (400) and (401) as

$$K\beta_1^{-\frac{1}{4}}(z_1 - z)^{-\frac{1}{4}} \exp\left\{\pm i[\alpha_1 - \tfrac{2}{3}\beta_1^{\frac{1}{2}}(z_1 - z)^{\frac{3}{2}}]\right\} \tag{406}$$

in terms of an arbitrary complex amplitude K and a constant of integration

$$\alpha_1 = \int_{z_0}^{z_1} m(z)\,dz. \tag{407}$$

When the Airy integral is used to 'heal' a ray-theory solution in this way, it gives always the unequivocal and highly useful result that, if the two solutions (406) are to add up to (405), then

$$\alpha_1 = n_1\pi + \tfrac{1}{4}\pi \tag{408}$$

with n_1 an integer. Also, for both solutions

$$K = \tfrac{1}{2}\pi^{-\frac{1}{2}} \beta_1^{\frac{1}{6}} Q_1(-1)^{n_1}. \tag{409}$$

These results show how ray-theory solutions are reflected at caustics: the amplitude is unchanged, while the phase relation is such that the wave and its reflection combine in a cosine as in (405) which reaches its maximum just before the caustic. Its phase has gone $\tfrac{1}{4}\pi$ past that maximum where the caustic is reached. The graph of $\operatorname{Ai}(X)$ itself (figure 97) reflects this behaviour.

Similarly, near a caustic forming a lower boundary to the wave region,

$$[m(z)]^2 = \beta_2(z - z_2) + O(z - z_2)^2 \tag{410}$$

with β_2 positive. A local solution of (398) is then

$$Q = Q_2 \operatorname{Ai}[\beta_2^{\frac{1}{3}}(z_2 - z)] \tag{411}$$

involving no waves where $\beta_2^{\frac{1}{3}}(z - z_2)$ is large and negative. Where that is large and positive, however, Q behaves like

$$Q_2 \pi^{-\frac{1}{2}} \beta_2^{-\frac{1}{12}} (z - z_2)^{-\frac{1}{4}} \cos\left[\tfrac{2}{3}\beta_2^{\frac{1}{2}}(z - z_2)^{\frac{3}{2}} - \tfrac{1}{4}\pi\right], \tag{412}$$

whereas the solutions (400) become

$$K\beta_2^{-\frac{1}{4}}(z-z_2)^{-\frac{1}{4}}\exp\{\pm i[\tfrac{2}{3}\beta_2^{\frac{1}{2}}(z-z_2)^{\frac{3}{2}}-\alpha_2]\}, \tag{413}$$

with

$$\alpha_2 = \int_{z_2}^{z_0} m(z)\,\mathrm{d}z. \tag{414}$$

Now the matching gives for both solutions

$$\alpha_2 = n_2\pi + \tfrac{1}{4}\pi \tag{415}$$

with n_2 another integer, and

$$K = \tfrac{1}{2}\pi^{-\frac{1}{2}}\beta_2^{\frac{1}{2}}Q_2(-1)^{n_2}, \tag{416}$$

indicating how ray-theory solutions are reflected at lower caustics.

One application of these results is to finding conditions for trapped waves. The sum of two solutions (400) can satisfy the conditions at *both* an upper and a lower caustic provided that both values of K are the same and the quantities α_1 and α_2 given by (407) and (414) satisfy (408) and (415). Adding the last two conditions, we obtain the simple rule

$$\int_{z_2}^{z_1} m(z)\,\mathrm{d}z = (n+\tfrac{1}{2})\pi \tag{417}$$

as the condition for a trapped wave. Here, $n = n_1 + n_2$ can be any integer. The condition (417) is famous also in quantum mechanics, in the theory of trapped-wave solutions of the Schrödinger equation (72).

Note that the $\tfrac{1}{2}\pi$ arises in (417) because $|Q|$ reaches its maximum just short of each caustic, with the phase going $\tfrac{1}{4}\pi$ past that maximum by the time the caustic is reached. The fundamental mode $n = 0$ (for internal waves, the mode of highest frequency, corresponding to sinuous oscillations of an ocean thermocline) has those two maxima *coinciding*. Ray theory, even when 'healed' by Airy integrals, may be only a crude representation of the waveform in that case, but the condition (417) for the frequency is often rather accurate.

Higher values of n correspond to the *number of crossings* of the axis by the graph of $Q(z)$, emphasised in the discussion of equation (71) as different for each trapped-wave solution. Waveforms with $n \geqslant 1$ are often quite well represented by using the two Airy-integral representations (404) and (411) up to the first axis crossing in each case; and, if $n > 1$ so that these do not coincide, ray theory in between (figure 100).

Sound waves may travel well over a calm cold lake in 'inversion' conditions, when the air temperature increases with height in a roughly linear

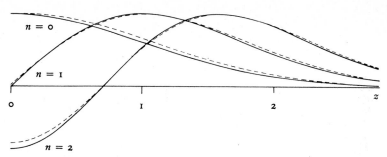

Figure 100. Full lines: trapped-wave solutions of (398) for the case $[m(z)]^2 + z^2 =$ constant. Broken lines: approximate solutions given by applying (417). These approximate solutions are calculated as proportional to $\mathrm{Ai}\left[-\left(\tfrac{3}{2}\int_z^{z_1} m(z)\,\mathrm{d}z\right)^{\frac{2}{3}}\right]$, which agrees with (404) near $z = z_1$ and may be shown by (374) to agree with ray theory for large $z_1 - z$. The constant value of $[m(z)]^2 + z^2$ happens to be exactly $n + \tfrac{1}{2}$ in both the exact and approximate theories. The solutions are even or odd functions of z according as n is even or odd. All solutions are here normalised to have the same maximum value.

manner. For this behaviour, a trapped wave may be responsible. The boundary condition, that $\partial p_e/\partial z$ vanishes on the lake surface $z = 0$, can be satisfied at the maximum ($X = -1.02$) of the Airy-integral solution (404). This makes

$$z_1 = 1.02\beta_1^{-\frac{1}{3}}, \quad \text{where} \quad \beta_1 = \omega^2 c_0^{-2}(T^{-1}\mathrm{d}T/\mathrm{d}z) \tag{418}$$

is the rate of *decrease* of (396) with height and z_1 is the height of the caustic. Wave energy for any given frequency, then, may be confined to a layer of height z_1, varying as $\omega^{-\frac{2}{3}}$. This tends to enhance the acoustic intensity at the surface for the sounds of relatively higher frequency.

4.12 Wave generation by travelling forcing effects

Waves generated by oscillating sources have been analysed above for homogeneous systems of three kinds: where the curvature (or Gaussian curvature) of the wavenumber curve (or surface) is nowhere zero (section 4.9), or is everywhere zero (section 4.10), or just vanishes locally (section 4.11). Now we show that this body of theory has many more applications than may be obvious at first.

This is because the theory can be immediately generalised (through, essentially, the Doppler relationship (143)) to wave generation by forcing effects travelling at a uniform velocity. Such forcing effects may, as before, be oscillating sources of fixed frequency ω_0. Particular importance may be attached, however, to travelling forcing effects that are *steady* (the case $\omega_0 = 0$); these include the waves made by the steady motion of a ship, described in a preliminary way in section 3.10.

In this section we give first the quite simple modifications that are needed in the theory of the last three sections to allow for forcing effects of frequency ω_0 travelling with uniform velocity. We take that velocity as $(-\mathbf{V})$ so that, as in section 4.6, $\mathbf{V}$ is the fluid's velocity relative to the source's motion. Ship waves (the case $\omega_0 = 0$ for gravity waves on deep water) are given as a principal example; though the theory is also used to indicate the geometry of wave patterns generated by oscillations of a moving ship at nonzero frequency. Lastly, the internal waves generated by an object moving steadily in stratified fluid are analysed at some length.

A source of waves oscillating at frequency ω_0 and travelling at velocity $(-\mathbf{V})$ may be represented in a differential equation or boundary condition by a forcing term

$$f(\mathbf{x} + \mathbf{V}t) \exp(\mathrm{i}\omega_0 t), \tag{419}$$

generalising equation (267). Then, if $f(\mathbf{x})$ has the Fourier transform $F(\mathbf{k})$ as in (268), the forcing term (419) becomes

$$\int_{-\infty}^{\infty} \int_{-\infty}^{\infty} \int_{-\infty}^{\infty} F(\mathbf{k}) \exp\{\mathrm{i}[\omega_0 t - k_j(x_j + V_j t)]\}\, \mathrm{d}k_1\, \mathrm{d}k_2\, \mathrm{d}k_3 \tag{420}$$

(where, however, the variable k_3 is suppressed in two-dimensional problems). In (420), the waves generated with wavenumber k_j have frequency

$$\omega_0 - k_j V_j = \omega_\mathrm{r} \tag{421}$$

relative to the fluid. Equation (421) coincides with the Doppler relationship (143) already established.

The corresponding wave motions can be written down immediately as

$$q = \int_{-\infty}^{\infty} \int_{-\infty}^{\infty} \int_{-\infty}^{\infty} \frac{F(\mathbf{k}) \exp\{\mathrm{i}[\omega_0 t - k_j(x_j + V_j t)]\}}{B(\omega_0 - k_j V_j, \mathbf{k})}\, \mathrm{d}k_1\, \mathrm{d}k_2\, \mathrm{d}k_3 \tag{422}$$

if, as in section 4.9, the governing differential equation or boundary condition acquires a right-hand side

$$B(\omega, \mathbf{k})\, a \exp[\mathrm{i}(\omega t - k_j x_j)] \tag{423}$$

for a sinusoidal wave

$$a \exp[\mathrm{i}(\omega t - k_j x_j)]. \tag{424}$$

In (422), the wavenumber surface S (or, in two dimensions, wavenumber curve) is defined as the locus

$$B(\omega_0 - k_j V_j, \mathbf{k}) = 0 \tag{425}$$

on which the integrand has poles; that is, where the dispersion relationship $B(\omega, \mathbf{k}) = 0$ for waves without forcing is satisfied by ω_r, the frequency (421) relative to the fluid.

All the theory needed to estimate the waves (422) is available in the previous three sections. We only need to

$$\text{replace} \quad B(\omega_0, \mathbf{k}) \quad \text{by} \quad B(\omega_0 - k_j V_j, \mathbf{k}) = B_r(\omega_0, \mathbf{k}), \qquad (426)$$

while at the same time we

$$\text{replace} \quad x_j \quad \text{by} \quad x_j + V_j t = X_j \qquad (427)$$

(a coordinate relative to the source), to make the integral (269) identical with (422). Therefore we can use any part of the general theory to estimate (422) if we simply make these substitutions throughout.

In particular, the substitution (426) replaces the equation (278) for the wavenumber surface (or curve) by (425). Even for waves isotropic relative to the fluid the motion of the source makes this latter equation (425) anisotropic. Therefore, the form of the general theory given in this chapter (a theory for anisotropic waves) is required even in isotropic systems to estimate waves made by travelling sources.

We found much earlier (section 3.9) that care is needed in correctly applying the radiation condition for travelling forcing effects. In the general theory for oscillating sources, the radiation condition locates waves of wavenumber $\mathbf{k}$ along a direction $\mathbf{n}$ normal to S, selecting from among the two normal directions the one along which ω is *increasing* above ω_0. A convenient way of expressing this is to use $S(\omega_0)$ to denote the wavenumber surface (425) for sources of frequency ω_0; then, the waves are found along that normal $\mathbf{n}$ to $S(\omega_0)$ which points in the direction of $S(\omega_0 + \delta)$ with $\delta > 0$.

The above rule has a simple physical interpretation. By (421), this normal direction $\mathbf{n}$ is that of the gradient vector

$$\partial \omega_0 / \partial k_j = V_j + \partial \omega_r / \partial k_j = V_j + U_j. \qquad (428)$$

Here, U_j is the group velocity relative to the fluid. One result of the substitution (427), however, is that the position of waves is described relative to the *source*; the vector (428) is, evidently, their energy propagation velocity in that frame of reference.

Two-dimensional waves generated by travelling forcing effects must take the asymptotic form given by equations (288), (289) and (290) after the substitutions (426) and (427). Thus, in an arbitrarily chosen direction L, they include contributions

$$q = F(k_1^{(0)}, k_2^{(0)}) [(\partial B_r/\partial n)^{(0)}]^{-1} (8\pi^3/\kappa^{(0)} X)^{\frac{1}{2}} \exp\left[i(\omega_0 t - k_j^{(0)} X_j + \Theta)\right] \quad (429)$$

from any points $(k_1^{(0)}, k_2^{(0)})$ on the wavenumber curve $S(\omega_0)$ where the normal **n**, pointing towards $S(\omega_0 + \delta)$, is in the direction of L. However, near a point of inflexion of $S(\omega_0)$, it is necessary, instead, to use a similarly modified version of (394).

We can easily obtain the power P_w needed to generate the waves in this case (for the similar expression in three dimensions, see (299) or (300)). If the energy per unit area for waves of wavenumber **k** propagating two-dimensionally is

$$W = W_0(\mathbf{k}) \overline{q^2}, \quad (430)$$

the directional distribution of wave energy is given by (429) as

$$W = W_0(\mathbf{k}^{(0)}) |F(\mathbf{k}^{(0)})|^2 [(\partial B_r/\partial n)^{(0)}]^{-2} (4\pi^3/\kappa^{(0)} X) \quad (431)$$

for each $\mathbf{k}^{(0)}$ satisfying (288). The energy flux, relative to the source, across unit transverse distance per unit time is then

$$\mathbf{I} = W(\mathbf{V} + \mathbf{U}); \quad (432)$$

here by (428), the vector $\mathbf{V} + \mathbf{U}$ points in the direction **n**, along a straight ray stretching outwards from the source. The transverse distance between rays from an element of the wavenumber curve of lengths ds is $\kappa^{(0)} X ds$ at distance X from the source. Therefore, the total power output can be written as an integral along the whole wavenumber curve S of the magnitude of **I** times this transverse distance. The integral, with the index (0) dropped, is

$$P_w = 4\pi^3 \int_S W_0(\mathbf{k}) |F(\mathbf{k})|^2 |\mathbf{V} + \mathbf{U}| (\partial B_r/\partial n)^{-2} ds. \quad (433)$$

Here, the magnitude of $\mathbf{V} + \mathbf{U}$ can be written

$$|\mathbf{V} + \mathbf{U}| = -(\partial B_r/\partial n)/(\partial B_r/\partial \omega) \quad (434)$$

since by (428) it is the rate of change of ω_0 in the direction **n** in changes keeping $B_r(\omega_0, \mathbf{k})$ equal to zero. Equation (434) leads to a form of P_w directly similar to (300). Alternatively, it leads to an often useful form

$$P_w = 4\pi^3 \int_S W_0(\mathbf{k}) |F(\mathbf{k})|^2 |\partial B_r/\partial \omega|^{-1} |\partial B_r/\partial n|^{-1} ds. \quad (435)$$

Here, for purposes of calculation we can use the identities

$$|\partial B_r/\partial n|^{-1} ds = |\partial B_r/\partial k|^{-1} |dl| = |\partial B_r/\partial l|^{-1} |dk| \quad \text{on} \quad S. \quad (436)$$

(For example, if the normal $\mathbf{n}$ makes an acute angle Φ with the x-axis, then $|\partial B_r/\partial k|$ is $|\partial B_r/\partial n|\cos\Phi$ while on S we have $|dl| = ds\cos\Phi$.)

As these forms for the power P_w put into the waves do not bring in the curvature of S explicitly, there is no infinity of the integrand at any point of inflexion of S. Accordingly, the very local modification made by (394) near the caustic (ray from the point of inflexion) has insignificant effect on the power output; its importance lies only in limiting the local amplitude on a caustic.

Surface gravity waves on deep water give an excellent illustration of these results. Consider forced oscillations of the free surface, in which its elevation ζ takes the form

$$\zeta = a\exp\left[i(\omega t - kx - ly)\right]. \tag{437}$$

In deep water, the corresponding velocity potential ϕ (satisfying Laplace's equation and having the surface value of $\partial\phi/\partial z$ equal to $\partial\zeta/\partial t$) is

$$\phi = i\omega a(k^2 + l^2)^{-\frac{1}{2}}\exp\left[i(\omega t - kx - ly) + (k^2 + l^2)^{\frac{1}{2}} z\right]. \tag{438}$$

Then the boundary value of the water pressure, relative to atmospheric pressure, is

$$\eta = p - p_a = p_e - \rho_0 g\zeta = -\rho_0[\partial\phi/\partial t]_{z=0} - \rho_0 g\zeta. \tag{439}$$

This vanishes in free gravity waves (section 3.1). For the forced motions (438), however, it takes the value

$$\eta = B(\omega, k, l)\, a\exp\left[i(\omega t - kx - ly)\right], \tag{440}$$

where

$$B(\omega, k, l) = \rho_0\,\omega^2(k^2 + l^2)^{-\frac{1}{2}} - \rho_0 g. \tag{441}$$

As expected, the equation $B = 0$ coincides with the dispersion relationship.

Before concentrating on waves generated by the *steady* motion of a ship, we consider the implications of these results for an oscillating source that produces a nonzero boundary value

$$\eta = f(x + Vt, y)\exp\left(i\omega_0 t\right) \tag{442}$$

travelling at velocity $(-V)$ where

$$\mathbf{V} = (V, 0). \tag{443}$$

Then equation (426) makes

$$B_r = \rho_0(\omega_0 - Vk)^2(k^2 + l^2)^{-\frac{1}{2}} - \rho_0 g, \tag{444}$$

and the equation $B_r = 0$ of the wavenumber curve $S(\omega_0)$ can be written

$$(\omega_0 - Vk)^4 = g^2(k^2 + l^2). \tag{445}$$

These curves $S(\omega_0)$ are plotted in figure 101 for various values of a nondimensional frequency $V\omega_0/g$. Arrows on curves show the direction of the normal $\mathbf{n}$ pointing towards curves with increasing ω_0. A 'c' on certain curves marks a point of inflexion, which must give rise (section 4.11) to a caustic.

The curve $S(0)$, with $V\omega_0/g = 0$, has two symmetrical branches; the normals to them fill a wedge of semi-angle $\sin^{-1}(\frac{1}{3}) = 19.5°$ representing, as we shall see, the ship waves of figure 70. The points of inflexion on $S(0)$ give rise to the caustics forming the boundary of that wedge.

When $V\omega_0/g$ takes small positive values (such as 0.125) the two branches of which $S(0)$ is composed are both displaced to the right. Certain waves with larger wavenumbers, found on the right-hand curve, fill a narrower wedge than before. Other waves with smaller wavenumbers, found on the left-hand curve, fill a wider wedge. In addition, $S(\omega_0)$ includes a very small oval branch involving wavenumbers of the order of $g^{-1}\omega_0^2$; the energy in these very long waves propagates so fast that they are disposed all round the travelling source.

Beyond a critical value 0.25 for $V\omega_0/g$ (when the curve $S(\omega_0)$ crosses itself) there are only two branches and the waves associated with each lie within a wedge. The boundary of such a wedge is formed in each case by caustics arising from points of inflexion on the curve. Figure 102 plots the semi-angles of both the wider wedge associated with the left-hand branch and the narrower wedge associated with the right-hand branch as a function of $V\omega_0/g$. (For $V\omega_0/g < 0.25$, a value of $180°$ is also included, to represent the fact that waves associated with the small oval branch are found in all directions around the forcing region.)

Figure 102 shows that waves are found in front of the source (that is, in the forward-facing semicircle) only if $V\omega_0/g < 0.27$. We may also note, before concentrating on the true ship-wave wedge (figure 70) due to the steady component ($\omega_0 = 0$) of a ship's motion, that waves associated with any unsteady components are confined within the *same* wedge if and only if $V\omega_0/g > 1.63$.

Any ship in steady motion can be regarded as producing a nonzero form of the boundary value η, which can generate waves. This is most obvious for a hovercraft, suspended above the water surface by the pressure of its 'air cushion'. Then equation (439) defines η as the excess of that pressure over atmospheric.

The distribution of η equivalent to an immersed ship may often be estimated to sufficient accuracy by writing the velocity potential ϕ as

$$\phi = \phi_h + \phi_w, \tag{446}$$

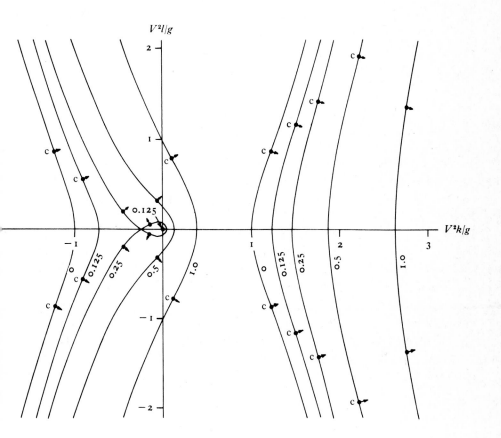

Figure 101. Wavenumber curves $S(\omega_0)$, representing surface gravity waves generated by a source oscillating with frequency ω_0 and travelling to the left at velocity V, for different values of $V\omega_0/g$ marked on the curves. The arrows show the directions in which waves of each wavenumber are found.

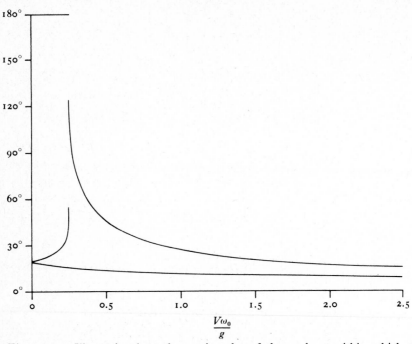

Figure 102. Illustrating how the semi-angles of the wedges, within which the waves represented in figure 101 lie, vary with $V\omega_0/g$.

where ϕ_h represents an irrotational flow that would be generated in unbounded fluid by the hull's motion, and ϕ_w is the velocity potential of the waves. Then the boundary value $\eta = \eta_w$ corresponding to the waves must satisfy the condition

$$\eta_w = -\eta_h \tag{447}$$

if the complete solution (446) is to satisfy the surface condition $\eta = 0$.

A 'double model' is often used, as described in section 3.10, for determining the frictional drag of a hull form. It is obtained by adding on to a model of the immersed part of the hull an inverted but otherwise identical shape which is its mirror image in the water surface. Such a double model may be used in a fully immersed condition for experimental determinations, not only of frictional drag but also of η_h. In steady motion of such a model the vertical displacement of fluid in the plane of symmetry is zero, so that equation (439) gives η_h as the distribution of excess pressure p_e measured in that plane of symmetry. (Alternatively, good computer programs exist for calculating this pressure distribution p_e in the plane of symmetry of such a double model.) Note, however, that smoothly extrapolated values must

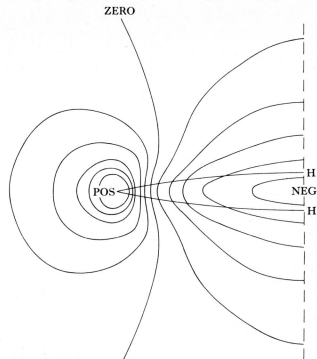

Figure 103. The distribution of excess pressure p_e in the plane of symmetry of a fully immersed double model of a ship hull H. These contours of equally spaced constant values of p_e are derived from a linear-theory computation for a parabolic ship with vertical sides and with draught–length ratio 0.25. The positive pressure peak (POS) near the bow is characteristic, however, of all sharp-bowed ship shapes. (This calculation, made to exhibit that intense localised pressure peak, pays no special attention to the stern and in fact takes the ship and the flow as symmetrical about the broken line.)

be used in that part of the plane of symmetry which actually intersects the hull form.

Evidently, the source region is wider than the ship itself (figure 103). The excess pressures p_e are spread out laterally in a way affected by both the ship's length and its draught (maximum depth). Nevertheless, the distribution of excess pressure is of limited extent, decaying as the inverse square of distance from the ship.

Whatever the form of

$$\eta_w = -\eta_h = f(x + Vt, y) = f(X, y) \tag{448}$$

so determined, the geometric pattern of wave crests is as in figure 70. This same pattern is derived whether we use the methods of section 3.10 or that of this section based on the wavenumber curve $S(o)$ of figure 101; in the latter case, the rule (319) is used for determining crest shapes.

From a knowledge of $f(X, y)$ and its two-dimensional Fourier transform $F(k, l)$, however, we can go further and determine the wavemaking power output (435). Using $W_0(\mathbf{k}) = \rho_0 g$ (the factor by which the mean square surface elevation must be multiplied to give the wave energy per unit area) and equations (436) and (444), we can write the wavemaking power output (435) as

$$P_\mathrm{w} = 4\pi^3 \int_{S(o)} \rho_0 g |F(k, l)|^2 \, |2\rho_0 \, Vk(k^2 + l^2)^{-\frac{1}{2}}|^{-1}$$
$$\times \, |\rho_0 \, V^2 k(k^2 + 2l^2) \, (k^2 + l^2)^{-\frac{3}{2}}|^{-1} \, |\mathrm{d}l|; \quad (449)$$

or, simplifying by use of the equation

$$V^4 k^4 = g^2(k^2 + l^2) \quad (450)$$

defining the curve $S(o)$, as

$$P_\mathrm{w} = 2\pi^3 V(\rho_0 g)^{-1} \int_{S(o)} [(k^2 + l^2)/(k^2 + 2l^2)] \, |kF(k, l)|^2 \, |\mathrm{d}l|. \quad (451)$$

This is quite a simple integral along the wavenumber curve of the quantity $|kF(k, l)|^2$, which may be called the *spectrum of* $\partial f/\partial x$, modified by a factor in square brackets which varies only between 1 and $\frac{1}{2}$. Wavemaking *resistance* D_w takes (section 3.10) the form $V^{-1} P_\mathrm{w}$. We now see that this can be estimated by (i) computing the pressure, $-f(X, y)$ in the plane of symmetry of a fully immersed double model in a stream of velocity V; (ii) integrating along the wavenumber curve (450) the spectrum of the pressure *gradient* $\partial f/\partial X$, modified by the said factor in square brackets; and (iii) multiplying by $2\pi^3(\rho_0 g)^{-1}$.

These facts are relevant to the search (section 3.10) for hull forms that will defer to relatively higher velocities a steep rise in the wavemaking resistance D_w. The lowest magnitude of the wavenumber on the curve $S(o)$ is gV^{-2}. Therefore, the wavemaking resistance begins to rise steeply when V reaches values where the gradient $\partial f/\partial X$ of pressure in the central plane of a fully immersed double model has significant spectral components around gV^{-2}. This steep rise may be deferred to higher velocities by choice of hull forms where spectral components of $\partial f/\partial X$ with relatively high wavenumber are 'ironed out'.

For example, figure 103 shows that the form of the excess pressure includes large gradients varying rather steeply in the region ahead of the bow of the double model where the excess pressure rises to its maximum

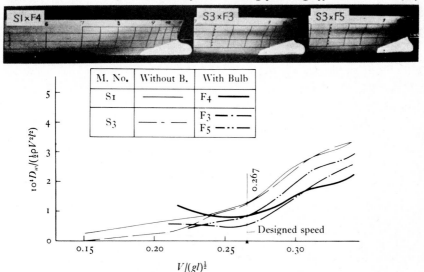

Figure 104. Results of Inui's classic experiments on ship models (of length 2.4 m) with bulbous bows designed so as virtually to eliminate the waves generated by the bow at the designed speed. Photographs confirmed the elimination of bow waves, which led to the large reductions in wavemaking drag D_w here shown.
[*Trans. Soc. Naval Architect. Mar. Eng.*]

value. These may be smoothed out by introducing an element of bulbous shape below the surface in this region (figure 104). The double model then includes two bulbous elements. By themselves they generate a pressure reduction on the central plane due to acceleration of the water motion relative to the model as it travels between them (Venturi effect). Careful design can make this distribution of negative excess pressure, due to the sub-surface bulbous element at the bow, cancel out enough of the distribution of positive excess pressure due to the main hull so that the spectral components of $\partial f/\partial X$ with higher wavenumbers are greatly reduced. The introduction of such 'bulbous bows' has deferred the steep rise of wavemaking resistance with speed (figure 104) enough to improve significantly the economics of shipping.

Wave patterns generated at somewhat higher speeds may be dominated by the waves near the caustic. These are given by equation (394) if special axes (x_1, x_2) are used so that the x_1-axis is along the caustic. For example, near the caustic
$$X = +8\tfrac{1}{2}y, \tag{452}$$
the required axes (x_1, x_2) are such that the wavenumber components become
$$k_1 = \tfrac{1}{3}(8\tfrac{1}{2}k+l), \quad k_2 = \tfrac{1}{3}(8\tfrac{1}{2}l-k). \tag{453}$$

At one of the two points of inflexion with normal **n** in the direction (452) we have

$$k = (\tfrac{3}{2})^{\frac{1}{2}}gV^{-2}, \quad l = -(\tfrac{3}{4})^{\frac{1}{2}}gV^{-2}, \quad \text{giving } k_1^c = (\tfrac{3}{4})^{\frac{1}{2}}gV^{-2}, \quad k_2^c = -(\tfrac{3}{2})^{\frac{1}{2}}gV^{-2}. \tag{454}$$

We readily calculate that

$$(\partial B_r/\partial k_1)^c = (\tfrac{4}{3})^{\frac{1}{2}}\rho_0 V^2, \quad (d^3k_1/dk_2^3)^c = (2^{\frac{5}{2}}V^4)/(27g^2); \tag{455}$$

the values required to write down equation (394) for $\omega_0 = 0$, as modified by the substitutions (426) and (427). However, we should use twice that value (where the real part is to be taken as usual) because the other point of inflexion $(-k_1^c, -k_2^c)$ with normal in the direction (452) makes a complex conjugate contribution to the wave amplitude.

Thus, in coordinates (x_1, x_2) such that the x_1-axis is the caustic,

$$\zeta = -\frac{8\pi^2 iF^c \exp[-i(k_1^c x_1 + k_2^c x_2)]}{(\tfrac{4}{3})^{\frac{1}{2}}\rho_0 V^2 (\tfrac{1}{3}\sqrt{2})(V^4 x_1/g^2)^{\frac{1}{3}}} \text{Ai}\left\{\frac{x_2}{(\tfrac{1}{3}\sqrt{2})(V^4 x_1/g^2)^{\frac{1}{3}}}\right\}. \tag{456}$$

Here, the wavenumber (k_1^c, k_2^c) given by (454) indicates that the wave crests are at $35°$ to the caustic (figure 70). Waves effectively disappear where the term in curly brackets is about 2.5; that is, at a perpendicular distance

$$x_2 = 1.2(V^4 x_1/g)^{\frac{1}{3}} \tag{457}$$

ahead of the caustic. The dotted lines in figure 70 indicate the significant enlargement which this makes to the Kelvin ship-wave pattern.

We conclude this section with a discussion of three-dimensional patterns of internal waves generated by steady motion of an obstacle. For internal waves generated by a source distribution

$$\partial Q/\partial t = f(x, y, z)\exp(i\omega_0 t), \tag{458}$$

equation (327) leads to a form

$$B(\omega_0, k, l, m) = [\omega_0^2 m^2 - (N^2 - \omega_0^2)(k^2 + l^2)]/(\omega_0 m) \tag{459}$$

for B.

For an obstacle moving vertically upwards with speed V we must take $\mathbf{V} = (0, 0, -V)$. Then, by (426),

$$B_r(\omega_0, k, l, m) = [(\omega_0 + Vm)^2(k^2 + l^2 + m^2) - N^2(k^2 + l^2)]/[(\omega_0 + Vm)m]; \tag{460}$$

and the wavenumber surface $S(\omega_0)$ is defined by the equation $B_r = 0$.

Steady disturbances are described by $S(0)$, a surface with equation

$$V^2 m^2(k^2 + l^2 + m^2) = N^2(k^2 + l^2) \tag{461}$$

plotted in figure 105(a), where the arrows indicate the normal direction pointing towards $S(+\delta)$. The vertical component of this direction is in all cases downwards, indicating that the waves generated trail behind the rising obstacle. Experimental results for a rising sphere are shown in figure 106, indicating the shapes of surfaces of constant phase. These may also be computed from equation (319) with the results shown in figure 105(b), where the comparison with the measured shapes is good.

When an obstacle moves *horizontally* through stratified fluid, at velocity $(-\mathbf{V})$ where $\mathbf{V} = (V, 0, 0)$, we have

$$B_r(\omega_0, k, l, m) = [(\omega_0 - Vk)^2 (k^2 + l^2 + m^2) - N^2(k^2 + l^2)]/[(\omega_0 - Vk)m] \quad (462)$$

so that $S(0)$ is the surface

$$V^2 k^2 (k^2 + l^2 + m^2) = N^2(k^2 + l^2). \quad (463)$$

The waves generated are now more complicated because $S(0)$ is not a surface of revolution. Figure 107 indicates its shape by plotting its intersections by different planes $l = $ constant. The direction of $\mathbf{n}$ (the normal to the surface in the direction of $S(+\delta)$) is indicated in each case by an arrow showing the *component of* $\mathbf{n}$ *in the plane* $l = $ constant. (Thus, the shorter arrows correspond to normal directions $\mathbf{n}$ with a substantial y-component.) The waves lie in the directions of the arrows; in all cases, in the right-hand half-plane (*behind* the moving obstacle).

The limiting case of two-dimensional obstacles (cylinders with generators in the y-direction) exhibits peculiar features of definite interest. Two-dimensional internal waves have $B(\omega, k, m)$ given by (330), leading to

$$B_r(\omega_0, k, m) = [(\omega_0 - Vk)^2 (k^2 + m^2) - N^2 k^2]/[(\omega_0 - Vk)m] \quad (464)$$

so that $S(0)$ is the curve

$$k^2[V^2(k^2 + m^2) - N^2] = 0. \quad (465)$$

Equations (464) and (465) are, of course, simply (462) and (463) with l put equal to zero. It is important, however, that the resulting form of $S(0)$ consists (figure 108) of a circle of radius N/V together with the straight line $k = 0$ *taken twice* (because the factor k appears squared in (465)).

The need to allow for the twofold nature of the straight-line branch $k = 0$ is clear when we draw the curve $S(+\delta)$ in order to determine the direction of the normal $\mathbf{n}$. The part of $S(+\delta)$ close to $k = 0$ is in two branches (figure 108). Outside the circle both branches are on the same side of $k = 0$,

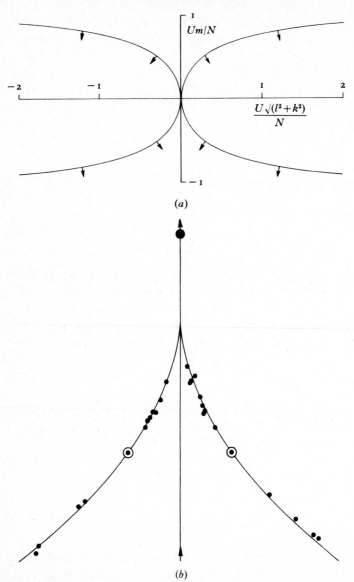

(a)

(b)

Figure 105. (a) Wavenumber surface $S(0)$ for internal waves generated by a steady disturbance moving vertically upwards with velocity V through stratified fluid with uniform Väisälä–Brunt frequency N.
(b) Full line: calculated shape of a surface of constant phase for internal gravity waves generated by steady vertical motion of a solid sphere (shown at top of diagram) through a uniformly stratified medium. Shape is normalised so that the points on the lines through the obstacle making an angle $\tan^{-1}(\frac{1}{4})$ with the vertical are in the ringed positions. Points: experimental results from figure 106 and from another similar photograph.

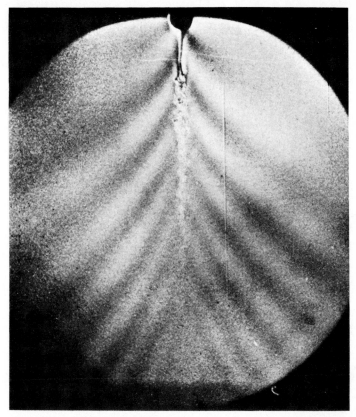

Figure 106. Schlieren photograph by Mowbray of internal gravity waves generated by steady vertical motion of a sphere (seen at top of diagram) of diameter 25.4 mm rising vertically at a speed 10.2 mm s⁻¹ in a salt solution whose density falls with height at a uniform rate of 0.2 kg m⁻³ per mm.

but inside it they are on opposite sides. This is why at any point of $k = 0$ *two* arrows should be drawn, and why one of them must point ahead of the obstacle at a point inside the circle (that is, for $|m| < N/V$). Indeed, it is evident from (464) that $B_r(\delta, k, m)$ vanishes for

$$Vk \doteq \delta \pm Nk|m|^{-1}, \quad \text{giving} \quad k \doteq \delta/(V \mp N|m|^{-1}); \quad (466)$$

these values of k are both positive if $|m| > N/V$, but one is negative if $|m| < N/V$.

When the integration with respect to k in the two-dimensional form of (422) is performed, subject to the radiation condition, contributions are

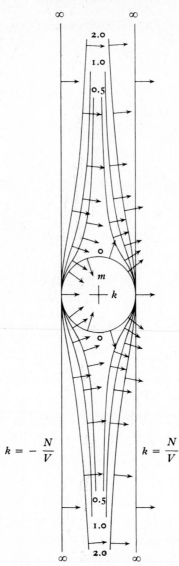

Figure 107. Wavenumber surface $S(0)$ for internal waves generated by a steady disturbance moving horizontally (to the left) with velocity V through stratified fluid with uniform Väisälä–Brunt frequency N. The different curves are the intersections of $S(0)$ by planes on which $V|l|/N$ takes different constant values, marked on the curves.

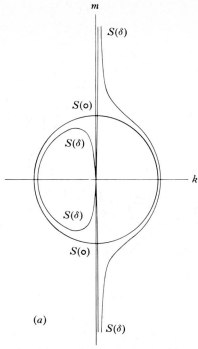

(a)

Figure 108. Two-dimensional internal waves generated by a stationary cylindrical obstacle (with generators perpendicular to the paper) in a horizontal flow of stratified fluid to the right (or due to the same obstacle moving to the left through stationary fluid).

(a) The wavenumber curves $S(\text{o})$ and $S(\delta)$. Near each point on the circular part (with radius N/V) of $S(\text{o})$, there is one point of $S(\delta)$, leading as usual to a single arrow on $S(\text{o})$. However, near each point on the straight part $k = \text{o}$ of $S(\text{o})$, there are two points of $S(\delta)$, leading to two arrows. Where $|m| > N/V$ these point in the *same* direction (both parts of $S(\delta)$ are on that side) but where $|m| < N/V$ they point in opposite directions.

made by two such poles (466), with δ taking (section 4.9) the pure imaginary value $(-i\epsilon)$. The residue at both is

$$\frac{F(\text{o}, m)\exp(-imz)}{[\partial B_r(\text{o}, k, m)/\partial k]_{k=\text{o}}} = \frac{Vm}{N^2 - V^2m^2}F(\text{o}, m)\exp(-imz). \qquad (467)$$

Physically, the poles reflect the values of the horizontal energy propagation velocity

$$\partial\omega_0/\partial k = V + \partial\omega_r/\partial k = V \mp N|m|^{-1} \qquad (468)$$

for internal waves with $k = l = \text{o}$ (the x-component of whose group velocity (91) is $\mp N|m|^{-1}$). One of the values (468) can be negative, corresponding

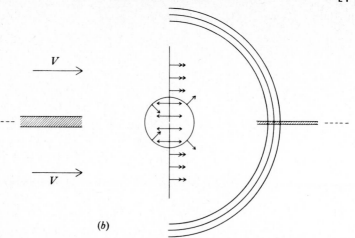

(b)

Figure 108.
(*b*) This arrangement of arrows is shown here, with the corresponding arrangement of waves. These include steady disturbances of relatively high vertical wavenumber downstream of the obstacle, and disturbances of relatively low vertical wavenumber |*m*| *upstream* of it. They also include 'lee waves' of wavelength $2\pi V/N$ on the lee side of the obstacle.

to energy propagation ahead of the obstacle, if $|m| < N/V$, making the group velocity exceed V.

Ahead of the obstacle, then, (422) is equal to an integral with respect to *m in this interval only* of $(-2\pi i)$ times the residue (467). This gives

$$q = -2\pi i \int_{-N/V}^{N/V} Vm(N^2 - V^2m^2)^{-1} F(\text{o}, m) \exp(-imz)\, \mathrm{d}m. \qquad (469)$$

Behind the obstacle, we have

$$q = -2\pi i \left(2\int_{-\infty}^{\infty} - \int_{-N/V}^{N/V} \right) Vm(N^2 - V^2m^2)^{-1} F(\text{o}, m) \exp(-imz)\, \mathrm{d}m, \qquad (470)$$

representing contributions from two poles (466) when $|m| > N/V$ and from only one when $|m| < N/V$.

We have shown that components of the source distribution on a relatively large vertical scale (corresponding to a vertical component of wavenumber less than N/V) generate an upstream disturbance. Fundamentally, this because the associated group velocity exceeds V.

Crudely, the mechanics of how 'two-dimensional' obstacles of relatively large vertical extent moving in stratified fluid can push fluid ahead of them is as follows. Some of the fluid in front, which is initially moving relative to the obstacle at velocity $(V, 0)$, may not have enough kinetic energy ($\frac{1}{2}\rho_0 V^2$ per unit volume) to yield the excess potential energy $\frac{1}{2}\rho_0 N^2 \zeta^2$ needed for vertical displacements ζ sufficient to circumvent the obstacle (either above or below). Accordingly, such fluid must pile up in a sheet ahead of the obstacle. This rough argument has the merit that it explains why the strong disturbance propagated ahead of a two-dimensional obstacle is completely absent for a three-dimensional one, which only generates waves behind it (figure 107). Evidently, fluid whose motion relative to a three-dimensional obstacle lacks the energy to circumvent it by vertical displacements can always make its way round by horizontal displacements.

The non-integrable singularity in the integral (469) at $|m| = N/V$ reflects a real physical phenomenon. The energy in Fourier components with $|m|$ just less than N/V can propagate forwards at speeds only just exceeding that of the obstacle. In the limit, it 'piles up' in front of the obstacle so that the disturbance is able to increase to very large values. This increase, however, is in practice limited by dissipation, by nonlinearity, or by finite duration of the forcing motion.

In addition to the sheet-like disturbances ahead of and behind the obstacle, figure 108 indicates the presence of cylindrical waves of length $2\pi V/N$, corresponding to wavenumbers on the circle. These waves are stationary relative to the obstacle, and found behind it at all angles θ to the horizontal less than $\frac{1}{2}\pi$. Their form is given by (429) as

$$q = F(V^{-1}N\cos\theta, V^{-1}N\sin\theta)(\tan\theta)(8\pi^3/NVX)^{\frac{1}{2}}\exp\left[-i(V^{-1}NX + \tfrac{1}{4}\pi)\right],$$
$$(471)$$

where X is distance from the obstacle. (In (471) we have used the facts that the circle's curvature κ is V/N while $(\partial B_r/\partial n)^{-1}$ is $(2N)^{-1}\tan\theta$; an extra factor 2 is again inserted to incorporate the complex conjugate term.) As $\theta \to \frac{1}{2}\pi$ there is a singular behaviour, for the same reason as before: these are waves with wavenumbers approaching $k = 0$, $|m| = N/V$, for which the magnitude of the energy propagation velocity $\mathbf{U} + \mathbf{V}$ again approaches zero. These waves too, then, take a very long time to escape from the forcing region; but the resulting build-up of waves vertically above the moving obstacle is limited in practice, once more, by dissipation, by nonlinearity, or by finite duration of the forcing motion.

4.13 Waveguides

An introduction to internal waves, like the present chapter, must avoid giving the impression that all interesting gravity waves in stratified fluids have wavelengths measured in metres (or, even, millimetres), rather than kilometres. To be sure, much of this chapter deals with ray theory and, therefore, with those relatively short waves to which that theory can be applied. Even the waves trapped on the ocean thermocline (section 4.3) have wavelengths only moderately greater. In this section, however, we describe waves with lengths of many tens of kilometres which, by combined acoustic and gravity-wave action, can propagate changes of pressure horizontally over great distances in the earth's stratified atmosphere.

Just as in other sections of this chapter, we embed this aspect of internal waves within the context of a general theory; this time, the theory of *quasi-one-dimensional waves in fluids*; or, more briefly, of waveguides. The strictly one-dimensional waves in fluids treated in chapter 2 involve fluid motions that to a close approximation are longitudinal; that is, in the direction of propagation. In waveguides, *propagation* in one dimension is brought about by fluid motions possessing substantial transverse as well as longitudinal components.

Good illustrations of the theory of dispersion and group velocity in one dimension are provided by waveguide propagation. Incidentally, waveguides for electromagnetic waves are much used in electrical engineering; but we discuss in this book only waveguides for waves in fluids.

We begin with acoustic waveguides, in which sound energy travels along ducts of constant cross-section. We show that, below a certain definite frequency ω_m, the wave motion must be strictly one-dimensional. At greater frequencies, however, it becomes highly three-dimensional. Details of the transition in a compact source's power output, as the frequency increases, from the strictly one-dimensional form to the three-dimensional form for effectively unbounded fluid (compare section 1.4) are found to be of particular interest.

Waveguides do not necessarily have boundaries all round: they can be *one-sided*. The atmosphere, only bounded at the bottom, operates as a one-sided 'gravity-acoustic waveguide' with nearly all the wave energy confined to within about 30 km of the boundary. We analyse this by use of the combined theory of sound and internal waves given in section 4.2.

From that standpoint, water with a free surface might be regarded as a 'one-sided waveguide' which allows one-dimensional propagation of surface waves with almost all their energy confined (chapter 3) to within about a quarter-wavelength of the free surface. True quasi-one-dimensional propagation is, however, illustrated better (since water waves, after all, spread in two horizontal dimensions) by the shoreline phenomenon of 'edge waves'. This section ends with an analysis of how 'edge waves' propagate along a beach signals of rather large wavelengths, their energy being confined to within a distance from the shoreline broadly comparable with that wavelength.

The acoustic waveguides we describe are hard-walled ducts of uniform cross-section A. We often use as an illustration the case when A is the rectangle

$$0 < y < b, \quad 0 < z < h \quad \text{with} \quad h \leqslant b. \tag{472}$$

In all cases, the x-axis is taken along the duct.

Sound of fixed frequency can travel along such a duct, not only in strictly one-dimensional propagation with

$$p_e = a \exp[i(\omega t - kx)], \quad \text{where} \quad \omega = c_0 k, \tag{473}$$

but also in 'waveguide modes'

$$p_e = a_{MN} \exp[i(\omega t - kx)] p_{MN}(y, z). \tag{474}$$

The expression (474) for p_e satisfies the wave equation if

$$\partial^2 p_{MN}/\partial y^2 + \partial^2 p_{MN}/\partial z^2 + \omega_{MN}^2 c_0^{-2} p_{MN} = 0, \tag{475}$$

where

$$\omega_{MN}^2 = \omega^2 - k^2 c_0^2; \tag{476}$$

also, it satisfies the boundary condition for the hard-walled duct if p_{MN} has zero normal derivative on the boundary. There is a doubly infinite sequence of 'eigenfunctions' $p_{MN}(y, z)$ for the cross-section A, each satisfying (475) with its associated 'eigenfrequency' ω_{MN} and satisfying also this boundary condition. By taking

$$p_{00}(y, z) = 1, \quad \omega_{00} = 0 \tag{477}$$

(a constant eigenfunction with zero eigenfrequency) we can include (473) as the special case $M = N = 0$ of (474).

For example, when A is the rectangle (472), these eigenfunctions are

$$p_{MN}(y, z) = \cos(M\pi y/b)\cos(N\pi z/h), \tag{478}$$

with eigenfrequencies satisfying

$$\omega_{MN}^2 = c_0^2 \pi^2 (M^2 b^{-2} + N^2 h^{-2}). \tag{479}$$

For any cross-section A, the eigenfrequencies are the frequencies at which two-dimensional waves of velocity c_0 can be excited into *resonance* within the area A.

Texts on eigenfunction expansions prove that for any A the eigenfunctions are a *complete set of orthogonal functions*. This means that any function $f(y, z)$ can be expanded as

$$f(y, z) = \sum_{M=0}^{\infty} \sum_{N=0}^{\infty} f_{MN} p_{MN}(y, z), \qquad (480)$$

where the equation

$$\iint_A f(y, z) p_{MN}(y, z) \, \mathrm{d}y \, \mathrm{d}z = C_{MN} f_{MN} \qquad (481)$$

defines the coefficients f_{MN}; here,

$$C_{MN} = \iint_A [p_{MN}(y, z)]^2 \, \mathrm{d}y \, \mathrm{d}z. \qquad (482)$$

Equation (481) is what would be obtained from (480) in term-by-term integration from the orthogonality property, that the integral of the product of two *different* p_{MN} over A vanishes. Completeness means that equation (480) represents any function $f(y, z)$ in A if the coefficients f_{MN} are so determined. In particular, the eigenfunctions (478) of the rectangle satisfy

$$C_{MN} = \tfrac{1}{4}A, \tfrac{1}{2}A, \text{ or } A \text{ if neither, just one, or both of } M \text{ and } N \text{ vanish}, \tag{483}$$

where $A = bh$ is the area of the rectangle; the expansion (480) is then just a double Fourier series.

Equation (476) shows that a waveguide mode can propagate only at frequencies $\omega > \omega_{MN}$. Therefore, if ω_m is the least positive eigenvalue ω_{MN}, sound propagation at frequencies less than ω_m is confined to the nondispersive mode $M = N = 0$; that is, to the strictly one-dimensional motion (473). For example, in the rectangle (472), equation (479) gives $\omega_m = c_0 \pi b^{-1}$.

All the other modes (the true waveguide modes) are dispersive, as (476) shows. For $k > 0$, this gives a positive phase velocity

$$\omega/k = c_0 \, \omega (\omega^2 - \omega_{MN}^2)^{-\frac{1}{2}} \qquad (484)$$

exceeding the sound speed c_0. However, wave energy travels along the duct at the positive group velocity

$$U_{MN} = \mathrm{d}\omega/\mathrm{d}k = (\mathrm{d}k/\mathrm{d}\omega)^{-1} = c_0 \, \omega^{-1}(\omega^2 - \omega_{MN}^2)^{\frac{1}{2}}, \qquad (485)$$

which is always *less* than c_0. As the frequency decreases to the 'cutoff frequency' ω_{MN} below which the mode cannot propagate, the group velocity (485) tends to zero.

It is interesting to study how a source of frequency ω_0 generates waves in modes with $\omega_{MN} < \omega_0$ and only those. Equation (301) for a general source distribution becomes

$$c_0^{-2}\,\partial^2 p_e/\partial t^2 - \nabla^2 p_e = f(x,y,z)\exp{(i\omega_0 t)} \qquad (486)$$

in terms of excess pressure p_e and the present (x,y,z) coordinates. When we apply the expansion (480) to $f(x,y,z)$, f_{MN} becomes a function of x, which we write as a Fourier integral

$$f_{MN}(x) = \int_{-\infty}^{\infty} F_{MN}(k)\exp{(-ikx)}\,dk. \qquad (487)$$

The solution of (486) is then

$$p_e = \sum_{M=0}^{\infty}\ \sum_{N=0}^{\infty} p_{MN}(y,z)\int_{-\infty}^{\infty} \frac{F_{MN}(k)\exp{[i(\omega_0 t - kx)]}}{B_{MN}(\omega_0, k)}\,dk, \qquad (488)$$

where

$$B_{MN}(\omega, k) = -\omega^2 c_0^{-2} + k^2 + \omega_{MN}^2\, c_0^{-2} \qquad (489)$$

represents for arbitrary ω and k the factor by which expression (474) is multiplied when substituted into the left-hand side of (486).

The integrand in (488) has no pole on the real k-axis if $\omega_{MN} > \omega_0$; then, the integral becomes exponentially small at large distances from the source region. For large positive x, for example, we prove this (sections 3.9 and 4.9) by *lowering* the path of integration in the complex k-plane. Physically, we can say that then all motion in the mode p_{MN} has its energy trapped around the source.

When $\omega_{MN} < \omega_0$, the waves propagating away from the source can be determined by evaluating (488) subject to the radiation condition. For large *positive* x, this picks out the poles of the integrand with k positive, corresponding to those waves for which the group velocity (485) is positive. Thus, the asymptotic form of the integral in (488) is the value of

$$-2\pi i\,\frac{F_{MN}(k)\exp{[i(\omega_0 t - kx)]}}{\partial B_{MN}(\omega_0, k)/\partial k} \quad \text{for} \quad k = +c_0^{-1}(\omega^2 - \omega_{MN}^2)^{\frac{1}{2}}. \qquad (490)$$

Adding up such a contribution to (488) for all modes with $\omega_{MN} < \omega_0$, we obtain

$$p_e = -\pi i \sum_{\omega_{MN} < \omega_0} \sum p_{MN}(y,z)\,\frac{F_{MN}[c_0^{-1}(\omega_0^2 - \omega_{MN}^2)^{\frac{1}{2}}]}{c_0^{-1}(\omega_0^2 - \omega_{MN}^2)^{\frac{1}{2}}}$$
$$\times\, \exp{\{i[\omega_0 t - (\omega_0^2 - \omega_{MN}^2)^{\frac{1}{2}} x c_0^{-1}]\}}. \qquad (491)$$

Note that in (491) the amplitude of each mode tends to infinity (unless $F_{MN}(0)$ happens to be zero) as $\omega_0 \downarrow \omega_{MN}$. This infinity might be expected, in a theory neglecting dissipation, as a resonant frequency is approached.

It is interesting to write down the wave energy W_{MN} per unit length of duct in one of the modes (474). This is obtained by integrating $\rho_0^{-1} c_0^{-2} \overline{p_e^2}$ over the cross-sectional area A. (There are no wave-energy contributions from products of two *different* eigenfunctions $p_{MN}(y, z)$, owing to their orthogonality.) In terms of the integral (482) of $[p_{MN}(y, z)]^2$, we obtain

$$W_{MN} = \tfrac{1}{2}\pi^2\rho_0^{-1} C_{MN}(\omega_0^2 - \omega_{MN}^2)^{-1} |F_{MN}[c_0^{-1}(\omega_0^2 - \omega_{MN}^2)^{\frac{1}{2}}]|^2. \tag{492}$$

The rate of transmission of wave energy in this mode is obtained by multiplying (492) by the group velocity (485). Adding this up for all modes propagating in the positive x-direction (modes with $k > 0$), and adding on a similar expression for modes with $k < 0$ propagating in the negative x-direction, we obtain the source's power output as

$$P = \tfrac{1}{2}\pi^2\rho_0^{-1} c_0\, \omega_0^{-1} \sum_{\omega_{MN} < \omega_0} \sum C_{MN}(\omega_0^2 - \omega_{MN}^2)^{-\frac{1}{2}} G_{MN}(\omega_0), \tag{493}$$

where

$$G_{MN}(\omega_0) = |F_{MN}[c_0^{-1}(\omega_0^2 - \omega_{MN}^2)^{\frac{1}{2}}]|^2 + |F_{MN}[-c_0^{-1}(\omega_0^2 - \omega_{MN}^2)^{\frac{1}{2}}]|^2. \tag{494}$$

When resonance is approached ($\omega_0 \downarrow \omega_{MN}$) the predicted power output (493) becomes infinite (again, unless $F_{MN}(0) = 0$). We can recognise this as analogous to cases at the end of section 4.12; where the velocity of energy propagation (485) tends to zero, fluctuations of pressure increase near the source region (because wave energy escapes only very slowly) and can extract a high power from the source.

At any frequency $\omega_0 < \omega_m$ there are no waveguide terms in (493); only the strictly one-dimensional term $M = N = 0$, giving

$$P = \tfrac{1}{2}\pi^2\rho_0^{-1} c_0\, \omega_0^{-2} C_{00}\{|F_{00}(c_0^{-1}\omega_0)|^2 + |F_{00}(-c_0^{-1}\omega_0)|^2\}. \tag{495}$$

Here, equations (477) and (482) give $C_{00} = A$.

For *compact* sources (those of dimension small compared with $c_0\omega_0^{-1}$) this expression (495) is effectively

$$P = \pi^2\rho_0^{-1} c_0\, \omega_0^{-2} C_{00}|F_{00}(0)|^2, \tag{496}$$

where, by (481) and (487),

$$2\pi C_{00}F_{00}(0) = C_{00}\int_{-\infty}^{\infty} f_{00}(x)\,\mathrm{d}x = \int_{-\infty}^{\infty}\mathrm{d}x \iint_A f(x, y, z)\,\mathrm{d}y\,\mathrm{d}z \tag{497}$$

is the amplitude of the *total source strength*, called $\dot{q}$ in chapter 1. Therefore, in terms of the mean square mass-outflow rate $\overline{q^2}$, the power output (496) can be written

$$P = \tfrac{1}{2}\rho_0^{-1} c_0 A^{-1} \overline{q^2}. \tag{498}$$

This value (498) coincides with results for a 'one-dimensional source' given in section 1.4, which however analyses sound radiated in one direction only by a fluctuating mass outflow $q(t)$ in that direction. Expression (498) agrees with the power radiated one-dimensionally in the positive x-direction by a fluctuating mass outflow $\tfrac{1}{2}q(t)$ in that direction, plus an equal amount radiated in the negative x-direction by a variable mass outflow $\tfrac{1}{2}q(t)$ in that direction.

Any compact source in a duct radiates according to one-dimensional theory, then, at frequencies less than ω_m. By contrast, we can expect that at very large frequencies the source will radiate as in unbounded fluid, with power output given (section 1.4) by

$$P = \overline{\dot{q}^2(t)}/(4\pi\rho_0 c_0); \tag{499}$$

essentially, because the boundaries of the duct are then 'in the far field' (at distances from the source *large* compared with c_0/ω_0). The approximation (499) for the power output assumes that in such a limit the source makes omnidirectional radiation which then bounces back and forth along the duct without significantly altering conditions near the source.

The process by which the limiting value (499) is reached for large ω is of considerable interest. For a compact source, (494) becomes

$$G_{MN}(\omega_0) \doteq 2|F_{MN}(0)|^2. \tag{500}$$

Here, by (481) and (487),

$$2\pi C_{MN} F_{MN}(0) = C_{MN} \int_{-\infty}^{\infty} f_{MN}(x)\, dx$$

$$= \int_{-\infty}^{\infty} dx \int\int_A f(x,y,z)\, p_{MN}(y,z)\, dy\, dz; \tag{501}$$

this integral, to close approximation, is the amplitude of the total source strength $\dot{q}$ times the value $p_{MN}(y_s, z_s)$ at the centre (y_s, z_s) of the source region (which is compact in relation to c_0/ω_0, and therefore also in relation to the greater scale c_0/ω_{MN} of variations of p_{MN}). Then the power (493) becomes

$$P = \tfrac{1}{2}\rho_0^{-1} c_0 \omega_0^{-1} \overline{\dot{q}^2(t)} \sum_{\omega_{MN} < \omega_0} \sum C_{MN}^{-1}[p_{MN}(y_s, z_s)]^2 (\omega_0^2 - \omega_{MN}^2)^{-\frac{1}{2}}. \tag{502}$$

When the frequency ω_0 is large enough, this sum (502) contains so many terms that it may be approximated by an integral. To do this we need to define $n(\omega)$ as a smooth function of ω giving approximately the *number* of eigenfunctions with eigenfrequency less than ω. For example, in the rectangle (472), that number of eigenfunctions (478) is the number of points (M, N) with integer coordinates inside the quarter-ellipse

$$M \geqslant 0, \quad N \geqslant 0, \quad M^2(\omega b/c_0 \pi)^{-2} + N^2(\omega h/c_0 \pi)^{-2} \leqslant 1. \qquad (503)$$

Evidently, $n(\omega)$ can be taken as the *area* of the quarter-ellipse (503); namely $\frac{1}{4}\pi$ times the product of its semi-axes $(\omega b/c_0 \pi)$ and $(\omega h/c_0 \pi)$. This is

$$n(\omega) = (\omega/c_0)^2 (A/4\pi), \qquad (504)$$

where A is the duct cross-sectional area.

Whatever the value of $n(\omega)$, we approximate the sum (502) for large ω_0 by

$$P \doteq \tfrac{1}{2}\rho_0^{-1} c_0 \omega_0^{-1} \overline{\dot{q}^2(t)} \int_0^{\omega_0} A^{-1}(\omega_0^2 - \omega^2)^{-\frac{1}{2}} n'(\omega) \, d\omega. \qquad (505)$$

Here, $n'(\omega) \, d\omega$ is a smoothed value of the number of terms in the sum with $\omega < \omega_{MN} < \omega + d\omega$, while by (482) a smoothed value of $C_{MN}^{-1}[p_{MN}(y_{\mathrm{s}}, z_{\mathrm{s}})]^2$ is its average A^{-1}. Expression (505) is immediately evaluated, with the value (504) of $n(\omega)$, as equal to the source's power output in an unbounded medium, (499). Furthermore, it is readily proved that equation (504) is the *only form of $n(\omega)$* which makes (505) independent of ω_0 and equal to (499). This fact can be regarded as a 'physical argument' for the necessary truth of a classical theorem: that equation (504) indeed holds for *any shape of cross-section* with area A.

It is interesting to trace details of the transition from (498) to (499) as ω_0 increases in a particular case: a rectangular duct (472) with a compact source at its *centre*. The expression (478) gives

$$[p_{MN}(y_{\mathrm{s}}, z_{\mathrm{s}})]^2 = [p_{MN}(\tfrac{1}{2}b, \tfrac{1}{2}h)]^2 = \begin{cases} 1 \text{ if both } M \text{ and } N \text{ are even,} \\ 0 \text{ otherwise;} \end{cases} \qquad (506)$$

so that the ratio of the power (502) to its low-frequency value (498) is

$$P/(\tfrac{1}{2}\rho_0^{-1} c_0 A^{-1} \overline{q^2}) = A\omega_0 \sum_{\substack{\omega_{MN} < \omega_0 \\ M, N \text{ even}}} \sum C_{MN}^{-1}(\omega_0^2 - \omega_{MN}^2)^{-\frac{1}{2}}, \qquad (507)$$

with C_{MN} as in (483).

Figure 109 shows how (507) makes the transition from a constant value 1 for $\omega_0 < \omega_{\mathrm{m}}$ to its quadratically increasing asymptotic value for high frequency given by (499) as

$$(\omega_0/c_0)^2 (A/2\pi). \qquad (508)$$

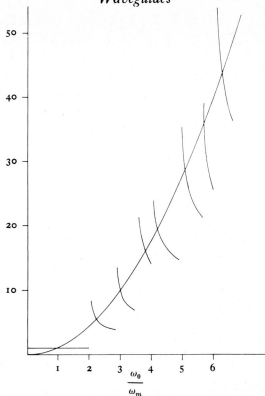

Figure 109. Power output P for a compact acoustic source of frequency ω_0 placed at the centre of a duct of rectangular cross-section with breadth–depth ratio $b/h = \sqrt{2}$. The ordinate gives the ratio of P to the value (498) which it always has for $\omega_0 < \omega_m$. (Actually, for a centrally placed source it has this value for $\omega_0 < 2\omega_m$.) The continuous curve represents the limiting form (508) taken when the duct is absent and the source radiates into free space. The discontinuous curve represents the exact expression (507).

Paradoxically, this nondimensional power output (507) is a *decreasing function* wherever it is continuous! At every eigenfrequency ω_{MN}, however, it shoots up (theoretically, up to $+\infty$), but extremely soon finds itself close to the asymptotic curve again. Only at frequencies which exceed just very marginally an eigenfrequency ω_{MN} is the departure pronounced; energy dissipation would still further smooth out those departures.

For sound waves in a duct, we have seen that propagation at specially low frequencies takes place through motions of a specially simple kind, as in equation (473). This, however, is not generally true in waveguides.

The atmosphere can act as a one-sided waveguide, in which greatest interest attaches to the low-frequency mode that can propagate pressure readjustments over the Earth's surface. Here, because the sound speed c_0 varies with altitude, no dispersion relationship as simple as (473) can apply. Furthermore, gravity waves are coupled with sound waves; indeed, the wavenumber is *small* compared with $[-\rho_0'(z)/\rho_0(z)]$, in contradiction to the condition (52) for sound and gravity waves to be decoupled.

Equations (35) and (36) describe gravity-acoustic waves in a stratified atmosphere. Propagation in the gravity-acoustic waveguide takes the form of waves

$$q = Q(z)\exp[\mathrm{i}(\omega t - kx)], \quad p_e = P(z)\exp[\mathrm{i}(\omega t - kx)], \tag{509}$$

where by (35) and (36) we have

$$\{[N(z)]^2 - \omega^2\}\, Q(z) = -\mathrm{i}\omega\{P'(z) + g[c_0(z)]^{-2}P(z)\}, \tag{510}$$

and

$$\{\omega^2[c_0(z)]^{-2} - k^2\}\, P(z) = \mathrm{i}\omega\{Q'(z) + g^{-1}[N(z)]^2 Q(z)\}. \tag{511}$$

Exponential integrating factors for these equations are

$$E_1(z) = \exp\left\{\int_0^z g[c_0(z)]^{-2}\,\mathrm{d}z\right\}, \quad E_2(z) = \exp\left\{\int_0^z g^{-1}[N(z)]^2\,\mathrm{d}z\right\}; \tag{512}$$

where it is important to note that their product

$$E_1(z)\,E_2(z) = [\rho_0(0)/\rho_0(z)] \tag{513}$$

varies inversely as the density; we can show, furthermore, that, under atmospheric conditions,

$$[E_2(z)/E_1(z)] = [c_0(z)/c_0(0)]^2 \exp\left\{-(2-\gamma)\int_0^z g[c_0(z)]^{-2}\,\mathrm{d}z\right\}; \tag{514}$$

a function which *decreases exponentially* with z (figure 110).

The upward component of mass flux q is zero at the ground, $z = 0$, so that the appropriate solution of (511) is

$$Q(z) = [\mathrm{i}\omega E_2(z)]^{-1}\int_0^z \{\omega^2[c_0(z_2)]^{-2} - k^2\}\, P(z_2)\,E_2(z_2)\,\mathrm{d}z_2. \tag{515}$$

Equation (510) can then be formally solved as

$$P(z) = [E_1(z)]^{-1}\left(P(0) + \int_0^z \{[N(z_1)]^2 - \omega^2\}\,[E_1(z_1)/E_2(z_1)]\,\mathrm{d}z_1\right.$$
$$\left. \times \int_0^{z_1} \{[c_0(z_2)]^{-2} - (\omega/k)^{-2}\}\, P(z_2)\,E_2(z_2)\,\mathrm{d}z_2\right): \tag{516}$$

an integral equation which lends itself to the determination of $P(z)$ by successive approximation. The first approximation is

$$P_1(z) = [E_1(z)]^{-1} P(0). \tag{517}$$

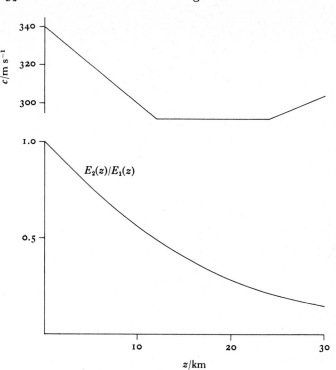

Figure 110. A typical variation of sound speed with height in the lowest 30 km of the atmosphere, and the corresponding ratio $E_2(z)/E_1(z)$ deduced from (514) with $\gamma = 1.4$.

This would only be exact in an isothermal atmosphere, with $c_0(z)$ constant so that the double integral in (516) can vanish if (ω/k) is equal to that constant. If this is not so, we go to a second approximation $P_2(z)$ given by (510) with $P(z)$ replaced by $P_1(z)$ on the right-hand side; and so on. At each stage, the $(n+1)$th approximation $P_{n+1}(z)$ is obtained from (516) with $P(z)$ replaced by $P_n(z)$ on the right-hand side.*

The associated value of ω/k must be such that the integral in (515) tends to zero as $z \to \infty$, giving

$$(\omega/k)^{-2} = \left\{ \int_0^\infty [c_0(z)]^{-2} P(z) E_2(z) \, dz \right\} \Big/ \left\{ \int_0^\infty P(z) E_2(z) \, dz \right\}. \quad (518)$$

* Note that the excess pressure takes values which are in phase at all heights above a given point on the Earth's surface. In fact, if the phase is chosen so that $P(0)$ is real, then all the $P_n(z)$ are also real.

Otherwise, an infinite amount of energy would be needed to set up the wave. Indeed, the potential energy density includes (see (37)) terms

$$\tfrac{1}{4}[N(z)]^2\,|Q(z)|^2\,[\omega^2\rho_0(z)]^{-1} \quad \text{and} \quad \tfrac{1}{4}|P(z)|^2\,[c_0(z)]^{-2}\,[\rho_0(z)]^{-1} \quad (519)$$

due to gravity and compressibility respectively, where $[\rho_0(z)]^{-1}$ by (513) is proportional to $E_1(z)\,E_2(z)$. Both terms (519) increase exponentially with height like $E_1(z)/E_2(z)$ (see (514)) unless the integral in (515), which is also the inner integral in (516), tends to zero for large z.

Equation (518) shows that the *inverse-square* wave speed $(\omega/k)^{-2}$, is a weighted mean of the inverse-square sound speed, $[c_0(z)]^{-2}$. The weighting function, $P(z)\,E_2(z)$ typically varies so that $c_0(z)$ takes a value intermediate between its ground-level and stratospheric values. For example, the first approximate form (517) gives a weighting function proportional to the function (514) plotted in figure 110. Typically, a stratospheric value of c_0 is around $290\,\mathrm{m\,s^{-1}}$ (corresponding to temperatures around $210\,\mathrm{K}$) and c_0 falls to this value from its ground-level value as z increases from 0 to around $12\,\mathrm{km}$. With a ground-level value of $340\,\mathrm{m\,s^{-1}}$ the corresponding weighted mean of $[c_0(z)]^{-2}$ leads to a wave speed of about $310\,\mathrm{m\,s^{-1}}$. This is close to wave speeds typically observed for long-distance propagation of pressure readjustments. (The first accurately measured disturbance of this kind was the air wave generated by the explosion of the island of Krakatoa in 1883.)

Such use of the first approximate form (517) of $P(z)$ in the weighting function gives a wave speed independent of frequency. More accurately, the $P(z)$ given by (516) depends on the frequency ω, losing such dependence only when ω is small compared with the Väisälä–Brunt frequency N. This frequency-dependence, through its effect on the weighting function $P(z)\,E_2(z)$, produces a quite small amount of dispersion; in fact, a very slight reduction in wave speed, proportional to ω^2, as ω increases from zero. This brings about, however, a reduction in *group velocity* three times as much.

We conclude chapter 4 by describing a marine 'one-sided waveguide'. This is associated with propagation along a shoreline of *long waves*; in the sense that most of their kinetic energy is in horizontal water motions varying little with depth. Although propagation is along the shoreline, the waves are *not* one-dimensional because the associated horizontal water motions are as much perpendicular to, as parallel to, the shoreline.

Propagation of such long waves in water of gradually varying depth $h(x,y)$ satisfies an equation of continuity,

$$\partial\zeta/\partial t + \partial(hu)/\partial x + \partial(hv)/\partial y = 0, \qquad (520)$$

that equates the rate of change of the free-surface elevation ζ to minus the two-dimensional divergence of the flow vector (hu, hv). With momentum equations

$$\partial u/\partial t = -g\,\partial\zeta/\partial x, \quad \partial v/\partial t = -g\,\partial\zeta/\partial y, \tag{521}$$

equation (520) gives

$$\partial^2\zeta/\partial t^2 = \partial(gh\,\partial\zeta/\partial x)/\partial x + \partial(gh\,\partial\zeta/\partial y)/\partial y. \tag{522}$$

This equation, of course, becomes the two-dimensional wave equation with wave speed $(gh)^{\frac{1}{2}}$ when the depth h is uniform.

When h depends only on a single coordinate y, representing distance from the shoreline, equation (522) can describe quasi-one-dimensional propagation along the shoreline if

$$\zeta = Z(y)\exp[\mathrm{i}(\omega t - kx)], \tag{523}$$

where

$$gh(y)\,Z''(y) + gh'(y)\,Z'(y) + [\omega^2 - gh(y)\,k^2]\,Z(y) = 0. \tag{524}$$

We are interested in cases when $h(y)$ increases with distance y from the shoreline. Then, where $h(y)$ becomes large, equation (524) becomes approximately

$$Z''(y) - k^2 Z(y) = 0; \tag{525}$$

and only a solution with, approximately,

$$Z(y) \quad \text{proportional to} \quad \exp(-ky) \tag{526}$$

has finite energy. Accordingly, we look for solutions of (524) that satisfy this condition as $h(y)$ becomes large.

There is one case when (526) is an exact solution of (524) for all y. This is the case $h = \beta y$ with uniform bottom slope β. Then (526) satisfies (524) everywhere provided that

$$\omega = (g\beta k)^{\frac{1}{2}}. \tag{527}$$

This is the classical edge-wave due to Stokes.*

In order to visualise the horizontal motions of fluid particles in the Stokes edge-wave, it is necessary only to look at figure 50; thinking of the

---

* Curiously.enough, Stokes did not need to make the long-wave approximation, because a velocity potential, proportional to $\exp(-\mathrm{i}kx - ky)$ and independent of z, exactly satisfies Laplace's equation. Also, it exactly satisfies the boundary condition on a uniformly sloping bottom if the y-axis is taken *along the bottom* and (as before) perpendicular to the shoreline. Finally, it exactly satisfies the free-surface condition on ϕ if (527) holds with β the *sine* (rather than, as above, the *tangent*) of the angle of the bottom to the horizontal; the difference is, of course, negligible for gradual slopes. None of these remarks apply, however, to edge-waves on *nonuniformly* sloping bottoms.

top of figure 50 as the shoreline and the particle paths as depicting the *horizontal* motions of fluid near the shoreline as the edge-wave propagates. That is because equations (521) make (u, v) an irrotational motion specified by a velocity potential $(ig\zeta/\omega)$ proportional, by (523) and (526), to

$$\exp\left[i(\omega t - kx) - ky\right];$$

this is just the same dependence on x and y as the dependence on x and z used to draw figure 50. Note also that equation (527) makes the wave speed $(g\beta/k)^{\frac{1}{2}}$ and the group velocity half as much.

The Stokes edge-wave solution for $h = \beta y$ is the simplest, but is in no way an isolated, solution of (524). In fact, edge-waves propagate along beaches of gradually *varying* slope $h'(y)$ according to equation (527) with β equal to some sort of intermediate value of the slope within the region where edge-wave solutions are significant.

For a rough demonstration of this, we characterise a gradually varying beach slope by (i) its value $h'(0)$ at the shoreline, and (ii) its value $h'(\infty)$ where ky is large, and (iii) a measure such as $h''(0)$ of the rate of change of slope with distance from the shoreline. Then we fit to (524) an approximate solution

$$Z(y) = Z(0)\left[\exp\left(-ky\right)\right]\left(1 + Cky\right)^D, \tag{528}$$

with the constants C and D chosen so that (528) satisfies (524) for both large and small values of ky.

We find that (524) is satisfied by (528) for large ky if

$$\omega^2 = (1 + 2D)gh'(\infty)k, \tag{529}$$

while the first two terms of its Taylor expansion for small ky give

$$\omega^2 = (1 - CD)gh'(0)k \tag{530}$$

and

$$2CD(C+1) - C^2D^2 = \left[-h''(0)/kh'(0)\right](1 - CD). \tag{531}$$

Figure 111 shows the results of using these equations to compute how the variable

$$\omega^2/gh'(0)k \tag{532}$$

(which is 1 for the Stokes solution) varies as a function of

$$h'(\infty)/h'(0) \tag{533}$$

(representing the reduction in slope as distance from the shoreline increases) for different values of

$$\left[-h''(0)/kh'(0)\right]. \tag{534}$$

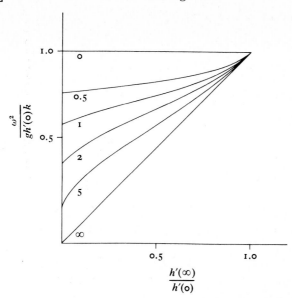

Figure 111. Edge-waves: a convenient approximate form of their dispersion relationship. Here, $h'(0)$ represents the bed slope near the beach and $h'(\infty)$ the smaller bed slope far from the beach, and the numbers on the curves are values of $[-h''(0)/kh'(0)]$ which measures the relative rate of initial fall in bed slope multiplied by $1/k$ (that is, by $\lambda/2\pi$).

The statement that the Stokes equation (527) is satisfied for a value of β intermediate between $h'(\infty)$ and $h'(0)$ is borne out by the fact that (532) *lies between* (533) and 1. It is closest to the latter value (making β close to $h'(0)$) when (534) is small, implying only a small reduction of slope within the region where wave motions are significant.

Edge-waves near a beach can often be identified in recorded components of the sea-wave spectrum having periods around a minute. Indeed, the common observation that ordinary beach waves (caused by swell propagating towards the shoreline) attain maximum amplitudes about once per minute (different legends make the 'seventh wave' or the 'ninth wave' the greatest) may result from the interaction between those surface waves and such an edge-wave.

EXERCISES ON CHAPTER 4

1. In a stratified fluid of uniform Väisälä–Brunt frequency N, show that the equations

$$p_e = (N \sin \theta)\, k^{-1} q_1 \exp\,[i(Nt \cos \theta - kx + kz \tan \theta)],$$

$$\rho_0\, \mathbf{u} = (\tan \theta,\, 0,\, 1)\, q_1 \exp\,[i(Nt \cos \theta - kx + kz \tan \theta)],$$

with $0 < \theta < \tfrac{1}{2}\pi$, represent plane internal waves which transmit an energy flux $\tfrac{1}{2}(\rho_0 k)^{-1} q_1^2 N \tan \theta$ in a direction upward and to the right specified by the unit vector $(\sin \theta,\, 0,\, \cos \theta)$.

This wave energy flux is incident upon a plane wall $x = z \tan \psi$, where $\theta - \pi < \psi < \theta$. Find the *reflected wave*; that is, a wave with the same frequency $\omega = N \cos \theta$ and with energy flux directed *away* from the wall, such that the sum of the incident and reflected waves has zero normal velocity on the wall itself. For the reflected wave it will be sufficient to find the values of the amplitude Q_1, the x-component of wavenumber K and the energy-flux direction Θ, corresponding to q_1, k and θ in the incident wave. In particular, show that

$$K = -k \sin (\theta - \psi)\, \operatorname{cosec} (\theta + \psi).$$

Give an alternative derivation of this equation by trigonometrical argument from a diagram showing the paths of energy flux in the incident and reflected rays. Finally, show that the energy flux in the reflected wave has a magnitude equal to that in the incident wave multiplied by a factor $\sin (\theta - \psi)\, |\operatorname{cosec} (\theta + \psi)|$. Is this consistent with the conservation of energy? Discuss in particular the limit as $\psi \to -\theta$.

2. There are suggestions in sections 4.2 and 4.5 that $[\rho_0(z)]^{\frac{1}{2}}$ (the square root of the undisturbed density in a stratified fluid) should be regarded as a naturally occurring factor both in the upward component of mass flux q and in the excess pressure p_e for small disturbances. Show, in fact, that if

$$q = [\rho_0(z)]^{\frac{1}{2}} q_s(x, y, z) \exp\,(i\omega t), \quad p_e = [\rho_0(z)]^{\frac{1}{2}} p_s(x, y, z) \exp\,(i\omega t),$$

then

$$\{[N(z)]^2 - \omega^2\}\, q_s = -i\omega[\partial p_s/\partial z + \sigma(z)\, p_s],$$

$$\partial^2 p_s/\partial x^2 + \partial^2 p_s/\partial y^2 + \omega^2[c_0(z)]^{-2} p_s = i\omega[\partial q_s/\partial z - \sigma(z)\, q_s],$$

where

$$\sigma(z) = \tfrac{1}{2} g[c_0(z)]^{-2} - \tfrac{1}{2} g^{-1}[N(z)]^2.$$

If the fluid is a perfect gas with undisturbed temperature $T_0(z)$ and with constant specific heats in the ratio γ, show that

$$\sigma(z) = (1 - \tfrac{1}{2}\gamma)\, g[c_0(z)]^{-2} - \tfrac{1}{2}[T_0'(z)/T_0(z)].$$

For an atmosphere at constant temperature T_0 with $\gamma = 1.4$, show that plane waves with both p_s and q_s proportional to $\exp\,[-i(kx + ly + mz)]$ are possible if

$$(N^2 - \omega^2)\,(k^2 + l^2) - \omega^2 m^2 = (1.225 N^2 - \omega^2)\,\omega^2 c_0^{-2}.$$

Show that this equation excludes frequencies ω between N and $1.11 N$. For $\omega < N$, sketch the departure from a conical wavenumber surface which prevents the horizontal component of wavenumber $(k^2 + l^2)^{\frac{1}{2}}$ from ever falling below $1.11 \omega c_0^{-1}$. For $\omega > 1.11 N$, show that $(k^2 + l^2)^{\frac{1}{2}}$ is at most equal to ωc_0^{-1}, and sketch the departure from a spherical wavenumber surface.

[The literature pays close, but perhaps misleading, attention to the exact nature of these departures from the conical wavenumber surface for internal waves and from the spherical wavenumber surface for sound waves. The reader will note that the departures are significant only for small wavenumbers of the order of

$$Nc_0^{-1} = 0.45[-\rho_0'(z)/\rho_0(z)],$$

corresponding to a wavelength equal to the altitude in which the density *falls by a factor of* more than 10^6. Atmospheres are never isothermal over such heights. Furthermore, even if they were, it might be impossible to excite a wave of this large wavelength.

The solutions for $(k^2+l^2)^{\frac{1}{2}}$ on the plane $m = 0$ are particularly misleading, because the idea of horizontally propagating waves of these great lengths is at first sight quite plausible. Nevertheless, such a wave with energy density uniform over a vast range of altitudes could not be excited. The true horizontally propagating gravity-acoustic waveguide mode (section 4.13) takes for the simple limiting case of an isothermal atmosphere the form $q_s = 0$, with p_s proportional to $\exp(-\sigma z)$. In other words, the vertical wavenumber m does *not* become zero in this case; it takes the imaginary value $(-i\sigma)$.]

3. The reader is invited to obtain a second approximation to the dispersion relationship (74) for the 'sinuous' mode of wave propagation on a steep ocean thermocline $z_2 < z < z_1$ separating regions $z < z_2$ and $z > z_1$ of negligible density gradient. From equation (71) for $Q(z)$ rewritten in terms of

$$Z(z) = Q'(z)/Q(z),$$

show that

$$Z(z) = k - k^2\omega^{-2}\int_{z_2}^{z}[N(z)]^2\,dz + \int_{z_2}^{z}\{k^2 - [Z(z)]^2\}\,dz.$$

Deduce that (74) is always an *overestimate* of ω^2. Obtain a second approximation for small $k(z_1 - z_2)$ as

$$\omega^2 = Gk(1 - kH),$$

where G as in (74) is $\frac{1}{2}g \ln(\rho_2/\rho_1)$, and where

$$H = 2\left(\ln\frac{\rho_2}{\rho_1}\right)^{-2}\int_{z_2}^{z_1}\left[\ln\frac{\rho_2}{\rho_0(z)}\right]\left[\ln\frac{\rho_0(z)}{\rho_1}\right]dz$$

is of the order of the layer thickness $z_1 - z_2$.

For these waves, is the ratio of group velocity to phase velocity greater or less than $\frac{1}{2}$?

4. In uniformly stratified fluid a small localised disturbance is made, as in figure 77, by the displacement of a cylinder normal to its axis. If this disturbance generated no waves of wavelengths exceeding a certain maximum λ_m, show that all waves after time t would be found within two cylindrical regions of radius $N\lambda_m t/4\pi$ which are tangent to each other at the position of the original disturbance.

5. Consider a nonhomogeneous wave system where the form (100) of the dispersion relationship is *slowly changing with time*. (For example, an external magnetic field which can influence waves as described in the epilogue may be slowly varying.) For waves whose frequency ω takes the form

$$\omega = \omega(k_1, k_2, k_3, x_1, x_2, x_3, t)$$

of a function varying slowly with respect to both x_i and t on a scale of wavelengths and wave periods, determine which one of the equations (103)–(108) is the only one to be modified. What essential change to the properties of rays does this modification make?

Demonstrate also the following extension of the stationary-phase principle to this very general case. Show that the change in phase between times $t = t_1$ and $t = t_2$ at a point moving along a path $x_i = x_i(t)$ on which the wavenumber takes the value $k_i = k_i(t)$ is

$$\int_{t_1}^{t_2} \{\omega[k_i(t), x_i(t), t] - k_i(t) \, \dot{x}_i(t)\} \, \mathrm{d}t.$$

Prove that this change in phase is stationary with respect to small variations $\delta x_i(t)$ and $\delta k_i(t)$ in $x_i(t)$ and $k_i(t)$ which leave unchanged the path's end-points (so that $\delta x_i(t_1) = \delta x_i(t_2) = 0$) if and only if the path is a ray.

6. For sound waves in a perfect gas with constant specific heats in the ratio γ, use the assumption that the volume-average of excess density is zero to show that the volume-average of excess pressure is equal to $\frac{1}{2}(\gamma - 1)$ times the wave energy per unit volume W. This may be regarded as suggesting that the full 'radiation stress' for sound waves should be written as the Reynolds stress (156) plus an isotropic term $\frac{1}{2}(\gamma - 1) \, W\delta_{ij}$. Note, however, that the associated mean motion can quickly adjust itself to generate an equal and opposite mean pressure, bringing to a halt any resulting movement. To what value is the volume-average of excess density then reduced?

7. Calculate the analytical form of the ray illustrated in figure 81; that is, a ray carrying small disturbances of frequency ω, *less* than the uniform Väisälä–Brunt frequency N, through stratified air which blows in the x-direction at a velocity τz increasing linearly with height. Show that the ray leaving the point $(x_0, 0, 0)$ in the plane $y = 0$ with x and z increasing has the shape

$$x = x_0 + z[N^2(\omega - k_{\mathrm{H}} \, \tau z)^{-2} - 1]^{\frac{1}{2}}$$

plotted in figure 81.

At what altitudes, if any, could internal waves be excited at a frequency ω *greater* than N so that the curve described by the above equation would represent the ray carrying their energy away in the plane $y = 0$ with x and z increasing? Furthermore, what would happen to any energy generated at such an altitude but moving away in the plane $y = 0$ with x increasing and z *decreasing*?

8. A large volume of liquid has undisturbed density distribution

$$\rho_0(z) = \rho_1(1 - \epsilon z)$$

with ρ_1 and ϵ constant, and the Väisälä–Brunt frequency can be taken as having the uniform value $N = (g\epsilon)^{\frac{1}{2}}$ (because ϵ is large compared with gc_0^{-2} but ϵz is small throughout the volume of liquid). At time $t = 0$, a region of fluid in the shape of a horizontal cylinder of radius a with axis $x = z = 0$ is *suddenly made homogeneous* with density ρ_1. (This is an idealisation of the sudden mixing effect due to the turbulence in the wake of a body moving fast through the liquid.) Solve the resulting *initial-value problem for two-dimensional internal waves*, assuming that the local vertical component of mass flux q is initially zero. (In other words, the turbulence has locally mixed up the fluid, which then is 'released from rest'.)

Express the initial value of ρ_e (the density excess over its undisturbed value) as

$$\rho_{eI} = \int_{-\infty}^{\infty} \int_{-\infty}^{0} F(k, m) \exp[-i(kx + mz)] \, \mathrm{d}k \, \mathrm{d}m,$$

and show that, with ρ_{eI} given as $+\rho_1 \epsilon z$ for $x^2 + z^2 < a^2$ and o elsewhere, the equation

$$F(k, m) = i\rho_1 \epsilon \pi^{-1} a^2 m (k^2 + m^2)^{-1} J_2[(k^2 + m^2)^{\frac{1}{2}} a]$$

expresses F in terms of the Bessel function J_2. Show that waves in the region $z > o$ take the form

$$q = \tfrac{1}{2} i g N^{-2} \int_{-\infty}^{\infty} \int_{-\infty}^{0} \omega(k, m) F(k, m) \exp [i\omega(k, m) t - kx - mz] \, dk \, dm,$$

with

$$\omega(k, m) = N|k| (k^2 + m^2)^{-\frac{1}{2}}.$$

Lastly, estimate these waves for large t at the point $x = r \sin \theta$, $z = r \cos \theta$ as

$$q = \tfrac{1}{2} \rho_1 N a^2 r^{-1} |\sin 2\theta| J_2(N t r^{-1} a \sin \theta) \exp (i N t \cos \theta).$$

[Here, the oscillation represented by the exponential has a form showing the radial character of wave crests. The Bessel function reflects the fact that waves of higher wavenumber, arriving at distant points relatively late, are generated with variable phase owing to the initial discontinuity in density at $r = a$.]

9. Use the approximate trapped-wave condition (417) to obtain the dispersion relationship for the sinuous mode of oscillation ($n = o$) on a particular thermocline. Take the Väisälä–Brunt frequency as zero outside the interval $z_2 < z < z_1$ and as

$$N(z) = N_m[1 - |z - z_m| (z_1 - z_m)^{-1}]$$

in the interval $z_2 < z < z_1$, where N_m is the maximum value, attained at

$$z = z_m = \tfrac{1}{2}(z_1 + z_2).$$

Show that the 'healed-ray-theory' dispersion relationship takes the form

$$k(z_1 - z_2) = \pi[(N_m^2 \omega^{-2} - 1)^{\frac{1}{2}} - \omega N_m^{-1} \cosh^{-1}(N_m \omega^{-1})]^{-1}.$$

This approximation is suitable for use in the frequency interval between $\omega = o.5 N_m$ and $\omega = N_m$, where it makes $k(z_1 - z_2)$ vary from 2.9 to ∞.

Conversely, show that the method of exercise 3, appropriate to small values of $k(z_1 - z_2)$, gives

$$\omega^2 = \tfrac{1}{6} N_m^2 k(z_1 - z_2) [1 - \tfrac{5}{28} k(z_1 - z_2)].$$

This approximation is suitable for use in the complementary range of values of $k(z_1 - z_2)$ between o and 2.9, where it makes ω vary from o to $o.48 N_m$.

Lastly, show that at frequencies ω for which the sinuous mode $n = o$ has its wavenumber k given accurately by healed ray theory, the varicose mode $n = 1$ has wavenumber just 3 times as great.

10. A horizontal cylinder with axis in the y-direction moves at constant velocity V in a direction specified by the unit vector $(\sin \psi, o, \cos \psi)$, where $o < \psi < \tfrac{1}{2}\pi$, through fluid with uniform Väisälä–Brunt frequency N. Find the associated wavenumber curve. Show that its two branches are asymptotic to the straight lines

$$V(k \sin \psi + m \cos \psi) = \pm N \cos \psi$$

for large values of k or m (either positive or negative), and also are tangent to the m-axis at the origin. Sketch the wavenumber curve for at least one value of ψ, and include arrows indicating in which direction waves of different wavenumbers are to be found. Lastly, prove that the coordinates (x, z) of points on any surface of constant phase take the form

$$x = CV^{-1}N(\cos \theta \sin \psi + \cos \psi \sin \theta \tan^2 \theta),$$
$$z = CV^{-1}N[\cos (\theta + \psi) + 2 \cos \psi \sin \theta \tan \theta],$$

where C is a constant and θ a parameter (actually, defined by the usual equation $m = -k \tan \theta$).

[Stevenson derived these equations, and for three fixed values of ψ ($45°$, $70°$ and $80°$) plotted the points (x, z) for varying θ, obtaining excellent agreement in each case with his experimentally observed shapes of surfaces of constant phase.]

11. Find the quasi-one-dimensional modes of waveguide propagation for water waves in a canal of uniform breadth b (with vertical sides) and of depth exceeding b. Determine the phase velocity and group velocity for each.

A solid body whose shape is symmetrical about the canal's central plane is pivoted about a vertical axis in that central plane. At time $t = 0$ it begins to make rotary oscillations of frequency ω and small amplitude about that vertical axis. Show that oscillations of frequency ω can first reach a great distance l from the body at a time approximately

$$t = 2\omega g^{-1}l(1 - \pi^2 g^2 b^{-2}\omega^{-4})^{-\frac{1}{2}}$$

if $\omega > (\pi g)^{\frac{1}{2}}b^{-\frac{1}{2}}$, and otherwise not at all.

12. A hardwalled acoustic duct has uniform cross-section in the shape of a square $0 < y < b$, $0 < z < b$. A loudspeaker diaphragm at the point $(0, \frac{1}{2}b, 0)$ generates by its vibrations a fluctuating volume flow

$$m_1 \exp (i\omega t).$$

The diaphragm is compact (small compared with c/ω). Show that the quantity $F_{MN}(k)$ defined in (481) takes a value independent of k, equal to

$$F_{MN} = (2\pi C_{MN})^{-1} i\omega\rho_0 m_1 \cos (\tfrac{1}{2}M\pi),$$

for all wavenumbers k which the diaphragm can excite. Determine from this the pressure fluctuations generated both for large positive x and for large negative x.

A similar diaphragm is now placed at the point $(l, \frac{1}{2}b, 0)$. It generates by its vibrations a fluctuating volume flow

$$m_2 \exp (i\omega t).$$

Show that a particular choice of m_2/m_1 will ensure that pressure fluctuations vanish for large positive x (appearing, in fact, only for large negative x) at all frequencies up to a certain limit, to be determined.

With the aim of increasing the range of frequencies for which the loudspeakers generate sound *only* in the negative x-direction, another similar diaphragm generating volume flow $m_1 \exp (i\omega t)$ is installed also at $(0, \frac{1}{2}b, b)$, while another generating volume flow $m_2 \exp (i\omega t)$ is installed also at $(l, \frac{1}{2}b, b)$. Show that for large positive x pressure fluctuations now vanish at all frequencies up to twice the previous limit.

EPILOGUE

Part 1   A variety of waves in fluids

In chapter 4 we described a number of important ideas and techniques that can be applied to any wave system satisfying linear equations, even if it is anisotropic as well as dispersive. We also analysed in depth one such system: internal gravity waves in stratified fluid.

This book's size, as explained in the prologue, forbids a similar treatment in depth of all systems of waves in fluids. It allows us, however, to sketch very briefly, in this first half of the epilogue, a variety of anisotropic and dispersive wave systems, with indications of how they can be analysed by the methods of chapter 4. For a similarly brief sketch of some important nonlinear effects on dispersive systems, see part 2.

The wave systems here discussed are those where propagation is controlled by either of two types of restoring forces not previously mentioned in this book: by Coriolis forces in rotating fluids, and by magnetic forces in electrically conducting fluids. Both types of system have in common one interesting feature: that their properties can be understood in terms of *field lines moving with the fluid*.

Texts on the dynamics of a homogeneous fluid show how vortex lines move with the fluid (except in so far as its viscosity causes some diffusion of the vorticity). A closely related result is that magnetic field lines move with the fluid (except in so far as its electrical resistivity causes some diffusion of the magnetic field). Propagation in each case depends on fluid motions in the wave distorting either the undisturbed parallel vortex lines of the rotating fluid, or the undisturbed magnetic field lines of the conducting fluid.

Another common feature of the two systems is that, although limited laboratory realisations of the wave motions are possible, their primary importance is geophysical. It is related, especially, to the effects of the Earth's rotation on oceanic or atmospheric movements, and the effects of its magnetic field on movements of the ionosphere or of the Earth's liquid-metal core.

The vorticity field $\mathbf{\Omega}$ in a uniformly rotating fluid takes a uniform value

$$\mathbf{\Omega}_0 = (0, 0, \Omega_0) \tag{1}$$

equal to *twice* the angular velocity of rotation (here, taken as a rotation about the z-axis). For a homogeneous fluid, with viscous dissipation neglected, the vortex lines move with the fluid. This means that any small fluid velocity field $\mathbf{u}$ (relative to the uniform rotation) causes a rate of change of vorticity

$$\frac{\partial}{\partial t} \operatorname{curl} \mathbf{u} = \Omega_0 \frac{\partial}{\partial z} \mathbf{u}, \tag{2}$$

proportional to the rate of change of $\mathbf{u}$ along a vortex line.

A more conventional approach to this fundamental equation (2) for the velocity field obtains it by taking the curl of a linearised momentum equation,

$$\rho_0 \, \partial \mathbf{u}/\partial t = -\nabla p_e - \rho_0 \mathbf{\Omega}_0 \times \mathbf{u}. \tag{3}$$

Here, p_e stands for any excess pressure distribution over that required to counteract centrifugal force in the uniformly rotating fluid, and the last term is the Coriolis force associated with any movement $\mathbf{u}$ relative to that uniform rotation.

When equation (2) has been derived by either of these means, the dispersion relationship for waves in a rotating fluid follows by applying a second time the operation on the left-hand side of (2). Since

$$\operatorname{curl} \operatorname{curl} \mathbf{u} = -\nabla^2 \mathbf{u} \tag{4}$$

for a divergenceless vector field $\mathbf{u}$, we obtain

$$-\frac{\partial^2}{\partial t^2} \nabla^2 \mathbf{u} = \Omega_0^2 \frac{\partial^2}{\partial z^2} \mathbf{u}, \tag{5}$$

a simple partial differential equation to be satisfied separately by each component of $\mathbf{u}$.

Very strong parallels between this equation and the corresponding equation for internal waves are evident from the resulting dispersion relationship. A plane-wave solution

$$\mathbf{u} = \mathbf{u}_1 \exp\left[\mathrm{i}(\omega t - kx - ly - mz)\right] \tag{6}$$

satisfies (5) if

$$\omega^2 = \Omega_0^2 m^2/(k^2 + l^2 + m^2). \tag{7}$$

Therefore, the frequency ω depends only on the direction and not on the magnitude of the wavenumber vector, and can take any value less than Ω_0 (all just as in internal waves but with Ω_0 replacing N).

In terms of the angle θ between the surfaces of constant phase and the z-axis (axis of rotation) equation (7) can be written

$$\omega = \Omega_0 \sin \theta; \tag{8}$$

this time, as in internal waves but with cos and sin interchanged. The group velocity

$$\mathbf{U} = (\partial\omega/\partial k,\, \partial\omega/\partial l,\, \partial\omega/\partial m) \tag{9}$$

can for positive ω/m be written

$$\mathbf{U} = \frac{\Omega_0(k^2+l^2)^{\frac{1}{2}}}{k^2+l^2+m^2} \left[\frac{(-km,\, -lm,\, k^2+l^2)}{(k^2+l^2)^{\frac{1}{2}}(k^2+l^2+m^2)^{\frac{1}{2}}} \right], \tag{10}$$

where the term in square brackets is a unit vector. Just as in section 4.4 we note that this velocity of wave-energy propagation is directed parallel to the surfaces of constant phase; in fact, perpendicular to the wavenumber vector and coplanar with it and the vertical. The magnitude of the group velocity can be written

$$U = (\Omega_0 \lambda/2\pi) \cos \theta \tag{11}$$

for waves of length λ. By contrast, the crests and other surfaces of constant phase are moving perpendicular to themselves at the wave speed,

$$c = (\omega\lambda/2\pi) = (\Omega_0 \lambda/2\pi) \sin \theta. \tag{12}$$

Equations (11) and (12) are again as in section 4.4 with Ω_0 replacing N and with cos and sin interchanged. For waves in rotating fluid, however, the z-component of wavenumber and group velocity have the *same* sign; therefore, the component of energy propagation along the axis of rotation has the *same sense* as the axial component of crest movement.

In rotating fluid, a source oscillating at frequency $\omega < \Omega_0$ generates wave energy travelling at the angle

$$\theta = \sin^{-1}(\omega/\Omega_0) \tag{13}$$

to the z-axis. These waves form a St Andrew's Cross in two-dimensional propagation; while, in three-dimensional propagation from a point source, they form a double cone. Just one or two crests are present, as analysed in section 4.10; only their *sense* of movement is opposite to that for internal waves.

The limiting case $(\omega/\Omega_0) \to 0$ is even more interesting than the corresponding limit for internal waves. The interchange of cos and sin means that the angle θ tends to *zero* in this case. Any solid body in very slowly changing motion, therefore, sends disturbances along a unique line $(\theta = 0)$

at a finite velocity* (which for components of length λ is $\Omega_0 \lambda/2\pi$). This implies that the steady motion of a solid body at velocity V along the axis of rotation can push a columnar signal ahead of it, consisting of all components of the disturbance with wavelength λ satisfying

$$\Omega_0 \lambda > 2\pi V. \tag{14}$$

A closely related phenomenon is the 'Taylor column', observed when a solid body moves at right angles to the axis of rotation at a *very* low speed, *small* compared with the group velocity $\Omega_0 \lambda/2\pi$ for waves of practically all lengths generated by a body of that size. This means that a columnar disturbance stretching in the $\theta = 0$ direction (parallel to the axis of rotation) must accompany that motion on both sides of the body.

The wave motions described so far are usually called 'inertial waves'; especially, because their energy is entirely kinetic: the Coriolis force is a nonconservative force incapable of storing, or otherwise changing, energy. Indeed, equation (3) implies that $\partial W/\partial t = -\nabla \cdot \mathbf{I}$ with $\mathbf{I} = p_e \mathbf{u}$ as usual but with W equal to the kinetic energy $\frac{1}{2}\rho_0(\mathbf{u} \cdot \mathbf{u})$. It also implies that, in plane waves (6), the fluid particles move with constant kinetic energy (around circular paths perpendicular to the wavenumber vector).

The above results for homogeneous fluid, although of obvious interest and well confirmed by laboratory experiments, are practically irrelevant to those notably *stratified* fluids, the ocean and the atmosphere. The value of Ω_0 for the rotation of the Earth is

$$\Omega_0 = 4\pi(\text{day})^{-1} = 1.45 \times 10^{-4} \text{s}^{-1}. \tag{15}$$

Corresponding values for the Väisälä–Brunt frequency N quoted in section 3.3 are centred around 10^{-2}s^{-1}. This suggests a strong tendency for internal wave propagation in the ocean and the atmosphere to be dominated by stratification effects; in other words, for the value of $\omega^2(k^2 + l^2 + m^2)$ to be dominated by the gravity contribution, equal (section 4.1) to

$$N^2(k^2 + l^2) \tag{16}$$

if the z-axis is vertical, rather than by the term suggested in equation (7); namely, $\Omega_0^2 m^2$ with the z-axis along the axis of rotation.

Detailed analysis confirms this conclusion in most cases, but an important exception (suggested by (16)) is made by very long waves with k and l extremely small. In the language of section 4.13, it is primarily the wave-

---

* By contrast, for internal waves the limiting angle $\theta = \frac{1}{2}\pi$ gives this result only for a two-dimensional source (section 4.12).

guide modes in the ocean (with wavelengths large compared to the ocean depth) that are significantly affected by the Earth's rotation. Among them, the lowest mode of all is called by oceanographers 'barotropic'; it is a 'long wave' in the sense of chapters 2 and 3 and shows no significant effect of stratification. Although higher waveguide modes, called 'baroclinic', are also of definite oceanographic interest, and are affected *both* by stratification and by Earth's rotation, we here concentrate on the long waves proper.

Fluid motions in such a long wave are, predominantly, horizontal motions showing insignificant variation with the vertical coordinate z; even though they allow small elevations ζ (positive or negative) of the free surface. Under these circumstances, the horizontal components of the vector equation (3) involve only the *vertical* component of the undisturbed vorticity $\mathbf{\Omega}_0$; this vertical component is always called f by oceanographers, and is equal to the magnitude (15) of $\mathbf{\Omega}_0$ times the sine of the latitude. Again, p_e can be taken as $\rho_0 g \zeta$ (where g, the observed gravitational acceleration on the rotating Earth, includes as always a small negative term due to centrifugal force). The linearised momentum equations for long waves, therefore, are

$$\partial u/\partial t = -g\,\partial\zeta/\partial x + fv, \quad \partial v/\partial t = -g\,\partial\zeta/\partial y - fu, \tag{17}$$

which must be solved with the same equation of continuity

$$\partial\zeta/\partial t + \partial(hu)/\partial x + \partial(hv)/\partial y = 0 \tag{18}$$

as was used in section 4.13.

An extensive theory has been developed for the long waves governed by equations (17) and (18). Their properties are simplest in those conditions when the variations of the undisturbed depth h with position, and of the 'Coriolis parameter' f with latitude, can to good approximation be neglected.

In that case when h and f can be regarded as constants, the additional vertical vorticity due to the wave motions takes the value

$$\partial v/\partial x - \partial u/\partial y = f\zeta/h. \tag{19}$$

This may be deduced from the above equations or interpreted as due to the undisturbed vortex lines changing their *vertical* extent from h to $h + \zeta$. Equation (18) then implies, with (17) and (19), that

$$\partial^2\zeta/\partial t^2 = gh(\partial^2\zeta/\partial x^2 + \partial^2\zeta/\partial y^2) - f^2\zeta, \tag{20}$$

so that the two-dimensional dispersion equation for long waves is

$$\omega^2 = gh(k^2 + l^2) + f^2. \tag{21}$$

Here, we note an interesting contrast: whereas rotation limits the frequency of inertial waves to values *less* than Ω_0, it has a precisely opposite effect on long waves, limiting ω to values *greater* than f. These long waves propagate isotropically but dispersively; the magnitude of the group velocity can be written

$$U = (gh)^{\frac{1}{2}} (\omega^2 - f^2)^{\frac{1}{2}} \omega^{-1}, \tag{22}$$

which tends to zero as ω falls to the cutoff value f.

Of course, many naturally occurring long waves, such as (for example) the 'tsunamis' generated by earthquake motions, have their important frequencies ω much larger than f. In any such case the effect of Earth's rotation on wave propagation is negligible.

The theory of that effect is, however, extremely important for one main reason: because the oceans are excited at frequencies comparable to f by *tide-raising forces*. The biggest of these is due to the Moon's excess of gravitational attraction on water nearer to it than the Earth's centre and its corresponding defect of attraction on water farther than the centre. The frequency of variation of that force at any fixed point of the rotating Earth, as the Moon pursues its orbit, takes a value $1.40 \times 10^{-4} \mathrm{s}^{-1}$ just less than Ω_0 (and corresponding to a period of 12 hours 25 minutes). This, however, exceeds f at all latitudes less than $75°$.

The Sun exerts somewhat smaller forces, with a frequency very close to $\Omega_0 = 1.45 \times 10^{-4} \mathrm{s}^{-1}$. Those are especially important every fourteenth day (at full moon or new moon) when the tide-raising forces due to the Sun and Moon reinforce one another.

The propagation properties of solutions of the long-wave equations have a particularly significant effect on the tides in shallow seas. These respond with a certain *delay* to the tidal rise and fall in an adjoining deep ocean, owing to the limited value of a propagation speed such as (22). The inevitable increase in amplitude of the wave as its energy moves into shallower waters and becomes confined within a reduced depth adds to the practical importance of precise knowledge of these effects.

Other modes of propagation besides that governed by the dispersion equation (21) are significant in this context. They include a quasi-one-dimensional mode of waveguide propagation known as the 'Kelvin wave' which may be thought of as an edge-wave (section 4.13) modified by the effects of the Earth's rotation.

Besides the predictable tide-raising forces of astronomical origin, certain other forces modify the tidal motions in shallow seas. These include the frictional action of the wind on the water surface. Very strong winds blowing at periods of maximum tidal motion can seriously threaten coastal defences. In order to study such threats to British coasts, the Institute of Oceanographic Sciences developed a computer program for solving equations (17), with the nonlinear terms restored and with arbitrary additional forcing terms on the right, throughout the shallow-sea region surrounding the British Isles, subject to equation (18) and to an input of known tidal movements of purely astronomical origin from the adjoining deep ocean. This program can be used with wind forecasts to determine when various emergency coastal defences need to be activated.

Long waves of frequency $\omega > f$ commonly behave in approximate accordance with the dispersion relationship (21) even though f and h are variable. Such variability is, nevertheless, found to make possible also some wavelike motions at much lower frequencies, called 'Rossby waves'. These may be especially important when the tidal response of a large ocean is studied. Such studies need, of course, to abandon Cartesian coordinates and adopt the spherical polar coordinates appropriate to the Earth's shape. At this stage, however, we ourselves must abandon our very brief sketch of waves in rotating fluids, referring readers to the bibliography for further information on the various topics mentioned.

In motions of homogeneous fluid, vortex lines and magnetic field lines share, as mentioned earlier, an important property. The full nonlinear equation for vorticity $\mathbf{\Omega}$ is

$$\partial\mathbf{\Omega}/\partial t = \operatorname{curl}(\mathbf{u} \times \mathbf{\Omega}) + \nu\nabla^2\mathbf{\Omega}, \tag{23}$$

where the last term, proportional to the kinematic viscosity ν, represents diffusion of vorticity. With that term neglected, equation (23) states that vortex lines move with the fluid; and its linearised form, where $\mathbf{\Omega}$ is replaced on the right-hand side by $\mathbf{\Omega}_0$, is the equation (2) governing inertial waves.

The corresponding equation for magnetic field $\mathbf{B}$ is

$$\partial\mathbf{B}/\partial t = \operatorname{curl}(\mathbf{u} \times \mathbf{B}) + (\tau/\mu)\nabla^2\mathbf{B}, \tag{24}$$

where τ is resistivity and μ magnetic permeability. Equation (24) arises because the electromotive force on a conductor moving at velocity $\mathbf{u}$ is the sum of the electrostatic force $\mathbf{E}$ and the Lorentz force $\mathbf{u} \times \mathbf{B}$. The current density $\mathbf{j}$ is accordingly determined by an equation

$$\mathbf{E} + \mathbf{u} \times \mathbf{B} = \tau\mathbf{j}. \tag{25}$$

Equation (24) then follows from the form

$$\partial\mathbf{B}/\partial t = -\operatorname{curl}\mathbf{E}, \quad \nabla \cdot \mathbf{B} = 0, \quad \operatorname{curl}\mathbf{B} = \mu\mathbf{j} \tag{26}$$

of Maxwell's equations appropriate to those processes, in a material of uniform permeability μ, which avoid such extreme frequencies that displacement current is significant in the last equation.

In (24) the diffusivity (τ/μ) is usually very large: far larger than common values of the kinematic viscosity ν. Thus, for mercury (an experimentally convenient electrically conducting liquid) $\tau/\mu = 0.75\,\text{m}^2\,\text{s}^{-1}$. Such an enormous diffusivity makes it almost impossible to reproduce in the laboratory effects like those to be described which depend on magnetic field lines being convected with the fluid. Even for liquid sodium, whose low resistivity τ makes it a common choice of fluid for cooling systems pumped electromagnetically, $\tau/\mu = 0.08\,\text{m}^2\,\text{s}^{-1}$ corresponding to a one-second diffusion distance of about 0.3 m.

Very high diffusivities may, however, become relatively unimportant where systems on the scale of the Earth are concerned, including for example those motions in the liquid core or in the ionosphere that may be responsible for parts of the observed fluctuations of the Earth's magnetic field. Even though typical diffusivities in the ionosphere may take still greater values, of order $10^5\,\text{m}^2\,\text{s}^{-1}$, they may well be relatively unimportant for phenomena on such large length scales. These are among the reasons for sketching below the remarkable properties of waves governed by the diffusionless form

$$\partial\mathbf{B}/\partial t = \text{curl}\,(\mathbf{u}\times\mathbf{B}) \tag{27}$$

of equation (24).

As the magnetic field lines are convected by the fluid, the force which, conversely, they exert on the fluid can be expressed in either of two ways: in terms of the aforementioned Lorentz force on moving charges as

$$\mathbf{j}\times\mathbf{B} = \mu^{-1}(\text{curl}\,\mathbf{B})\times\mathbf{B} = \mu^{-1}[\mathbf{B}\cdot\nabla\mathbf{B} - \nabla(\tfrac{1}{2}B^2)] \tag{28}$$

or in terms of the *equivalent stress system* of Maxwell. This consists of an isotropically acting pressure $\tfrac{1}{2}\mu^{-1}B^2$, together with a uniform tension $\mu^{-1}B^2$ per unit cross-sectional area acting along the magnetic field lines.

The latter form allows a rapid inference of the nature of the wave propagation possible in a uniform magnetic field $\mathbf{B}_0$ when compressibility is insignificant. A purely physical argument suffices: under conditions of incompressibility, fluid pressures can adjust themselves to balance exactly the variations in magnetic pressure; then, each magnetic field line and the fluid moving with it behave like a simple string in tension. In fact, fluid of cross-sectional area A surrounding the field line has mass per unit length

$\rho_0 A$ and tension $\mu^{-1}B_0^2 A$. Therefore, transverse motions travel one-dimensionally along it, like waves along a stretched string, at a wave speed

$$c_A = \left(\frac{\text{tension}}{\text{mass per unit length}}\right)^{\frac{1}{2}} = \frac{B_0}{(\mu\rho_0)^{\frac{1}{2}}}, \tag{29}$$

known as the Alfvén wave speed.

The surprising feature of this conclusion is that, in three-dimensionally unbounded fluid, waves may propagate one-dimensionally along magnetic field lines. The same conclusion follows from a direct small-perturbation analysis. Small changes in **B** from a uniform undisturbed magnetic field

$$\mathbf{B}_0 = (0, 0, B_0) \tag{30}$$

satisfy

$$\partial\mathbf{B}/\partial t = B_0\,\partial\mathbf{u}/\partial z \tag{31}$$

(a direct analogy of equation (2) for rotating fluid). The fluid momentum equation, with fluid pressures and magnetic pressures in balance, becomes

$$\rho_0\,\partial\mathbf{u}/\partial t = \mu^{-1}B_0\,\partial\mathbf{B}/\partial z. \tag{32}$$

From (31) and (32), every component of the velocity field (or of the magnetic field) satisfies the equation

$$\partial^2\mathbf{u}/\partial t^2 = (B_0^2/\mu\rho_0)\,\partial^2\mathbf{u}/\partial z^2 \tag{33}$$

describing one-dimensional propagation in the z-direction at the Alfvén wave speed (29).

We note in particular that not *only* transverse components of **u** are so propagated. Evidently, the component of **u** along the field lines is directly coupled to the transverse components in a divergenceless motion, and therefore satisfies the same law of propagation.

For radiation from a source of fixed frequency ω_0, equation (33) is a special case of those systems considered at the beginning of section 4.10. Thus, the wavenumber surface S is a pair of parallel *planes*. Any such case, as shown there, corresponds to one-dimensional wave propagation along the direction normal to those planes. Thus, the usual reduction of amplitude as wave energy spreads out in two or three dimensions is avoided. When wave energy dissipation is not large, such modes of propagation may allow the maintenance of significant amplitudes over exceptionally long distances.

Again, curved magnetic field lines may act as rays in the sense of section 4.5. Any disturbance with characteristic wavenumber small compared to their curvature can be expected to travel along them.

The possible importance of such conclusions is an incentive to careful study of how far effects neglected above may alter them. Here, we do not pursue further the earlier comments regarding dissipative processes (viscosity and, especially, electrical resistivity), but indicate some competing nondissipative effects.

Compressibility was neglected in the above analysis. However, much as in sections 3.1 and 4.2, we can estimate simply that the neglect of compressibility should be justified if the Alfvén wave speed (29) is small compared with the speed of sound, c_0. Then the density changes due to fluid pressure changes should occur at too slow a rate to affect the assumption of a divergenceless velocity field.

For a gas (such as the part of the upper atmosphere known as the ionosphere, rendered electrically conducting by photo-ionisation) the Alfvén speed (29) can be small compared with the sound speed $(\gamma p_0/\rho_0)^{\frac{1}{2}}$ if and only if the magnetic pressure $\frac{1}{2}\mu^{-1}B_0^2$ is small compared with the gas pressure p_0. The magnitude B_0 of the Earth's magnetic field is typically around 5×10^{-5} T in middle geomagnetic latitudes (where T, the SI unit of magnetic flux intensity, is the tesla $= 10^4$ gauss; note also that the permeability μ takes the free-space value 1.26×10^{-6} T^2 N^{-1} m^2). The resulting magnetic pressure, 10^{-3} N m^{-2} or 10^{-8} bar, is small compared with gas pressures in the lower part of the ionosphere, known as the D region, at an altitude of 80 to 100 km. On the other hand, gas pressures are comparable with 10^{-8} bar in the E region, around 130 km altitude; whereas in the F regions (from 180 km upwards) gas pressures are much *less* than the magnetic pressure.

There may, therefore, be a strong interest in analysing the interactive propagation of Alfvén waves and sound waves, both in cases when c_A and c_0 are comparable and in cases when the Alfvén speed greatly exceeds the sound speed. Such an analysis uses the linearised equations of continuity and of momentum for a compressible fluid, with a linearised form

$$\mu^{-1}[\mathbf{B_0} \cdot \nabla \mathbf{B} - \nabla(\mathbf{B_0} \cdot \mathbf{B})] \tag{34}$$

of the magnetic force (28), and with equation (31) for the rate of change of **B**. We omit details and give only the results.

The equation for the component w of velocity parallel to the magnetic field lines has plane-wave solutions

$$w = w_1 \exp[i(\omega t - kx - ly - mz)], \tag{35}$$

where

$$k^2 + l^2 = (\omega^2 - c_0^2 m^2)(\omega^2 - c_A^2 m^2)[(c_0^2 + c_A^2)\omega^2 - c_0^2 c_A^2 m^2]^{-1}. \tag{36}$$

Equations for the components (u, v) of velocity perpendicular to field lines

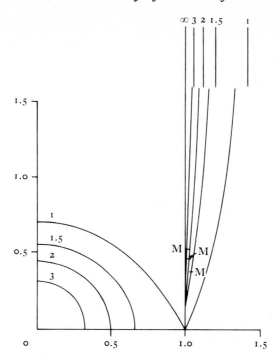

Figure 112. Wavenumber surfaces S for magnetohydrodynamic waves in a compressible fluid with sound speed c_0 and Alfvén speed c_A. The magnetic field is in the z-direction. The abscissa is m, the z-component of wavenumber, while the ordinate is $(k^2 + l^2)^{\frac{1}{2}}$. The points marked M are points of inflexion giving rise to caustics.
(a) When c_0/c_A takes the marked values 1, 1.5, 2, 3 and ∞ (corresponding to $c_A/c_0 = 1$, $\frac{2}{3}$, $\frac{1}{2}$, $\frac{1}{3}$ and 0) the surfaces are as shown with ω_0/c_0 as the unit of wavenumber.
(b) However, when c_A/c_0 exceeds 1, taking the marked values 1, 1.5, 2, 3 and ∞, the surfaces are as shown with ω_0/c_A as the unit of wavenumber. [*Phil. Trans. Roy. Soc.*]

are more complicated, but become simple if we concentrate on the two-dimensional divergence and curl of that two-dimensional velocity field:

$$\Delta_{II} = \partial u/\partial x + \partial v/\partial y \quad \text{and} \quad \Omega_{II} = \partial v/\partial x - \partial u/\partial y. \tag{37}$$

The values of Δ_{II} and Ω_{II} determine uniquely a velocity field (u, v) vanishing at large distances. Of these quantities (37), Δ_{II} satisfies

$$\Delta_{II} = i(m - \omega^2 c_0^{-2} m^{-1}) w_1 \exp[i(\omega t - kx - ly - mz)], \tag{38}$$

so that its variations are coupled to (indeed, directly proportional to) those of $\partial w/\partial z$. By contrast, Ω_{II} satisfies an independent equation

$$\partial^2 \Omega_{II}/\partial t^2 = c_A^2 \, \partial^2 \Omega_{II}/\partial z^2. \tag{39}$$

Equation (39) means that the z-component of vorticity Ω_{II} is propagated one-dimensionally at the Alfvén velocity c_A along the magnetic field lines. In other words, compressibility does not eliminate such one-dimensional propagation; it merely confines it to the z-component of vorticity.

The wavenumber surface (36) for the propagation of both w and Δ_{II} is illustrated in figure 112 for various values of c_A/c_0 between 0 and 1. Except in the 'incompressible' limit ($c_A/c_0 \to 0$), wave propagation along the normals to this wavenumber surface spreads out in three dimensions the w and Δ_{II} signals radiated by a source. When those signals have become fully separated from the one-dimensional propagation of Ω_{II}, the motion propagated along field lines must be purely transverse and divergenceless.

Admittedly, for small values of c_A/c_0, the wavenumber surface S consists essentially of a sphere of radius ω/c_0 representing the isotropic propagation of sound waves and a pair of planes representing a quite independent one-dimensional propagation at the Alfvén speed. However, for larger values such as $c_A/c_0 = \frac{1}{2}$, although the propagation at velocity near to c_0 is still almost isotropic, the propagation at velocity near to c_A spreads out conically; it fills a cone of semi-angle $4.7°$ whose boundary is a caustic arising from the point of inflexion in figure 112.

Note that, because c_0 and c_A occur symmetrically in equation (36) for S, the surface takes the same forms again when $c_A \geqslant c_0$. Thus, the forms of S for the values $\frac{1}{3}$, $\frac{1}{2}$, $\frac{2}{3}$ and 1 of c_0/c_A are as shown in figure 112 if the unit of wavenumber is taken as ω_0/c_0 instead of ω_0/c_A.

The limit of large c_A/c_0 is very interesting, and may be relevant to propagation in the F regions of the ionosphere. In that limit, the wavenumber surface reduces to a sphere of radius ω_0/c_A, corresponding to *isotropic* propagation at the Alfvén velocity c_A, together with two planes,

$$m = \pm \, \omega_0 c_0^{-1}, \qquad\qquad (40)$$

corresponding to *one-dimensional* propagation at the velocity of sound c_0.

Such an unexpected possibility of one-dimensional *sound* propagation can be understood in terms of the fact that in this limit the tension in magnetic field lines greatly exceeds the gas pressures. This makes possible a strictly one-dimensional propagation in the sense of chapter 2: sound waves may travel along, effectively, *rigid tubes* provided by the tightly stretched magnetic field lines. The longitudinal nature of the wave is shown by the fact that equations (35) and (38) make Δ_{II} vanish, while (39) makes Ω_{II} propagate at the much faster Alfvén speed. In the sound wave itself, then, they are both zero so that u and v vanish; only the velocity component w in the direction of propagation is nonzero.

If c_A/c_0 is large, there may be sources with a spectrum of frequencies too low (in relation to the source size) to generate any disturbances propagating at velocity c_A; whether the spherical waves propagating Δ_{II}, or the one-dimensional waves propagating Ω_{II}. From such sources, the only waves generated will be the above-mentioned sound waves propagated one-dimensionally along effectively rigid magnetic field tubes.

Even in a brief sketch of waves in ionised gases we should mention the existence of various nondissipative effects causing departures from the ideal equation (27) for conducting fluids. From those we select the one that typically becomes significant first as the wave frequency increases: the Hall effect.

The electrons in an ionised gas, because of their exceedingly small mass, contribute negligibly to the gas *velocity* $\mathbf{u}$, defined as a local average momentum per unit mass. However, any electric current density $\mathbf{j}$ in the gas depends on the electrons (of charge $-e$ and, say, number density n_{el}) moving at a velocity $(-\mathbf{j}/en_{el})$ relative to the positive ions. In particular, when positive ions and neutral particles have the same mean velocity $\mathbf{u}$, that of the electrons is

$$\mathbf{u}_{el} = \mathbf{u} - (\mathbf{j}/en_{el}). \tag{41}$$

On the other hand, field lines tend to move not with the gas as a whole but with the electron gas, so that (27) is replaced by

$$\partial \mathbf{B}/\partial t = \text{curl}\,(\mathbf{u}_{el} \times \mathbf{B}). \tag{42}$$

This is because the inertia of the electrons is negligibly small at the frequencies under discussion. Accordingly, when the 'collisional' effects contributing to electrical resistivity are also neglected, the mean force on electrons, $-e(\mathbf{E} + \mathbf{u}_{el} \times \mathbf{B})$, equilibrates either to zero, or to a term involving the gradient of electron pressure which disappears when curl $\mathbf{E}$ is calculated. In either case the equations (41) and (42) replace the former equation (27).

The Hall effect briefly described above is found to destroy the one-dimensional propagation of transverse motions along magnetic field lines at wave speed c_A. It converts it into a conical propagation confined within a cone of semi-angle $\sin^{-1}(\omega\rho_0/B_0\,en_{el})$.

By contrast, the other one-dimensional propagation, which occurs at the sound speed c_0 when that is small compared with c_A, turns out to be uninfluenced by Hall effect. This longitudinal form of one-dimensional propagation may persist, therefore, up to higher frequencies than the transverse form.

Very brief sketches of a variety of waves in fluids have been given in this part 1 of the epilogue, mainly in order to help indicate the width of the field. For detailed information on any of the types mentioned, see the texts referred to in the bibliography.

Part 2   Nonlinear effects on dispersive wave propagation

By nonlinear effects on waves we mean any features of real wave motions which cannot be reproduced in a linear analysis; that is, in an analysis neglecting the squares of the disturbances. In the core of this book (chapters 1–4) nonlinear effects are described only for nondispersive waves (sections 2.8–2.14); mainly for sound waves, and also for long waves in channels.

There are some good reasons for thus giving prominence to nonlinear effects in the *nondispersive-wave* context, where they can take such 'spectacular' forms as shock waves and hydraulic jumps; and where, as shown in section 2.11, quite small nonlinear terms in the equations of motion can, given a long enough time, change ordinary waveforms into forms so spectacularly different. These are systems where changes of waveform in successive instants can 'add up' until they have become very large changes.

By contrast, the dispersive property (wave speed varying with wavelength) can significantly limit the possibilities for such cumulative consequences of small nonlinear effects. No longer is there a single basic wave speed c_0 at which the whole waveform travels while it becomes progressively 'sheared' as a result of small variations of wave speed from that basic speed c_0. Instead, every new Fourier component which nonlinear modifications may generate has a different basic wave speed. Accordingly, the modifications produced become spatially separated, which hinders their progressive accumulation.

The ordinary dynamics of homogeneous fluid is another field where small nonlinear effects have a cumulative action. The basic wave speed is zero for vorticity (section 1.1); the familiar statement that vortex lines move with the fluid incorporates both this fact and the essentially nonlinear action of a vorticity field on itself. Its accumulated effect produces, for example, the complicated properties of those fluid flows with highly contorted vortex lines called turbulence.

The cumulative action of nonlinear effects wherever there is a single wave speed (zero for that vorticity field, or nonzero for a nondispersive system) is absent for a dispersive system. Dispersion prevents accumulation because different Fourier components in a wave system become displaced from each other. This can mean that the dynamics of a 'simple' homogeneous fluid is actually more complicated than that of a stratified fluid, or than that of a homogeneous fluid in uniform rotation; essentially, because disturbances in these systems, instead of staying together, are dispersed three-dimensionally: the group velocity is different for each wavenumber vector. Similarly, much useful knowledge about water waves can be derived from a purely linear theory like that of chapter 3: in fact, wherever there is no tendency for small nonlinear effects to act cumulatively, such a linear analysis has a good chance of producing useful results.

In this part 2 of the epilogue we take up three questions suggested by the above discussion. First, we concentrate on waves such as gravity waves on deep water for which a linear analysis (section 3.2) is relatively simple and predicts a high degree of dispersion. We sketch an account of how far the properties of these waves may be altered by substantial nonlinear effects such as may accompany large disturbances of the water surface.

Secondly, we consider waves with only a small degree of dispersion, such as 'fairly long' water waves; these are waves whose length λ falls only a little short of those values (exceeding 20 times the water depth) for which dispersion is quite negligible. In 'fairly long' waves there can be rather an equal balance, as we shall see, between the influence of small nonlinear effects and that small degree of dispersion.

Thirdly, we sketch very briefly an approach to analysing a statistical assemblage of dispersive waves. Especially, we seek to indicate why certain properties of such an assemblage may be simpler than those of its non-dispersive analogues including turbulence.

Unity is given to the subject by the concept of 'wave action', first shown by G. B. Whitham to be of general significance for nonlinear dispersive waves. The analysis of ray tracing in a wind (section 4.6) has suggested already the importance of wave action for a limited range of linear systems, but we shall find that it has a much wider significance.

For gravity waves on deep water, a periodic wave of length λ is predicted by a linear analysis to be exactly sinusoidal in shape and to travel with a wave speed

$$c = (g\lambda/2\pi)^{\frac{1}{2}}. \tag{43}$$

We begin by asking whether periodic waves of length λ continue to be possible for wave amplitudes so great that the squares and higher powers of disturbances can no longer be neglected.

We might expect such periodic waves to be possible, even though they need not be exactly sinusoidal and even though their wave speed c need not exactly satisfy (43), for the following reason. Suppose that a large cylindrical obstacle moves horizontally through deep water at a velocity V perpendicular to its axis. Then it should experience a wavemaking resistance even if the linear analysis (section 3.9) of that resistance is not exact. Thus, we expect energy to be carried away by gravity waves behind the cylinder. On the other hand, the flow relative to the cylinder should be a steady flow at velocity V past a stationary obstacle, in which all wave crests must be motionless. Therefore, those crests must move forward relative to the undisturbed water at one and the same wave speed

$$c = V; \tag{44}$$

a criterion which takes the same form as in the linear analysis (section 3.9).

Note also that if W is the wave energy per unit horizontal area and I is the energy flux in the waves per unit length of crest,[*] then

$$cW - I \tag{45}$$

represents the flux of energy backwards relative to a cylinder moving forward at velocity $c = V$. The expression (45) represents the *wavemaking power* per unit length of cylinder (presumed large enough so that it generates only gravity waves). With energy attenuation neglected, the waves behind the cylinder can be expected to be of uniform amplitude so that the flux (45) is uniform. Thus, they should be periodic waves whose length and amplitude are defined by the value (44) for wave speed and the value (45) for flux of energy backward relative to crests.

Such a picture has been used successfully in the computation of the periodic wave, calculated as a steady flow in a frame of reference in which the crests are fixed. Then the velocity potential ϕ satisfies a free-surface condition $\quad \frac{1}{2}(\partial\phi/\partial x)^2 + \frac{1}{2}(\partial\phi/\partial z)^2 + g\zeta = \text{constant on } z = \zeta \tag{46}$

given by Bernoulli's equation. At great depths, the motion relative to the crests is a uniform motion with velocity c, giving

$$\phi \sim cx \quad \text{as} \quad z \to -\infty. \tag{47}$$

The solution is facilitated by the introduction of a stream-function ψ so that the free surface becomes $\psi = 0$. A form

$$\left.\begin{aligned} x &= c^{-1}\phi - \sum_{n=1}^{\infty} b_n e^{2n\pi\psi/c\lambda} \sin\left(2n\pi\phi/c\lambda\right), \\ z &= c^{-1}\psi + \sum_{n=1}^{\infty} b_n e^{2n\pi\psi/c\lambda} \cos\left(2n\pi\phi/c\lambda\right), \end{aligned}\right\} \tag{48}$$

[*] In the linear theory of section 3.9, I is UW, where U is the group velocity.

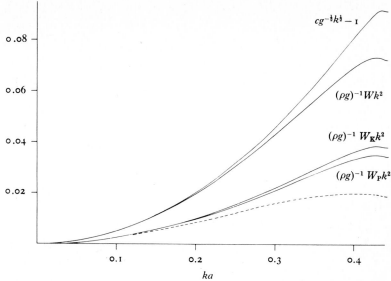

Figure 113. Periodic gravity waves of large amplitude on deep water. Full lines: values of the wave speed c, of the wave energy W and of its kinetic and potential components W_K and W_P for waves of length $2\pi/k$ and varying amplitude a (defined in (49)). Broken line: see (76) below.

in which x and z are written as Fourier series in ϕ satisfying Laplace's equation with respect to ϕ and ψ, corresponds to waves of length λ (so that the period for ϕ is $c\lambda$).

The computation of how these waves vary with amplitude, defined as

$$a = \tfrac{1}{2}(\zeta_{\text{crest}} - \zeta_{\text{trough}}),\tag{49}$$

involves a very intricate evaluation of Fourier coefficients and the use of ingenious methods of high accuracy for accelerated summation of series. We give only the results.

Waves are possible for all values of a up to a maximum value

$$a_{\text{max}} = 0.0706\lambda.\tag{50}$$

The wave velocity varies from the linear-theory value (43) to a value just 1.092 times greater when $a = a_{\text{max}}$ (figure 113). The waveform remains symmetrical about the crest, but increasingly departs from the sinusoidal form as a increases towards a_{max} (figure 114). In the limit $a \to a_{\text{max}}$ the crest becomes a 120° angle; relative to the crest, the water at the apex of the 120° angle is at rest, with the square of its speed increasing linearly with distance from the apex so that equation (46) can be satisfied.

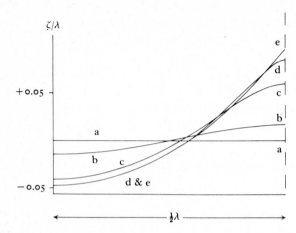

Figure 114. Waveforms of periodic gravity waves of amplitude a on deep water. The scales used enhance all vertical distances by a factor of 2 in comparison with horizontal distances. The broken line marks the plane of symmetry at each crest. (a) Undisturbed water surface. (b) Waveform for $a = 0.015\lambda$. (c) Waveform for $a = 0.050\lambda$. (d) Waveform for $a = 0.065\lambda$. (e) Waveform for the maximum amplitude $a = 0.0706\lambda$.

Figure 113 plots also a nondimensional form

$$(\rho g)^{-1} W k^2 = (\rho g)^{-1} W_{\mathrm{K}} k^2 + (\rho g)^{-1} W_{\mathrm{P}} k^2 \qquad (51)$$

of the wave energy per unit area W, and of its kinetic-energy and potential-energy components W_{K} and W_{P}, as a function of ka. Here, $k = 2\pi/\lambda$ as usual, so that $ka_{\max} = 0.444$. We see that waves possess not only a maximum amplitude but also a maximum energy (attained, actually, at a slightly lower amplitude with $ka = 0.429$). Additional modes of dissipation of energy, characterised by the appearance of foam ('white horses') due to entrapping of air by water, are brought into being at energies above this value.

Instead of pursuing this breakdown of the integrity of the water surface above a certain value of wave energy density, we use the properties indicated in figures 113 and 114 to introduce some general theory. This theory applies to any conservative system allowing unidirectional propagation in which periodic waves can exist for any wavelength within some given range (here, $\lambda > 0.1$ m for surface tension to be negligible) and for any amplitude within another given range (here, $0 < a < a_{\max}$).

Figure 113 shows that although the kinetic and potential energies W_{K}

and W_P per unit area, are necessarily equal in linear theory, nevertheless nonlinear effects cause them to become different. The difference

$$\mathscr{L} = W_K - W_P,\tag{52}$$

which from figure 113 might appear to be of only minor importance, proves the key to a deeper understanding of nonlinear effects in dispersive systems.

The curly $\mathscr{L}$ is used in (52) to denote *Lagrangian density*. Texts on dynamics show that the properties of any conservative dynamical system can be expressed by *Lagrange's equations*, entirely in terms of the Lagrangian function L. This is defined as the kinetic energy of the system minus its potential energy, regarded as a function of the generalised coordinates q_n of the system and their time-derivatives $\dot{q}_n$ (the generalised velocities). Hamilton's principle states that Lagrange's equations are equivalent to a variational statement; namely, that a time-integral

$$\int_{t_1}^{t_2} L \, \mathrm{d}t\tag{53}$$

of the Lagrangian is stationary in any small variation of the generalised coordinates as functions of time with zero variation at the end-points t_1 and t_2.

We sketch briefly the extension of this theory to wave motions in a conservative system. We limit the analysis to one-dimensional propagation in a spatially homogeneous system, although the theory is readily extended to propagation in two or three dimensions and to systems with properties varying gradually on a scale of wavelengths.

For describing a wave system, we retain the use of discrete generalised coordinates q_n, which however are now functions *both* of the time t *and* of the distance x in the direction of propagation. The Lagrangian

$$L = L(q_n, \partial q_n/\partial t, \partial q_n/\partial x) = L(q_n, \dot{q}_n, q'_n)\tag{54}$$

is then defined as the difference between the kinetic energy and potential energy *per unit distance*. It depends not only on the q_n and their time-derivatives $\dot{q}_n$ but also on their space-derivatives q'_n.

As in classical dynamics, the potential energy must be independent of the generalised *velocities* $\dot{q}_n$, while the kinetic energy is a homogeneous function of the second degree in those velocities. This means that the kinetic energy per unit distance can always be written as

$$\tfrac{1}{2}\dot{q}_n \, \partial L/\partial \dot{q}_n.\tag{55}$$

Hamilton's principle now requires that an integral

$$\int_{t_1}^{t_2} dt \int_{x_1}^{x_2} L\, dx \tag{56}$$

be stationary in small variations of $q_n(t,x)$ vanishing at the end-points $t = t_1$, t_2 and $x = x_1$, x_2. Since L is here a Lagrangian per unit distance, this integral, like (53), is the time-integral of a Lagrangian.

For application to periodic waves, it is sometimes convenient to take the integral (56) over a single period and a single wavelength. It can be shown that the principle continues to hold in that case if the variations of q_n are simply constrained to remain periodic with that period and wavelength. Thus, among all undulations with a given period and wavelength, the actual waveform is that which gives a stationary value of the integral (56) taken over a wavelength. This is the same as saying that the *average* Lagrangian density $\mathscr{L}$ (averaged over a wavelength) is stationary.

In a purely linear theory this can only mean that $\mathscr{L}$ is zero. All energies, indeed, are exactly quadratic functions of all the disturbance variables in a linear theory; therefore, when each q_n is varied to $(1+\epsilon)q_n$ (a specially simple small variation), $\mathscr{L}$ changes to $(1+\epsilon)^2\mathscr{L}$. This can be a zero change only if $\mathscr{L}$ itself is zero: the familiar linear-theory result that average kinetic and potential energies are equal.

Note that a nonlinear theory may predict a unique wave of given wavelength λ and period λ/c. For example, figure 113 implies that for given λ the choice of c (within certain limits) fixes the amplitude, and thence also the form, of the wave. In a nonlinear theory, however, energies can depend both on quadratic *and* on higher powers of the coordinates; accordingly, no requirement for the stationary value of $\mathscr{L}$ to be zero exists.

In a periodic wave all the generalised coordinates must take a form

$$q_n = f_n(\omega t - kx), \tag{57}$$

where as usual

$$k = 2\pi/\lambda, \quad \omega = 2\pi c/\lambda. \tag{58}$$

Then the quantity stationary for a real waveform can be written as

$$\mathscr{L}(\omega, k) = (2\pi)^{-1} \int_0^{2\pi} L(f_n(\alpha),\, \omega f_n'(\alpha),\, -k f_n'(\alpha))\, d\alpha; \tag{59}$$

that is, as an average of the Lagrangian (54) with respect to the phase $\alpha = \omega t - kx$.

We can now define a quantity of great significance in dispersive-wave theory: the action density

$$\mathscr{A} = \partial\mathscr{L}/\partial\omega. \tag{60}$$

This is the rate of change of Lagrangian density with frequency, keeping the wavenumber constant. When (60) is evaluated using the expression (59) for $\mathscr{L}$, we do not have to allow for the small change in waveform $f_n(\alpha)$ which accompanies a small change in ω, because $\mathscr{L}$ is stationary with respect to any small changes in waveform! Therefore, we can evaluate (60) for fixed waveform $f_n(\alpha)$ as

$$\partial\mathscr{L}/\partial\omega = (2\pi)^{-1}\int_0^{2\pi} f_n'(\alpha)\,(\partial L/\partial\dot{q}_n)\,\mathrm{d}\alpha. \tag{61}$$

This makes $\omega\,\partial\mathscr{L}/\partial\omega$ the average of $\dot{q}_n\,\partial L/\partial\dot{q}_n$ which by (55) is $2W_K$: twice the average kinetic energy per unit distance. Accordingly, the action density $\mathscr{A}$ can be written

$$\mathscr{A} = \partial\mathscr{L}/\partial\omega = 2W_K\,\omega^{-1}; \tag{62}$$

it is merely for linear systems that this takes the form $W\omega^{-1}$ (total wave energy density divided by frequency) noted in section 4.6.

When the above theory is applied to long-crested water waves, all the energies are taken per unit length of crest so that the above energies per unit distance become energy densities per unit horizontal area. Then equation (62) quantifies, for waves of given length, how the excess $\mathscr{L}$ of kinetic over potential energy rises as ω increases above its linear-theory value $(gk)^{\frac{1}{2}}$:

$$W_K - W_P = \mathscr{L} = \int_{(gk)^{\frac{1}{2}}}^{\omega} 2W_K\,\omega^{-1}\,\mathrm{d}\omega. \tag{63}$$

The accuracy of the computations on which figures 113 and 114 are based was checked by the fact that they satisfied (63) to within a large number of significant figures.

One of the advantages of variational theory is that it leads to useful identities such as (63); later, we find also an expression for the wave energy flux I. A still greater advantage, however, is that it enables the theory of the dispersion process itself to be extended to include nonlinear effects.

We sketch this for the simple theory (section 3.6) of how an extended group of waves propagating one-dimensionally evolves in time when the wave properties (wavelength and amplitude) vary only gradually on a scale of wavelengths. The nonlinear theory still assumes, as in section 3.6, that a local phase α can be defined; essentially, so that it increases smoothly by 2π between successive crests (even though the *waveform* may be quite different from a sine wave). Then the definitions of ω and k give

$$\omega = \partial\alpha/\partial t, \quad k = -\partial\alpha/\partial x, \tag{64}$$

as before.

Next, we consider the integral (56) to which Hamilton's principle applies, but with t_1, t_2, x_1 and x_2 chosen so that the integral is taken over an extended time and over the whole group. We assume that, wherever properties vary slowly on a scale of wavelengths, the wave will locally be very like a periodic wave with the local values (64) of ω and k. In (56), then, where the Lagrangian L is integrated over very many wavelengths and very many periods, it can be replaced at each point by its value averaged over a wavelength or a period; namely, $\mathscr{L}(\omega, k)$.

The integral (56) then becomes

$$\int_{t_1}^{t_2} \mathrm{d}t \int_{x_1}^{x_2} \mathscr{L}(\partial\alpha/\partial t, -\partial\alpha/\partial x)\,\mathrm{d}x. \tag{65}$$

This integral (65) is stationary for variations in α vanishing at the end-points if and only if the Euler condition

$$\frac{\partial}{\partial t}\left(\frac{\partial \mathscr{L}}{\partial \omega}\right) = \frac{\partial}{\partial x}\left(\frac{\partial \mathscr{L}}{\partial k}\right) \tag{66}$$

is satisfied.

Equation (66) is Whitham's fundamental equation for nonlinear one-dimensional dispersion. It can be interpreted most directly as a *wave-action conservation law*

$$\partial\mathscr{A}/\partial t = -\partial\mathscr{B}/\partial x \tag{67}$$

in terms of the wave-action density (60) and a wave-action flux

$$\mathscr{B} = -\partial\mathscr{L}/\partial k. \tag{68}$$

Alternatively, it can be used to yield an energy conservation law,

$$\partial W/\partial t = -\partial I/\partial x, \tag{69}$$

where the energy flux is

$$I = -\omega\,\partial\mathscr{L}/\partial k. \tag{70}$$

This is because (62) gives the total wave energy density $2W_{\mathrm{K}} - \mathscr{L}$ as

$$W = \omega\,\partial\mathscr{L}/\partial\omega - \mathscr{L}, \tag{71}$$

so that

$$\begin{aligned}
\frac{\partial W}{\partial t} &= \left[\omega\frac{\partial}{\partial t}\left(\frac{\partial\mathscr{L}}{\partial\omega}\right) + \frac{\partial\omega}{\partial t}\frac{\partial\mathscr{L}}{\partial\omega}\right] - \left[\frac{\partial\omega}{\partial t}\frac{\partial\mathscr{L}}{\partial\omega} + \frac{\partial k}{\partial t}\frac{\partial\mathscr{L}}{\partial k}\right] \\
&= \omega\frac{\partial}{\partial x}\left(\frac{\partial\mathscr{L}}{\partial k}\right) + \frac{\partial\omega}{\partial x}\frac{\partial\mathscr{L}}{\partial k} = -\frac{\partial I}{\partial x}
\end{aligned} \tag{72}$$

by (66), (64) and (70).

In the particular case of gravity waves on deep water these equations for density and flux of action and energy simplify further because $(\rho g)^{-1}\mathscr{L}k^2$, as figure 113 shows, is a function of ω^2/gk. This implies that

$$k\,\partial\mathscr{L}/\partial k + \tfrac{1}{2}\omega\,\partial\mathscr{L}/\partial\omega = -2\mathscr{L}; \tag{73}$$

which means, for example, that the energy flux I can be written in terms of the quantities plotted in figure 113 as

$$c(2\mathscr{L} + \tfrac{1}{2}\omega\,\partial\mathscr{L}/\partial\omega) = c(2\mathscr{L} + W_{\mathrm{K}}) = c(3W_{\mathrm{K}} - 2W_{\mathrm{P}}). \tag{74}$$

Again, the backward flux (45) of energy relative to crests is

$$cW - I = c(3W_{\mathrm{P}} - 2W_{\mathrm{K}}). \tag{75}$$

This formula is used in figure 113 (broken line) to plot a nondimensional form

$$gP_{\mathrm{w}}/\rho bV^5 = g(3W_{\mathrm{P}} - 2W_{\mathrm{K}})/\rho c^4 \tag{76}$$

of the wavemaking power of a cylinder of length b moving at speed V perpendicular to its generators and making waves of speed $c = V$ without energy dissipation by foaming. Here, the maximum of 0.0200 implies a critical value $0.02\rho bV^5/g$ for wavemaking power (or $0.02\rho bV^4/g$ for the wavemaking drag D_{w}) above which there must be foaming in the waves close to the cylinder.

The relation between wave action and wave energy is somewhat different for waves propagated through fluid moving with velocity V. Then, the Lagrangian of the motions relative to the undisturbed flow depends on generalised coordinates which in a periodic wave vary as

$$q_n = f_n(\omega t - k(Vt + x)) \tag{77}$$

in terms of a displacement x relative to the undisturbed flow. This gives

$$\dot{q}_n = (\omega - kV)f_n' = \omega_{\mathrm{r}} f_n', \tag{78}$$

where ω_{r} is the relative frequency as defined in section 4.6. Accordingly, the action density (61) has $\omega_{\mathrm{r}}\,\partial\mathscr{L}/\partial\omega$ equal to the average of $\dot{q}_n\,\partial L/\partial\dot{q}_n$ which is $2W_{\mathrm{K}}$: twice the kinetic energy of the wave motions relative to the undisturbed flow. For the special case of infinitesimal waves, this is W_{r}, giving the formula

$$\mathscr{A} = W_{\mathrm{r}}/\omega_{\mathrm{r}} \tag{79}$$

for action density. The fundamental wave-action conservation law (66) for a system varying gradually on a scale of wavelengths can then be shown to have the interpretation used in section 4.6.

Even without any undisturbed flow, however, the clearest indication of nonlinear effects on wave dispersion comes from the wave-action conservation law (66). Here we give just one indication of this. Expanding the derivatives in (66) when ω and k satisfy (64), we obtain

$$\frac{\partial^2 \mathscr{L}}{\partial \omega^2}\frac{\partial^2 \alpha}{\partial t^2} - 2\frac{\partial^2 \mathscr{L}}{\partial \omega\, \partial k}\frac{\partial^2 \alpha}{\partial t\, \partial x} + \frac{\partial^2 \mathscr{L}}{\partial k^2}\frac{\partial^2 \alpha}{\partial x^2} = 0; \tag{80}$$

a second-order equation for the phase α, with coefficients functions of the first derivatives (64). Among treatments of this much studied equation we single out the following.

Consider waves generated by a wavemaker oscillating with fixed frequency ω_0 at $x = 0$. Suppose that its amplitude of oscillation departs from a large constant value by a very small amount which varies periodically at a frequency δ far smaller than ω_0. This gives conditions of the form

$$\alpha = \omega_0 t, \quad \partial \alpha / \partial x = -k_0 + \epsilon \exp{(it\delta)} \tag{81}$$

at the wavemaker $x = 0$; where the frequency $\partial \alpha / \partial t$ takes the constant value ω_0 but the amplitude, and therefore also the wavenumber $(-\partial \alpha / \partial x)$, is modulated at frequency δ.

To the first order in ϵ, departures of the phase α from the linear form $\omega_0 t - k_0 x$ must satisfy (80) with the coefficients replaced by constants $\mathscr{L}_{\omega\omega 0}$, $\mathscr{L}_{\omega k 0}$ and $\mathscr{L}_{kk 0}$ equal to the values of the three second derivatives of $\mathscr{L}$ at $\omega = \omega_0$, $k = k_0$. The solution is a linear combination of solutions

$$\exp{[i(t\delta - x\kappa)]}, \tag{82}$$

where κ is either of the roots of the quadratic

$$\mathscr{L}_{kk 0}\kappa^2 + 2\mathscr{L}_{\omega k 0}\kappa\delta + \mathscr{L}_{\omega\omega 0}\delta^2 = 0. \tag{83}$$

These roots may be real or imaginary; for gravity waves on deep water, their computation from the data given above shows them to be real when $a > 0.054\lambda$ and imaginary when $a < 0.054\lambda$.

When they are imaginary, the combination of solutions (82) which satisfies the boundary conditions (81) is

$$\alpha = \omega_0 t - k_0 x + (\beta^{-1}\sinh \beta x)\epsilon \exp{[i(t - xU_0^{-1})\delta]}; \tag{84}$$

here, U_0 is an 'effective group velocity' (velocity of propagation of the amplitude modulation) given by

$$U_0 = -\mathscr{L}_{kk 0}/\mathscr{L}_{\omega k 0}, \tag{85}$$

and β is a rate of exponential increase of modulation with distance given by

$$\beta/\delta = (\mathscr{L}_{\omega\omega 0\, kk 0} - \mathscr{L}_{\omega k 0}^2)^{\frac{1}{2}}/\mathscr{L}_{kk 0}. \tag{86}$$

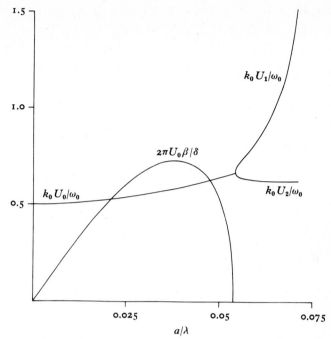

Figure 115. Waves generated by a wavemaker oscillating with fixed frequency ω_0, and with amplitude departing from a large constant value by a small amount which varies periodically at a frequency δ far smaller than ω_0. The rate β of exponential increase of modulation with distance is plotted, together with the effective group velocity U_0, as functions of the amplitude a. Above $a = 0.054\lambda$ the effective group velocity splits into two (U_1 and U_2) while β vanishes.

The hyperbolic sine in (84) indicates that a periodic wave of finite amplitude $a < 0.054\lambda$ on deep water is weakly *unstable*, in the sense that a slow modulation of it tends to become more pronounced the farther the wave travels.

The instability disappears above $a = 0.054\lambda$. When the roots of (83) are real, equation (84) is still the solution of the problem but the quantity defined by (86) is a pure imaginary number. The modulation then splits into two modulations by constant, equal amounts propagated at two different 'effective group velocities' U_1 and U_2 defined by equations

$$U_1^{-1}\delta = U_0^{-1}\delta + |\beta|, \quad U_2^{-1}\delta = U_0^{-1}\delta - |\beta|. \tag{87}$$

Whitham indicated such a 'splitting of the group velocity' as a possible nonlinear effect on dispersion in his earliest work on the variational method.

Figure 115 shows how in the amplitude range $a < 0.054\lambda$ the effective group velocity U_0 rises only slightly above the linear-theory value $0.5\omega_0/k_0$;

while the parameter $2\pi U_0 \beta/\delta$ rises to a maximum of 0.7 and then falls to zero. This parameter is the logarithmic increase in depth of modulation in the distance $2\pi U_0/\delta$ between successive maxima of wave amplitude. A value 0.7, therefore, means that the depth of modulation associated with a particular (travelling) amplitude peak will have doubled approximately by the time the next amplitude peak is generated at the wavemaker . . . Above $a = 0.054\lambda$, it is interesting that while one modulation is propagated at an effective group velocity U_2 close to the linear-theory value, the other small modulation of this large-amplitude basic wave runs ahead at a substantially greater effective group velocity U_1.

Several more intricate calculations related to the weak instability of waves of moderate amplitude on deep water have been made. When the depth of modulation has increased so that the coefficients in (80) vary significantly, these calculations indicate a changed shape of modulation. Crudely speaking, the crests of largest amplitude travel forward fastest, which reduces the wavelength in front of the amplitude peak and increases it behind; then, the energy behind moves forward faster and the energy in front slower, leading to a highly localised enhancement of the amplitude peak. After a finite time this takes a cusped form, and then the assumptions of the theory (amplitude varying gradually) break down. In practice this limits the growth of peak amplitude: in a variety of calculations it was predicted to grow to between 24 % and 57 % above its initial value.

Work of Benjamin and Feir first demonstrated experimentally the growth of a small, slow modulation to waves of moderate amplitude. In the later stages of development, a complicated change in the modulation curve was found to succeed the cusped shape. By contrast, modulations that were insufficiently gradual did *not* grow. Recently, improved dispersion theories have allowed for an extra term in the Lagrangian density (52) proportional to the square of the amplitude *gradient*, and obtained much better agreement with these and similar experiments.

Whitham developed a variational theory also for waves on shallow water. In these, the velocity potential may contain a slowly varying nonperiodic mean portion Φ whose gradient $\partial\Phi/\partial x = s$ represents a mean horizontal stream velocity induced by the waves. The averaged Lagrangian density takes the form

$$\mathscr{L}(\omega, k, \eta, s), \quad \text{where} \quad \eta = -\partial\Phi/\partial t, \tag{88}$$

leading to a double system of Euler equations

$$\frac{\partial}{\partial t}\left(\frac{\partial\mathscr{L}}{\partial\omega}\right) = \frac{\partial}{\partial x}\left(\frac{\partial\mathscr{L}}{\partial k}\right), \quad \frac{\partial}{\partial t}\left(\frac{\partial\mathscr{L}}{\partial\eta}\right) = \frac{\partial}{\partial x}\left(\frac{\partial\mathscr{L}}{\partial s}\right). \tag{89}$$

Instability for waves of moderate amplitude was found if and only if the ratio λ/h of wavelength to depth was less than 4.6.

We pass now to waves with λ/h in the range between (say) 10 and 20, sometimes described as 'fairly long' waves. For those, because dispersion is small, and yet not wholly negligible, the nonlinear theory of long-wave propagation as described in chapter 2 may be modified only to a moderate extent.

The linear-theory value

$$c = (gk^{-1}\tanh kh)^{\frac{1}{2}} \tag{90}$$

for the wave speed is well represented for such 'fairly long waves' (with $kh < 0.63$) by two terms in its Taylor expansion:

$$c = c_0(1 - \tfrac{1}{6}k^2h^2), \tag{91}$$

where $c_0 = (gh)^{\frac{1}{2}}$. Often, wave systems with *small* dispersion exhibit such a k^2 term in the departure of the linear-theory wave speed from a constant value c_0. It means that waves propagating in the positive x-direction are close to solutions of

$$\partial v/\partial t + c_0\,\partial v/\partial x + \sigma\partial^3 v/\partial x^3 = 0; \tag{92}$$

here, σ is a constant taking the value

$$\sigma = \tfrac{1}{6}c_0 h^2 \tag{93}$$

for 'fairly long' waves.

These remarks suggest how the nonlinear theory of one-dimensional wave propagation (chapter 2) can be modified for systems with a small amount of dispersion, such that the linear-theory wave speed is $c_0 - \sigma k^2$ for wavenumber k. In chapter 2, excess wave speed due to nonlinear effects is called v, and an equation

$$\partial v/\partial t + (c_0 + v)\,\partial v/\partial x = 0 \tag{94}$$

is used for a nondispersive nonlinear system. (For long waves $v = \tfrac{3}{2}u$, where u is the fluid velocity.) When the ideas in equations (92) and (94) are combined to make allowance for small wave-speed modifications due *both* to nonlinear and to dispersive effects, we obtain the famous Korteweg–de Vries equation

$$\partial v/\partial t + (c_0 + v)\,\partial v/\partial x + \sigma\,\partial^3 v/\partial x^3 = 0; \tag{95}$$

which those authors proposed in the late nineteenth century for the analysis of fairly long waves. Since then, it has been used to treat many nonlinear wave systems with a small amount of dispersion. Sometimes a coordinate

$X = x - c_0 t$ is used (chapter 2) to give a frame of reference moving with the basic wave speed c_0; then the equation becomes

$$\partial v/\partial t + v \, \partial v/\partial X + \sigma \, \partial^3 v/\partial X^3 = 0. \tag{96}$$

Although the theory of the solutions to the Korteweg–de Vries equation (96) is very extensive, we include less of it than of the nonlinear deep-water theory; partly, because an equation ignoring dissipation is relatively less appropriate in shallow water owing to bottom friction. Here, we concentrate on the main changes which dispersion makes to the conclusions of chapter 2.

The significance of the nonlinear term $v \, \partial v/\partial X$ in (96) relative to the dispersive term $\sigma \, \partial^3 v/\partial X^3$ depends on a non-dimensional parameter $v_1 \lambda^2/\sigma$, where v_1 is the peak value of the excess wave speed and λ is a measure of the length of the wave. This parameter is proportional, by (93), to

$$a\lambda^2/h^3 \tag{97}$$

since v_1/c_0 is proportional to the ratio of a wave amplitude a (defined as in (49)) to the mean depth h.

The solutions of (96) show that, when this parameter $a\lambda^2/h^3$ is large enough (greater than about 16), waveform shearing by nonlinear effects can proceed to steepen the front face (compressive phase) of a wave just as in chapter 2. The consequent hydraulic jump is further discussed below.

By contrast, a balance between waveform steepening and dispersion turns out to be possible for values of $a\lambda^2/h^3$ below about 16 (for example, with $\lambda/h = 12$ and $a < \frac{1}{9}h$). Under those circumstances, indeed, periodic waves of unchanging form become possible once more due to a balance of those two effects.

Crudely, we can see that such a balance might occur as follows: if the 'fundamental' (lowest-order Fourier component) of a waveform is $a_1 \cos kx$, then the first harmonic $a_1 \cos 2kx$ propagates more slowly owing to dispersion. Evidently, its lagging phase corresponds to a negative rate of introduction of a $\sin 2kx$ component; on the other hand, shearing of the fundamental is easily shown to give a positive rate of introduction of such a $\sin 2kx$ component. This suggests that a cancellation may be possible, leading to a travelling wave of unchanging form.

More rigorously, we seek such a solution by substituting

$$\partial v/\partial t = -v_0 \, \partial v/\partial X \tag{98}$$

in (96). Here, v_0 is the wave speed in the (X, t) coordinate system; thus, the wave speed is $c_0 + v_0$ in the original (x, t) coordinates. It is easy to compute

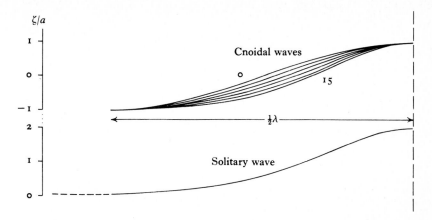

Figure 116. Waveforms for cnoidal waves of length λ and amplitude a on water of depth h for six values of $a\lambda^2/h^3$: o (sinusoidal waveform), 3, 6, 9, 12 and 15. Below, the solitary wave is plotted in the same way by taking its effective length λ as in equation (100).

the solutions of the resulting ordinary differential equation. They consist of a singly infinite family of periodic-wave solutions and one limiting solution. The periodic-wave solutions are called 'cnoidal waves' because they take the form of the square of the Jacobian elliptic function cn; a circumstance useful when they were first calculated, although unimportant when modern computational aids are available.

These cnoidal waveforms are plotted in figure 116. For the smaller values of $a\lambda^2/h^3$ the nonlinearity causes little departure from a sinusoidal form; as $a\lambda^2/h^3$ increases, the first harmonic is rather obvious initially (as the crest becomes peakier and the trough shallower), followed by higher harmonics. For $a\lambda^2/h^3 = 15$ the periodic wavetrain has almost degenerated into a sequence of isolated 'humps' with a flat water surface between them.

This fact makes less astonishing than it, perhaps, might otherwise be the famous limiting solution

$$v = 3v_0\{\cosh\left[\tfrac{1}{2}(v_0/\sigma)^{\frac{1}{2}}(X - v_0 t)\right]\}^{-2} \tag{99}$$

of the Korteweg–de Vries equation (96). This represents, *not* a periodic wave but the propagation of a single isolated 'hump' of unchanging form. Scott Russell in the first half of the nineteenth century found experimentally that such a 'solitary wave' with length, depth and amplitude properly matched can propagate virtually unchanged (except for a very gradual attenuation due to bottom friction); later, Rayleigh found the solution (99), before Korteweg and de Vries embedded it in a general approximate theory of fairly long waves.

The solitary wave can be claimed to possess a wavelength λ, *not* in the sense of a spatial period, but in the sense of the distance within which the surface elevation exceeds (say) 3 % of its maximum value. In this sense,

$$\lambda = 4(\sigma/v_0)^{\frac{1}{2}} \cosh^{-1}[(0.03)^{-\frac{1}{2}}], \quad \text{giving} \quad a\lambda^2/h^3 = 16. \quad (100)$$

In figure 116 the solitary wave is plotted (lower curve) using this value for the wavelength. It is seen as something of a limiting case of the cnoidal waveform.

Thus, dispersion effects can counteract waveform steepening up to the limiting value (100) of the parameter $a\lambda^2/h^3$. Beyond that limiting value it is impossible for a waveform to be unchanging. In practice, its front face steepens as described in chapter 2 and forms a hydraulic jump.

At last, we are in a position to discuss more fully than was possible in section 2.12 the hydraulic jump; and, especially, its need to get rid of energy at a specified rate as it propagates. For a jump where the water depth rises from a constant value h_0 to another value h_1, the required rate of energy loss is

$$\rho_0 Ug(h_1 - h_0)^3/4h_1 \quad (101)$$

per unit breadth. Here,

$$U = [gh_1(h_1 + h_0)/2h_0]^{\frac{1}{2}} \quad (102)$$

is the speed of propagation of the jump into still water.

A hydraulic jump with strength $(h_1 - h_0)/h_0$ less than around 0.3 exhibits rather little foaming (if any) but is followed by a train of periodic waves with crests stationary relative to the jump. We analyse these first by linear, and then by nonlinear, theory.

The water following the jump moves relative to it at a velocity equal, by mass-conservation, to Uh_0/h_1. On linear theory, this must be the wave speed

$$Uh_0/h_1 = (gk^{-1}\tanh kh_1)^{\frac{1}{2}}, \quad \text{with} \quad k = 2\pi/\lambda, \quad (103)$$

if λ is the wavelength of waves behind the jump with crests stationary relative to it. Their rate of energy leakback relative to the crests, (45) per unit breadth, can be balanced against the required rate of energy loss (101). On linear theory, this gives

$$\rho_0 Ug(h_1 - h_0)^3/4h_1 = \frac{1}{4}\rho_0 ga^2(Uh_0/h_1)[1 - (2kh_1)(\sinh 2kh_1)^{-1}] \quad (104)$$

for waves of amplitude a. Numerical solution of equations (102)–(104) gives results which to 2 significant figures take the simple form

$$kh_1 = 3[(h_1 - h_0)/2h_0]^{\frac{1}{2}}, \quad a/h_1 = 3^{-\frac{1}{2}}(h_1 - h_0)/h_0. \quad (105)$$

Thus, the amplitude a is predicted to be around 0.6 times the height $h_1 - h_0$ of the jump.

Observed amplitudes are centred about $0.6(h_1 - h_0)$ but vary somewhat randomly from $0.3(h_1 - h_0)$ to $0.9(h_1 - h_0)$. This variability may be thought to need explanation, since the measurements are of only moderate difficulty.

Actually, the results (105) give a value

$$a\lambda^2/h_1^3 = 8\pi^2/3^{\frac{5}{2}} = 5.1. \tag{106}$$

Figure 116 shows that this is *not* small enough for nonlinear effects to be neglected. Therefore, the waves behind a jump *must* be cnoidal. In fact, they are observed to have the cnoidal waveforms and the cnoidal-wave relationships between length, wave speed, and amplitude.

Rather surprisingly, a recalculation using the full nonlinear cnoidal-wave theory shows that the rate of leakback of energy relative to crests cannot, for any wave amplitude, take the *full* value (101). However, it can take *any* value *less* than that! Specifically, values of $a/(h_1 - h_0)$ equal to 0.6 (as given by linear theory, or by the average of the observations) correspond to a rate of leakback 0.8 times the required energy loss. Therefore, they correspond to a hydraulic jump which dissipates 20% of the required energy loss by viscous action or by foaming.

On this nonlinear theory, the large spread in observed values of $a/(h_1 - h_0)$, from 0.3 to 0.9 in undular jumps, corresponds to a large spread in the amount of viscous and/or foaming loss of energy occurring under different experimental conditions: from 75% to 5% of the required total. For still stronger jumps, foaming is intense and accounts for almost all the required energy loss.

Hydraulic jumps are often found as stationary phenomena in streams. Hydraulic theory shows that a steady stream of speed exceeding the long-wave speed can slow down to a speed below the long-wave speed only by discontinuous deceleration at a hydraulic jump. This is usually of the foaming variety (figure 49), although an undular jump can occur if the initial speed excess was only slight.

The above sketch of fairly long waves in channels gives little prominence to the solitary wave; which, in *this* context, is little more than an experimental curiosity, hardly ever observed in nature. Texts on the Korteweg–de Vries equation show, however, that in applications where no dissipation is present the solution (99), often called a 'soliton', can play a key role; indeed, it can be proved that an arbitrary initial motion breaks up, ultimately, into an 'ensemble of solitons' . . .

Figure 117. The Sierra Wave. Air flowing down the mountain range (here, from right to left), acquires great speed and is then suddenly decelerated at something like a stationary hydraulic jump (visible at the left of the photograph). [*Photograph by R. Symons, from R. S. Scorer: Clouds of the World.*]

To every property of surface gravity waves noted in this epilogue there are corresponding properties of internal waves. Interfacial waves between two fluids exist up to a certain amplitude beyond which energy is rapidly dissipated by an interpenetration of the two fluids analogous to foaming. In continuously stratified fluid, it is possible for dispersive and nonlinear effects to be in balance; in particular, powerful techniques of nonlinear analysis have been used to prove that, once more, a solitary wave can exist.

Above all, the hydraulic jump can rather spectacularly be present in stratified fluid. Where a stream of cold air flowing down the lee side of a mountain range is accelerated to a speed exceeding the wave speed for waves long compared with the stream's depth, its deceleration may take place at a vast discontinuity like the famous Sierra Wave (figure 117).

In conclusion, we note a still more advanced area of nonlinear dispersion theory where researches of great potential importance are in progress. This treats the nonlinear processes by which the wavenumber spectrum of a statistical assemblage of waves evolves in time.

The theory has been developed primarily for weak nonlinearities. These allow exchange of wave energy between the wavenumbers $\mathbf{k_1}$, $\mathbf{k_2}$ and $\mathbf{k_3}$ in a 'triad' of values which, with the corresponding linear-theory frequencies ω_1, ω_2 and ω_3, satisfies

$$\mathbf{k_1} + \mathbf{k_2} + \mathbf{k_3} = 0, \quad \omega_1 + \omega_2 + \omega_3 = 0. \tag{107}$$

(Sometimes, tetrads also are considered.)

The theory has much in common with the statistical theory of turbulence. In one respect, however, it is less complex. The relatively simple statistical properties of a Gaussian distribution are guaranteed us by the central limit theorem in any case when the random variable is a sum of a large number of statistically independent random variables. In a dispersive wave system, this is assured by the fact that signals at any point include a large number of weak wave packets travelling at different group velocities from widely different locations.

Thus, although statistical theories of turbulence are bedevilled by 'closure' problems associated with the non-Gaussian statistics of vortex lines moving with the fluid, a rational analysis of the statistics of a weakly interacting assemblage of dispersive waves has been found to be more feasible. In particular, it is beginning to throw light (in a quantitative fashion) both on ocean-wave spectra and on an important matter hinted at in section 4.6: how vertical transport of horizontal mean momentum in stably stratified fluid is achieved through the agency of a statistical assemblage of internal waves interacting with each other and with a sheared mean flow.

BIBLIOGRAPHY

Part 1. Some basic texts

For appreciating the material in this book, some prior knowledge is desirable in four fields: basic mechanics including the mechanics of vibrations; basic dynamics of fluids; the elementary theory of Fourier series and integrals; and the elementary theory of functions of a complex variable. We first note several texts in these fields; secondly, we mention texts in other subjects related to material in the book. Although readers will be able to proceed comfortably without knowledge in this second group of subjects, they will notice that the book includes many statements related to these subjects which they may wish to follow up in more detail.

A good text covering all the basic mechanics used in this book is

Greenwood, D. T. *Principles of Dynamics*. Englewood Cliffs, NJ: Prentice-Hall (1965).

This includes the general linear theory of vibrations. A more elementary treatment of vibrations, which usefully brings out the analogies between mechanical and electrical oscillations emphasised in sections 2.3–2.5, is

Jones, D. S. *Electrical and Mechanical Oscillations*. London: Routledge & Kegan Paul (1961).

The theory of vibrations is clearly described, and related to the theory of sound, in

Morse, P. M. *Vibration and Sound*, 2nd edn. New York: McGraw-Hill (1948).

An advanced mechanics text, particularly useful for matters referred to in the epilogue (part 2), is

Goldstein, H. *Classical Mechanics*. Reading, Mass.: Addison-Wesley (1950).

All of the basic dynamics of fluids used in this book is excellently covered in

Batchelor, G. K. *An Introduction to Fluid Dynamics*. Cambridge University Press (1967).

Another very suitable, and slightly shorter, text is

Curle, N. & Davies, H. J. *Modern Fluid Dynamics*. London: Van Nostrand (1968).

A more advanced account of boundary-layer effects, especially relevant to the material of sections 2.7, 3.5, and 4.7, is

Rosenhead, L. (ed.). *Laminar Boundary Layers*. Oxford University Press (1964).

Note that the method of dimensional analysis, as applied, for example, in section 3.10, is fully explained in all of these texts.

A good text covering the whole mathematical background used in this book is

Apostol, T. M. *Mathematical Analysis*. Reading, Mass.: Addison-Wesley (1957).

This includes both the necessary Fourier analysis and the knowledge assumed on functions of a complex variable. A short, but clear, elementary account of Fourier series and integrals is

Stuart, R. D. *An Introduction to Fourier Analysis*. London: Methuen (1961).

Equally, a good brief account of complex-variable theory is

Phillips, E. G. *Functions of a Complex Variable*, 8th edn. Edinburgh: Oliver & Boyd (1957).

An advanced text covering both subjects and giving more detail on many matters treated in chapters 3 and 4 (for example, an alternative real-variable approach to the method of stationary phase) is

Jeffreys, H. & Jeffreys, B. S. *Methods of Mathematical Physics*. Cambridge University Press (1950).

No special knowledge in any other fields has been assumed. Nevertheless, we note below a few basic texts in each neighbouring area of science to which substantial reference is made in this book.

An excellent account of partial differential equations is

Garabedian, P. R. *Partial Differential Equations*. New York: Wiley (1964).

Here, for example, the reader will find a general analysis of characteristic curves for nonlinear hyperbolic equations (section 2.8) and a proof of the theorem on numbers of eigenvalues referred to in section 4.13. For less advanced matters, see

Sneddon, I. N. *Elements of Partial Differential Equations*. New York: McGraw-Hill (1957).

For the differential geometry of surfaces, a few ideas from which are used in sections 4.9 and 4.10, see

Willmore, T. J. *An Introduction to Differential Geometry*. Oxford University Press (1964).

Aspects of the theory of elasticity, mentioned especially in discussing distensible tubes (section 2.2), can be followed up at length in

Sokolnikoff, I. S. *Mathematical Theory of Elasticity*. New York: McGraw-Hill (1956).

For a rather more elementary treatment, see

Southwell, R. V. *Introduction to the Theory of Elasticity*. Oxford University Press (1936).

For the biomechanical and physiological background to blood-pulse propagation in arteries, see

McDonald, D. A. *Blood Flow in Arteries*, 2nd edn. London: Edward Arnold (1974).

A rather fuller account of the associated fluid dynamics is given by

Lighthill, M. J. *Mathematical Biofluiddynamics*. Philadelphia: Society for Industrial and Applied Mathematics (1975).

A good text covering all the basic thermodynamics to which we refer is

Andrews, F. C. *Thermodynamics: Principles and Applications*. New York: Wiley (1971).

For readers wishing to follow up the statistical-physics foundations of the subject, a suitable text is

Pointon, A. J. *An Introduction to Statistical Physics for Students*. London: Longmans (1967).

An outstanding, though rather more advanced, account of both topics, is

Sommerfeld, A. *Thermodynamics and Statistical Mechanics*. New York: Academic Press (1956).

Those aspects of electromagnetic theory to which reference is made (especially, in sections 2.3–2.5 and in the epilogue, part 1) are all excellently expounded in

Clemmow, P. C. *An Introduction to Electromagnetic Theory*. Cambridge University Press (1973).

Another good text is

Reitz, J. R. & Milford, F. J. *Foundations of Electromagnetic Theory*. Reading, Mass.: Addison-Wesley (1969).

A first-rate account of the interactions between a magnetic field and the motions of a conducting fluid, including those aspects dealt with in the epilogue, is

Shercliff, J. A. *A Textbook of Magnetohydrodynamics*. Oxford: Pergamon (1965).

The present book points out relationships to several important results from quantum mechanics (sections 4.2, 4.5 and 4.11). Readers will find more information on these in

Merzbacher, E. *Quantum Mechanics*, 2nd edn. New York: McGraw-Hill (1961).

Another good, slightly fuller, treatment is

Schiff, L. I. *Quantum Mechanics*, 3rd edn. New York: McGraw-Hill (1968).

A fine general account of the oceans, and of the classical methods of oceanographic science, is

Sverdrup, H. U., Johnson, M. W. & Fleming, R. H. *The Oceans*. Englewood Cliffs, NJ: Prentice-Hall (1942).

Aspects of the interaction between oceanography and mathematical analyses of waves in fluids are well described in

Neumann, C. & Pierson, W. J. *Principles of Physical Oceanography*. Englewood Cliffs, NJ: Prentice-Hall (1966).

A more extensive treatise, also of great value, is

Defant, A. *Physical Oceanography* (2 vols.). Oxford University Press (1961).

An excellently written elementary text on the atmosphere is

Riehl, H. *Introduction to the Atmosphere*, 2nd edn. New York: McGraw-Hill (1972).

Basic analytic methods in the field are well set out in

Hess, S. L. *Introduction to Theoretical Meteorology*. New York: Henry Holt (1959).

For aspects of the interaction between meteorology and analysis of waves in fluids, see

Holton, J. R. *An Introduction to Dynamical Meteorology*. New York: Academic Press (1972).

An article surveying waves in fluids in general, which depicts rather briefly almost all the whole subject matter of this book, is

Lighthill, M. J. *Commun. Pure & Appl. Math.* **20**, 267–93 (1967).

Part 2. Acoustic literature

In part 2 we list some of the acoustic literature, relevant to the material of chapter 1 and also to the portions of chapters 2 and 4 concerned with linear and nonlinear acoustics. Among general books on sound we must recommend above all

Rayleigh, Lord. *Theory of Sound* (2 vols.), 2nd edn. London: Macmillan (1896); also, New York: Dover (1945).

This book, a delight to read, is crammed with perennially significant information and ideas; as in the Appendix with its remarkable general derivation of the energy propagation velocity (see our section 3.8). Acoustic streaming (our section 4.7) is another topic penetratingly analysed.

Modern texts which similarly give excellent accounts of the basic science of the kind of sound to which our ears are sensitive are

Beranek, L. L. *Acoustics*. New York: McGraw-Hill (1954),

and

Stephens, R. W. B. & Bate, A. E. *Acoustics and Vibrational Physics*. 2nd edn. London: Edward Arnold (1966).

Sound waves as a phenomenon in physics, with the main emphasis on ultrasonics (frequencies above the audible range) and their applications, are treated in the monumental work

Mason, W. P. (ed.). *Physical Acoustics: Principles and Methods* (10 vols.). New York: Academic Press (1964–73).

Note, in particular, vol. 2, where chapter 3 deals with the 'lag' effects on acoustic attenuation discussed in our section 1.13, and where chapter 11 deals with acoustic streaming (our section 4.7). Again, vol. 4 contains a chapter 12 on the interactions between sound waves and electromagnetic effects in ionised gases (our epilogue, part 1).

The subject known as acoustical engineering involves the application of acoustical knowledge of the kind described in our sections 1.12, 2.5, 4.13, etc., to design. An excellent basic textbook is

Olson, H. F. *Elements of Acoustical Engineering.* London: Van Nostrand (1940).

For loudspeakers and horns, see

Jordan, E. J. *Loudspeakers.* London: Focal Press (1963).

An older, but still very useful, text is

McLachlan, N. W. *Loudspeakers.* Oxford University Press (1934).

A good modern treatment placing this subject in a wider context is

Moir, J. *High Quality Sound Reproduction.* London: Chapman & Hall (1961).

A major area of practical application of acoustical theory is to noise abatement. An important general text in this field is

Beranek, L. L. (ed.). *Noise Reduction.* New York: McGraw-Hill (1960).

For an account of the theories underlying work on the reduction of aircraft noise, see

Goldstein, M. E. *Aeroacoustics.* Washington: Nat. Aeron. & Space Admin. (1974).

Nonlinear acoustics and shock-wave theory are among the many important matters admirably discussed in

Whitham, G. B. *Linear and Nonlinear Waves.* New York: Wiley (1974).

This book includes also a full account of nonlinear dispersive waves (our epilogue, part 2).

An outstanding account of shock waves in general and of their importance in gas dynamics is

Raizer, Yu. P. & Zeldovich, Ya. B. *Physics of shock waves and high-temperature hydrodynamic phenomena* (2 vols.). New York: Academic Press (1966–7).

The importance of the shock wave in physical chemistry as a method of generating a known and practically instantaneous rise in temperature, with chemical-kinetic consequences of great interest, is well brought out in

Bradley, J. N. *Shock Waves in Chemistry and Physics.* London: Methuen (1962).

For the case when the thermal changes in a consequent chemical reaction greatly modify the propagation of the shock wave, see especially

Zeldovich, Ya. B. & Kompaneets, A. S. *Theory of Detonations.* New York: Academic Press (1960).

We give also some more detailed acoustical references, as follows. For the history (section 1.2) of sound-speed theory, see p. 477 of

Lamb, H. *Hydrodynamics*, 6th edn. Cambridge University Press (1932); also, New York: Dover (1945).

This, indeed, is a text full of valuable information regarding very many matters concerned with waves in fluids.

The decibel, emphasised in section 1.3 as a measure of acoustic intensity (energy flux), is supplemented by other units defined with the object of estimating the effect of sound on human ears. The widely used perceived-noise decibel (PNdB) was introduced by

Kryter, K. D. *J. Acoust. Soc. Amer.* **31**, 1415–29 (1959),

who compares it with previously available measures of the subjective effects of sounds. All are based on a nonuniform weighting of the intensity spectrum as a function of frequency.

For a general account of sound generation (sections 1.4–1.10) see

Lighthill, M. J. *Proc. Roy. Soc.* A, **267**, 147–82 (1962).

An excellent verification that the rules of section 1.6 can be applied even to so complex a phenomenon as sound generated by irregular combustion in an open flame is in

Hurle, I. R., Price, R. B., Sugden, T. M. & Thomas, A. *Proc. Roy. Soc.* A, **303**, 409–27 (1968).

They measured the variation in total rate of reaction photochemically, and deduced from it the fluctuations in the associated rate of volume outflow, $\dot{V}(t)$, into the surrounding atmosphere. They demonstrated that the observed acoustic waveform followed the form of $\ddot{V}(t)$ with the expected time-lag.

Propeller noise associated with a *steady* distribution of force around a rotating blade was calculated with dipole strength equated to force by

Garrick, I. E. & Williams, C. E. *NACA Report* no. 1198. Washington: Nat. Adv. Comm. Aeronautics (1954);

and with the additional volume term in the dipole strength (section 1.7) by

Arnoldi, R. A. *Propeller noise caused by blade thickness.* East Hartford, Conn.: United Aircraft Corp. Rep. no. R-0896-1 (1956).

For many-bladed propellers and fans, the considerable cancellation in the dipole radiation fields associated with the different blades may create a situation when the quadrupole fields associated with wake turbulence become the dominant noise source, as shown by

Ffowcs Williams, J. E. & Hawkings, D. L. *J. Sound Vib.* **10**, 10–21 (1969).

For the scientific study of ripple-tank representations of sound-generation phenomena (section 1.8), see

Ffowcs Williams, J. E. & Hawkings, D. L. *J. Fluid Mech.* **31**, 779–88 (1968).

Their application to jet-noise phenomena is given in

Webster, R. B. *J. Fluid Mech.* **40**, 423–32 (1970).

Those theories of scattering of acoustic or electromagnetic waves that go beyond scattering by compact bodies (section 1.9; that is, beyond the Rayleigh/Born approximation) are described by

King, R. W. P. & Wu, T. T. *Scattering and Diffraction of Waves.* Harvard University Press (1959).

This includes interaction of waves both with finite obstacles (scattering) and with infinite or semi-infinite screens (diffraction).

Quadrupole analysis of aerodynamic sound generation (section 1.10) was introduced by

Lighthill, M. J. *Proc. Roy. Soc.* A, **211**, 564–87 (1952).

For its application to jet noise, see especially

Lighthill, M. J. *Amer. Inst. Aeron. Astron. Journ.* **1**, 1507–17 (1963).

A useful compendium of material on flow-generated noise is

Schwartz, I. R. (ed.). *Basic Aerodynamic Noise Research.* Washington: Nat. Aeron. & Space Admin. (1969).

A fine general survey of this field is

Ffowcs Williams, J. E. *Ann. Rev. Fluid Mech.* **1**, 197–222 (1969).

A still more recent and comprehensive survey, also admirable, is

Crighton, D. G. *Prog. Aerospace Sci.* **16**, 31–96 (1975).

A later paper of particular importance is

Howe, M. S. *J. Fluid Mech.* **71**, 625–73 (1975).

The mechanisms of attenuation of sound in the body of the fluid (section 1.13) are described at length by M. J. Lighthill in pp. 250–351 of

Batchelor, G. K. & Davies, R. M. (eds.). *Surveys in Mechanics.* Cambridge University Press (1956).

That survey includes also a detailed analysis of the nonlinear propagation of simple waves of moderate strength with attenuation taken into account. This approach uses the 'Burgers equation', and the theory of section 2.11 appears as a limiting case when the diffusivity of sound δ is small.

The original 1859 paper on the Riemann theory (section 2.8) is most easily accessible in pp. 156–75 of

Riemann, B. *Gesammelte Mathematische Werke*, 2nd edn. Leipzig: Teubner (1892).

His transformation of the equations of motion is on p. 159 (where our P is his f).

The conditions governing changes at a shock wave, in forms allowing for an entropy increase (section 2.10), were given by

Rankine, W. J. M. *Phil. Trans. Roy. Soc.* **160**, 277–88 (1870);

and, for a rather general fluid, by

Hugoniot, A. *J. de l'Ecole Polytech.* **58**, 1–125 (1889).

In 1910, analyses of a shock wave's internal structure were given simultaneously by

Rayleigh, Lord. *Proc. Roy. Soc.* A, **84**, 247–84 (1910),

and by

Taylor, G. I. *Proc. Roy. Soc.* A, **84**, 371–7 (1910).

The latter gives an account quite close to that in section 2.10; for the recent work in this field, see references given above. We note also that a good discussion of the possibility, not ruled out by thermodynamics, that there can be fluids in which expansive waves would become discontinuous, is in

Thompson, P. A. & Lambrakis, K. C. *J. Fluid Mech.* **60**, 187–208 (1973).

The general law for embedding shock waves in a developing weak simple wave was put forward by

Whitham, G. B. *J. Fluid Mech.* **1**, 290–318 (1956),

who later made the extensions to nonuniform media given in our sections 2.13 and 2.14; see

Whitham, G. B. *J. Fluid Mech.* **4**, 337–60 (1958).

For recent surveys of supersonic boom theory, see the article by M. J. Lighthill in

Stollery, J. L., Gaydon, A. G. & Owen, P. R. (eds.). *Shock Tube Research.* London: Chapman & Hall (1971);

and, also,

Hayes, W. D. *Ann. Rev. Fluid Mech.* **3**, 269–90 (1971).

A rather intricate theory of propagation of *strong* simple waves under conditions of gradually varying composition and cross-section is given by

Varley, E. & Cumberbatch, E. *J. Inst. Math. Applics.* **2**, 133–43 (1966) and *J. Fluid Mech.* **43**, 513–37 (1970).

This rigorous large-amplitude theory can be regarded as justifying the treatment given in sections 2.13 and 2.14 (to which it reduces in cases of small amplitude). It has not, however, been extended so as to treat propagation after a strong shock wave has appeared.

The other aspect of nonlinear acoustics treated in this book is acoustic streaming (section 4.7). The theory, in a form general except that terms in the streaming velocity of higher order than the second are neglected, is set out by

Westervelt, P. J. *J. Acoust. Soc. Amer.* **25**, 60–7 (1953),

and by

Nyborg, W. L. *J. Acoust. Soc. Amer.* **25**, 68–75 (1953).

Since then, extensive developments in slow viscous flow theory occurred, as described by

Happel, J. & Brenner, H. *Low Reynolds Number Hydrodynamics*. Englewood Cliffs, NJ: Prentice-Hall (1965),

so that our description by means of a stokeslet distribution now seems appropriate. The work of

Stuart, J. T. *J. Fluid Mech.* **24**, 673–87 (1966)

established clearly that the standard theory is valid only when the Reynolds number based on the streaming velocity itself is rather small; by contrast, high Reynolds number hydrodynamics is applicable to strong acoustic streaming. This suggests ideas such as the application of the jet solution of

Squire, H. B. *Quart. J. Mech. Appl. Math.* **4**, 321–9 (1951)

made in our section 4.7.

Vibration of a compact body in fluid at rest generates a streaming motion essentially the same as that generated when plane sound waves are incident on the stationary body. For a general review of streaming in such cases, see

Riley, N. *J. Inst. Math. Applics.* **3**, 419–34 (1967).

Part 3. Water-wave literature

Discussion of water waves in this book begins with the analysis of long waves in open channels in chapter 2. An impressive general account of water motions in open channels is

Chow, V. T. *Open-Channel Hydraulics*. New York: McGraw-Hill (1959).

Another good text is

Henderson, F. M. *Open Channel Flow*. New York: Macmillan (1966).

Note that the effect of turbulence on long-wave attenuation (an important topic omitted from chapter 2) is well covered in both these books.

Our chapter 3 gives the linear theory of water waves across the whole spectrum of wavelengths, while the epilogue sketches something of the nonlinear theory. A mathematically oriented text with a comparably wide scope is

Stoker, J. J. *Water Waves*. New York: Interscience (1953).

An excellent treatment of this subject in its relationship to oceanography is given by

Phillips, O. M. *The Dynamics of the Upper Ocean*. Cambridge University Press (1966).

For applications of water-wave knowledge to the study of environmental stresses on offshore structures, see

McCormick, M. E. *Ocean Engineering Wave Mechanics*. New York: Wiley (1973).
An outstanding account of the basic physical chemistry behind phenomena involving surface tension is

Defay, R. & Prigogine, I. *Surface Tension and Adsorption*. London: Longmans (1966).
This may be found to give useful background to the discussions in sections 3.4 and 3.5.

As in part 2, we follow the above bibliography of general texts with some more specialised references. A fine application of Helmholtz resonator theory (section 2.5) to long-wave resonance in harbours is made by

Miles, J. W. & Lee, Y. K. *J. Fluid Mech.* **67**, 445–64 (1975).
For the structure of hydraulic jumps (section 2.12 and epilogue) see

Benjamin, T. B. & Lighthill, M. J. *Proc. Roy. Soc.* A, **224**, 448–60 (1954).
A demonstration that viscous action alone is able (as those authors predict) to dissipate enough energy to allow the formation of a jump with a train of cnoidal waves behind it is given in the analysis of

Byatt-Smith, J. G. B. *J. Fluid Mech.* **48**, 33–40 (1971).
A detailed investigation of those conditions under which the Severn bore forms is given by

Abbott, M. R. *Proc. Camb. Phil. Soc.* **52**, 344–62 (1956).
This obtains good agreement between the observations and a theory using the methods described in chapter 2 as modified by the effects of turbulent friction.

For the attenuation of surface waves, particularly as affected by variability of surface tension, see the excellent review by

Lucassen-Reynders, E. H. & Lucassen, J. *Adv. Colloid & Interface Sci.* **2**, 347–95 (1969).
Also, an important bibliography 'Oil on Troubled Waters' covering the vast literature of this field is shortly to be published by J. C. Scott of the Fluid Mechanics Research Institute, University of Essex. Stokes' idea underlying the treatment given in our section 3.5 is indicated on p. 73 of his famous pioneering paper (1851) on viscosity, reprinted as vol. 3, pp. 1–141 of

Stokes, G. G. *Mathematical and Physical Papers* (5 vols.). Cambridge University Press (1880–1905).
Observations of ocean swell at enormous distances from the storm that can be shown to have generated it are in

Munk, W. H., Miller, G. R., Snodgrass, F. E. & Barber, N. F. *Phil. Trans. Roy. Soc.* A, **255**, 505–84 (1963), and in

Snodgrass, F. E., Groves G. W., Hasselmann, K. F., Miller, G. R., Munk, W. H. & Powers, W. M. *Phil. Trans. Roy. Soc.* A, **259**, 431–97 (1966).
A survey of group velocity and its properties, covering much of the general theory in chapters 3 and 4, is given by

Lighthill, M. J. *J. Inst. Math. Applics.* **1**, 1–28 (1965).
The simple Stokes argument for two waves of nearly equal frequency and amplitude is in vol. 5, p. 362 of his *Papers* just referred to.

Application of the kinematics of wave crests as in sections 3.6 and 3.8 to group-velocity theory is given, for example, by

Lighthill, M. J. & Whitham, G. B. *Proc. Roy. Soc.* A, **229**, 281–345 (1955).
For the extension (sections 4.4 and 4.5) to anisotropic waves, see

Whitham, G. B. *J. Fluid Mech.* **9**, 347–52 (1960).
An early analysis of stationary waves on a stream (section 3.9) is

Rayleigh, Lord. *Proc. London Math. Soc.* **15**, 69–78 (1883).
This uses a hypothetical attenuation mechanism (as in Rayleigh's argument on energy propagation velocity given at the end of section 3.8) to overcome the 'important and famous' difficulty regarding the poles. For the poetry of stationary waves, see pp. 257–60 of

Latham, E. C. (ed.). *The Poetry of Robert Frost.* London: Jonathan Cape (1971).
Scott Russell's work on waves was published mainly as the two Reports on Waves in

Russell, J. Scott. *British Assoc. Reports* **6**, 417–96 (1838) and **13**, 311–90 (1845).
They are notable for many important observations, including the first detailed description of the solitary wave (epilogue, part 2).

An outstanding general survey of ship waves (our sections 3.10 and 4.12), with a very full bibliography, is

Wehausen, J. V. *Adv. Appl. Mech.* **13**, 93–245 (1973).
For the important early work of William Froude, see his collected papers:

Froude, W. *Papers.* London: Institution of Naval Architects (1955).
Lord Kelvin (then Sir William Thomson) first gave his theoretical ship-wave pattern in an 1883 lecture reprinted in

Thomson, W. *Popular Lectures*, vol. 3, pp. 450–500. London: Macmillan (1891).
A wealth of information on wavemaking resistance is in

Havelock, T. H. *Collected Papers.* Washington: Office of Naval Research (1966).
An early paper proposing the use of data from 'double-model' experiments or computations in the prediction of wavemaking resistance is

Gadd, G. E. *Trans. Roy. Inst. Nav. Arch.* **112**, 335–45 (1970).
Against the background of the existing very wide range of intricate procedures aimed at calculating wavemaking resistance, the crude and imprecise method of estimation suggested in section 3.12 may appear very limited. Nevertheless, it is put forward in the hope of stimulating physical thinking in that field. For a rather accurate calculation procedure which uses a similar distribution of sources in the plane $z = 0$, together with a distribution of sources on the hull surface, and goes far towards satisfying exactly the full boundary condition on the hull surface and on the displaced position of the free surface, see

Gadd, G. E. *Trans. Roy. Inst. Nav. Arch.* **118**, 207–19 (1976).
Inui's classic work on bulbous bows, from which figure 104 is taken, is described in

Inui, T. *Trans. Soc. Nav. Arch. Mar. Eng.* **70**, 282–353 (1962).
The application of caustic theory to Kelvin's ship-wave pattern is given by

Ursell, F. *J. Fluid Mech.* **8**, 418–31 (1960).
For extensions to cases of wave patterns made by ships in relatively shallow water, see, for example,

Everest, J. T. & Hogben, N. *Trans. Roy. Inst. Nav. Arch.* **112**, 319–29 (1970).
Interactions between ship waves and the *wake* of the ship are studied by

Peregrine, D. H. *J. Fluid Mech.* **49**, 353–60 (1971).
Modifications to the Kelvin pattern due to nonlinear effects are analysed by

Hogben, N. *J. Fluid Mech.* **55**, 513–28 (1972).
The earliest account of edge-waves (section 4.13) is in the 1846 paper reprinted in vol. 1, pp. 157–87 (see p. 167) of

Stokes, G. G. *Mathematical and Physical Papers* (5 vols.). Cambridge University Press (1880–1905).

The subject was revived by
Ursell, F. *Proc. Roy. Soc.* A, **214**, 79–97 (1952).
Excellent modern accounts are by
Mysak, M. E. *J. Marine Res.* **26**, 24–42 (1968),
and by
Grimshaw, R. *J. Fluid Mech.* **62**, 775–91 (1974).

Part 4. Stratified-fluids literature

We can recommend several basic texts on the dynamics of stratified fluids. A fairly early book strong on the physical and mathematical background to the subject is
Eckart, C. *Hydrodynamics of Oceans and Atmospheres*. New York: Pergamon (1960).
The general flow behaviour of stratified fluids, as well as the resulting generation of internal waves (including *lee waves*), is well described by
Yih, C. S. *Dynamics of Nonhomogeneous Fluids*. New York: Macmillan (1965).
    A book which surveys a wide range of experimental and observational data against a background of clear theoretical analysis is
Turner, J. S. *Buoyancy Effects in Fluids*. Cambridge University Press (1973).
This book exhibits the many remarkable properties (briefly referred to in our section 4.3) of sea-water so stratified that the gradients of temperature and of salinity have the same sign. Also, it gives an excellent account of the structure and generation of thermoclines (our section 4.3).
    For data and theory on waves in the atmosphere, see especially
Gossard, E. E. & Hooke, W. H. *Waves in the Atmosphere*. Amsterdam: Elsevier (1975).
This includes an extensive description of 'infrasound' (the acoustic-gravity waveguide mode of propagation described in section 4.13).
    Passing to detailed references, we note first the classic account of ship-generated interface waves in an estuary with fresh water overlying salt water (section 4.1):
Ekman, V. W. *Scientific Results of the Norwegian North Polar Expedition* **5**, no. 15, 1–152 (1904).
    For the origins of the Boussinesq approximation in the study of internal waves, see
Boussinesq, J. *Théorie Analytique de la Chaleur* (2 vols.). Paris: Gauthier-Villars (1901–3).
    For the work of Väisälä and Brunt, see
Väisälä, V. *Soc. Sci. Fenn. Commentat. Phys.-Math.* **2**, no. 19, 1–46 (1925)
and
Brunt, D. *Quart. J. Roy. Met. Soc.* **53**, 30–2 (1927).
    A useful general review of waves in stratified fluids (relevant to sections 4.2 and 4.13 in particular) is
Tolstoy, I. *Rev. Mod. Phys.* **35**, 207–30 (1963).
    A description of overturning motions in the Gulf of Toulon (section 4.3) is in
Anati, D. A. *Cahiers Océanographique* **23**, 427–43 (1971).
    Waves on the ocean thermocline are admirably described in chapter 5 of
Phillips, O. M. *The Dynamics of the Upper Ocean*. Cambridge University Press (1966).
    For a case study of observed internal-wave propagation in the atmosphere (our section 4.3), compared with a theoretical analysis, see
Kjelaas, A. G., Gossard, E. E., Young, J. M. & Moninger, W. R. *Tellus*, **27**, 25–33 (1975).

Note also that the role of internal waves in the formation and generation of Clear Air Turbulence is extensively discussed in

Pao, Y. H. & Goldburg, A. (eds.). *Clear Air Turbulence and its Detection.* New York: Plenum Press (1969).

The University of Manchester Department of Mechanics of Fluids has developed highly effective 'schlieren' techniques, described by

Mowbray, D. E. *J. Fluid Mech.* **27**, 595–608 (1967),

for making visible the internal waves in stratified salt solution. The St Andrew's Cross (section 4.4, figure 76) is taken from an account of waves generated by oscillating bodies in

Mowbray, D. E. & Rarity, B. S. H. *J. Fluid Mech.* **28**, 1–16 (1967).

Internal waves generated by steadily moving bodies (section 4.12) are described in

Mowbray, D. E. & Rarity, B. S. H. *J. Fluid Mech.* **30**, 489–95 (1967),

from which figure 106 is taken, and by

Stevenson, T. N. *J. Fluid Mech.* **33**, 715–20 (1968).

The effect of viscous attenuation on the St Andrew's Cross (section 4.10) is analysed experimentally and theoretically in

Thomas, N. H. & Stevenson, T. N. *J. Fluid Mech.* **54**, 495–506 (1972),

from which figures 93 and 94 are taken. The pictures of transient motions (figure 77) are from

Stevenson, T. N. *J. Fluid Mech.* **60**, 759–67 (1973).

I am indebted to Professor N. H. Johannesen, the Head of Department, for supplying me with excellent prints of all the above photographs and to the above-named authors for permission to publish.

A general account of wave refraction in horizontally stratified media (section 4.5), with however the main emphasis on electromagnetic waves and on elastic waves, is

Brekovskikh, L. M. *Waves in Layered Media.* New York: Academic Press (1960).

A general survey of sound propagation through moving fluids (section 4.6), including moving stratified fluids, is

Lighthill, M. J. *J. Sound Vib.* **24**, 471–92 (1972).

This subject owes much to the 1945 paper of Blokhintzev available in English as

Blokhintzev, D. I. *Acoustics of a Nonhomogeneous Moving Medium.* Tech. Memor. no. 1399. Washington: Nat. Adv. Comm. Aeronautics (1956).

Ray-tracing techniques for sound waves in an atmosphere with stratified temperature and wind are described, for example, by

Thompson, R. J. *J. Acoust. Soc. Amer.* **51**, 1675–82 (1972).

Such techniques are compared with the results of using more exact equations by

Pridmore-Brown, D. C. *J. Acoust. Soc. Amer.* **34**, 438–43 (1962).

The concept of the Reynolds stress was first expounded for turbulence by

Reynolds, O. *Phil. Trans. Roy. Soc.* **186**, 123–64 (1895).

It was applied later to the hydrodynamic-instability waves that may precede transition to turbulence, and then more and more to waves in general; for example, by

Longuet-Higgins, M. S. & Stewart, R. W. *J. Fluid Mech.* **13**, 481–504 (1962).

A general derivation of a wave-action conservation law for nonhomogeneous moving media is in

Bretherton, F. P. & Garrett, C. J. R. *Proc. Roy. Soc.* A, **302**, 529–54 (1968).

Critical-layer absorption of internal waves was first clarified by

Booker, J. R. & Bretherton, F. P. *J. Fluid Mech.* **27**, 513–39 (1967):

and more fully explored by

Bretherton, F. P. *Quart. J. Roy. Met. Soc.* **95**, 213–43 (1969) and *J. Fluid Mech.* **36**, 785–803 (1969).

The importance of internal waves as a means for vertical transfer of horizontal momentum, referred to at the ends of sections 4.6 and 4.7, is well brought out in several recent papers; see, for example,

Andrews, D. G. & McIntyre, M. E. *J. Atmos. Sci.* **33**, 2031–53 (1976).

The general theory of stationary phase in three dimensions and of waves from an oscillating source given in sections 4.8 and 4.9 is taken from

Lighthill, M. J. *Phil. Trans. Roy. Soc.* A, **252**, 397–430 (1960).

He analyses at length various aspects described only briefly in this book, such as the determination of surfaces of constant phase as 'reciprocal polars' of wavenumber surfaces. Applications to magnetohydrodynamic waves are given in some detail.

The first use of the Airy integral to study caustics (section 4.11) was made, for the case of light waves, by

Airy, G. B. *Trans. Camb. Phil. Soc.* **6**, 379–402 (1838).

The method has continued to thrive ever since! Excellent tables of the Airy integral are in

Miller, J. C. P. *The Airy Integral.* British Association Mathematical Tables, Part-Volume B. Cambridge University Press (1946).

For waves generated by travelling forcing effects in general (section 4.12), see

Lighthill, M. J. *J. Fluid Mech.* **27**, 725–52 (1967).

A good account of the upstream influence of a cylindrical body moving horizontally through stratified fluid (or, what is the same thing, the blocking effect of a stationary cylinder on the flow of stratified fluid) is given by

Long, R. R. *Tellus*, **22**, 471–80 (1970).

This paper includes a rather full discussion of the nonlinear effects.

Waveguides in general are described by

Budden, K. G. *The Waveguide Mode Theory of Wave Propagation.* London: Logos (1961).

For an application of the rather elementary theory of the acoustic case given in our section 4.13 to flow-noise generation in ducts, see

Davies, H. G. & Ffowcs Williams, J. E. *J. Fluid Mech.* **32**, 765–78 (1968).

Extensions of waveguide mode theory, needed to analyse sound propagation through ducts including a nonuniform steady flow, were ingeniously made by

Swinbanks, M. A. *J. Sound Vib.* **40**, 51–76 (1975).

The analysis in section 4.13 of the gravity-acoustic waveguide in the atmosphere follows the general spirit of the papers by

Garrett, C. J. R. *Quart. J. Roy. Met. Soc.* **95**, 731–53 (1969) and

Bretherton, F. P. *Quart. J. Roy. Met. Soc.* **95**, 754–57 (1969).

In particular, it does not enquire closely into the possible influence of the thermal structure of very high parts of the atmosphere. Actually, certain computations suggest that peculiar resonance-like effects could result from apparently well-founded detailed assumptions about that upper-atmosphere structure; see, for example,

Press, F. & Harkrider, D. *J. Geophys. Res.* **67**, 3889–908 (1962).

Nevertheless, the observed propagation properties of the fundamental mode, including voluminous data summarised by

Pierce, A. D., Posey, J. W. & Iliff, E. F. *J. Geophys. Res.* **76**, 5025–41 (1971) on waves generated by nuclear explosions, fail to show such peculiar features and tend to confirm the views of Garrett and of Bretherton.

Part 5. A bibliography for the epilogue

Because the epilogue sketches only briefly several extensive areas of knowledge and research, it demands a relatively full bibliography. Once more, we begin with basic texts: an excellent general account of the dynamics of rotating fluids is

Greenspan, H. P. *The Theory of Rotating Fluids.* Cambridge University Press (1968).
     There are several good basic texts on magnetohydrodynamics and its geophysical and astrophysical applications, including

Cowling, T. G. *Magnetohydrodynamics.* New York: Interscience (1957), and

Ferraro, V. C. A. & Plumpton, C. *An Introduction to Magneto-Fluid Mechanics*, 2nd edn. Oxford University Press (1966).
     The latter gives an introduction to nonlinear effects in waves. These are further pursued in

Anderson, J. E. *Magnetohydrodynamic Shock Waves.* Cambridge, Mass.: M.I.T. Press (1963).
     For texts emphasising many important effects associated with the statistical mechanics of electrons and ions in plasmas (ionised gases), see

Holt, E. H. & Haskell, R. E. *Plasma Dynamics.* New York: Macmillan (1965); and, also,

Stix, T. H. *The Theory of Plasma Waves.* New York: McGraw-Hill (1962).
     Ionospheric waves, covering much of the spectrum from magnetohydrodynamic to electromagnetic waves, are treated by

Helliwell, R. A. *Whistlers and Related Ionospheric Phenomena.* Stanford University Press (1965).
     Another stimulating account of these matters together with many astrophysical topics is

Dungey, J. W. *Cosmic Electrodynamics.* Cambridge University Press (1958).
     An important collective work on nonlinear wave theory, with the main emphasis on dispersive waves, is

Leibovich, S. & Seebass, A. R. (eds.). *Nonlinear Waves.* Cornell University Press (1974).
     An excellent specialised text on nonlinear dispersive waves is

Karpman, V. I. *Nonlinear Waves in Dispersive Media.* Oxford: Pergamon (1975).
     Another work of great importance in this field, mentioned already in the context of nonlinear acoustics, demands a second mention in relation to nonlinear dispersive waves:

Whitham, G. B. *Linear and Nonlinear Waves.* New York: Wiley (1974).
     We turn to specialised references. An introduction to the dynamics of rotating fluids is given by

Lighthill, M. J. *J. Fluid Mech.* **26**, 411–31 (1966).
     This seeks to relate their properties as closely as possible to the classical hydrodynamic theorems concerning vorticity.
     Inertial waves were treated first by Kelvin in the 1880 paper reprinted as vol. 4, pp. 152–65 of

Kelvin, Lord. *Mathematical and Physical Papers* (6 vols.). Cambridge University Press (1882–1911).
For experimental confirmations of their properties, see
Oser, H. *Z. angew. Math. Mech.* **38**, 386–91 (1958),
and
Fultz, D. *J. Meteorol.* **16**, 199–208 (1959).
The generation of tsunamis is well described by
Braddock, R. D., van den Driessche, P. & Peady, G. W. *J. Fluid Mech.* **59**, 817–28 (1973).
For both their generation and their subsequent propagation, see
Hammack, J. L. *J. Fluid Mech.* **60**, 769–99 (1973).
The properties of Taylor columns were theoretically and experimentally investigated by
Taylor, G. I. *Proc. Roy. Soc.* A, **102**, 180–9 (1922) and **104**, 213–18 (1923).
He was following up a previous theoretical suggestion by
Proudman, J. *Proc. Roy. Soc.* A, **92**, 408–24 (1916).
Nonlinear distortions of the Taylor column were analysed by
Lighthill, M. J. *Proc. Camb. Phil. Soc.* **68**, 485–91 (1970),
who obtained good agreement with experiments by
Hide, R. & Ibbetson, A. J. *J. Fluid Mech.* **32**, 251–72 (1968).
Experimental studies of Taylor columns in density-stratified rotating fluids were made by
Davies, P. A. *J. Fluid Mech.* **54**, 691–717 (1972).
The corresponding column found ahead of an axially moving obstacle was also observed by Taylor, and investigated further by
Long, R. R. *J. Meteorol.* **10**, 197–203 (1953).
The phenomenon was explained rather clearly by
Stewartson, K. *Quart. J. Mech. Appl. Math.* **11**, 39–51 (1958);
and, for a wider range of cases, by
Trustrum, K. *J. Fluid Mech.* **19**, 415–32 (1964)
in a paper which also brought out clearly the analogy to motions in stratified fluid (our section 4.12).
An excellent introduction to tides is given by
Hendershott, M. & Munk, W. H. *Ann. Rev. Fluid Mech.* **2**, 205–24 (1970).
For an example of the use of the theory to compute the tidal system in a major ocean (the Atlantic), see
Pekeris, C. L. & Accad, Y. *Phil. Trans. Roy. Soc.* A, **265**, 413–36 (1969).
Good observational agreement with these computations was found in the work of
Cartwright, D. E. *Phil. Trans. Roy. Soc.* A, **270**, 603–49 (1971).
An excellent modern account of the 'Kelvin-wave' mode of tidal propagation is
Miles, J. W. *J. Fluid Mech.* **55**, 113–27 (1972).
A computational procedure for following tidal propagation over a large continental-shelf area, as modified by wind stress *and* nonlinear effects, is described in highly interesting detail by
Flather, R. A. & Davies, A. M. *Institute of Oceanographic Sciences Report* no. 16. London: Her Majesty's Stationary Office.
Among the different types of tidal wave propagation, the nature and importance of the low-frequency 'Rossby waves' linked closely to the variation of Coriolis parameter with latitude was first clearly brought out by

Rossby, C. G. *J. Marine Res.* **2**, 38–55 (1939).

A full and clear analysis of such waves is in

Longuet-Higgins, M. S. *Proc. Roy. Soc.* A, **279**, 446–73 (1964).

Waveguide modes in the ocean include, not only the 'barotropic' mode (with fluid motions almost independent of depth) as found in the tides proper, but also 'baroclinic' modes. These modes, dominated by the density stratification in the ocean, involve fluid motions which vary so with depth that they actually change sign; also, they propagate relatively slowly. Usually, baroclinic modes are strongly affected by the variation in the Coriolis parameter *f* with latitude. The theory of these modes is given, for example, in the Appendix to

Lighthill, M. J. *Phil. Trans. Roy. Soc.* A, **265**, 45–92 (1969).

This paper shows them to be particularly important in equatorial regions. It indicates their relevance and that of the barotropic Rossby waves to the remarkably sensitive response of the Indian Ocean to onset of the Southwest Monsoon; for recent work on this topic, see

Anderson, D. L. T. & Rowlands, P. B. *J. Marine Res.* **34**, 395–417 (1976).

An important broad survey of ocean currents and their dynamics was published by

Deacon, G. E. R. (ed.). *Phil. Trans. Roy. Soc.* A, **270**, 349–465 (1971).

The original work of Alfvén on magnetohydrodynamic waves is in

Alfvén, H. *Ark. Mat. Astr. Fys.* **29B**, no. 2, 1–7 (1942);

see also

Alfvén, H. & Fälthammar, C. G. *Cosmical Electrodynamics*, 2nd edn. Oxford University Press (1963),

where many of their applications are described. Combined acoustic and magnetohydrodynamic waves were first studied by

Herlofson, N. *Nature*, **165**, 1020–1 (1950).

Their form as quoted in the epilogue, and the associated physical interpretation, were given by

Lighthill, M. J. *J. Fluid Mech.* **9**, 465–72 (1960).

He applied this analysis to a discussion of the waves generated when a spacecraft penetrates the upper ionosphere.

Within the liquid core of the earth, another blend of two types of wave is important; in fact, magnetohydrodynamic waves in a rotating (as well as conducting) fluid. A good general survey of this field is by

Hide, R. & Stewartson, K. *Rev. Geophys. Space Phys.* **10**, 579–98 (1972).

An important discussion meeting on the subject is reviewed by

Moffatt, H. K. (ed.). *J. Fluid Mech.* **57**, 625–49 (1973).

Relationships between the waves and the fluctuating component in the Earth's magnetic field are much studied. For recent work on the other important question, how far a random assemblage of such waves can act as a dynamo generating the main magnetic field of the Earth, see, for example, the excellent work of

Soward, A. M. *J. Fluid Mech.* **69**, 145–77 (1975);

and, especially, the fine recent monograph by

Moffatt, H. K. *Magnetic Field Generation in Electrically Conducting Fluids*. Cambridge University Press (1977).

The epilogue's part 2 begins by noting a feature of dispersive waves which makes their nonlinear theory less complicated than that of, say, turbulence. For a good recent survey of turbulence data and theory, see

Hinze, J. O. *Turbulence*, 2nd edn. New York: McGraw-Hill (1975).

All but one of the results quoted in the epilogue for nonlinear periodic waves on deep water are taken from

Longuet-Higgins, M. S. *Proc. Roy. Soc.* A, **342**, 157–74 (1975).

He gives full references to the earlier work, dating back to Stokes. He also discusses the implications for wave breaking of the important discovery that the maximum of wave energy is found at an amplitude *below* its maximum value. The computed waveforms (figure 114) are from

Schwartz, L. W. *J. Fluid Mech.* **62**, 553–78 (1974).

The variational theory for a slowly varying train of nonlinear dispersive waves is due to

Whitham, G. B. *J. Fluid Mech.* **22**, 273–83 (1965) and **44**, 373–95 (1970).

The application to waves on deep water was made by

Lighthill, M. J. *J. Inst. Math. Applics.* **1**, 269–306 (1965).

Both authors contributed also to the major discussion meeting on nonlinear dispersive waves, published by

Lighthill, M. J. (ed.). *Proc. Roy. Soc.* A, **299**, 1–145 (1967).

This includes the Benjamin and Feir theoretical and experimental work on the stability of such waves, and complementary work of importance on that subject by O. M. Phillips. It includes also a valuable survey article by K. Hasselmann on statistical assemblages of dispersive waves.

The wave-action interpretation of Whitham's results is noted in his early papers. In addition, this interpretation, with its numerous applications, is emphasised, and explained very interestingly, by

Hayes, W. D. *Proc. Roy. Soc.* A, **320**, 187–208 (1970) and **332**, 199–221 (1973).

Refinements of the Whitham theory have been made by incorporating into the equations additional terms representing a dependence on rate of change of amplitude. One of the first papers along these lines was by

Chu, V. H. & Mei, C. C. *J. Fluid Mech.* **47**, 337–51 (1971).

They obtained improved agreement with the experimental results of Feir. A very interesting general treatment was given by

Davey, A. *J. Fluid Mech.* **53**, 769–81 (1972),

and further extended by

Davey, A. & Stewartson, K. *Proc. Roy. Soc.* A, **338**, 101–10 (1974).

Still more recently, good agreement for waves on deep water between theory and an extensive set of new experiments has been achieved by

Yuen, H. C. & Lake, B. M. *Physics of Fluids* **18**, 956–60 (1975).

Their analysis is based on a variational principle using a calculated Lagrangian density with a dependence on amplitude gradients.

For the case of 'fairly long' waves, the competition between small nonlinear effects and small dispersive effects was first clearly exhibited by

Korteweg, D. J. & de Vries, G. *Phil. Mag.* (5), **39**, 422–34 (1895),

through their introduction of the equation which bears their names. They developed the theory of cnoidal waves, deriving as a limiting case the equation for Scott Russell's solitary wave given by

Rayleigh, Lord. *Phil. Mag.* (5) **1**, 257–79 (1876).

For recent, much more exact, theories of the solitary wave, see, for example,

Longuet-Higgins, M. S. & Fenton, J. D. *Proc. Roy. Soc.* A, **340**, 471–93 (1974), and

Byatt-Smith, J. G. B. & Longuet-Higgins, M. S. *Proc. Roy. Soc.* A, **350**, 175–89 (1976).

A method for obtaining very general solutions to the Korteweg–de Vries equation was obtained by

Gardner, C. S., Greene, J. M., Kruskal, K. D. & Miura, R. M. *Phys. Rev. Lett.* **19**, 1095–7 (1967).

This has been widely applied (see Whitham's book) to the solution of particular initial-value problems. A careful experimental evaluation of the extent to which solutions of the Korteweg–de Vries equation represent actual 'fairly long' waves is given by

Zabusky, N. J. & Galvin, C. J. *J. Fluid Mech.* **47**, 811–24 (1971).

Note also the equation's application to quite different wave systems; for example, to nonlinear waves in rotating fluids by

Leibovich, S. *J. Fluid Mech.* **42**, 803–22 (1970).

In that system, the existence of a solitary wave had been shown experimentally by

Pritchard, W. G. *J. Fluid Mech.* **42**, 61–83 (1970).

Internal waves of large amplitude are analysed by

Benjamin, T. B. *J. Fluid Mech.* **29**, 559–92 (1967).

A combined experimental and theoretical account in considerable detail is given by

Thorpe, S. A. *Phil. Trans. Roy. Soc.* A, **263**, 563–614 (1968).

For the development of a train of internal waves in fluid of finite depth, see

Long, R. R. *Tellus*, **24**, 88–99 (1972).

A remarkable application of modern functional analysis to obtain an existence proof for the solitary wave in stratified fluid is in

Benjamin, T. B. *An Exact Theory of Finite Steady Waves in Continuously Stratified Fluids*. University of Essex: Fluid Mechanics Research Institute Report no. 48 (1973).

The nonlinear theory of propagation of internal waves in a gradually varying medium is given by

Grimshaw, R. *J. Fluid Mech.* **54**, 193–207 (1972).

The general theory of statistical assemblages of nonlinearly interacting waves owes much to work of

Phillips, O. M. *J. Fluid Mech.* **9**, 193–217 (1960),
and of

Longuet-Higgins, M. S. & Phillips, O. M. *J. Fluid Mech.* **12**, 333–6 (1962).

A key role was played by

Hasselmann, K. *J. Fluid Mech.* **12**, 481–500 (1962); also, **15**, 273–81 and 385–98 (1963).

Furthermore, he later gave the subject great strength and generality, by application of powerful techniques from solid-state physics theory, in

Hasselmann, K. *Rev. Geophys.* **4**, 1–32 (1966).

These methods are applied to interactions between surface waves and internal waves by

Kenyon, K. E. *J. Marine Res.* **26**, 208–31 (1968).

For fuller references, see the short survey article by

Phillips, O. M. *Ann. Rev. Fluid Mech.* **6**, 93–110 (1974).

An indication of the power of the Hasselmann technique is given by

Müller, P. *J. Fluid Mech.* **77**, 789–823 (1976).

Using a 'closure' system appropriate to the essentially Gaussian statistics of an assemblage of internal waves, he actually calculates the effective vertical diffusion coefficient for horizontal momentum which they must generate in the ocean . . . Similar achievements still lie beyond the reach of turbulence theorists.

NOTATION LIST

We list only symbols that appear frequently. The first column gives the meanings which they usually have. The second column notes any local meaning which they may be given in some specialised discussion.

Symbol	Usual meaning	Any exceptions
a	wave amplitude (ch. 3ff)	sphere radius (ch. 1)
b	breadth (ch. 2ff)	sphere displacement (ch. 1)
c	wave speed	
c_p; c_v	specific heats at constant pressure; at constant volume	
f	function	Coriolis parameter (epil.)
g	gravitational acceleration	occasionally, another function
h	depth	
i	square root of -1	
k	wavenumber in one-dimensional propagation	thermal conductivity (1.13)
l	length (ch. 1, 2, 3)	
$\mathbf{k} = (k, l, m)$	three-dimensional vector wavenumber (ch. 4)	m volume outflow from source (1.4)
$\mathbf{n}$	unit normal vector	
n	an integer	
o	of order smaller than	
p	pressure	
q	{ mass outflow from source (ch. 1); vertical component of mass flux (ch. 4)	
$\mathbf{r}$	position vector	
r	distance from origin	
t	time	
u	fluid velocity in one-dimensional motions	
v	excess signal speed in one-dimensional propagation	
$\mathbf{u} = (u, v, w)$	vector velocity in three-dimensional motions	
x, y, z	coordinates	
A	area	
B	forcing coefficient (3.9ff)	$\mathbf{B}$ magnetic field (epil.)
C	capacity (ch. 1 and 2 only)	thereafter, occasionally used for some *constant*

Symbol	Usual meaning	Any exceptions
D	drag (ch. 3ff)	distensibility (2.2)
E	specific internal energy	Young's modulus (2.2)
		E electric field (epil.)
F	force	
F	Fourier transform of forcing function	
G	dipole strength	
I	intensity (wave energy flux)	
J	volume flow (ch. 2)	Jacobian (4.8)
K	compressibility (ch. 1 and 2 only) Gaussian curvature of wavenumber surface (4.9ff)	magnitude of wavenumber vector (4.8)
M	mass	
M, N	integer subscripts	components of source wavenumber (1.12)
O	of order at most	
P	power	integral occurring in the Riemann theory (2.8, 2.9)
Q	mass outflow per unit volume (ch. 1 and 4 only)	waveform area (ch. 2)
$P(z), Q(z)$	giving z-dependence of excess pressure and upward mass flux (ch. 4)	
R	gas constant	
S	specific entropy	wavenumber surface (4.9ff)
T	temperature surface tension	a subsidiary time-variable (1.13)
U	shock-wave velocity (ch. 2) group velocity (ch. 3 and 4)	velocity of centroid of body $=$ **U** (ch. 1)
V	volume (ch. 1 and 2); velocity of fluid relative to an obstacle or other source of waves (ch. 3 and 4)	potential in Schrödinger's equation (4.5)
W	wave energy density	
Y	admittance (ch. 2)	
Z	impedance (ch. 2)	
X, Y, Z	elsewhere, used as coordinates of various special kinds	
α	local phase	coefficient of expansion (1.2, 4.3)
γ	ratio of specific heats	
δ	diffusivity of sound (ch. 1 and 2); a small quantity with the dimensions of frequency (3.7ff)	boundary-layer thickness (3.5)
ζ	vertical displacement (especially, of water surface)	(ξ, η, ζ) direction cosines (1.1)
η	a boundary value (3.9ff)	

Symbol	Usual meaning	Any exceptions
θ	angle to axis	
κ	curvature	a small quantity with the dimensions of wavenumber (1.13, 3.9 and epil.)
λ	wavelength	
μ	viscosity	magnetic permeability (epil.)
ν	kinematic viscosity	
ρ	density	
ϕ	velocity potential	
χ	salinity	
ω	frequency	
Δ	divergence	
Θ	phase-shift	
Ω	vorticity	

Subscripts

a	atmospheric	
e	excess over undisturbed value	
ex	external to boundary layer	'excluded' (1.3)
r	relative to undisturbed motion	
w	associated with waves	

AUTHOR INDEX

In this Author Index, all references are to pages A to Q of the Bibliography. An asterisk means that an author's name is listed also in the Subject Index.

[490]

SUBJECT INDEX

Numbers (1 to 469) refer to pages of the main text, while capital letters (A to Q) refer to pages of the Bibliography.

simple waves 143–56, *DF*
  incorporating weak shock waves 165–75
  in open channel 148–9, 182–3
  in perfect gas 148–50, 166–71
sinuous waves on thermocline 305–6, 398, 433, 435, *J*
sinusoidal waves 79, 84–8, 111–20, 124, 126–8, 130–6, 200
  on deep water 208–14, 223–5, 235–6
  on water of uniform depth 214–21, 225–7, 229–35, 256–7
  theory for general systems 258–60, 316, 352, 362–3
siren 31, 87–8
sky, blue light of the 56, *E*
'smoothed' number of eigenfunctions in interval of eigenfrequencies 424, *B*
Snell's law 323, 333
solids, mechanics of 96–9, 104, *AB*
solid surface, boundary condition at 128–31, 209, 214, 233, *A*
solitary wave 465–7, *IPQ*
  in stratified fluid 468, *Q*
'soliton' 467, *Q*
'sonic wind' (acoustic streaming) 339–45, *DG*
sound speed 4–11
  'frozen' 84–5
  in perfect gases 6–8
  in stratified fluid 291–8
  in water 10–11, 95, 206
  local 138–41, 148
  Newtonian 5, 8–10, 299–300
source regions, acoustically compact 25, 51, 64, *CDEF*
  cases with dipole far fields 35–41, 371
  cases with monopole far fields 31–5, 45, 66, 73, 370–1, 422–5, 436
  case with dissipation 76–7
  with nonlinear effects included 194–6
source regions, compact (for waves in general) 371, 380–1
source regions, noncompact 65–76, 371–2
spatial distortion of temporal waveforms 151–2, 174–5, 187–99
spatial waveforms, temporal distortion of 150–1, 165–74
specific heats 6–10, 82–4, 148, 161–2, 185–7, 202, 295, 300, 307, 426–7, 432–4, *B*
speedboats generating waves 274, 279
spheres, radiation from 33–5, 39–40, 65–70, 87–8
  scattered 50–7, 87, *E*
spherical attenuation factor 19, 24, 26

spherical pulse (nonlinear theory) 194–6, 203, *DF*
splitting of general standing wave into travelling waves 379, 384
standing waves 279
  of sound in a tube, generating streaming 347–8
  splitting of, into travelling waves 379, 384
stationary phase 70–1, 76, 88, 191, 197, 247–53, 270, 351–7, 366–9, 383–4, 386–93, 434, *BHL*
  to second approximation 281–2
stationary waves 260–70, 282–3, 452–4, *IP*
statistical assemblage of waves 468–9, *Q*
steady streaming generated by wave attenuation 337–51, *G*
steady stream, stationary waves on a 260–70, 282–3, 452–4, *IP*
steam siren 87–8
steepening of waveform's compressive phase 150–6, 161, 169–74
steepest ascent (direction of particle motion in surfaces of constant phase) 289–90, 350
steepest descent, method of 251–2, 387–90, *B*
  second approximation 281–2
step in bed of stream, generating stationary waves 261
Stevenson's studies of internal-wave generation 315–16, 379–82, 433, 436, *K*
Stokes boundary layer 130–6, 229–32, 281, *A*
Stokes calculation of internal dissipation in water waves 232–6, *H*
Stokes drift 280
Stokes edge-wave 429–30, *I*
'stokeslet' velocity field 341–3, *G*
stone thrown into pond 239–40, 245, 393–4
Straits of Gibraltar 200
stratification 286–98, 308–9, 312–16, 322–3, 332, 358–60, 376–85, 395–8, 410–17, *CFJKLMO*
  of atmosphere 11, 197–9, 205, 295, 306–8, 323–5, 332–7, 398–9, 425–8, 432–4, 468
  of ocean 205, 298–306, 324
  stability of 287, 301–2, 307–8
stratified-fluids experiments 313–16, 380–2, 411–13, *JKL*
stratosphere 307, 428, *C*
stream, stationary waves on a 260–70, 282–3, 452–4, *IP*
stream-function 452–3, *A*